AF606504

GEOSTATISTICS FOR THE NEXT CENTURY

Quantitative Geology and Geostatistics

VOLUME 6

The titles published in this series are listed at the end of this volume.

GEOSTATISTICS FOR THE NEXT CENTURY

An International Forum
in Honour of Michel David's Contribution
to Geostatistics, Montreal, 1993

Edited by

ROUSSOS DIMITRAKOPOULOS
McGill University,
Department of Mining and Metallurgical Engineering,
Montreal, Canada

KLUWER ACADEMIC PUBLISHERS
DORDRECHT / BOSTON / LONDON

Library of Congress Cataloging-in-Publication Data

Geostatistics for the next century : an international forum in honour of Michel David's contribution to geostatistics, Montreal, 1993 / editor, Roussos Dimitrakopoulos.
p. cm. -- (Quantitative geology and geostatistics ; v. 6)
ISBN 0-7923-2650-4 (acid-free)
1. Geology--Statistical methods--Congresses. I. David, Michel, 1945- . II. Dimitrakopoulos, Roussos. III. Series.
QE33.2.S82G45 1994
550'.72--dc20 93-44969

ISBN 0-7923-2650-4

Published by Kluwer Academic Publishers,
P.O. Box 17, 3300 AA Dordrecht, The Netherlands.

Kluwer Academic Publishers incorporates
the publishing programmes of
D. Reidel, Martinus Nijhoff, Dr W. Junk and MTP Press.

Sold and distributed in the U.S.A. and Canada
by Kluwer Academic Publishers,
101 Philip Drive, Norwell, MA 02061, U.S.A.

In all other countries, sold and distributed
by Kluwer Academic Publishers Group,
P.O. Box 322, 3300 AH Dordrecht, The Netherlands.

Printed on acid-free paper

Printed in the Netherlands

TABLE OF CONTENTS / TABLE OF MATIERES

Information Measures, Integration and Geostatistical Analysis / Mesures du Niveau d'Information, Intégration et Analyse Géostatistique

Conditional Simulations / Simulations Conditionnelles

FOREWORD

To honour the remarkable contribution of Michel David in the inception, establishment and development of Geostatistics, and to promote the essence of his work, an international Forum entitled *Geostatistics for the Next Century* was convened in Montreal in June 1993. In order to enhance communication and stimulate geostatistical innovation, research and development, the Forum brought together world leading researchers and practitioners from five continents, who discussed-debated current problems, new technologies and futuristic ideas.

This volume contains selected peer-reviewed papers from the Forum, together with comments by participants and replies by authors. Although difficult to capture the spontaneity and range of a debate, comments and replies should further assist in the promotion of ideas, dialogue and criticism, and are consistent with the spirit of the Forum.

The contents of this volume are organized following the Forum's thematic sessions. The role of theme sessions was not only to stress important topics of today but in addition, to emphasize common ground held among diverse areas of geostatistical work and the need to strengthen communication between these areas. For this reason, any given section of this book may include papers from theory to applications, in mining, petroleum, environment, geohydrology, image processing.

New advances and ideas, some of which may be considered controversial, are part of the present volume. These include: developments in dealing with uncertainty, advances in sampling, fuzzy set and Bayesian frameworks for information and data integration, algorithmic frameworks, fractal and multifractal approaches, neural network based simulation, optimization based conditional simulations, spatiotemporal modelling, issues of support change and upscaling, new stochastic fluid flow related formulations, as well as applications of new technologies and new solutions to old problems. Notably, conditional simulation appears as an exciting and prominent approach and is featured in most parts of the volume.

Having instigated and worked on the Forum for the last two years, I consider it an outstanding success due to the enthusiastic response of the international geostatistical community, and the ingenuity, quality and leadership of its participants. The present volume should further expand this success and communicate substantial results and ideas to a wider audience.

Lastly, I hope that David's spirit will continue to inspire events that are imaginative, non-conventional and daring, thereby promoting the free and unrestricted interaction of people and ideas. One's concerns for enhancing the recovery of metals, petroleum or pollutants should also stimulate a process of *Enhanced Ideas Recovery* - Isn't this what the rewarding growth of geostatistics calls for.

Montreal, August 1993

Roussos Dimitrakopoulos
Editor

AVANT-PROPOS

Dans le but de souligner la remarquable contribution de Michel David à l'introduction, à l'établissement et au développement de la géostatistique, et afin de promouvoir l'essence de son travail, un colloque international, intitulé *Geostatistics for the Next Century*, a eu lieu à Montréal en juin 1993. Ayant pour objet d'accroître la communication et de stimuler l'innovation, la recherche et le développement dans le domaine de la géostatistique, le colloque a réuni des chercheurs et des praticiens de premier plan au niveau international, venant des cinq continents, qui ont discuté des problèmes actuels, des nouvelles technologies et des idées futuristes.

Le présent volume contient des articles choisis parmi ceux présentés lors du colloque et révisés par des pairs, ainsi que certains commentaires des participants et les réponses des auteurs. Malgré qu'il soit difficile de rendre la spontanéité et l'éventail des débats, les commentaires et les réponses devraient aider à promouvoir les idées, le dialogue et les critiques et, dans ce sens, sont conformes à l'esprit du colloque.

Le contenu de ce volume est organisé suivant les sessions thématiques du colloque. Ces sessions thématiques avaient pour but, non seulement de souligner les sujets actuels importants, mais aussi de mettre l'accent sur les points communs à divers secteurs de la géostatistique et sur le besoin de raffermir la communicaton entre ces secteurs. C'est pourquoi un chapitre donné de ce livre peut comprendre des articles allant de la théorie aux applications possibles, aussi bien dans le domaine des mines, du pétrole, de l'environnement ou de l'hydrogéologie que dans celui du traitement d'image.

De nouveaux développements et de nouvelles idées, dont certains peuvent paraître controversés, apparaissent dans le présent volume. Parmi ceux-ci, l'on retrouve : les développements dans la façon de traiter l'incertitude, les progrès dans le domaine de l'échantillonage, les ensembles flous et la logique Baysienne pour l'intégration des données et des différents types d'information, les logiques algorithmiques, les approches fractales et multifractales, les simulations basées sur les réseaux neuroniques, les simulations conditionnelles basées sur l'optimisation, la modélisation dans l'espace et le temps, les problèmes de changement de support et de changement d'échelle, les nouvelles formulations stochastiques de la mécanique des fluides, ainsi que l'application de nouvelles technologies et de solutions nouvelles à de vieux problèmes. La simulation conditionnelle, notamment, semble une approche dominante et excitante, et se retrouve dans la plupart des sections du volume.

Étant à l'origine du colloque et y ayant travaillé au cours des deux dernières années, je considère personnellement qu'il a connu un succès remarquable, et ce en raison de la réponse enthousiaste de la communauté internationale oeuvrant dans le domaine de la géostatistique, ainsi que de l'ingéniosité, de la qualité et du dynamisme des participants. Le présent volume devrait accroître ce succès et communiquer des résultats et des idées importantes à un auditoire plus vaste.

Pour terminer, j'espère que l'esprit de Michel David continuera à susciter des événements innovateurs, non conventionnels et audacieux, encourageant ainsi la libre interaction des personnes et des idées. Les préoccupations de chacun pour améliorer la récupération des métaux, du pétrole ou des polluants devrait aussi stimuler un processus de *d'amélioration de la récupération des idées.* N'est-ce pas là ce que demande la croissance de la géostatistique?

Montréal, août 1993

Roussos Dimitrakopoulos
Editeur

MICHEL DAVID

Listed in the "Who's Who" of America, Michel David is credited as one of the first to put geostatistics on the scientific world map. Having started his career in his native country of France, Michel received his Bachelor's degree of Ingenieur civil des mines from the University of Nancy in 1967. He then went on to receive both his M.Sc.(1969) and Ph.D. (1973) in operations research from the University of Montreal. Early in his career, Michel made it apparent that the best word to describe him in his academic and professional life, is that of a teacher. He began as an assistant professor at Ecole Polytechnique de Montreal before he completed his Ph.D. and quickly became a full professor in the Department of Mineral Engineering at the same university. It soon became evident that Michel felt that his theoretical and academic approach to the subject should be tested in "real world" situations and thus, he began his long romance with consulting in 1971. Beginning with ore reserve estimation and ore body models, Michel has over the years applied his work to numerous mines worldwide.

Of most significance in his earlier years, is the publication of the first english language book on geostatistics in 1977. This book became an "overnight sensation" and set the pace for the rapidly emerging field. It highlighted Michel's insightful ideas on relative variograms, proportional effects and the log-normal short cut. The book has been translated into Russian, while the newly revised edition of the book is due out in Spanish in 1994. A second book dealing with advanced techniques was published in 1988. In order to set the "geostatistical wheel" in motion, Michel toured the world and taught over 40 short courses to approximately 1500 professionals, starting at the University of Nevada, Reno in 1973. His international acclaim was recognized immediately and he became one of the directors of N.A.T.O. Advanced Study Institutes in Rome (1975) and Tahoe (1982), a visiting professor at the Colorado School of Mines, and an invited speaker at many universities in Europe, North America, South America, and Australia.

Having a natural flair for bringing experts together, Michel became the director of the Mineral Exploration Research Institute in Montreal in 1978. Later he used his leadership skills to create Geostat Systems International Inc. (Denver, 1980, and Montreal, 1981). Throughout this, Michel continued to teach at the university and to supervise numerous graduate students from around the world.

Diverting more of his professional energy and zeal into his private life these days, Michel has been awarded the W.C.Krumbein medal from the International Association for Mathematical Geology (1988), and the Blaylock medal of the Canadian Institute of Mining, Metallurgy and Petroleum. In addition, Michel has been confirmed as a fellow of the Royal Society of Canada. His interest and influence in the field of geostatistics is commendable and is reflected in the experts and scholars that attended the Forum. His strive for excellence and education should be an inspiration for generations of geostatisticians to come.

MICHEL DAVID

Mentionné dans le «Who's Who» américain, Michel David est reconnu comme étant l'un des premiers à faire connaître la géostatistique dans le monde scientifique. Michel David a débuté sa carrière dans son pays natal, la France, où il a obtenu en 1967 son diplôme d'ingénieur civil des mines de l'École des mines de Nancy. Il a ensuite obtenu sa maîtrise (1969) et son doctorat (1973) en recherche opérationnelle à l'Université de Montréal. Les talents de professeur de Michel David se sont manifestés très tôt dans sa carrière, à la fois dans sa vie universitaire et professionnelle. Avant la fin de son doctorat, il a commencé à titre de chargé de cours à l'Ecole Polytechnique de Montréal; il est ensuite rapidement devenu professeur titulaire au département de génie minier de cette même école. Il est bientôt devenu évident que Michel croyait que son approche théorique et académique du sujet devait être mise à l'épreuve dans des situations réelles. En 1971, il a donc entrepris sa longue histoire d'amour avec la consultation. Commençant avec l'estimation des réserves et les modèles de gisements, Michel a, au cours des années, travaillé sur de nombreuses mines à travers le monde.

L'un des gestes très importants posés au début de sa carrière a été la publication en 1977 du premier livre de géostatistique en anglais. Ce livre a connu un succès immédiat et a tracé la voie dans un domaine qui évoluait rapidement. Il soulignait la perspicacité des idées de Michel sur les variogrammes relatifs, les effets proportionnels et le raccourci log-normal. Ce livre a été traduit en russe, et une nouvelle édition révisée doit être publiée en espagnol en 1994. Un deuxième livre, qui traite des techniques avancées, a été publié en 1988. Pour mettre en route la «roue géostatistique», Michel a parcouru le monde et a donné plus de 40 séminaires à environ 1500 professionnels, dont le premier a été donné à l'Université du Nevada à Reno en 1973. Son succès international a été immédiatement reconnu et il est devenu l'un des directeurs des conférences de l'Advanced Study Institute de l'OTAN à Rome (1975) et à Tahoe (1982). Il a aussi été professeur invité au Colorado School of Mines, et orateur invité dans plusieurs universités d'Europe, d'Amérique du nord, d'Amérique du sud et d'Australie.

Possédant un flair naturel pour réunir les experts, Michel est devenu directeur de l'Institut de recherche en exploration minérale de Montréal en 1978. Il a ensuite mis à profit ses talents de meneur pour créer Systèmes Géostat International Inc., d'abord à Denver en 1980 puis à Montréal en 1981. À travers cela, Michel a continué à emseigner à l'École Polytechnique et à superviser plusieurs étudiants de deuxième et de troisième cycle venant de partout à travers le monde.

Aujourd'hui, Michel consacre une plus grande partie de son énergie à sa vie privée. Il s'est vu attribuer pour son travail la médaille W.C. Krumbein de l'International Association for Mathematical Geology (1988), ainsi que la médaille Blaylock de l'Institut Canadien des Mines. En plus, Michel a été reçu membre de la Société royale du Canada. Son intérêt pour la géostatistique et son influence dans ce domaine sont louables et se reflètent dans le nombre d'experts et d'érudits qui ont assisté au colloque. Sa recherche de l'excellence et ses efforts dans le domaine de l'éducation devraient inspirer plusieurs générations futures de géostatisticiens.

ACKNOWLEDGEMENTS

The overwhelming success of this Forum would not have been possible without the support of several sponsors (please refer to the List of Sponsors), the dedication of its contributors and reviewers, the concern, ideas and suggestions of numerous colleagues, and the support of the international geostatistical community as a whole.

Acknowledgements are also in order to the personel of McGill's Conference Office, as well as Karen Richardson, Anne-Marie Czitrom-Dagbert, Bruce Robins and Xiaochun Luo for their assistance.

Montreal, August 1993

The Organizing Committee of the Forum

RoussosDimitrakopoulos,Chairperson(Canada)
George Christakos (USA)
Peter Dowd (England)
Ricardo Olea (USA)
Shahrokh Rouhani (USA)
Michel Soulie (Canada)

REMERCIEMENTS

L'immense succès de ce colloque n'aurait pu être rendu possible sans l'appui des organismes qui l'ont parrainé (veuillez vous reporter à la Liste des organismes parrainant le colloque). Ce succès est aussi dû au dévouement des collaborateurs et des réviseurs, à l'intérêt démontré par de nombreux collègues, qui nous ont fourni des idées et des suggestions, ainsi qu'à l'appui de toute la communauté internationale oeuvrant en géostatistique.

Nous tenons aussi à remercier le bureau des conférences de l'université McGill, ainsi que Karen Richardson, Anne-Marie Czitrom-Dagbert, Bruce Robins et Xiaochun Luo, pour l'aide qu'ils nous ont apportée.

Montréal, août 1993

Le comité organisateur du colloque

Roussos Dimitrakopoulos, président (Canada)
George Christakos (Etats-Unis)
Peter Dowd (Angleterre)
Ricardo Olea (Etats-Unis)
Shahrokh Rouhani (Etats-Unis)
Michel Soulié (Canada)

LIST OF SPONSORS / LISTE DES ORGANISMES PARRAINANT LE COLLOQUE

BP Exploration

Cameco Corporation

Canadian Institute of Mining, Metallurgy and Petroleum

Centre de Geostatistique, Ecole des Mines de Paris

De Beers

Ecole Polytechnique de Montreal

Geological Survey of Canada

GEOSTAT Systems International Inc.

INCO Limited

International Association for Mathematical Geology

C. Lemmer

McGill University

North American Council on Geostatistics

Snowden Associates Pty Ltd.

Stanford Centre for Reservoir Forecasting, Stanford University

University of Leeds

URANERZ Exploration and Mining Limited

LIST OF PARTICIPANTS / LISTE DES PARTICIPANTS

Frits P. Agterberg, Geological Survey of Canada
601 Booth Street, Ottawa, Ontario, Canada K1A 0E

Abdullah Alattas, P.O. Box 345 DGMR, Jeddah,Saudi Arabia 21191

Marco Alfaro, Universite de Chile, Mining Department
Tupper 2069, Casilla, Santiago, Chile 2777

Margaret Armstrong, Centre de Gostatistique
35, rue Saint-Honore, Fontainebleau, France 77305

Bruce Bancroft, Consolidation Coal Company
1800 Wasington Road, Consol Plaza, Pittsburgh, Pennsylvania, U.S.A. 15241

Marc F.P. Bierkens, University of Utrecht, Department of Physical Geography
P.O. Box 80.115, Utrecht, Netherlands 3508 TC

Richard Bilonick, Consolidation Coal Co.
Consol Plaza, Pittsburgh, Pennsylvania, U.S.A. 15241

Dominique F.-Bongarçon, MRD, Bayshore Corporate Center,
1710 So. Amphlett Blvd., Suite 302, San Mateo, California, USA 94402

Luiz Braga, Federal University of Rio de Janeiro, I. Mathematica
C.P. 68530, Rio de Janeiro, Brazil 21945-970

John, A. Brunette, Placer Dome Inc.
P.O. Box 49330, Vancouver, B.C., Canada V7X 1P1

Bruce E. Buxton, Battelle Memorial Institute
505 King Avenue, Columbus, Ohio, U.S.A. 43201-2693

Richard Chambers, Amoco Production Company
P.O. Box 3385, Tulsa, Oklahoma, U.S.A. 74102

George Christakos, University of North Carolina, Dept. of Environmental
Science & Eng., CB #7400, 1316 Rosenau, Chapel Hill, NC, U.S.A. 27599

Gerard Conan, Marine Research Laboratory, University of Moncton
Moncton, NB, Canada E1A 3E9

Marc V. Cromer, Montgomery Watson
365 Lennon Lane, Walnut Creek, California, U.S.A. 94598

Michel Dagbert, Geostat Systems International Inc.
800 boul. Chomedey, Suite A-240, Laval, Quebec, Canada H7V 3Y4

Kadri Dagdelen, Colorado School of Mines, Mining Engineering Dept. ESM, Golden, Colorado, U.S.A. 80401

Colin Daly, B.P, Research Centre
Chertsey Road, Sunbury-on-Thames, Middlesex, United Kingdom TW16 7LN

Michel David, Ecole Polytechnique
630 Terrasse D'Auteuil, Laval, Quebec, Canada H7L 1K5

Robert de l'Etoile, Geostat Systems International Inc.
800 Boul. Chomedey #A240, Laval, Quebec, Canada H7V 3Y4

Pierre Delfiner
7 rue Guynemer, L'Hay Les Roses, France

Jean-Pierre Delhomme, Schlumberger E.P.S.
26 rue de la Cavee, B.P. 202, Clamart Cedex, France F92140

Claude Demange, COGEMA
2 rue Paul Dautier, B.P. 4, Velizy-Villa Coublay Cedex, France 78141

Alexandre Desbarats, Geological Survey of Canada
601 Booth Street, Ottawa, Ontario, Canada K1A 0E8

Clayton Deutsch, Exxon Production Research Company
P.O. Box 2189, Houston, Texas, U.S.A. 77252-2189

Roussos Dimitrakopoulos, Dept. of Mining and Met. Eng., McGill University
3480 University Street, Montreal, PQ, Canada H3A 2A7

Peter Dowd, University of Leeds
Dept. of Mining and Mineral Engineering, Leeds, United Kingdom LS2 9JT

Olivier Dubrule, ELF Aquitaine Production
Pau Cedex, France 64018

Marie-Josee Fortin, Centre D'Etudes Nordiques, Universite Laval
Sainte-Foy, Quebec, Canada G1K 7P4

Alain Galli, Ecole Nationale Superieure des Mines, de Paris Ecole,
Centre de Geostatistique, 35 rue Saint-Honore, Foutainebleau, France 77305

Gary Giroux, Montgomery Consultants Ltd.
701 - 675 W. Hastings Street, Vancouver, B.C., Canada V6B 1N2

Pierre Goovaerts, Stanford University, Dept. of Applied Earth
Sciences, Mitchell Building, Stanford, California, U.S.A. 94305-2225

William R. Green, Placer Dome Inc.
P. O. Box 49330, Bentall Station, Vancouver, B.C., Canada V7X 1P1

Dominique Guerillot, IFP c/o Elf Geoscience Research Centre
114 A Cromwell Road, United Kingdom SW7 4EU

Kateri Guertin, Kateri Guertin Engineering
244 Stoneheng, Beaconsfield, Quebec, Canada H9W 3X9

Allan Gutjahr, Dept of Mathematics, New Mexico Tech.
Socorro, New Mexico, U.S.A. 87801

Mark S. Handcock, New York University
44 West W 4th #860, New York, New York, U.S.A. 10012-1126

Goosa Hedhili, Societe Italo Tunisienne D'Exploration, Petroliere
Centre Urbain Nord, B.P. 424, Tunis Cedex, Tunis, Tunisia 1080

Steve Hoerger, Newmont Gold Company
P.O. Box 669, Carlin, NV, U.S.A. 89822

Dominique Jeulin, Ecole Nationale Superieure, des Mines de Paris
Centro de Geostatistique, 35 rue St. Honore, Fontainebleau, France 77305

Andre G. Journel, Applied Earth Sciences Department
Stanford University, Stanford, California, U.S.A. 94305-2225

Mark Jutras,Placer Dome Inc.
P.O. Box 49330, Bentall Station, Vancouver, B.C., Canada V7X 1P1

Marek Kacewicz, ARCO Exploration and Production Tech.
2300 West Plano Parkway, Plano, Texas, U.S.A. 75024

Wynand Johannes Kleingeld, De Beerserican Corp.
Ore Evaluation Department, P.O. Box 41848, Craighall, South Africa 2024

Danie Krige, University of Witwatersrand
P.O. Box 121, Florida Hills, South Africa 1716

Gregory Kushnir, GeoGraphix Inc.
1860 Blake Street, Suite 900, Denver, Colorado, U.S.A. 80202

Carina Lemmer, Eagle House 10th Floor
70 Fox Street, Johannesburg, South Africa 2001

Oddvar Lia, Norwegian Computing Centre
P.O. Box 114, Blindern, Oslo, Norway N-0314

Thorsten Liebers, Technische Universitat Dresden, Institut fur
Planetare Geodasie, Mommsenstr. 13, D-O-8027 Dresden, Germany,

Xiaochun Luo, Dept. of Mining and Met. Eng.
McGill University, 3480 University Street, Montreal, PQ, Canada H3A 2A7

Denis Marcotte, Department de Genie Mineral, Ecole Polytechnique
C.P. 6079 Succursale A, Montreal, Quebec, Canada H3C 3A7

Pierre Montes, Ecole Polytechnique
Dept. de Genie Civil, C.P. 6079, Succ. A, Montreal, QC, Canada H3C 3A7

Pierre Mousset-Jones, Mackay School of Mines
University of Nevada, Reno, Nevada, U.S.A. 89557-0047

Roger Murray, Western Geophysical Company
3600 Briarpark, Houston, Texas, U.S.A., 77042

Donald E. Myers, University of Arizona
Dept, of Mathematics, Tucson, Arizona, U.S.A. 85721

Jeff Myers, Estox
P.O. Box 1365, Golden, Colorado, U.S.A., 80402

Brent Noland, Esso Resources Canada Limited
237 Fourth Avenue SW, Calgary, Alberta, Canada T2P 0H6

Ramagopal Nutakki, Computer Modelling Group
Office 200 3512-33 Street NW, Calgary, Alberta, Canada T2L 2A6

Ricardo Olea, Kansas Geological Survey
1930 Constant Ave. Campous West, Lawrence, Kansas USA 66044-2598

Dean Oliver, University of North Carolina, Dept. of Environmental
Science & Eng., CB #7400, 131 Rosenau, Chapel Hill, NC, U.S.A. 27599

Tinus Oosterveld, De Beers Consolidated Mines
P.O. Box 41848, Craighall, South Africa 2024

Yvan Pannatier, Institute of Mineralogy
University of Lausanne, BFSH 2, Lausanne, Switzerland, 1015

Harry M. Parker,, Mineral Resources Dept, Inc. Bayshore Corporate Centre,
1710 S. Amphlett Blvd. Suite 302, San Mateo, CA, U.S.A. 94402

Vera Pawlowsky, Univ. Polit. Catalunya
Dpto. Matem. Aplicada III, Gran Capitan, s/n, Barcelona, Spain, 08034

Jurgen Pilz, Bergakademie Freiberg, Fachbereich Geowissenschaften,
Gustav-Zeuner-Strabe 12, Freiberg, Saxony, Germany, D-9200

Francis Pitard, Francis Pitard Sampling Consultants
14710 Tejon Street, Broomfield, Colorado, U.S.A. 80020

Carlos E. Puente, Hydrologic Science, Department of Land and Water
Resources, University of California, Davis, Davis, California, U.S.A. 95616

Wen Renjun, Department of Geology, Norwegian Institute of Technology, University of Trondheim, Trondheim, Norway N-7034

Charles Rose,
P.O. Box 4344, Monroe, Louisiana, U.S.A. 71211-4344

David Rossi, Schlumberger - Doll Research
Old Quarry Road, Ridgefield, CT, U.S.A. 06877

Mario Rossi, Mineral Resources Development, Inc., Bayshore Corporate Center, 1710 S. Amphlett Blvd. Suite 302, San Mateo, CA, U.S.A. 94402

Maria-Theresia Schafmeister, Freie University Berlin, Institute for Applied Geology, Malteserstr. 74-100, Haus B, Berlin 46, Germany D-1000

Edmund Sides, International Institute for Aerospace
Survey and Earth Sciences, Kanaalweg 3, Delft, Netherlands 2628 EB

Alastair J. Sinclair, University of British Columbia
Dept. of Geological Sciences, Vancouver, BC, Canada V6T 1Z4

Steve Smithies, Genmin - General Mining Metals and
Minerals, P.O. Box 61820, Marshalltown, Johannesburg, South Africa 2107

Vivienne Snowden, Snowden Associates Pty Ltd
P.O. Box 77, West Perth, W.A., Australia WA6872

Andrew Solow, Woods Hole Oceanographic Center
Marine Policy Div., Woods Hole, MA, U.S.A. 02543

Philippe Sonnet, Universite Catholique de Louvain, Laboratoire de Geotechnique et, Mineralogie, 3 Place L. Pasteur, Louvain-a-Neuve, Belgium 1348

Michel Soulie,, Ecole Polytechnique, Section Geotechnique
C.P. 6079, Succ. A., Montreal, QC, Canada H3C 3A7

Mohan R. Srivastava, FSS International
800 Millbank, False Creek South, Vancouver, BC, Canada V5Z 3Z4

Francois E. Steffens, University of South Africa
Statistics Dept., P.O. Box 392, Pretoria, Transvaal, South Africa 0001

Dimitri Stolyarenko, University of Moncton
School of Engineering, Moncton, New Brunswick, Canada E1A 3E9

Ali M. Subyani, Colorado State University
1440 Edora, Apt. # 28, Fort Collins, Colorado, U.S.A. 80525

Suresh Thadani, Stochastic Systems International
1524 Mockingbird, Plano, Texas, U.S.A. 75093

Kelly Tyler, Statoil, Dept. of Research Technology
P.O. Box 300, Stavanger, Norway N-4001

Marcel Vallée, Marcel Vallee Geolonseuil Inc.
706, Ave. Routhier, Sainte Foy, QC, Canada G1X 3J9

Arthur Van Beurden, National Institute Public Health & Envir Ant. van Leeuwenhoeklaan 9, P.O. Box 1, Bilthoven, Utrecht, Netherlands 3720 BA

Georges Verly, Reservoir Mechanisms and Simulation, BP Research Centre, Chertsey Road, Sunbury-on-Thames, Middlesex, United Kingdom TW16 7LN

Jeffrey M. Yarus, Marathon Oil Company
P.O. BOx 269, Littleton, Colorado, U.S.A. 80160

Evangelos A. Yfantis, University of Nevada, Las Vegas, Computer Science Dept.
4505 S. Maryland Parkway, Las Vegas, NV, U.S.A. 89120

Li-Ping Yuan, Alberta Geological Survey - Alberta Research Council, Box 8330, Stn. F, Edmonton, Alberta, Canada T6H 5X2

MODELLING PRACTICES, NEEDS, DIRECTIONS AND CONCERNS

ESTIMATING OR CHOOSING A GEOSTATISTICAL MODEL*

OLIVIER DUBRULE
Elf Aquitaine Production
Avenue Larribau, 64000 Pau, FRANCE

Geostatisticians base their estimations, simulations, uncertainty evaluations on the definition of a model. The uncertainty figures they produce are a function of the model chosen and of the parameters of this model. In mining applications, where data are plenty, this issue may not be as important as in petroleum applications, where data are very scarce. Computations derived from a model based on few data must be carefully analysed. For instance, geostatistical computations are often obtained under the assumptions that some parameters of the model (nugget effect..) are perfectly known. These assumptions must be clearly identified, especially when uncertainty evaluations are derived.

Introduction

Geostatistical models are experiencing great success in the oil industry. The growing use of these models has provided new solutions to the problem of representing the spatial distribution of variables such as lithology, porosity or permeability. This success also highlights the difficulties associated with the use of these models, when such issues as uncertainty quantification are addressed. It will be shown that the bayesian formalism can help understand the assumptions that are made, sometimes explicitly but in many cases implicitly, by many geostatistical studies.

Nugget Effect and Variogram Model at the origin

Geostatistics books are full of examples showing experimental variograms, together with the model fitted to them. The spherical model is used most often. A common approach seems to be, given an experimental variogram, to choose as theoretical model the spherical variogram that best fits the values at the first lags. The "nugget effect" is then derived as the intersection of the variogram model with the Y-axis (Figure 1). It is often

---

* Title inspired from Matheron's book, 1988

R. Dimitrakopoulos (ed.), Geostatistics for the Next Century, 3–14.

recommended to interpret the nugget effect as a measurement error, or a microstructure, or an undetermined mixture of both. Unfortunately, the derivation of the nugget effect, and its interpretation either way are often arbitrary. The data themselves cannot give any information for distances smaller than their minimum spacing. And cross-validation cannot give any further information since, being based on the data themselves, it cannot validate a model over distances smaller than the minimum data spacing. Unfortunately, in many cases, the arbitrariness affecting the determination of the nugget effect is ignored. This is not always the case, and many nice examples can be found that explain how short-distances sampling (De Fouquet et al, 1989) or considerations on the likely origin of measurement errors (Chiles, 1977) can help reduce uncertainty on the evaluation of the nugget effect.

This uncertainty on the nugget effect can be of great consequence on subsequent geostatistical computations, such as kriging. Take the simplistic case of a variogram that appears to be flat at all distances greater than the minimum data spacing. Such a variogram can be modelled by:

- a pure nugget effect
- a model with a range that is just equal to the minimum data spacing.

Both models are compatible with the data (Figure 2). Note that an infinity of models, taken between these two extremes, could also be fitted to a flat experimental variogram. Now use both variograms for interpolating the variable of interest: with the first variogram the interpolation will be simply equal to the mean of the data (assuming a global neighborhood is used), and, at the data points, it will be equal to (Figure 3b):

- the mean if the nugget is interpreted as a measurement error
- the measured value if the nugget is interpreted as a microstructure with a very small range.

Between these two extremes, and depending on the assumption made about the relative importance of microstructure and measurement error, intermediate interpolations at the data points will be possible.

However, away from the data, the interpolated value will always be the mean.

Now take the other model: interpolation (Figure 3c) will be equal to the mean at a distance greater than the range from all the data points, then, as the distance from a data point decreases from the range value to zero, the interpolated value will tend towards the measured value.

Other examples could be given (estimation variances, variances of average block values ...) of the great importance of the nugget effect, which is in most cases, the least-known part of the variogram.

Therefore it is, in many cases, wrong to assume that a reliable variogram model can be fitted to the available data. Note here that, in the first mining applications of geostatistics (Matheron, 1970, Journal and Huijbregts, 1977), recommendations were often made by geostatisticians to drill extra closely spaced drill-holes, with the goal of better understanding the variogram behaviour at the origin. With the spread of geostatistical applications to other fields such as petroleum, where the cost of drilling new wells is huge, the importance of this wise recommendation has often been forgotten. So far, only the nugget effect has been discussed. The choice of the variogram's behaviour at the origin (linear, parabolic) can also greatly impact geostatistical computations.

The conclusion is that, in many cases, the behaviour of a variogram model at the origin is not derived from the data themselves. Experience shows that it is often derived on the basis of subjective knowledge that the geostatistician has of the variable of interest. For instance, the thickness of a geological layer is expected to show better horizontal continuity than an average permeability.
Therefore, significant uncertainty may affect the parameters of a variogram model: these parameters (nugget, behaviour at the origin) are often "chosen" for conveniency reasons rather than "estimated" from the data (this terminology is taken from Matheron, 1988). However, not only the parameters of the model may result from a subjective decision: the choice of using a variogram-based approach may also be arbitrary, albeit reasonable. This issue is not often mentioned in mining geostatistics, but appears unescapable in the data-scarce world of petroleum.

Petroleum Applications: Choice of a Probabilistic Model and its Parameters

Choice of a Probabilistic Model

In the petroleum industry, geostatistics is now widely used for quantitative reservoir characterisation: two main categories of approaches exist, either "pixel-based" or "object-based". Approaches based on variograms (De Fouquet et al, 1989, Journel and Gomez Hernandez, 1989), Fractals (Hewett, 1986) or Markov random fields (Farmer, 1992, Ripley, 1993) belong to the former category, whereas boolean-like models (Haldorsen and MacDonald, 1987) belong to the latter.

In the worst scenarios, the choice of which approach to use is based on software availability. In other applications, this choice is controlled by the geological knowledge: for instance, object-based models will be preferred in the case of a fluvio-deltaic reservoir composed of channel sands distributed within a shaly floodplain, whereas pixel-based models will be chosen in cases where geometries are harder to define, such as carbonate environments.

Determining the parameters

Once the geostatistical model has been chosen, its parameters are determined: these can be the size of genetic units, the variogram, or the Markov transition probabilities. Their choice is rather important, since they summarise, together with the model itself, the quantitative geological knowledge.

In petroleum applications, parameters of the model can be, in favorable cases, calculated from the data (Alabert and Massonnat, 1990). As was discussed earlier about the nugget effect, well data, no matter how numerous, will always leave uncertainty about the variogram's behaviour at the origin. In most petroleum applications, because of the lack of wells, these parameters are inferred from outcrop studies, or simply chosen because they result in three-dimensional representations that "look" right. The use of outcrop data requires the strong assumption that the outcrop used for deriving statistical parameters is

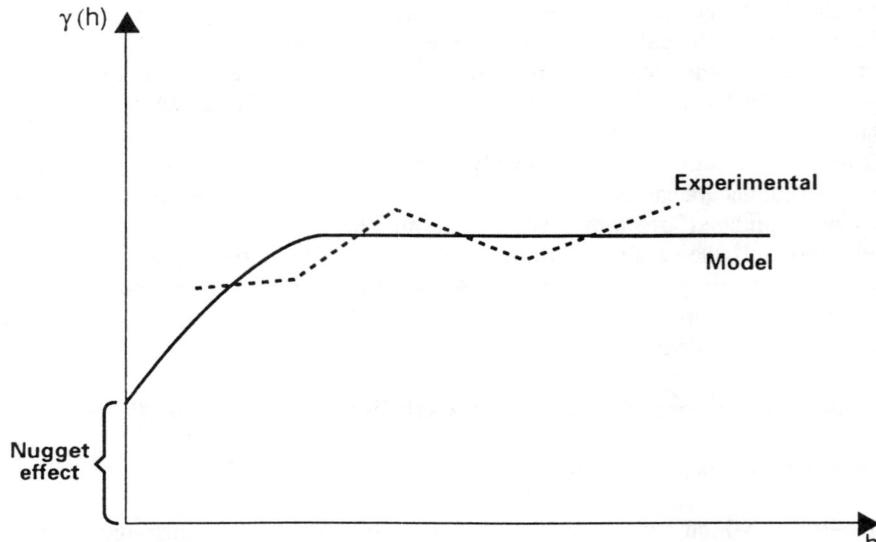

Figure 1 : The Nugget effect is often chosen as the intersection between the Y-axis and the model that best fits the experimental variogram at sampled distances

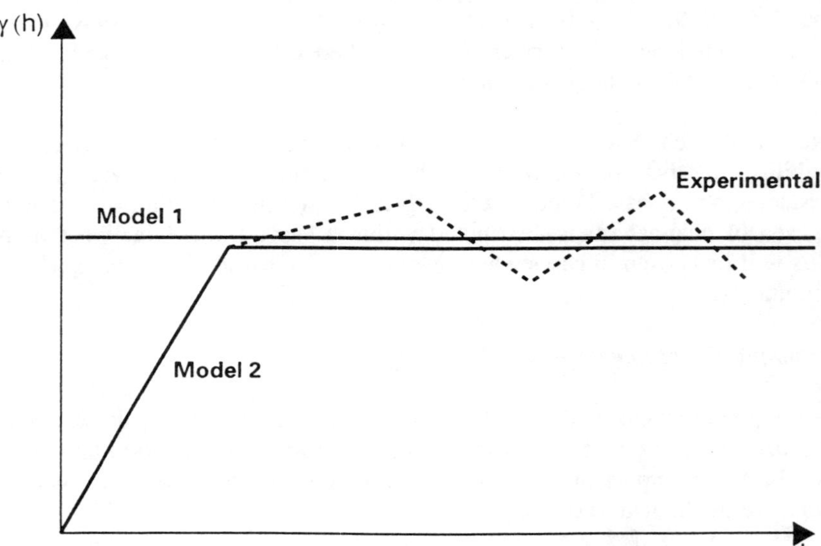

Figure 2 : Two models that both provide a good fit of the experimental variogram at sampled distances

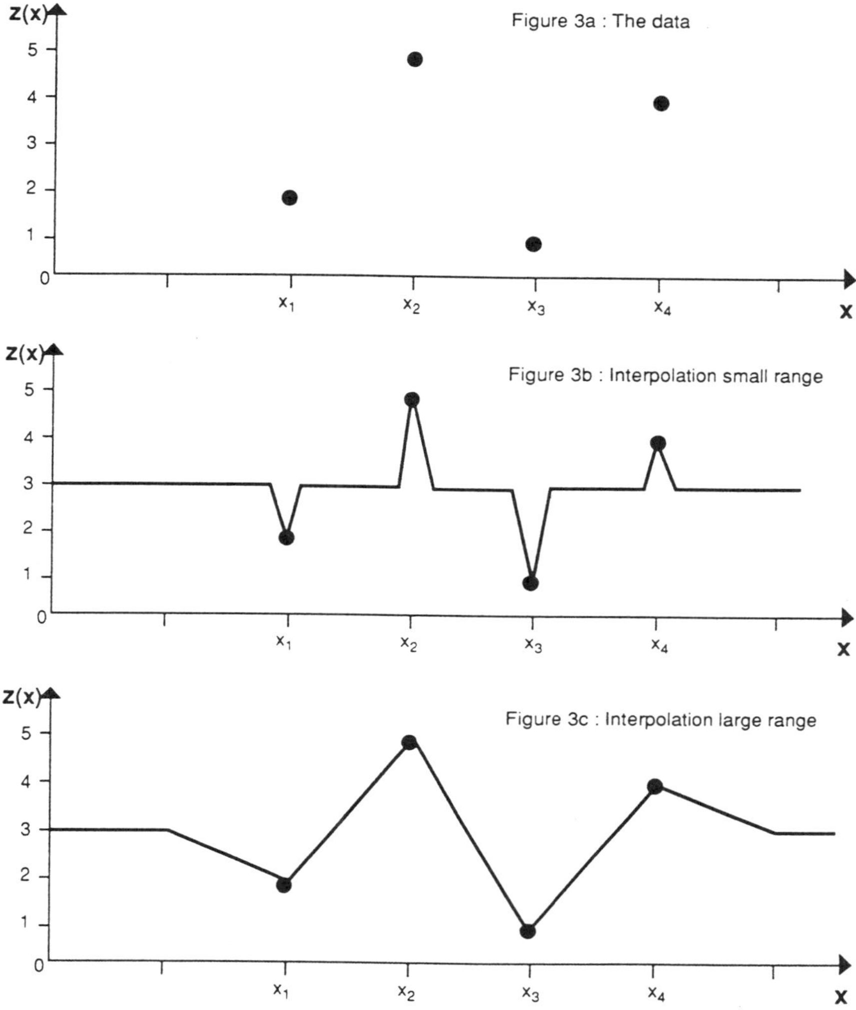

Figure 3 : Comparing the impact of kriging with two variogram models, both compatible with the experimental variogram at sampled points

geologically and statistically "analogous" to the studied reservoir. A non-trivial aspect of the use of "external" information, such as that derived from outcrops, is to make sure that the parameters of the model are consistent with the small number of available data. For instance, object-based models often use size information derived from outcrop studies, which may be inconsistent with the few available observations (closely-spaced wells, geological correlations..). Rather than calculating a parameter from the data, through a standard variogram calculation and fitting approach, the question becomes that of making sure that one chosen parameter is not incompatible with these data. This issue constitues an area of research that deserves to be more investigated.

In spite of this uncertainty affecting parameters of the model, examples exist where authors have derived uncertainty figures for petroleum reservoirs simply from the variability between realisations of conditional simulations obtained at fixed parameters. This approach, neglecting the lack of knowledge about the parameters themselves, can lead to a significant underestimation of the uncertainties.

All uncertainties must be quantified

In conclusion from the above, a certain degree of uncertainty always exists, both on the validity of the model itself and on the parameters of this model. An attractive solution seems to be, through a bayesian apprach, to also attach uncertainty to the parameters, as recommended by Omre et al, 1993. The few examples presented below will show that this approach is not as simple as it may appear at first sight.

Assume that we are faced with the problem of estimating sand connectivity within a layer of sand percentage say 40%, deposited within a fluvial environment. It is assumed for simplification that the only reservoir bodies are channels distributed within a shaly background. These are often modelled with an object-based technique, using the approach described by Haldorsen and Macdonald, 1987: a set of channels is distributed by simulation in 3D. Their shape can be modelled using simple elongate rectangular bars, the width and thickness of which following the size estimate given by the geologist. But what about the direction of these channels? Since, in some practical applications, nothing is known about paleocurrent direction, it can be decided to account for this state of ignorance by drawing the direction of each individual channel from an uniform distribution between 0 and 180°.

The result is that realisations produced with such input parameters will show an excellent connectivity in 3D (Figure 4)! Therefore, although nothing was initially known about channel direction, the very "deterministic" conclusion was drawn that connectivity is excellent! The flaw in this approach is of course that the quantitative assumption "nothing is known about channel direction" was wrongly translated into an assumption of the type "for each possible representation of the reservoir, channel directions are very scattered and independent from each other". These two implicit assumptions result in the excellent connectivity observed within each realisation. Note that it would have been very difficult, if not impossible, to generate a spectrum of realisations based on no assumption at all about possible channel directions: this would have required the generation of as many

realisations as there were possible scenarios for the distribution of channel directions: realisations with all channels having the same direction, realisations with independent directions, intermediate scenarios...
This simply means, as already emphasised in the literature (Jaynes, 1968) that it may be very difficult, in practical applications, to quantify a state of complete ignorance about a parameter. Jaynes has illustrated this problem with the example of the uniform distribution: assuming nothing is known about the value of a parameter such as porosity, except that this porosity is between 0 and 30%, how could this state of complete ignorance be represented? The use of a uniform distribution between 0 and 30% can lead to inconsistencies since, with such a model, the square of porosity would not be uniformly distributed: therefore, the square of porosity would not be completely unknown, in spite of the fact that porosity itself is unknown. Jaynes concludes , as in the example of channel connectivity, that a state of complete ignorance on one parameter may not be preserved through a non-linear transform on this parameter. This constitutes one of the obstacles to real-life applications of the bayesian formalism.

On the same subject, Journel and Deutsch (1993) have shown that various geostatistical assumptions, such as those of gaussian simulation, could have a significant impact as far as uncertainty quantification is concerned. The argument for using a gaussian model is often that is maximum entropy or "least committal": it maximizes uncertainty beyond the statistics that are considered known. However, a maximum entropy permeability or porosity distribution may not necessarily generate a maximum entropy distribution of reservoir simulation outputs. In fact, as quoted from Journel and Deutsch, "Maximum entropy of the random function model does not entail maximum entropy of the response distributions; in fact, the contrary is observed for the flow performances studied".

In spite of the above difficulties, the issue of quantifying uncertainties on the parameters of a variogram model (in particular nugget and behaviour at the origin) deserves to be better investigated on both a theoretical and a practical point of view. Surprisingly, to the author's knowledge, very few efforts have been made in that direction (see however the interesting paper by Handcock and Stein, 1993). Such an approach would certainly be preferable to making arbitrary assumptions about a nugget effect when little is known about the geology (approaches such as the choice of the "highest possible" or "worst scenario" nugget can lead, as mentioned earlier, to unrealistic geostatistical computations).

Various levels of uncertainty

Once a probabilistic model and its parameters have been identified, the model predicts that, as the number of data points increases, the variability from one realisation to another also decreases. This is true, of course, as long as the model and its parameters are left unchanged. However, experience shows that, especially for applications where few data are available to start with, uncertainty can increase as the number of data increases! This simply illustrates the fact that new data can require a change in the parameters of the model, or worse, in the model itself.
This frequently happens in petroleum applications, the knowledge about an oil field

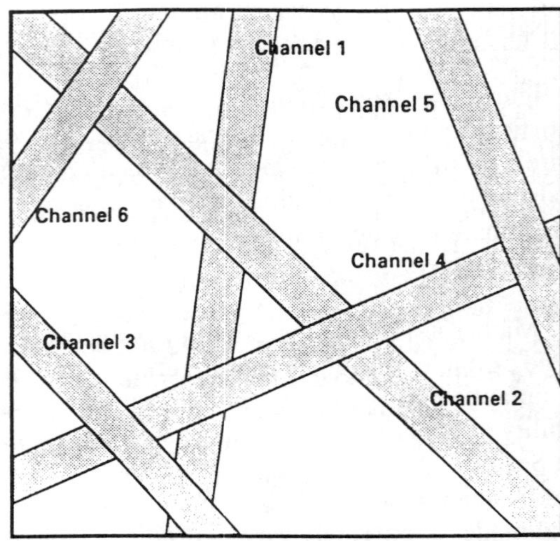

Figure 4 : Illustration of the excellent connectivity obtained when assuming uniformly distributed direction of channels (map view)

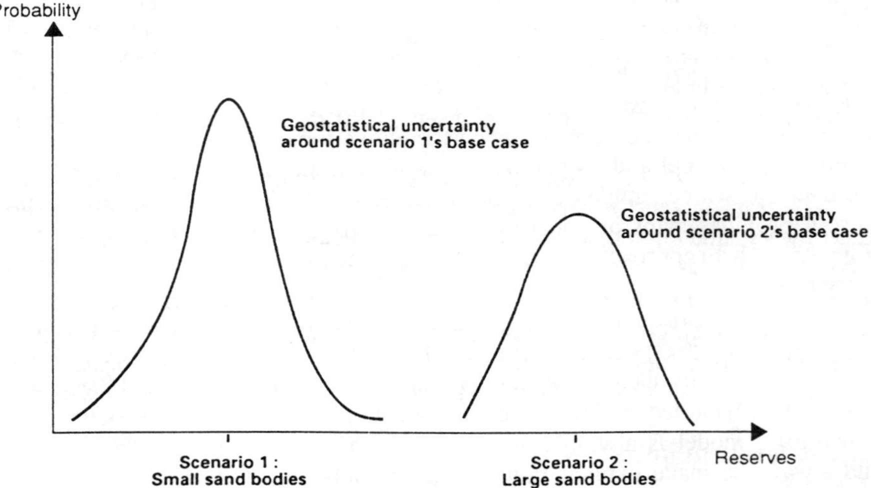

Figure 5 : Illustrating the risk of quantifying unimportant uncertainties using geostatistics

evolving from appraisal stage, where a few wells are available, to later development stages, where information from tens of wells, 3D seismics and production data may be at hand. Uncertainty figures may have to be increased as more data are gathered because, although the geostatistical model was right, the uncertainty on its parameters was underestimated (Figure 5). Omre et al (1993) have shown (Figure 6) that predicting the fraction of water produced over an oil field using fixed values of parameters would lead to a significant underestimation of uncertainty, compared to an approach accounting for uncertainties on these parameters.

As pointed out by G. Matheron (1988), there will also always be a risk of radical error on the model itself. The example of Ruijtenberg (Figure 7) demonstrates how a model can change from almost unfaulted to extremely faulted within a matter of ten years, just because of the availability of better seismic and more wells. Corrigan, 1988, also presents "post mortem" North Sea examples, showing how the knowledge about various oil fields has changed over time. A quantification of uncertainties due to potentially radical changes of the initial geological model is extremely hard to produce. However, these uncertainties can be essential. Therefore, geostatistical uncertainty evaluations should always carefully mention the assumptions under which the results are obtained (fixed probabilistic model, fixed parameters of this model...).

Conclusion

This paper may seem like a mere collection of remarks on the various degrees of uncertainty affecting geostatistical models and their parameters. Indeed, this work has for only ambition to generate discussions on the above topics.

The choice of a model and its parameters is often, especially - but not only - in cases where data are scarce, guided by subjective considerations: a model may be preferred because it provides maps or conditional simulations that "look" right. There is nothing wrong with such an approach, and the success of geostatistical techniques in the petroleum industry proves it.

Nevertheless, geostatisticians should be extremely cautious when assuming that they can quantify uncertainty, especially when very few data are available to start with. If this is done, the assumptions made by the model must be clearly identified, and these may prove wrong as more data become available. In the data-scarce world of petroleum, the use of a probabilistic model is always associated with many assumptions. These assumptions should always be made explicit rather than left implicit, as it is too often the case. It should always be kept in mind that the main uncertainty may not lie in the variability between realisations at fixed model and fixed parameters but rather on the validity of the model itself, or of the parameters of this model.

Acknowledgements

The author thanks the management of Elf Aquitaine Production for permission to present this paper.

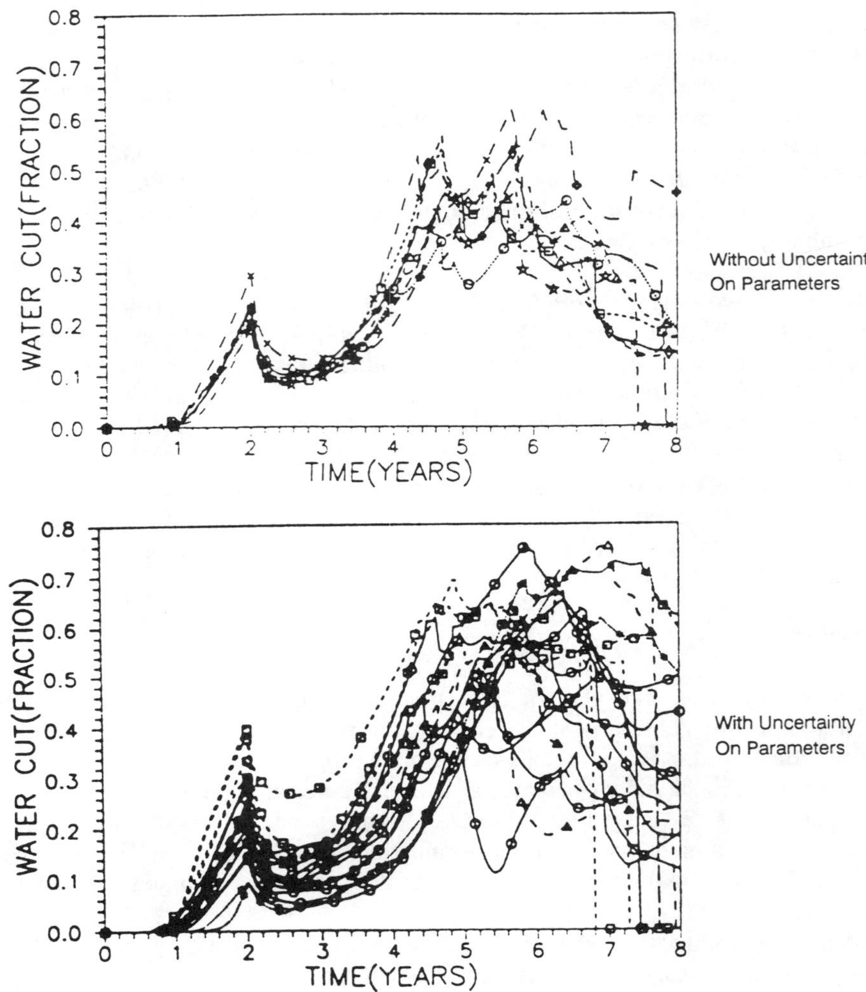

Figure 6: Impact of accounting for uncertainty on parameters of the geostatistical model when predicting water production over 8 years of an oil field's life (from Omre et al, 1993). The upper plot (8 realizations) is obtained by assuming statistical parameters such as those controlling the areal extent of large shales are fixed (standard approach). The lower plot (20 realizations) takes into account the uncertainty on these statistical parameters (bayesian approach). The variability between oil production characteristics changes dramatically from one plot to the other.

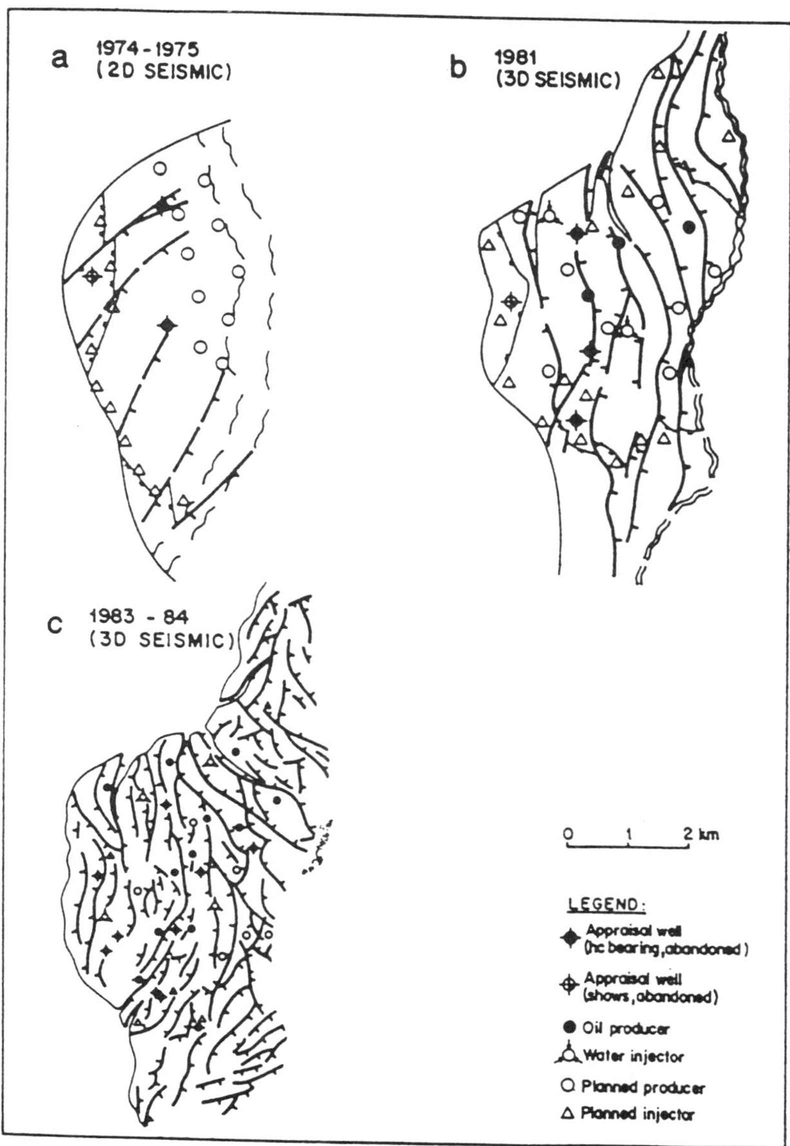

Figure 7: Comparison of 2D and 3D structural maps at Block IV, Cormorant Field, UK, North Sea (from Ruijtenberg et al, 1990).

References

Alabert, F.G. and G.J. Massonnat, 1990, Heterogeneity in a Complex Turbiditic Reservoir: Stochastic Modelling of Facies and Petrophysical Variability, SPE Paper 20604, p. 775-790.

Chiles, J.P., 1977, Géostatistique des Phénomènes non-stationnaires, Thèse de Docteur-Ingénieur, Ecole des Mines de Paris.

Corrigan, T., 1988, Factors controlling successful reservoir prediction: a cautionary tale from the UK North Sea, presented at the 2nd Conference on Resetrvoir Management in Field Development and Production, Stavanger.

Farmer, C.L, 1992, Numerical Rocks, in Mathematics of Oil Recovery, King ed, Oxford University Press.

Fouquet, C. de, H. Beucher, A. Galli, C. Ravenne, 1989, Conditional Simulation of Random Sets - Application to an Argilaceous Sandstone Reservoir, Geostatistics, Kluwer Publishers, Armstrong ed, Vol. 2, pp. 517-530.

Haldorsen, H.H. and C.J. Macdonald, 1987, Stochastic Modelling of Underground Reservoir Facies, SPE Paper 16751.

Handcock, M.S. and M.L. Stein, 1993, A Bayesian Analysis of Kriging, this volume.

Hewett, T.A., 1986, Fractal Distributions of Reservoir Heterogeneity and Their Influence on Fluid Transport, SPE Paper 15386

Jaynes, E.T., 1968, Prior Probabilities, IEEE Tansactions, Systems Science and Cybernetics SSC-4, 227-241. Reprinted in V.M. Rao Tummala and R.C. Hensha eds, Concepts and Applications of Modern Decision Models (Michigan State University Business Studies Series).

Journel, A.G. and C.V. Deutsch, 1993, Entropy and Spatial Disorder, Mathematical Geology, Vol. 25, Nuber 3, April 1993, pp. 329-355.

Journel, A.G. and C. Huijbregts, 1977, Mining Geostatistics, Academic Press, New York.

Journel, A.G. and J.J. Gomez Hernandez, 1989, Stochastic Imaging of the Wilmington Clastic Sequence, SPE Paper 19857.

Matheron, G., 1970, The Theory of Regionalized Variables and its Applications, Fasc. 5, Ecole des Mines de Paris, Fontainebleau.

Matheron G., 1988, Estimating and Choosing, Springer Verlag, Berlin.

Omre, H., H. Tjelmeland, Y. Qi, L. Hinderaker, 1993, Assessment of Uncertainty in the Production Characteristics of a Sandstone Reservoir, Reservoir Characterization III, Penwell, Linville ed, pp.556-603 .

Ripley, B.D, 1993, Stochastic Models for the Distribution of Rock Types in Petroleum Reservoirs, "Statistics in the Environmental and Earth Science", ed. A. Walden and P. Guttoep, Griffin.

Ruijtenberg, P.A., R. Buchanan, and P. Marke, 1990: Three-Dimensional Data Improve Reservoir Mapping, Journal of Petroleum Technology 42, January 1990, p. 22-61.

COMMENT ON "ESTIMATING OR CHOOSING A GEOSTATISTICAL MODEL" BY OLIVER DUBRULE

MARK S. HANDCOCK
Department of Statistics & Operations Research
Leonard N. Stern School of Business
New York University
New York, NY 10012-1126 U. S. A.

Oliver Dubrule has presented an insightful and articulate view of the uncertainty quantification issues facing geostatisticians. These comments are intended to provide additional discussion.

The key feature of Bayesian approaches is to choose to quantify all uncertainty via probability distributions. This not only includes the variation in the phenomena, but extends to the uncertainty in the model and other "states of nature". This allows the geostatistician to quantify uncertainty in a natural and powerful way. Often the uncertainty represented is not solely the variability in the process of geological deposition, but also that of the geostatistician's lack of knowledge of that process. For example, the porosity of the rock at an given location is usually regarded as fixed and the model expresses the uncertainty in the geostatistician's knowledge of that value. Quantifying geological knowledge can be tricky and explicitly demands the involvement of the geostatistician and the geologist. Often this is because we are required to provide knowledge about aspects of the model or the phenomena that are not easy to express.

We do not get something for nothing, as Dubrule's sand connectivity example shows. Expressing lack of knowledge about the direction of the channels by a uniform distribution over the possible directions leads to simulated data with a unreasonably highly level of connectivity. The key point here is that we are able to recognize that the implications of this particular quantification of knowledge are not consistent with our geological knowledge: our intuition about the resultant map-view is much better than our intuition about the distributions of channel directions. We can exploit this to ensure our quantification of geological knowledge are much more consistent with our actual knowledge. Let X be the map-view of the sand connectivity and θ the direction of the channels. Let $f(X \mid \theta)$ be the distribution of map-views given (a distribution of) directions θ. In the example θ is taken to be uniform and the resulting map-views ($f(X \mid \theta)$) were overly connected relative to map-views consistent with our geological knowledge. However, note that

$$P(\,X \mid \pi) \;=\; \int_{\Theta} f(X \mid \theta)\pi(\theta)d\theta$$

where $P(\,X \mid \pi)$ is the (marginal) distribution of the map-view given the chosen

R. Dimitrakopoulos (ed.), Geostatistics for the Next Century, 15–16.

distribution $\pi(\theta)$ of θ. This expression makes it clear how the distribution of directions effects the resulting map-views. We can use it to play "what if" scenarios with particular distributions for the direction by choosing $\pi(\theta)$ and seeing if the resulting map-view is consistent with our knowledge. For example we could choose a distribution with subsequent directions highly correlated rather than independent. This would lead to less connected map-views. The distribution could be fine-tuned by comparing its implications in terms of map-views with our intuition about how the map-views should look.

In this example we see that "lack of knowledge" can be a slippery concept. Apparent ignorance from one perspective appears to be knowledge from another. The issue is how to elicit the knowledge. By "lack of knowledge" we often mean that it is difficult to specify any knowledge beyond that in the data. The objective then is to choose a model for model uncertainty that has minimal impact on the subsequent inference relative to the information in the data. This is often difficult, as **any** quantification of geological knowledge will say something precise about the phenomena (as this example shows). The solution is to check that the conclusions from the analysis are robust to the choice of quantification of knowledge. If they are not, then we have learned that the present data can not answer the question and we can not reach an unambiguous conclusion unless we augment the data by more information in the form of additional data or geological knowledge.

REPLY TO "COMMENT ON ESTIMATING OR CHOOSING A GEOSTATISTICAL MODEL" BY M.S. HANDCOCK

Olivier Dubrule
EFL Aquitaine Production, France

I would like to thank Mark Handcock for his comments and emphasise two of his sentences: "We do not get something for nothing" and "Apparent ignorance from one perspective appears to be knowledge from another". Uncertainty quantification is usually achieved through the use of a probabilistic model. Too often, the geologist is not asked the right questions by the geostatistician. For instance, if the problem is probabilistic heterogeneity modelling, instead of being asked whether an object–based, a pixel–based or a gaussian approach is realistic for the depositional environment considered, the geologist may only be asked to comment on the range value of a variogram model: he only has control of limited aspects of model definition. As a result, realisations of the chosen model may not span the whole range of possible geological scenarios, which results, in most cases, into an underestimation of uncertainty. As with the example of channel directions in the paper, the geologist must be told about all the assumptions of the model: assuming that channel directions are independent is as important as deriving the shape and size of the same channels from outcrop observations. As argued by M. Handcock you get something from nothing by translating a lack of knowledge about channel orientation into the strong assumption of statistical independence, and ignorance about channel direction wrongly appears as knowledge about connectivity. This was recently illustrated by Journal and Deutsch (1993): In case of ignorance about the pdf of a random function, the choice of the Maximum Entropy Gaussian distribution tends to artificially decrease the variability between flow simulations performed on various realisations.
Because of this limited communication between geologist and geostatistician, the heterogeneity distribution may appear better known than it really is. For solving this communication problem, it is the geostatistician's responsibility to train the geologist about all the assumptions – explicit or implicit – made by a model. Geostatistics is not black magic, it is simply a method of quantifying geology, and the associated uncertainty.

REFERENCE

JOURNEL, A. G. and DEUTSCH, C. V. (1993), Entropy and Spatial Disorder, Mathematical Geology **25**, 3, 329-355.

R. Dimitrakopoulos (ed.), Geostatistics for the Next Century, 17.

COMPARISON OF COIK, IK AND MIK PERFORMANCES FOR MODELING CONDITIONAL PROBABILITIES OF CATEGORICAL VARIABLES

P. GOOVAERTS
Geology and Environmental Sciences Department
Stanford University
Stanford CA 94305
U.S.A.

A performance comparison of three algorithms (coindicator kriging, multiple indicator kriging, median indicator kriging) for estimating conditional probabilities of categorical variables is presented. The reference soil data set includes 2649 locations at which the soil type was determined. Three subsets of 50, 100 and 500 sample locations were randomly selected and used to estimate the probability for the different soil types at the remaining locations. In all cases the variograms required were modeled using the complete data set. The comparison of estimated vs true probabilities shows that, whatever the number of conditioning data, the theoretically better coIK does not provide more accurate re-estimates than the two other algorithms. The latter two algorithms involve less variogram modeling effort and smaller computational cost. Furthermore, the number of order relation deviations is showed to be significantly higher for the coIK results.

INTRODUCTION

In many areas, one has to deal not only with quantitative variables but very often with categorical variables. Then a common problem consists of estimating the probability for each category to prevail at any particular location. Using an indicator approach these probabilities can be established by three classes of algorithms: coindicator kriging (coIK), multiple indicator kriging (IK), and median indicator kriging (mIK). The "theoretically" better algorithm is coIK which accounts for the transition probabilities between different categories and yields the smallest estimation variance. Its implementation, however, requires inference and modeling all category indicator (cross) variograms, and solving large and possibly unstable cokriging systems. This explains why IK which involves only modeling the direct indicator variograms has been the preferred algorithm so far. Further simplification consists of considering the same model for all sample indicator variograms, this is referred to as median indicator kriging. Many variants to these three basic algorithms (Suro-Pérez and Journel, 1990; Soares, 1992; Dimitrakopoulos and Dagbert, 1993) have been developed but will not be considered in this study.

The recent development of iterative procedures (Goulard and Voltz, 1992) or graphical procedures (Chu, 1993) for fitting a linear model of coregionalization, coupled with the increase in computing power have rendered coindicator kriging practical. There remains to assess the actual practical advantage brought by coIK.

R. Dimitrakopoulos (ed.), Geostatistics for the Next Century, 18–29.

The fact that the cokriging estimation variance is smaller than the kriging estimation variance does not necessarily entail that the cokriging estimate is in practice the most accurate. Practice has recurrently ıown (J urnel and Huijbregts, 1978, p. 326) that cokriging improves little over kriging when il the covariables involved are equally sampled; such is the case in most indicator kriging applications. The reconciliation between the theoretical promise of cokriging and its lesser performance in practice remains an open question in geostatistical theory, yet it has major practical implications.

This paper adds another case study to the file "cokriging vs kriging". It presents a performance comparison utilizing a reference soil data set comprising 2649 locations at which the soil type was determined. Three subsets of 50, 100 and 500 sample locations were randomly selected and used to re-estimate by coIK, IK and mIK the probability for the different soil types at the remaining locations. The comparison criteria account for the following characteristics : number and magnitude of order relation deviations, and differences between true and estimated vectors of probabilities.

INDICATOR ALGORITHMS FOR ESTIMATING CONDITIONAL PROBABILITIES OF CATEGORICAL VARIABLES

Let $\{s_k \; ; \; k = 1, \ldots, K\}$ be a set of K mutually exclusive categories observed over a field $\mathcal{D}$. At each sample location $\mathbf{u}_\alpha$, a vector of K indicator values $i(\mathbf{u}_\alpha; s_k)$ can be formed where

$$i(\mathbf{u}_\alpha; s_k) = \begin{cases} 1 & \text{if } \mathbf{u}_\alpha \in s_k \\ 0 & \text{otherwise} \end{cases} \tag{1}$$

The different algorithms aim at estimating the conditional probabilities of the K categories at any location $\mathbf{u} \in \mathcal{D}$, i.e.

$$p(\mathbf{u}; s_k) = \text{Prob}\,\{\mathbf{u} \in s_k | (n)\} = \text{Prob}\,\{I(\mathbf{u}; s_k) = 1 | (n)\} \tag{2}$$

where the notation (n) represents the conditioning information utilized.

The estimated probability for a given category s_{k_0}, $p^*(\mathbf{u}; s_{k_0})$, is computed as a linear combination of neighboring indicator data $i(\mathbf{u}_\alpha; s_k)$.

Coindicator kriging

The coindicator kriging (coIK) estimate utilizes indicator data related to any category :

$$p^*_{coIK}(\mathbf{u}; s_{k_0}) = \sum_{k=1}^{K} \sum_{\alpha=1}^{n} \lambda_k(\mathbf{u}_\alpha; s_{k_0}).i(\mathbf{u}_\alpha; s_k) \tag{3}$$

The weights $\lambda_k(\mathbf{u}_\alpha; s_{k_0})$ are obtained by solving the following ordinary cokriging system of K(n+1) equations

$$\sum_{k'=1}^{K} \sum_{\beta=1}^{n} \lambda_{k'}(\mathbf{u}_\beta; s_{k_0}) \;\; \gamma_I(\mathbf{u}_\alpha - \mathbf{u}_\beta; s_k, s_k') - \mu_k \quad = \quad \gamma_I(\mathbf{u}_\alpha - \mathbf{u}_0; s_k, s_{k_0})$$

$$\forall\, \alpha = 1 \;\text{ to }\; n, \quad k = 1 \text{ to } K$$

$$\sum_{\beta=1}^{n} \lambda_k(\mathbf{u}_\beta; s_{k_0}) = \delta_{kk_0}$$
$$\forall \ k = 1 \ \text{ to } \ K \tag{4}$$

where μ_k is the kth Lagrange multiplier and δ_{kk_0} is the Kronecker delta.

This cokriging system calls for the joint modeling of K(K+1)/2 indicator (cross) variograms $\gamma_I(\mathbf{h}; s_k, s_{k'})$. This large number of variograms to model is the main reason why coIK is rarely used in practice. In fact, the difficulty does not lie in the number of models to infer but in the fact that the matrix of variogram models, $\mathbf{\Gamma}_I(\mathbf{h})$, must be conditionally negative semi-definite (Journel and Huijbregts, 1978, p. 171),

$$\mathbf{\Gamma}_I(\mathbf{h}) = \begin{bmatrix} \gamma_I(\mathbf{h}; s_1, s_1) & \cdots & \gamma_I(\mathbf{h}; s_1, s_K) \\ \vdots & & \vdots \\ \gamma_I(\mathbf{h}; s_K, s_1) & \cdots & \gamma_I(\mathbf{h}; s_K, s_K) \end{bmatrix}$$

The only model that is of wide usage so far is the linear model of coregionalization which involves modeling all the variograms as linear combinations of the same set of L basic variogram functions $g_l(\mathbf{h})$:

$$\gamma_I(\mathbf{h}; s_k, s_{k'}) = \sum_{l=1}^{L} b^l_{kk'} g_l(\mathbf{h}) \qquad \forall k, k' \tag{5}$$

where $b^l_{kk'}$ is the weight applied to the function $g_l(\mathbf{h})$ in the model for $\gamma_I(\mathbf{h}; s_k, s_{k'})$.

Using matrix notation, Eq. (5) is rewritten :

$$\mathbf{\Gamma}_I(\mathbf{h}) = \sum_{l=1}^{L} \mathbf{B}^l g_l(\mathbf{h}) \tag{6}$$

where each $\mathbf{B}^l = \left[b^l_{kk'}\right]$ is a K×K symmetric matrix called a coregionalization matrix.

This matrix formulation allows the conditionally negative semi-definite condition for $\mathbf{\Gamma}_I(\mathbf{h})$ to be expressed in practical terms: the condition holds true if the $g_l(\mathbf{h})$ are licit variogram models and each matrix $\mathbf{B}^l$ is positive semi-definite (Journel and Huijbregts, 1978, p. 172). Therefore, modeling the coregionalization involves, in practice, the following steps:

- choosing the number (L) and characteristics (type, range) of the basic variogram functions $g_l(\mathbf{h})$, and
- fitting the selected model to the experimental values, i.e. estimating the coefficients $b^l_{kk'}$, under the constraint of positive semi-definiteness of the coregionalization matrices $\mathbf{B}^l$.

Goulard and Voltz (1992) and Goovaerts (1992) describe in some detail the practice of choosing the number and type of models. Once selected the basic models $g_l(\mathbf{h})$

are fitted iteratively by weighted least squares.

Multiple indicator kriging

Unlike the coIK estimate, the multiple indicator kriging (IK) estimate utilizes only indicator data on the category s_{k_0} being considered :

$$p^*_{IK}(\mathbf{u}; s_{k_0}) = \sum_{\alpha=1}^{n} \nu_{k_0}(\mathbf{u}_\alpha; s_{k_0}).i(\mathbf{u}_\alpha; s_{k_0}) \tag{7}$$

The system to be solved is less cumbersome and requires only the K direct indicator variogram models $\gamma_I(\mathbf{h}; s_k, s_k)$.

$$\begin{aligned} \sum_{\beta=1}^{n} \nu_{k_0}(\mathbf{u}_\beta; s_{k_0}) \;\; \gamma_I(\mathbf{u}_\alpha - \mathbf{u}_\beta; s_{k_0}, s_{k_0}) - \mu_k &= \gamma_I(\mathbf{u}_\alpha - \mathbf{u}_0; s_{k_0}, s_{k_0}) \\ &\forall \quad \alpha = 1 \text{ to } n \\ \sum_{\beta=1}^{n} \nu_{k_0}(\mathbf{u}_\beta; s_{k_0}) &= 1 \end{aligned} \tag{8}$$

Unlike the linear model of coregionalization, there is no requirement for every indicator variogram to be modeled with the same set of basic variogram functions. Moreover, there is no constraint on the coefficients of the different models. This greater flexibility of modeling leads to models that generally fit the experimental values better.

Median indicator kriging

Similarly to IK, the median indicator kriging (mIK) estimate accounts only for the indicator information about the category s_{k_0} being considered :

$$p^*_{mIK}(\mathbf{u}; s_{k_0}) = \sum_{\alpha=1}^{n} \tau(\mathbf{u}_\alpha).i(\mathbf{u}_\alpha; s_{k_0}) \tag{9}$$

In addition, the mIK algorithm assumes that all K direct indicator variograms are proportional to each other (Deutsch and Journel, 1992, p. 74). In this paper, the common standardized indicator variogram $\gamma_{mI}(\mathbf{h})$ was computed as the mean of the K rescaled indicator variograms :

$$\gamma_{mI}(\mathbf{h}) = \frac{1}{K} \sum_{k=1}^{K} \frac{\gamma_I(\mathbf{h}; s_k, s_k)}{p_k(1 - p_k)}, \tag{10}$$

where $p_k = E[I(\mathbf{u}; s_k)]$ is the global proportion of facies k, and $p_k(1 - p_k)$ is the corresponding indicator variance.

Since the indicator data configuration is the same for all categories, one single IK system needs to be solved with the resulting weights $\tau(\mathbf{u}_\alpha)$ being used in Eq. (9) for all categories,

$$\sum_{\beta=1}^{n} \tau(\mathbf{u}_\beta) \;\; \gamma_{mI}(\mathbf{u}_\alpha - \mathbf{u}_\beta) - \mu \;\; = \;\; \gamma_{mI}(\mathbf{u}_\alpha - \mathbf{u}_0)$$
$$\forall \quad \alpha = 1 \text{ to } n$$
$$\sum_{\beta=1}^{n} \tau(\mathbf{u}_\beta) \;\; = \;\; 1 \qquad (11)$$

PERFORMANCE COMPARISON : A CASE STUDY

The data set

This study relates to a survey undertaken over a 3500 km^2 area located in south-east Scotland (McBratney et al., 1982). The data set comprises 2649 locations at which the soil type was determined according to morphological soil properties. Six different soil types, denoted s_k (k=1, ...,6), were identified. Most locations (61.6%, exactly) pertain to the soil types s_3 or s_4. The four other soil types represent fairly similar proportions of the data.

An important feature of each soil type is its spatial distribution. Figure 1 shows that locations of type s_1 are scattered in space whereas the other soil types form more spatially compact groups oriented NE-SW. Therefore, estimating the probabilities for soil s_1 is expected to be more difficult than for other soils.

General procedure

To assess the performances of coIK, IK and mIK and investigate their dependence on the number N_c of conditioning data, we proceeded as follows:

1. Three subsets of $N_c = 50, 100$ and 500 sample locations were randomly selected.

2. Each subset was used to estimate by the three algorithms the probability for the different soil types at the N_t remaining locations ($N_t = 2649 - N_c$). In each case, the search radius was 25 km and the nearest data up to a maximum of 32 were retained.

3. For each algorithm and set of N_t data locations,

    - the number and magnitude of order relation deviations were computed, see later definitions (12) and (13),
    - the departures between true and estimated vectors of probabilities were assessed using two criteria, defined hereafter in relations (14) and (15).

Variogram modeling

Whatever the number of conditioning data, the variograms required by the different algorithms were modeled using the complete reference data set. We decided to work in each case with the best structural information available because our primary interest is in comparing the performances of the algorithms, independently of questions of inference. Note that this decision privileges coIK because it is the algorithm that involves the most difficult inference.

For coIK, a linear model of coregionalization was fitted to the 21 indicator (cross) variograms, see Figures 2 and 3 (solid line), using the iterative procedure developed by Goulard and Voltz (1992). The fit appears quite satisfactory apart from the cross-variograms $\gamma_{s_3 s_5}$ and $\gamma_{s_3 s_6}$. These two cross variograms are zero up to 10 or 15 km, which indicates that the two soil types have no common border.

For IK, the 6 indicator variograms were modeled independently, see Figure 2 (dashed line). Since there is no constraint on the parameters of the models, the fits are seen to be better. Note the small range of the indicator variograms γ_{s_1} and γ_{s_2}, which reflects the greater scattering of the two soil types in the region.

The median indicator variogram required by mIK was computed from the mean of the 6 direct standardized indicator variograms, see Figure 4.

So far only isotropic models have been considered. According to Figure 1, an anisotropic behaviour is expected for most categories. Indeed, except from soil types s_1 and s_2, the indicator variograms are anisotropic, especially for lags greater than 5 km. To investigate the possible loss of accuracy due to the assumption of isotropy, an anisotropic model was fitted to the four direct indicator variograms (s_3, s_4, s_5, s_6) and to the median indicator variogram, see Figure 5. As the categorical variables do not share the same directions of anisotropy, it was not possible to fit an anisotropic linear model of coregionalization. Therefore the comparison between isotropic and anisotropic modeling is performed only for the IK and mIK algorithms.

Result 1 : Order relation deviations

At each location $\mathbf{u}_\alpha$, the set of K estimated probabilities $\{p^*(\mathbf{u}_\alpha; s_k)\ ;\ k = 1, \ldots, K\}$ must satisfy the two following order relations :

$$p^*(\mathbf{u}_\alpha; s_k) \in [0, 1] \tag{12}$$

$$\sum_{k=1}^{K} p^*(\mathbf{u}_\alpha; s_k) = 1 \tag{13}$$

The first condition may not be satisfied because kriging weights can be negative, therefore the kriging estimate is a non-convex linear combination of the conditioning data. For each algorithm, the percentage of probabilities valued outside the interval $[0, 1]$ as well as the magnitude of these order relation deviations were computed, see Table 1. Whatever the number of conditioning data, coIK has about two to four times more order relation problems than the two other algorithms. This can be explained by the coIK constraint that the sum of the weights applied to secondary variables is zero, which implies a larger number of negative weights and so a greater risk of getting probabilities outside the interval $[0, 1]$.

Except for anisotropic indicator kriging ($N_c = 50$), the magnitude of order relation deviations is small.

The first constraint (12) was met by resetting the faulty probabilities to the nearest bound, 0 or 1. Once this correction has been performed, each value $p^*(\mathbf{u}_\alpha; s_k)$ was restandardized by the sum $\sum_{k=1}^{K} p^*(\mathbf{u}_\alpha; s_k)$ to meet the second condition (13).

Result 2 : Differences between true and estimated vectors of probabilities

For the N_t test locations, both true indicator values $i(\mathbf{u}_\alpha; s_k)$, with $i(\mathbf{u}_\alpha; s_k) = 0$ or 1, and estimated probabilities $p^*(\mathbf{u}_\alpha; s_k)$, with $p^*(\mathbf{u}_\alpha; s_k) \in [0, 1]$, are available. These were compared using the two following criteria.

The first criterion denoted C_1 is defined as :

$$C_1 = \frac{1}{N_t} \sum_{\alpha=1}^{N_t} C_1(\mathbf{u}_\alpha) \ , \tag{14}$$

where

$$C_1(\mathbf{u}_\alpha) = \sum_{k=1}^{K} \omega_k \left| p^*(\mathbf{u}_\alpha; s_k) - i(\mathbf{u}_\alpha; s_k) \right|$$

The weights ω_k were computed as

$$\omega_k = \frac{\left(\sum_{k=1}^{K} 1/p_k(1-p_k)\right)^{-1}}{p_k(1-p_k)} \quad , \quad \sum_{k=1}^{K} \omega_k = 1$$

The index C_1 measures the *average absolute deviation between the true and estimated vectors of probabilities.* Its value is close to zero if the algorithm is accurate. The weighting system adopted gives more weight to categories with small indicator variance, i.e. to soils of small areal extent. Indeed, categories that are rare are often of higher interest.

According to criterion C_1, IK and mIK perform similarly and slightly better than coIK, whatever the number of conditioning data, see Table 2. The accuracy of the estimation increases with the number of conditioning data but is not improved by taking into account anisotropy.

For $N_c = 50$, Table 3 gives the values of C_1 for each soil type. There is a strong relation between accuracy of the estimation and the extent (proportion) of the corresponding soil type. The lowest (best) C_1 values are observed for the two most frequent categories, s_3 and s_4. The worst results are obtained for the least frequent categories, s_1 and s_2. For these last two categories, IK and especially mIK do a better job than coIK. The anisotropic modeling improves the accuracy of the estimation only for soil type s_5.

The second criterion denoted C_2 is defined as :

$$C_2 = 1 - \frac{1}{N_t} \sum_{\alpha=1}^{N_t} C_2(\mathbf{u}_\alpha) \ , \tag{15}$$

where

$$C_2(\mathbf{u}_\alpha) = \sum_{k=1}^{K} p^*(\mathbf{u}_\alpha; S_k).i(\mathbf{u}_\alpha; S_k)$$

The index C_2 equals one minus the average estimated probability for the correct soil type. C_2 measures the *average probability to draw a wrong category when using the estimated probabilities in the simulation process.* Its value is close to zero if the algorithm is accurate.

When only 50 conditioning data are available, the average probability for misdrawing is 0.45, see Table 2. This probability decreases with increasing number of conditioning data. According to C_2, no algorithm performs better than the others.

For $N_c = 50$, Table 3 gives the values of C_2 for each soil type. As for C_1, there is a strong correlation between the probability of misdrawing and the smaller extent of the corresponding soil type. Again, the highest values (0.7-0.8) are observed for the least frequent categories s_1 and s_2, with IK and especially mIK performing better than coIK. For soil type s_5, the anisotropic modeling allows reducing the probability of misdrawing from 0.7 to 0.6.

CONCLUSIONS

In theory, coindicator kriging should outperform the other algorithms because it accounts for the cross relations, i.e. the transition probabilities between the different categories. In practice, however, coIK does not appear to do a better job than IK or mIK.

- The number of order relation deviations is found to be higher for coIK.
- The comparison of true vs estimated probabilities shows that coIK does not provide more accurate estimations than the other algorithms. It underperforms IK and mIK for the categories of smaller extent which are potentially of greater interest.

Even if better software makes coindicator kriging less demanding, this study draws our attention to the possibility that putting more partially redundant information into a (co)kriging system does not necessarily improve the accuracy of the estimation. It is conjectured that the linear model of coregionalization may artificially limit the potential of cokriging.

ACKNOWLEDGEMENTS

The author thanks the East Scotland College of Agriculture, especially Mr H. M. Gray, Dr R. G. McLaren, and Dr R. B. Speirs, for the data from their survey.

REFERENCES

- Chu, J. (1993) "XGAM user's guide", Stanford Center for Reservoir Forecasting, Stanford University, Unpublished Annual Report n^0 6.
- Deutsch, C.V. and Journel, A.G. (1992) GSLIB : Geostatistical Software Library and User's Guide, Oxford University Press, New-York.
- Dimitrakopoulos, R. and Dagbert, M. (1993) "Sequential modelling of relative indicator variables : dealing with multiple lithology types", in A. Soares (ed.), Geostatistics Tróia '92, Kluwer Academic Publishers, Dordrecht, pp. 413-424.
- Goovaerts, P. (1992) Multivariate geostatistical tools for studying scale-dependent correlation structures and describing space-time variations, Ph. D. thesis, U.C.L., Belgium.

- Goulard, M. and Voltz, M. (1992) "Linear coregionalization model; tools for estimation and choice of cross-variogram matrix", Math. Geol. 24, 269-286.
- Journel, A.G. and Huijbregts, C.J. (1978) Mining Geostatistics, Academic Press, London.
- McBratney, A. B., Webster, R., McLaren, R. G. and Spiers, R. B. (1982) "Regional variation of extractable copper and cobalt in the topsoil of south-east Scotland", Agronomie 2, 969-982.
- Soares, A. (1992) "Geostatistical estimation of multi-phase structures", Math. Geol. 24, 149-160.
- Suro-Pérez, V. and Journel, A.G. (1990) "Stochastic simulation of lithofacies: an improved sequential indicator approach", in Guerillot and Guillon (eds), Proc. of the 2nd European Conf. on the Math. of Oil Recovery, Publ. Technip, pp. 3-10.

Table 1. Percentage and magnitude of order relation deviations for different number of conditioning data.
The deviation magnitude is defined as : $(p^* - 1)$ if $p^* > 1$, and $|p^*|$ if $p^* < 0$.

	Percentage					Magnitude				
	Isotropic			Anisotropic		Isotropic			Anisotropic	
	coIK	IK	mIK	IK	mIK	coIK	IK	mIK	IK	mIK
$N_c = 50$	29.4	13.5	8.8	10.3	10.0	0.026	0.031	0.012	0.092	0.018
$N_c = 100$	29.4	15.0	17.3	19.8	15.9	0.025	0.017	0.014	0.029	0.015
$N_c = 500$	35.5	6.6	4.5	9.4	7.6	0.011	0.010	0.005	0.030	0.007

Table 2. Values of criteria C_1 and C_2 for different number of conditioning data (best if values small).

	Criterion C_1					Criterion C_2				
	Isotropic			Anisotropic		Isotropic			Anisotropic	
	coIK	IK	mIK	IK	mIK	coIK	IK	mIK	IK	mIK
$N_c = 50$	0.129	0.126	0.126	0.129	0.125	0.446	0.443	0.452	0.450	0.446
$N_c = 100$	0.112	0.110	0.110	0.110	0.109	0.381	0.376	0.381	0.378	0.377
$N_c = 500$	0.086	0.080	0.079	0.081	0.080	0.287	0.271	0.269	0.274	0.270

Table 3. Values of criteria C_1 and C_2 for each category ($N_c = 50$) (best if values small).

	Criterion C_1					Criterion C_2					
	Isotropic			Anisotropic		Isotropic			Anisotropic		Ref.
	coIK	IK	mIK	IK	mIK	coIK	IK	mIK	IK	mIk	prop.
s_1	0.224	0.210	0.202	0.210	0.202	0.782	0.734	0.713	0.718	0.712	10.4
s_2	0.349	0.330	0.313	0.323	0.309	0.803	0.757	0.724	0.736	0.717	5.4
s_3	0.062	0.062	0.060	0.067	0.061	0.262	0.264	0.265	0.278	0.269	37.7
s_4	0.095	0.091	0.101	0.108	0.100	0.361	0.375	0.408	0.435	0.400	23.9
s_5	0.214	0.206	0.208	0.188	0.188	0.730	0.716	0.732	0.618	0.664	10.4
s_6	0.152	0.161	0.162	0.156	0.170	0.495	0.509	0.534	0.509	0.549	12.2

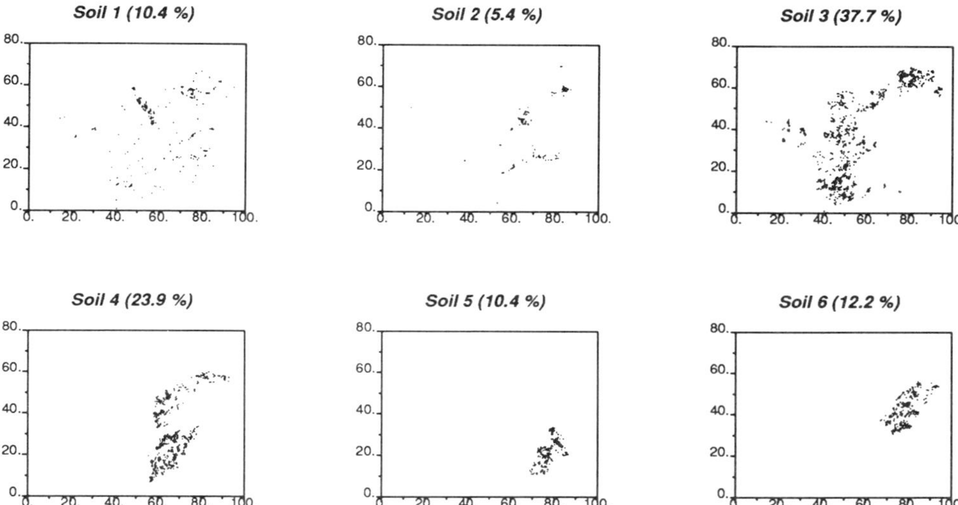

Figure 1: Location of the samples for the different soil types the proportion of which is indicated into brackets (units = km).

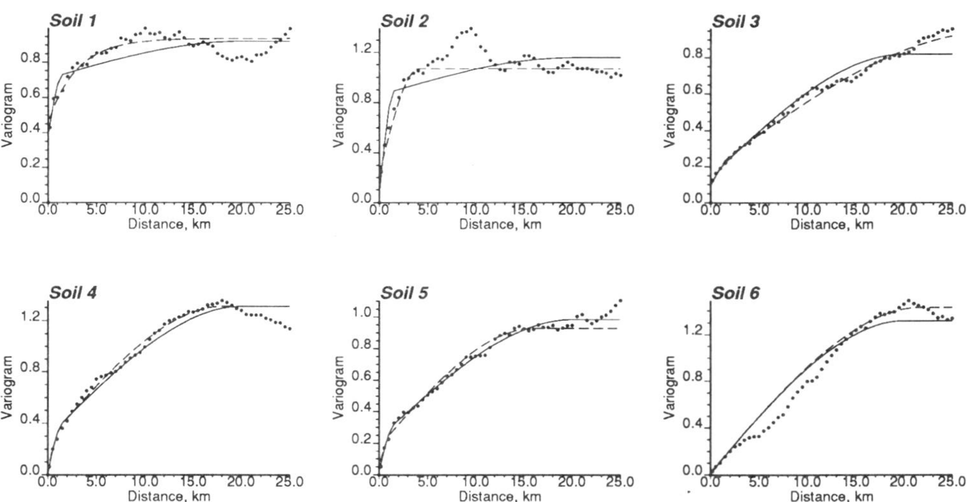

Figure 2: Indicator variograms and models, as used in CoIK (solid line) or IK (dashed line).

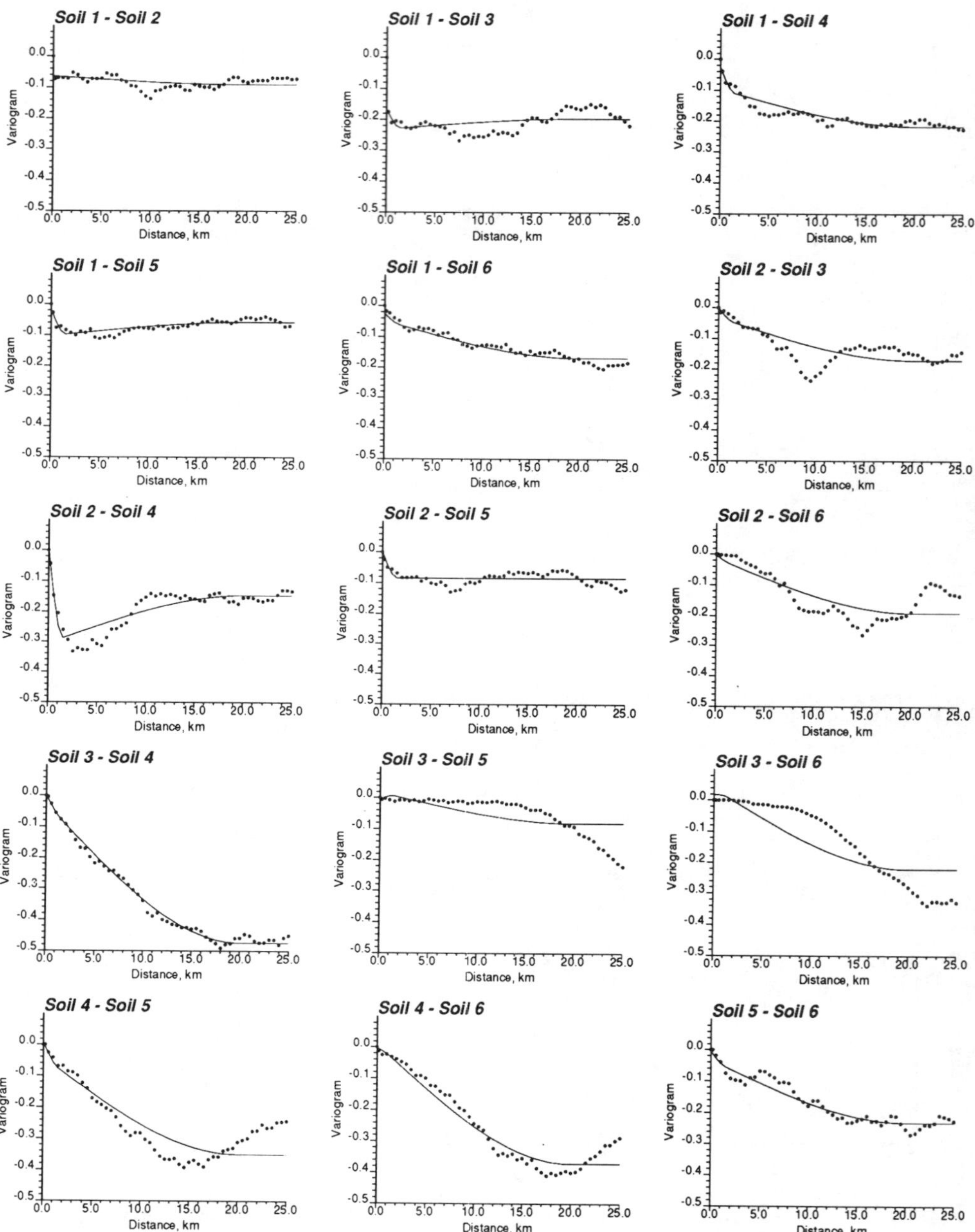

Figure 3: Cross indicator variograms and the linear model of coregionalization fitted.

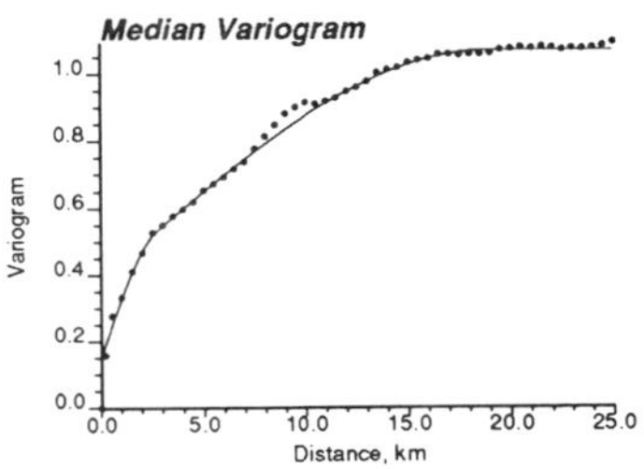

Figure 4: Median indicator variogram.

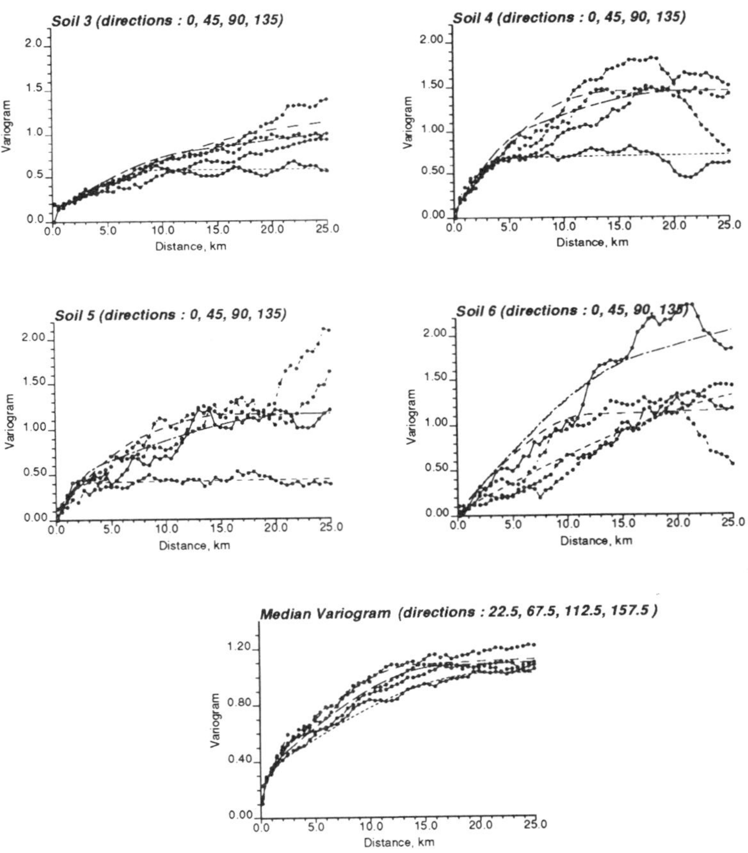

Figure 5: Anisotropic models for the direct indicator variograms and median indicator variogram.

MODELING UNCERTAINTY: SOME CONCEPTUAL THOUGHTS

ANDRÉ G. JOURNEL
Geology and Environmental Sciences Department
Stanford University
Stanford CA 94305

ABSTRACT

The place of random experiments and stochastic models in experimental sciences is discussed first. Models of uncertainty are necessarily based on prior decisions about the randomization of either the unknown itself and/or its estimate and, as such, cannot claim to any objectivity. A safeguard for building such models is to charge them maximally with data deemed relevant to the "unknown" at hand.

Another look at Bayes' relation shows that its practice requires the transfer of statistics from a calibration data set to the (prior) environment of the updating process. The IK paradigm is shown to be such transfer of calibration likelihood statistics. The paper ends with a generalization of the sequential Gaussian simulation algorithm allowing for a sequence of parametric non-Gaussian conditional distributions yet with means and variances given by SK.

Keywords: Stochastic simulation, random models, Bayes updating, likelihood functions, sequential simulation, non-Gaussian models.

THE PLACE OF RANDOM EXPERIMENTS

The utilization of Monte Carlo techniques, and more generally any algorithm that relies on the processing of random numbers to simulate experiments or numerical models otherwise unaccessible, has been a puzzle to some earth scientists. How could random numbers be re-organized to mimick even remotely the complexity of natural phenomena? The answer lies in the revolution brought forward by computers with their ever increasing speed and, possibly, more importantly RAM memory.

Graduating from the Bernouilli jar where random numbers had only in common their originating single-point distribution (histogram), we can now ask the computer to organize these random numbers according to specific two-point, three-point, multiple-point statistics or, better and if available, according to deterministic laws that control aspects of the joint distribution of variables in space. Beyond these input data, the stochastic model allows the resulting images to explore the remainder uncertainty, that which is very precisely the focus of investigation.

There are many advantages to such random, brute force, approach to modeling or exploring natural processes, most prominently,

R. Dimitrakopoulos (ed.), Geostatistics for the Next Century, 30–43.

• it is much easier, faster and less expensive that actual field or lab experiments, which is *not* saying that it should replace such experiments. I boldly liken the stochastic physicist of the late twentieth century to his prestigious forefathers of the nineteenth century whose discoveries were often associated to innovative experiments to collect relevant data. The stochastic physicist will build a numerical model that integrates (freezes) any prior information and relies on raw computer power to explore the remaining space of uncertainty. That exploration seeks structures that may lead to an improved understanding of the natural process under study.

• there is no need to fully understand a priori the algorithm through which the original random numbers are being re-assembled. I would even go as far as to say that potential for discovery decreases as the algorithm (not the physics!) is better understood and, thus, tend to cool down the random mobility too early beyond the unquestionable (frozen) data.

Rigid models, such as the infamous iid model for multivariate or multiple-point distributions or the over famous multivariate Gaussian model, were useful in a time when scientists had only calculus and a slide-rule to process data. It was then important that prior models be parametric, more exactly parameter-poor: a single univariate distribution suffices to characterize multivariate distribution of iid models, that distribution plus a mere covariance suffice to characterize multivariate Gaussian-related models and their close cousins the factorable-type models. Also, because hand-performed calculus was tedious and does not allow for random perturbations, prior internal consistency of these parameter-poor models was necessary.

Stochastic models can afford to be parameter-rich, a better qualifier than "non-parametric", including, e.g., all or part of the two-point distribution (beyond the mere covariance), plus parts of specific multiple-point distributions plus constraints as generated by known physical laws. The problems associated with using such parameter-rich models are twofold,

(1) **inference:** these multiple parameters and/or conditioning data must often be inferred from different sources. For example, there may be direct samples on the phenomenon (field) being investigated whereas statistics are inferred from a similar but different training image (outcrop). There may be also interpretative expert information, e.g., geological interpretation leading to a model of anisotropy.

(2) **consistency:** these multiple parameters originating from different sources of information may be in part inconsistent. Computing power allows finding a balance between conflicting input parameters: some conditions are allowed to be relaxed, some others are strictly enforced (deeply frozen). Typically in geostatistics, reproduction of well/borehole data is enforced, reproduction of variogram model(s) is more lax.

Problems of prior inconsistency are solved as the stochastic simulation progresses. Too strict attention to consideration of prior consistency leads to using parameter-poor models that are inconsistent with information they cannot account for. In an era of computer imaging, it often pays to see first then to understand rather than understand before seeing. Of course, the computer images must be reliable in that they incorporate a maximum of the information deemed relevant.

MODELING UNCERTAINTY

A trademark of geostatistics and more generally of probabilistic prediction is the modeling of uncertainty. No one argues the fact that in any estimation there is uncertainty involved, what is arguable is the definition of uncertainty itself.

Any "true value" z which is the goal of an estimation endeavor is unique,[1] it is either known and the estimation problem vanishes, or it is unknown and for each possible estimated value z^* there is an error, unfortunately equally unique and unknown. Assessment of the error $z^* - z$ calls for a randomization of either z^*, z or $r = z^* - z$. "Randomization" refers to a process allowing the definition of other true values $z^{(l)}$, or other estimated values $z^*_{(k)}$, or other error values $r^{(j)}$. Then, observation of the distribution of these substitutes allows an assessment of the actual error $z^* - z$. Since this randomization process is not unique there cannot be a fully objective assessment of uncertainty, unless a specific randomization process is agreed upon, then, rather than "objective" the correct adjective should be "conventional".

Examples of randomization processes that would lead to different uncertainty models are:

- The same data set may be given to K independent laboratories / consultants who will return K estimated values $z^*_{(k)}, k = 1, \ldots, K$. Note that the distribution of the $z^*_{(k)}$'s involve all K estimated values, it is not specific to anyone in particular. The problem here lies with the arbitrary range and selection of independent consultants or, more generally, of different massaging of the same data set [Englund, 1990].
- A specific laboratory/consultant is given L sets of data related to L known calibration values $z^{(l)}, l = 1, \ldots, L$, and is asked to provide the corresponding L estimates $z^{(l)*}$. The distribution of the L observed errors $z^{(l)*} - z^{(l)}$ is used to model the uncertainty of that lab in evaluating an unknown value z. The problem now lies with the availability and representativity of the calibration set, $\left\{z^{(l)}, l^{\underline{th}} \text{ data set}\right\}, l = 1, \ldots, L$. Is that set representative of future errors?

This second approach is similar to cross-validation in geostatistics, whereby known values $z^{(l)} = z(\mathbf{u}_l)$ are re-estimated from neighboring values. Unfortunately, when the true value z refers to a global property it is usually difficult to find representative substitutes $z^{(l)}$, unless a random variable or function model is used, see hereafter.

- L known values $z^{(l)}$, similar in some sense to the unknown value z, are retained and their distribution is used to provide a model of uncertainty about z. Note again that such distribution is not specific to any particular estimated value z^*; in fact it may be established prior to the derivation of any estimate for z. Yet data are needed to determine whether the environment of a known value $z^{(l)}$ qualifies it to belong to the previous distribution.

This last approach is similar to that underlying the choice of a random variable or function model Z. A random variable (RV) can be seen as a set of outcomes $\left\{z^{(l)}, l = 1, \ldots, L\right\}$ associated with a corresponding set of probabilities of occurence p_l (in the case of a discrete distribution). Data are needed to "condition" the distribution, i.e., to define the RV model. Here the data are used *not* to build an estimate but to define the randomization process of the true value z; then from the RV model Z an "optimal" estimate $z^*_{(k_0)}$ can be derived specific to a particular utility or loss function [Srivastava, 1987].

Some guidelines

In the absence of an objective or agreed-upon definition of the model of uncertainty,

[1]There is not even a "true value" z. The image z of any phenomenon is necessarily related to a specific measurement device and to its unique instance of application. Estimation would then be the exercise of building an alternative image z^* of z from related available images called "data".

one can limit the impact of subjectivity with the following guidelines:

1. First and foremost is honesty: *accept that all assessments of uncertainty are but models and there is no "best" model.* The randomization process from which the uncertainty assessment is derived should be stated clearly, i.e. in layman terms.

2. The second guideline relates to the definition of the randomization process: it should be such as to *freeze* (i.e. reproduce) *any piece of information deemed intrinsic to the estimation at hand* and make the rest variable.

3. The third guideline is to *perform systematic calibration of any data used.*

When dealing with estimation in space, one would typically "freeze" the sample values at their locations, but not necessarily the statistics derived from these data. Indeed, if samples can often be considered as representative of the corresponding local values, their histogram or variogram may not be considered as fully representative of the underlying population because of sparsity and possible bias (wells tend to be located in high pay zones). It is perfectly licit to condition a random function (RF) model to local sample values and, simultaneously, to statistics (histogram, variogram, multiple points covariances) taken from a different data set: the conditional simulation algorithm should reconciliate the two sets of information. If there is a conflict (inconsistency) between the two sets of information, some tradeoff is obtained usually by privileging the local information.

Although there is no best model of uncertainty, a good model is one that accounts for (freeze) the data deemed relevant to the estimation at hand. In other words, a "good" model of uncertainty is one that is data-charged, which implies that one cannot lose by getting more data and more relevant data.

The third guideline is a safeguard against subjectivity in retaining "deemed" relevant information. Consider, for example, the case of seismic data as used to evaluate petrophysical rock properties (porosity, permeability). Field experience and/or lab data may indicate a correlation between bright spots of seismic travel time and local concentrations of gas or oil. No matter how critical is that information, it should be quantitatively calibrated prior to being used as conditioning data in a numerical reservoir model. For example, that calibration can take the form of a prior conditional distribution of the type [Zhu and Journel, 1993]:

$$\text{Prob}\left\{Z(\mathbf{u}) \leq z \mid \text{seismic hot spot around } \mathbf{u}\right\},$$

where $Z(\mathbf{u})$ is the RF modeling oil concentration at location $\mathbf{u}$.

Short of actual field data to perform such calibration, the previous conditioning may be introduced as a prior expert judgement but it should be clearly stated as such. A "good" reservoir model depends on the accuracy of such prior deterministic judgements that need only stem from relevant field experience.

Although calibration should rely on statistical data at the scale at which the information is to be used, beware that statistical correlation does not provide causal interpretation and can be blurred by poor data. A "good" calibration must be backed by sound physical interpretation based, possibly, on rigorous lab data.

ANOTHER LOOK AT BAYESIAN UPDATING

Some naive non-statisticians, like this author, may have wondered at the essence of Bayes' relation and the reason for its unquestionable efficiency beyond what, at first look, may appear as a tautology. A dual look at Bayesian updating allows an interesting and potentially fruitful new interpretation of generalized indicator kriging.

Some recalls

Consider two random events A and B, where A models the unknown and B models the information set. Realization of A and B are denoted with corresponding small case letters a and b. The conditional probability definition is written:

$$P\{A=a|B=b\} = \frac{P\{A=a,B=b\}}{P\{B=b\}} \tag{1}$$

or

$$\begin{aligned} P\{A=a,B=b\} &= P\{A=a|B=b\}\cdot P\{B=b\} \\ &= P\{B=b|A=a\}\cdot P\{A=a\} \end{aligned} ,$$

from which Bayes' relation is derived:

$$P\{A=a|B=b\} = \frac{P\{B=b|A=a\}}{P\{B=b\}}\cdot P\{A=a\} \tag{2}$$

with in addition: $P\{B=b\} = \sum_{(L)} P\{B=b|A=a_l\}\cdot P\{A=a_l\}$,
where (L) is the set of all possible realizations of A.

Bayes' relation (2) appears as a rewriting of the conditional probability definition (1); indeed, the likelihood probabilities $P\{B=b|A=a_l\}$ can be retrieved from the joint probabilities $P\{A=a_l,B=b\}$. The practice of Bayes' relation escapes this tautology because the likelihood function, $P\{B=b|A=a_l\}$ as a function of a_l, is imported from a data environment different from that of the evaluation of the prior probability $P\{A=a\}$. The practice of Bayes' relation is more fairly represented by notations that would reveal that import process:

$$P\{A=a|B=b\} = \frac{P_1\{B=b|A=A\}}{P_1\{B=b\}}\cdot P_0\{A=a\} \tag{3}$$

The subscript $_0$ refers to the information available prior to knowledge of $B=b$, and $_1$ refers to the calibration set from which the updating factor is determined.

Bayes' relation (3) can also be written as:

$$P\{A=a|B=b\} = (1+\lambda_1)\cdot P_0\{A=a\} \tag{4}$$

with:

$$\lambda_1 = \frac{P_1\{B=b|A=a\} - P_1\{B=b\}}{P_1\{B=b\}}$$

λ_1 appears as a relative likelihood probability or a relative updating of the probability of the data event $B=b$ by the information $A=a$. Again, the subscript $_1$ recalls that the factor λ_1 originates from a calibration data set different from that used to determine the prior probability $P_0\{A=a\}$.

The formulation (4) allows an interesting interpretation of the practice of Bayesian updating:

$$\frac{P\{A=a|B=b\} - P_0\{A=a\}}{P_0\{A=a\}} = \frac{P_1\{B=b|A=a\} - P_1\{B=b\}}{P_1\{B=b\}} = \lambda_1$$

The relative updating ratio of the probability of event $A=a$ is identified to the relative updating ratio λ_1 of the probability of the data event $B=b$, the latter being inferred from a calibration data set. Identifying ratios of change is, indeed, a common very simple yet very effective inference paradigm.

An inference problem:

The relative likelihood λ_1 can be derived from a prior parametric model for either the likelihood function, $P\{B = b|A = a_l\}$ as a function of a_l, or for the joint probability $P_1\{B = b, A = a\}$. The parameters of such model are typically inferred from the calibration set (subscript $_1$). The most critical decision in such approach is the choice of the model, not so much the algorithm for determination of its parameters.

The likelihood factor λ_1 can also be derived from experimental frequencies directly read from the calibration data set. That data set, however, may not be exhaustive enough to include enough realizations of the joint event $\{A = a, B = b\}$, particularly if the data event $B = b$ is complex, e.g., it includes data values at many specific locations in space. One way to circumvent such problem is to split the information $B = b$ into less complex component data events and consider separate updatings from each of the component data events.

The indicator of occurrence of any event, say $B = b$, can always be expressed as a product of a, possibly very large, number of more basic indicator random variables:

$$I(B;b) = \Pi_{k=1}^{K} I_k = \begin{cases} 1 & , \text{ if } B = b \\ 0 & , \text{ if not} \end{cases} \tag{5}$$

The I_k's are binary indicator RV's valued 0 or 1. The realization $B = b$ calls for all indicator I_k's to be valued 1. The I_k's are "basic" in the sense that they represent occurrence of events elementary enough to be observable often enough in the calibration data set, more precisely, they are such that the following statistics can be inferred:

$$\begin{aligned} E\{I_k\} &= P_1\{I_k = 1\} \\ \text{Cov}_1\{I_k, I_{k'}\} &= P_1\{I_k I_{k'} = 1\} - P_1\{I_k = 1\} \cdot P_1\{I_{k'} = 1\} \\ \text{Cov}_1\{I_k, I(A;a)\} &= P_1\{I_k = 1, A = a\} - P_1\{I_k = 1\} \cdot P_1\{A = a\} \\ &\forall k, k' = 1, \ldots, K. \end{aligned}$$

The exact Bayes' relation (4) can be used for each of the K partial updatings:

$$P\{A = a|I_k = 1\} = (1 + \lambda_{1,k}) \cdot P_1\{A = a\} \tag{6}$$

with:

$$\lambda_{1,k} = \frac{P_1\{I_k = 1|A = a\} - P_1\{I_k = 1\}}{P_1\{I_k = 1\}} = \frac{\text{Cov}_1\{I_k, I(A;a)\}}{E_1\{I(A;a)\} \cdot E_1\{I_k\}} \tag{7}$$

for all $k = 1, \ldots, K$.

The problem now is to recombine these K partial updatings into an estimate of the probability $P\{A = a|B = b\}$ updated by the global data event $B = b$, i.e., $I(B;b) = \Pi_k I_k = 1$.

An identification paradigm

Starting from the exact Bayes' relation (4),

$$P\{A = a|B = b\} = (1 + \lambda_1) \cdot P_0\{A = a\},$$

the idea is to replace the global likelihood factor λ_1 by a linear combination of the available partial factor $\lambda_{1,k}$'s:

$$\lambda_1^* = \sum_{k=1}^{K} \omega_k \lambda_{1,k} \tag{8}$$

then:

$$P^*\{A=a|B=b\}=(1+\lambda_1^*)\cdot P_0\{A=a\} \tag{9}$$

The problem now is one of determining the K weights ω_k's. The criterion proposed here is identification of known results, a criterion used in dual kriging [Journel 1989, p.14 and Appendix A]. The ω_k's should be such that the resulting estimator (9) identifies the following K limit results:

$$P^*\{I_k=1|B=b\}=(1+\lambda_1^*)\cdot P_0\{I_k=1\}=1,\quad \forall k=1,\ldots,K \tag{10}$$

When the event $A=a$ is identified to the elementary event $I_k=1$, the definition (7) gives for each of the K partial updating factors

$$\lambda_{1,k'}=\frac{\mathrm{Cov}_1\{I_k,I_{k'}\}}{E_1\{I_k\}\cdot E_1\{I_{k'}\}},\quad k'=1,\ldots,K$$

The K constraints (10) are then written, accounting for the definition (8) of the estimate λ_1^* :

$$\left[1+\sum_{k'}\omega_{k'}\frac{\mathrm{Cov}_1\{I_k,I_{k'}\}}{E_1\{I_k\}\cdot E_1\{I_{k'}\}}\right]\cdot E_0\{I_k\}=1,\quad \forall\ k=1,\ldots,K$$

For consistency, identify $E_0\{I_k\}$ to the calibration value $E_1\{I_k\}$, the previous constraints are simplified into the linear system:

$$\sum_{k'=1}^{K}\nu_{k'}\ \mathrm{Cov}_1\{I_k,I_{k'}\}=1-E_1\{I_k\},\quad k=1,\ldots,K \tag{11}$$

with: $\omega_{k'}=\nu_{k'}E_1\{I_{k'}\}$.

In summary, the exact Bayes' relation (4) with likelihood factor λ_1 is estimated by the following relation using a linear combination of partial likelihood factors $\lambda_{1,k}$:

$$P^*\{A=a|B=b\}=(1+\lambda_1^*)\cdot P_0\{A=a\} \tag{12}$$

where: $\lambda_1^*=\sum_{k=1}^{K}\nu_k\cdot P_1\{I_k=1\}\cdot\lambda_{1,k}$,
the weights ν_k's are given by the linear system (11), and the partial likelihood factors $\lambda_{1,k}$ are determined from the calibration data set, see relation (7).

Remarks:

- The estimator (12) of the conditional probability $P\{A=a|B=b\}$ is none other than that provided by indicator kriging (IK), and the system (11) is none other than the dual IK system; for a proof see Appendix A.

The solution to system (11) exists and is unique as long as the left hand side matrix is correctly inferred or modeled as a covariance matrix.

- The weight applied to each partial likelihood factor $\lambda_{1,k}$ is proportional to the probability of occurrence $P_1\{I_k=1\}$ of the corresponding elementary data event, which makes sense.

If the K elementary data events I_k are independent one from another:

$$\mathrm{Cov}_1\{I_k,I_{k'}\}=\begin{cases}0,\ \forall k\neq k'\\ E_1\{I_k\}\cdot[1-E_1\{I_k\}],\ \text{for } k=k',\end{cases}$$

the system (11) becomes diagonal with for solution:

$$\nu_k = 1/E_1\{I_k\}, \text{ thus: } \omega_k = 1, \quad \forall k = 1, \ldots, K, \text{ hence: } \quad \lambda_1^* = \sum_k \lambda_{1,k}$$

All *relative* partial likelihood factors are equally weighted, which again makes sense.

Order relations:

Each of the K partial updatings, being obtained from the exact Bayes' relation (2), results in a licit posterior probability verifying the constraint interval:

$$P\{A = a|I_k = 1\} \in [0, 1], \quad \forall k = 1, \ldots, K \tag{13}$$

Indeed, the constraint interval (13) calls for:

$$1 + \lambda_{1,k} \in [0, 1/P_1\{I_k = 1\}] \tag{14}$$

which, in its turn, calls for: $P_1\{I_k = 1|A = a\} \geq 0$, which is trivial, and: $P_1\{I_k = 1, A = a\} \leq P_1\{I_k\}$, which is also trivial. Similarly, for the probability (12) resulting from the approximate updating process to be valued in [0,1], the global likelihood factor λ_1^* should be such that:

$$1 + \lambda_1^* \in [0, 1/P_0\{A = a\}] \tag{15}$$

Unfortunately, neither the expression (12) for λ_1^*, nor the system (11) guarantees those constraints. One solution, actually that provided by quadratic programming, would be to keep the value $1 + \lambda_1^*$ resulting from system (11) if it falls into the interval (15), if not reset it to the closest bound.

Order relation deviations is the most nagging problem of the indicator approach in geostatistics. Many ad hoc corrections are possible; however, a worthwhile research would be to define a probability updating algorithm that ascertains licit probability values yet does not call for crippling prior parametric models that cannot account for the large variety of conditioning information available. A first avenue for such research would be to find alternatives to the IK-related identification algorithm (10).

A GENERALIZATION OF SEQUENTIAL GAUSSIAN SIMULATION

The sequential simulation principle [Johnson, 1987] has proven widely successful in enlarging the geostatistical toolbox for stochastic simulation. It is used either in a parametric multivariate Gaussian framework [Isaaks, 1990; Gomez-Hernandez and Journel, 1993] or in an indicator framework [Journel, 1989; Gomez-Hernandez and Srivastava, 1990].

The sequential simulation approach is particularly suited to the Gaussian model because all successive conditional distributions are then Gaussian with parameters fully determined by a simple kriging (SK) system. It is shown hereafter that the conditional distributions can be of any other parametric type as long as its mean and variance are determined by SK; this generalization preserves the honoring of the prior covariance model. The implicit random function (RF) model is not anymore Gaussian and may be difficult to identify a priori, but as in the many algorithm-driven simulation algorithms the properties of that implicit RF can be observed on the simulated realizations.

The sequential simulation principle

Consider a stationary RF $Z(\mathbf{u})$ not necessarily Gaussian. Without loss of generality assume:

$$E\{Z(\mathbf{u})\} = 0, \quad \text{Var}\{Z(\mathbf{u})\} = 1$$

The stationary correlogram of $Z(\mathbf{u})$ is:

$$E\{Z(\mathbf{u})Z(\mathbf{u}+\mathbf{h})\} = \rho(\mathbf{h})$$

Given n original data $\{Z(\mathbf{u}_\alpha) = z(\mathbf{u}_\alpha), \alpha = 1, \ldots, n\}$, sequential simulation proceeds with the following steps:

1. Define a random path through all nodes $\mathbf{u}'_l, l = 1, \ldots, L$, to be simulated.
2. At each node $\mathbf{u}'_l$, build the cumulative conditional distribution function (*ccdf*) of $Z(\mathbf{u}'_l)$ given the n original data and all previously simulated values $z_s(\mathbf{u}'_{l'}), l' < l$. Denote that *ccdf* by

$$F(\mathbf{u}'_l; z|(n+l-1)) = P\{Z(\mathbf{u}'_l) \leq z|(n+l-1)\}.$$

3. Draw a realization $z_s(\mathbf{u}'_l)$ from that *ccdf*, that realization becomes a conditioning datum for all subsequent nodes $\mathbf{u}'_{l''}, l'' > l$.
4. Loop until all L nodes are visited and simulated.

This sequential algorithm amounts to sample from the following L-variate distribution conditional to the original n data

$$F(\mathbf{u}'_1, \ldots, \mathbf{u}'_L; z_1, \ldots, z_L|(n)) = P\{Z(\mathbf{u}'_1) \leq z_1, \ldots, Z(\mathbf{u}'_L) \leq z_L|(n))$$

with density equal to the product of the L single-variate conditional probability density functions (*cpdf*'s):

$$f(\mathbf{u}'_1, \ldots, \mathbf{u}'_L;\ z_1, \ldots, z_L|(n)) = \tag{16}$$

$$f(\mathbf{u}'_L; z_L|(n+L-1)) \cdot f(\mathbf{u}'_{L-1}; z_{L-1}|(n+L-2)) \cdot \cdots \cdot f(\mathbf{u}'_1;\ z_1|(n))$$

The L-variate *pdf* (16) resulting from the sequential simulation approach thus depends on the sequence of L univariate *cpdf*'s $f(\mathbf{u}'_l; z_l|(n+l-1))$, $l = 1, \ldots, L$.

The simple kriging principle:

Generalizing from the Gaussian case, consider for each of the L univariate *cpdf*'s $f(\mathbf{u}'_l;\ z_l|(n+l-1))$ a two-parameter distribution (not necessarily Gaussian) fully characterized by

~ its conditional mean, identified to the linear regression (i.e. SK) of the RV $Z(\mathbf{u}'_l)$ on the $(n+l-1)$ conditioning data:

$$E\{Z(\mathbf{u}'_l)|(n+l-1)\} \equiv z(\mathbf{u}'_l)^*_{SK} = \sum_{\alpha=1}^{n+l-1} \lambda_\alpha(\mathbf{u}'_l) \cdot z(\mathbf{u}_\alpha) \tag{17}$$

~ its conditional variance, *assumed independent of the data values* (homoscedasticity property), and identified to the SK variance:

$$\text{Var}\{Z(\mathbf{u'}_l)|(n+l-1)\} \equiv \sigma^2_{SK}(\mathbf{u'}_l) = 1 - \sum_{\alpha=1}^{n+l-1} \lambda_\alpha(\mathbf{u'}_l) \cdot \rho(\mathbf{u'}_l - \mathbf{u}_\alpha) \tag{18}$$

The weights $\lambda_\alpha(\mathbf{u'}_l)$'s are given by the SK system:

$$\sum_{\beta=1}^{n+l-1} \lambda_\beta(\mathbf{u'}_l) \cdot \rho(\mathbf{u}_\beta - \mathbf{u}_\alpha) = \rho(\mathbf{u'}_l - \mathbf{u}_\alpha),\ \alpha = 1,\ldots,n+l-1 \tag{19}$$

Not only the L *cpdf*'s $f\,(\mathbf{u'}_l; z_l|(n+l-1))$ need not be Gaussian, they need not be all the same. For example, one could choose these *cpdf*'s from a 3-parameters distribution family where the 3rd parameter changes from one node $\mathbf{u'}_l$ to another $\mathbf{u'}_{l'}$; this 3rd parameter may be a vehicle for entering soft prior information specific to each node. Although the resulting L-variate *cpdf*, as written in (16), is perfectly licit, one important question arises:

> "Does that distribution (16) identify the prior covariance model $\rho(\mathbf{h})$ for $Z(\mathbf{u})$?"

The answer is, remarkably, yes, no matter the choice for the L successive *cpdf* types.

Covariance reproduction:

The RV corresponding to the realization $z_s(\mathbf{u'}_l)$ drawn at the lth node can be written as:

$$Z_s(\mathbf{u'}_l) = Z(\mathbf{u'}_l)^*_{SK} + R_s(\mathbf{u'}_l) \tag{20}$$

where $R_s(\mathbf{u'}_l)$ is a residual RV with mean zero and variance $\sigma^2_{SK}(\mathbf{u'}_l)$ as given by expression (18). The critical point here is the independence of $R_s(\mathbf{u'}_l)$ on the SK estimator $Z(\mathbf{u'}_l)^*_{SK}$ which allows the realization $r_s(\mathbf{u'}_l)$ to be drawn independently of the SK value $z(\mathbf{u'}_l)^*_{SK}$. Then, the simulated value at $\mathbf{u'}_l$ is:

$$z_s(\mathbf{u'}_l) = z(\mathbf{u'}_l)^*_{SK} + r_s(\mathbf{u'}_l)$$

Consider the next node $\mathbf{u'}_{l+1}$ and the SK estimate using the $(n+l)$ data including $z_s(\mathbf{u'}_l)$:

$$z(\mathbf{u'}_{l+1})^*_{SK} = \sum_{\alpha=1}^{n+l-1} \lambda_\alpha(\mathbf{u'}_{l+1})\ z(\mathbf{u}_\alpha) + \lambda_{n+l}(\mathbf{u'}_{n+l}) \cdot\ z_s(\mathbf{u'}_l) \tag{21}$$

with the $(n+l)$ weights given by the SK system:

$$\begin{cases} \sum_{\beta=1}^{n+l-1} \lambda_\beta(\mathbf{u'}_{l+1})\ \rho(\mathbf{u}_\beta - \mathbf{u}_\alpha) + \lambda_{n+l}(\mathbf{u'}_{l+1})\ \rho(\mathbf{u'}_l - \mathbf{u}_\alpha) = \rho(\mathbf{u'}_{l+1} - \mathbf{u}_\alpha) \\ \qquad\qquad \alpha = 1,\ldots,n+l-1 \\ \sum_{\beta=1}^{n+l-1} \lambda_\beta(\mathbf{u'}_{l+1})\ \rho(\mathbf{u}_\beta - \mathbf{u'}_l) + \lambda_{n+l}(\mathbf{u'}_{l+1}) = \rho(\mathbf{u'}_{l+1} - \mathbf{u'}_l) \end{cases} \tag{22}$$

and:

$$\sigma^2_{SK}(\mathbf{u'}_{l+1}) = 1 - \sum_{\alpha=1}^{n+l-1} \lambda_\alpha(\mathbf{u'}_{l+1})\ \rho(\mathbf{u'}_{l+1} - \mathbf{u}_\alpha) - \lambda_{n+l}(\mathbf{u'}_{l+1}) \cdot\ \rho(\mathbf{u'}_{l+1} - \mathbf{u'}_l)$$

A realization $z_s(\mathbf{u'}_{l+1})$ is drawn from the RV

$$Z_s(\mathbf{u'}_{l+1}) = Z(\mathbf{u'}_{l+1})^*_{SK} + R_s(\mathbf{u'}_{l+1}) \tag{23}$$

where $R_s(\mathbf{u}'_{l+1})$ is a residual RV with mean zero and variance $\sigma^2_{SK}(\mathbf{u}'_{l+1})$ independent of the SK estimator $Z(\mathbf{u}'_{l+1})^*_{SK}$.

The covariance of the two simulated values is:

$$E\{Z_s(\mathbf{u}'_l)\ Z_s(\mathbf{u}'_{l+1})\} = \tag{24}$$
$$E\{Z(\mathbf{u}'_l)^*_{SK} \cdot\ Z(\mathbf{u}'_{l+1})^*_{SK}\} +\ E\{Z(\mathbf{u}'_l)^*_{SK} \cdot\ R_s(\mathbf{u}'_{l+1})\}$$
$$+E\{Z(\mathbf{u}'_{l+1})^*_{SK} \cdot R_s(\mathbf{u}'_l)\} + E\{R_s(\mathbf{u}'_l) \cdot\ R_s(\mathbf{u}'_{l+1})\}$$

The four component terms of expression (24) are developed in Appendix B. Their regrouping yields for the covariance of simulated values:

$$E\{Z_s(\mathbf{u}'_l) \cdot\ Z_s(\mathbf{u}'_{l+1})\} = \sum_{\beta=1}^{n+l-1} \lambda_\beta(\mathbf{u}'_{l+1})\ \rho(\mathbf{u}'_l - \mathbf{u}_\beta) +\ \lambda_{n+l}(\mathbf{u}'_{l+1}) \tag{25}$$

$= \ \rho(\mathbf{u}'_{l+1} - \mathbf{u}'_l)$, per the last equation of the SK system (22).
Thus, and per recurrence: $\mathrm{Cov}\{Z_s(\mathbf{u}'),\ Z_s(\mathbf{u}'+\mathbf{h})\} =\ \rho(\mathbf{h})$ (26)

The marginal covariance $\rho(\mathbf{h})$ is indeed reproduced by the simulated values $Z_s(\mathbf{u})$.

Remarks:

• The proof of the covariance reproduction (26) never called for any restriction on the distribution types of the L random deviates $R_s(\mathbf{u}'_l)$; they only need to be of mean zero and variance equal to the SK variances $\sigma^2_{SK}(\mathbf{u}'_l)$.

• In practice, the data retained at each note $\mathbf{u}'_l$ is limited to a search neighborhood. Consequently, the covariance model $\rho(\mathbf{h})$ is reproduced only up to the radii of that neighborhood.

• In theory, any path can be used to visit the simulation nodes. In practice, however, if the path follows a sequence of very close nodes the simulated features tend to be overly spread along that path creating artifact structures [Isaaks, 1990]. Therefore a random path is commonly used, forfeiting the CPU advantage associated to a sequential and regular path, see hereafter.

• This sequential simulation algorithm is similar to an auto-regressive (AR) process. Indeed, each successive conditional mean is obtained by a regression involving all previously simulated values. Also, the random deviate $r_s(\mathbf{u}'_l)$ added to the regression value $z(\mathbf{u}'_l)^*_{SK}$ is independent of the data values, as in a AR process. However, and because a random path is used, the weighting system $\{\lambda_\alpha(\mathbf{u}'_l),\ \alpha = 1, \ldots n+l-1\}$ and the variance $\sigma^2_{SK}(\mathbf{u}'_l)$ of the random deviate change from one node to another, as opposed to an AR process.

The additional degrees of freedom given by the free choice of the *cpdf* type at each node could be put to use to input additional information and to control statistics of the resulting L-variate distribution other than the covariance $\rho(\mathbf{h})$. The sequential simulation approach with SK need not be constrained by the carcan of Gaussian models with their maximum entropy character [Journel and Deutsch, 1993].

REFERENCES

[1] Englund, E. J. (1990) - "A variance of geostatisticians", in Math Geology **22**(4), p. 417-456

[2] Gomez-Hernandez, J. J. and Journel, A. G. (1993) - "Joint sequential simulation of multiGaussian fields", in Proc. of the 4th Int. Geostat. Congress, Troia 1992, ed. Soares, publ. Kluwer.

[3] Gomez-Hernandez, J. J. and Srivastava, R. M. (1993) - "ISIM3D: an Ansi-C 3 dimensional multiple indicator simulation", in Computers & Geosciences, **16**(4), p. 395-440.

[4] Isaaks, E. H. (1990) - "The application of Monte Carlo methods to the analysis of spatially correlated data", Ph.D. thesis, Stanford University.

[5] Johnson, M. E. (1987) - **Multivariate statistical simulations**, Wiley & Sons, 230 p.

[6] Journel, A. G. (1989) - **Fundamentals of Geostatistics in Five Lessons**, Short course in Geology, **8**, American Geophy., Union Press, Washington DC, 40 p.

[7] Journel, A. G. and Deutsch, C. V. (1993) - "Entropy and spatial disorder", in Math Geology **25**(3), p. 329-355.

[8] Solow, A. (1986) - "Mapping by simple indicator kriging", in Math Geology **18**(3), p. 335-352.

[9] Srivastava, R. M. (1987) - "Minimum variance or maximum profitability", in CIMM, **80**(901), p 63-68.

[10] Zhu, H. and Journel, A. G. (1993) - "Formatting and integrating soft data: Stochastic imaging via the Markov-Bayes algorithm", in Proc. of the 4th Int. Geostat. Congress, Troia 1992, ed Soares, publ., Kluwer.

APPENDIX A: Dual indicator kriging (IK)
Consider the indicator of occurence of the random event $A = a$

$$I(A;a) = \begin{cases} 1 & \text{if } A = a \\ 0 & \text{if not} \end{cases}$$

The prior probability of occurence $P\{A = a\} = E\{I(A:a)\}$ is to be updated by the information event $B = b$, i.e., $I(B;b) = 1$. The simple IK estimate of the conditional probability $P\{A = a|B = b\}$ is written, [Journel, 1989, p. 35]:

$$E^*\{I(A;a)|I(B;b) = 1\} = \\ E\{I(A;a)\} + \tau \cdot [1 - E\{I(B;b)\}]$$

where τ is the simple kriging (SK) weight given by the SK system, here limited to a single equation since there is a single global piece of information $B = b$:

$$\tau \cdot \ \text{Var}\{I(B;b)\} = \ \text{Cov}\{I(B;b), I(A;a)\}$$

The SK estimate is thus written:

$$E^*\{I(A;a)|I(B;b) = 1\} =$$

$$E\{I(A;a)\} + \frac{\text{Cov}\{I(B;b), I(A;a)\}}{E\{I(B;b)\}} =$$

$$P\{A = a\} + \frac{P\{A = a, B = b\} - P\{A = a\} \cdot P\{B = b\}}{P\{B = b\}} = \frac{P\{A = a, B = b\}}{P\{B = b\}}$$

Thus, simple IK with one single piece of information identifies exactly Bayes' relation (1) and (2). This is a classical result [Solow, 1986].

This exact result, however, requires inference of the covariance $\text{Cov}\{I(A;a), I(B;b)\}$, i.e. and de facto, inference of the joint probability $P\{A = a, B = b\}$. Such inference may be difficult if the data event $B = b$ is complex, hence unlikely to be observed repetitively.

The idea then is to decompose the global data event $B = b$ or $I(B; b)$, into the intersection of K more elementary event as in relation (5):

$$I(B; b) = \Pi_{k=1}^{K} I_k$$

The simple IK estimate is then: (A-1)

$$E^* \{I(A; a)|I_k = 1, k = 1, \ldots, K\} = E\{I(A; a)\} + \sum_{k=1}^{K} \tau_k [1 - E\{I_k\}],$$

where the SK weights τ_k are given by the SK system:

$$\sum_{k'=1}^{K} \tau_{k'} \operatorname{Cov}\{I_k, I_{k'}\} = \operatorname{Cov}\{I_k, I(A; a)\}, \quad \forall k = 1, \ldots, K,$$

or in matrix notations:

$$E^* \{I(A; a)|I_k = 1, k = 1, \ldots, K\} - E\{I(A; a)\} = \boldsymbol{\tau}' \cdot \mathbf{d}$$

with the SK system: $\boldsymbol{\Sigma} \cdot \ \boldsymbol{\tau} = \mathbf{C}_k$
then:

$$\begin{aligned} \boldsymbol{\tau} &= \ \boldsymbol{\Sigma}^{-1} \cdot \mathbf{C}_k, \quad \boldsymbol{\tau}' = \mathbf{C}'_k \cdot \ \boldsymbol{\Sigma}^{-1} \\ \boldsymbol{\tau}' \cdot \mathbf{d} &= \mathbf{C}'_k \ \boldsymbol{\Sigma}^{-1} \mathbf{d} = \mathbf{d}' \ \boldsymbol{\Sigma}^{-1} \mathbf{C}_k = \boldsymbol{\nu}' \mathbf{C}_k \\ \text{with: } \boldsymbol{\nu}' &= \mathbf{d}' \cdot \ \boldsymbol{\Sigma}^{-1}, \text{ i.e. } \boldsymbol{\Sigma} \cdot \boldsymbol{\nu} = \mathbf{d} \end{aligned}$$

The dual version of the SK estimate (A1) and the corresponding dual SK system are thus written:

$$\begin{aligned} & E^* \{I(A; a)|I_k = 1, k = 1, \ldots, K\} \\ = \ & E\{I(A; a)\} + \textstyle\sum_{k=1}^{K} \nu_k \operatorname{Cov}\{I_k, I(A; a)\} \\ = \ & P\{A = a\} \cdot \left[1 + \textstyle\sum_k \nu_k \dfrac{\operatorname{Cov}\{I_k, I(A; a)\}}{P\{A = a\}}\right] \\ = \ & P\{A = a\} \cdot [1 + \textstyle\sum_k \nu_k \cdot P\{I_k = 1\} \cdot \lambda_{1,k}] \end{aligned} \tag{A-2}$$

where the weights are given by the dual SK system $\boldsymbol{\Sigma} \cdot \ \boldsymbol{\nu} = \mathbf{d}$, interpreted as the K identification relations:

$$\begin{aligned} & E\{I_k|I_{k'} = 1, k' = 1, \ldots, K\} \\ = \ & E\{I_k\} + \textstyle\sum_{k'=1}^{K} \nu_{k'} \operatorname{Cov}\{I_{k'}, I_k\} = 1 \quad, \forall\, k \end{aligned}$$

i.e.

$$\sum_{k'=1}^{K} \nu_{k'} \operatorname{Cov}\{I_{k'}, I_k\} = 1 - E\{I_k\} \quad, \forall\, k \tag{A-3}$$

The dual IK **estimate** expression (A2) identifies relation (12), and the dual IK system (A3) identifies **system** (11). Thus, the IK algorithm has been shown to identify the approximate Bayes' updating relation (9).

APPENDIX B: Covariance of Simulated values

The four terms of the covariance expression (24) are developed into:

(1) $E\{Z(\mathbf{u'}_l)^*_{SK} \cdot Z(\mathbf{u'}_{l+1})^*_{SK}\} =$

$\sum_{\beta=1}^{n+l-1} \lambda_\beta(\mathbf{u'}_{l+1}) \sum_{\alpha=1}^{n+l-1} \lambda_\alpha(\mathbf{u'}_l)\rho(\mathbf{u}_\beta - \mathbf{u}_\alpha)$

$+ \lambda_{n+l}(\mathbf{u'}_{l+1}) \sum_{\alpha=1}^{n+l} \lambda_\alpha(\mathbf{u'}_l) E\{Z_s(\mathbf{u'}_l) \cdot Z(\mathbf{u}_\alpha)\}$

Now, per expression (20) and the SK system (19):

$$E\{Z_s(\mathbf{u'}_l)Z(\mathbf{u}_\alpha)\} = E\{Z(\mathbf{u'}_l)^*_{SK} Z(\mathbf{u}_\alpha)\} + E\{R_s(\mathbf{u'}_l)Z(\mathbf{u}_\alpha)\}$$

$$= \sum_{\beta=1}^{n+l-1} \lambda_\beta(\mathbf{u'}_l) \rho(\mathbf{u}_\beta - \mathbf{u}_\alpha) + 0 = \rho(\mathbf{u'}_l - \mathbf{u}_\alpha)$$

Thus:

$$\sum_{\alpha=1}^{n+l} \lambda_\alpha(\mathbf{u'}_l) E\{Z_s(\mathbf{u'}_l) Z(\mathbf{u}_\alpha)\} = \sum_{\alpha=1}^{n+l} \lambda_\alpha(\mathbf{u'}_l) \rho(\mathbf{u'}_l - \mathbf{u}_\alpha) = 1 - \sigma^2_{SK}(\mathbf{u'}_l)$$

Hence:

$$E\{Z(\mathbf{u'}_l)^*_{SK} \cdot Z(\mathbf{u'}_{l+1})^*_{SK}\} =$$

$$\sum_{\beta=1}^{n+l-1} \lambda_\beta(\mathbf{u'}_{l+1}) \rho(\mathbf{u'}_l - \mathbf{u}_\beta) + \lambda_{n+l}(\mathbf{u'}_{l+1}) [1 - \sigma^2_{SK}(\mathbf{u'}_l)]$$

(2) $E\{Z(\mathbf{u'}_l)^*_{SK} \cdot R_s(\mathbf{u'}_{l+1})\} = 0$, since $R_s(\mathbf{u'}_{l+1})$ is independent of the $(n+l)$ data values, hence of the previous SK estimator $Z(\mathbf{u'}_l)^*_{SK}$.

(3) $E\{Z(\mathbf{u'}_{l+1})^*_{SK} \cdot R_s(\mathbf{u'}_l)\} =$

$$\sum_{\alpha=1}^{n+l-1} \lambda_\alpha(\mathbf{u'}_{l+1}) \cdot E\{Z(\mathbf{u}_\alpha) \cdot R_s(\mathbf{u'}_l)\} + \lambda_{n+l}(\mathbf{u'}_{l+1}) \cdot E\{Z_s(\mathbf{u'}_l) \cdot R_s(\mathbf{u'}_l)\}$$

with: $E\{Z(\mathbf{u}_\alpha) \cdot R_s(\mathbf{u'}_l)\} = 0, \quad \forall\, \alpha = 1, \ldots, n+l-1$ and:

$$E\{Z_s(\mathbf{u'}_l) \cdot R_s(\mathbf{u'}_l)\} = E\{Z(\mathbf{u'}_l)^*_{SK} R_s(\mathbf{u'}_l)\} + E\left\{R_s^2(\mathbf{u'}_l)\right\} = \sigma^2_{SK}(\mathbf{u'}_l),$$

since $R_s(\mathbf{u'}_l)$ is independent of the SK estimator $Z(\mathbf{u'}_l)^*_{SK}$.
Hence:

$$E\{Z(\mathbf{u'}_{l+1})^*_{SK} \cdot R_s(\mathbf{u'}_l)\} = \lambda_{n+l}(\mathbf{u'}_{l+1}) \cdot \sigma^2_{SK}(\mathbf{u'}_l)$$

(4) $E\{R_s(\mathbf{u'}_l) \cdot R_s(\mathbf{u'}_{l+1})\} = 0$, since the two deviates are drawn independently. Regrouping the four results (1), (2), (3), and (4) yields the covariance expression (25) in text.

COMMENT ON "MODELING UNCERTAINTY: SOME CONCEPTUAL THOUGHTS" BY A.G. JOURNEL

R. MOHAN SRIVASTAVA
FSS International
800 Millbank
Vancouver, BC
Canada V5Z 3Z4

It strikes me that there is an inconsistency between the first and third guidelines given by Journel for modelling uncertainty.

His first guideline is to "accept that all assessments of uncertainty are but models and there is no 'best' model". I entirely agree with this principle and wish that all geostatisticians would acknowledge that they are tackling real world problems with probabilistic models. If they did make such an acknowledgement, they would be more likely to realize that all of the parameters they are choosing for their random function—its expected value, its variogram and its stationarity (or lack of it)—are choices that they are making about their model and are not properties of the reality they are studying. Since the probabilistic model is a figment of our imagination (a convenient and practical figment, but a figment nonetheless), its parameters are not real facts of the world.

In the same way that it is convenient to think of planets as having circular orbits, geostatisticians find it convenient to think of a regionalized variable as an outcome of a random function. The circle of a planet's orbit has a radius but this is purely a mental construct; a planet does not follow a circular path (an ellipse would be a better, but not perfect, approximation) and its "true" radius does not exist. For certain problems that we might want to solve, our circular orbit model will do a good enough job and, even though a true radius does not exist, we can still choose a value for this parameter of our circular orbit model and use this model to get a reasonably accurate answer. The parameters of our random function model are as imaginary as the non-existent radius of a planet's orbit. If we deem the model to be reasonable, then we have to take responsibility for the choice of parameter values and we should not try to divest ourselves of that responsibility by using the fig-leaf of actual data.

Journel's third guideline, unfortunately, encourages the belief that we can (and should) calibrate everything using available data: "The third guideline is to perform systematic calibration of any data used". It is always necessary to show that a model is appropriate, and that the parameter choices are reasonable; in this sense, data are an

R. Dimitrakopoulos (ed.), Geostatistics for the Next Century, 44–45.

invaluable source of supporting information. But we should not fool ourselves into thinking that the only way to defend a parameter choice is through actual data.

The most classic of probabilistic processes, the throwing of dice, provides a good example of how we normally proceed with the use of our probabilistic models. When asked to bet on the outcome, with our winnings coming only when our guess for the sum of the two dice is exactly correct, most of us would bet on number 7. This reflects our belief that an appropriate model for this process is a random variable[1] whose mean, median and mode are all 7. We would make our bet and play the game without a shred of actual data to support our mental construct. If, as we played the game, we started to realize that our initial model was wrong, and that we were playing with loaded dice, we might, at that point, try to use the available data to build a revised model in which one particular value was more likely than all others.

This same basic philosophy should apply to the use of probability models in geostatistics. Based on our familiarity with the problem at hand, and with a good dose of common sense and good judgement, we should choose an appropriate probabilistic model and reasonable parameters. In reporting our study, we owe it to the readers of our report to explain why our model is appropriate and why the parameter choices are reasonable. We may use actual data to support these arguments, but the final choice is entirely our responsibility. No calibration plot will absolve us of this fundamental responsibility by proving that our parameter choice is correct.

While I would agree that calibration of data is an illuminating exercise for practitioners who are trying to understand the data and their limitations, I do not feel that it is necessary, as Journel's example of seismic data suggests, to prove the degree of correlation between hard and soft data through statistical calibration. It is definitely necessary to state what degree of correlation is being assumed in the model, and to defend this choice through qualitative (and, perhaps, quantitative) arguments. But, having stated and defended the choice, we should feel free to use geostatistics to explore the consequences of this model, and should not feel embarassed if we do not have calibration data to support our choice.

The Bayesian framework that Journel embraces is the ideal framework for living with the consequences of our assumptions. It begins with a prior model (that includes parameter choices) and then updates this using a consistent set of rules. Journel's vision of how we should model uncertainty would be improved by more frankly acknowledging that proving the correctness of the prior has no place in a Bayesian framework. The prior cannot be proved correct (or incorrect); it serves merely as a starting point from which updating takes place.

---

[1] We find this model appropriate despite the fact that we know that the sum of the two dice is not truly random but is merely the result of some very complicated combination of physical principles that are completely deterministic and well understood.

THE ROLE OF MASSIVE GRADE DATA BASES IN GEOSTATISTICAL APPLICATIONS IN SOUTH AFRICAN GOLD MINES

D.G. KRIGE and
Honorary Professorial Research Fellow
Department of Mining Engineering
University of the Witwatersrand
P.O. WITS, Johannesburg, 2050
South Africa

C.E. DOHM
Consulting Ore Evaluation Analyst to
Anglo American Corporation of South
Africa and De Beers Cons Mines Ltd.
P.O. Box 41848, Craighall, 2024
South Africa

The critical and indispensable role played by the massive data bases of the South African gold mines in the birth and early development of geostatistics is covered briefly. The continuation of this role throughout and particularly during the last decade is dealt with in more detail. It is shown how this has led to the development of new distribution models and also to studies of the regional variations in the parameters of the distributions and spatial models concerned. The presence of spatial structures for the parameters as such, is also studied. Based on these studies, the advantages of applying Bayesian principles via macro-kriging are demonstrated. Regional studies of the available massive data bases are continuing and will inevitably contribute to further developments.

INTRODUCTION

It is no coincidence that geostatistics originated in South Africa and specifically on the Witwatersrand goldfields. The crucial factor in this regard was the massive records of sampling data collected and maintained from the very early days of mining operations a full century ago. The first attempts to analyse the large numbers of gold values statistically dates back to 1919 and 1929 but were not successful. In 1947, however Sichel suggested the lognormal frequency model. But possibly of greater significance were the detailed records kept consistently on the gold mines of comparisons, via the so-called block factors, of the annual reserve block estimates based on peripheral samples and the sampling results from the stope faces advancing within these blocks. Thus, the block factors provided a practical follow-up monitoring of the orthodox routine block valuations, i.e. comparisons between original block estimates and the subsequent 'real'(or close to real) values found inside the blocks. These block records showed serious conditional, or grade-related

R. Dimitrakopoulos (ed.), Geostatistics for the Next Century, 46–54.

biases with blocks valued as low grade being seriously undervalued on average and blocks valued as high grade being overvalued.

This complicated grade control and gave misleading grade-tonnage trends over a range of cutoff grades. No rational explanation for this feature was forthcoming and no action was taken until 1951 when it was shown (Krige, 1978 and David, 1977) with the use of the lognormal model, to be a straightforward and unavoidable regression effect present when any estimates subject to error, are correlated with the corresponding actual values. The practical solution implemented on many mines in the 50's and 60's, was to use regression corrections that, effectively, correspond to block estimates based on weighted averages of the peripheral estimates and of the mean grade for the mine section concerned. In this regard the block grade was seen as a member of a population of block grades within the mine section; the section mean grade then provides a second estimate of the specific block grade subject to an error variance equal to the variance of the block distribution. Also, the relevant correlations between peripheral and internal values provided the first evidence of spatial correlations. The regressed estimates can therefore be regarded as the first application of what later became known as kriging; more specifically simple or elementary kriging.

This historical fact stresses the need to use all available data in practical follow-up studies to monitor and assess the quality of ore reserve block valuations whatever techniques are used. It is a serious reflection on many practising ore evaluators and geostatisticians in particular that, even today on several diverse mines, the main author has encountered an absence of this common sense procedure. In several cases sophisticated modelling of spatial structures and the use of kriging procedures were in use but on implementation of the up to then absent follow-up correlations, significant conditional biases were still present. This was generally due to the use of ordinary kriging with too limited a search range and/or the neglect of vital geological information.

Against this background this paper will attempt to highlight the role which massive data bases have played, are still playing, and will continue to play in the development of improved geostatistical techniques in the South African goldfields; also wherever extensive databases are available.

FREQUENCY DISTRIBUTION MODELS

Suitable frequency distribution models are essential to geostatistical ore evaluations for the following main reasons:

(i) to provide a suitable base for all analyses of variability, error distributions, confidence limits, correlations and regressions, spatial structures, etc.

(ii) where possible to provide improved estimates of the unknown population mean, given a limited number of available observations.

Based on analyses of large numbers of gold values, Sichel suggested the lognormal model to cater for the skew nature of the distributions observed. He subsequently provided the solution to the small-sampling theory for the lognormal via the so-called Sichel's-'t' estimator with confidence limits.

In 1960 the main author (Krige, 1960) analyzed 43 sets of data with up to 33031 values per set, and totalling nearly 300000 observations. It was found that virtually all the logarithmic distributions were negatively skew and resulted in significant biasses of up to 118% in estimates of the mean when using the 2-parameter lognormal. The 3-parameter lognormal with a constant ß added before taking logs was consequently proposed and reduced the biases to a maximum of 2.6%. This was not significant as biases were measured relative to the corresponding arithmetic means which themselves were subject to some error. This model was introduced on the mines and is still in extensive use throughout the gold mining industry.

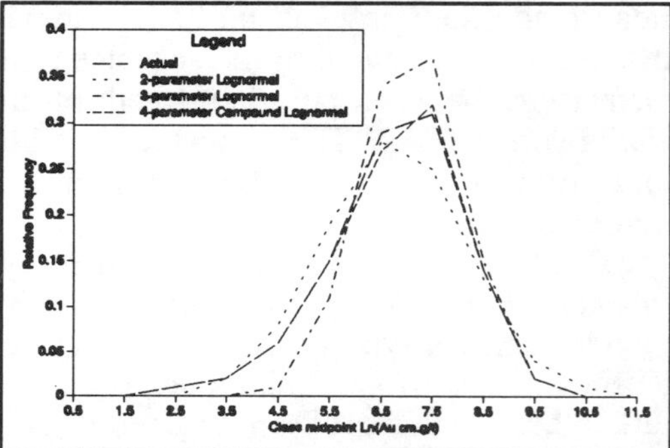

Figure 1 Elandsrand Gold Mine - VCR Frequency Distribution Models Fitted to 21412 Values

More recently in 1992, 22 distributions representing four reefs and some 150000 values were analyzed (Sichel et al, 1992), and showed that, particularly where geological facies and hence the lognormal populations are mixed, significant deviations from the 3-parameter model can occur in the left and right-hand tails of the distribution resulting in biases of up to 4,9% when using that model. Two more general models, i.e. the compound lognormal and the log-generalised inverse Gaussian were proposed and gave somewhat better results. This comparison has now been carried further by the coauthor for a distribution of 21412 values from the mixed facies VCR reef on the Elandsrand Gold Mine on the far West rand. Figure 1 shows the histogram of the actual values and the three models fitted i.e. the 2-parameter lognormal, 3-parameter lognormal and compound lognormal. The models as fitted show means of 2122 cm.g/t, 1693 cm.g/t and 1720 cm.g/t respectively compared to the arithmetic mean of 1716 cm.g/t. The indicated biases are +23,66%, -1.34% and +0.23% and with the fits shown demonstrate the advantage of the new model.

A further important aspect of the frequency distribution studies over the last 40 years has been the fact that the massive data bases allowed for groupings of the individual or 'point' values into large support areas representing ore blocks, mine sections and even sets of adjacent mines. Thus the effects of changes of support on the distribution models applicable could be studied in a practical way and compared with those predicted from point distributions by geostatistical theory. In general it has been found that, as the support size increases, the variance, skewness and kurtosis of the log-values tend towards normality and the patterns change gradually from compound lognormal and the 3-parameter lognormal to the 2-parameter lognormal. This is shown in Table 1 for the Hartebeestfontein and Elandsrand mines. For the former mine the trend for skewness is not confirmed, but the number of large blocks was limited to only 20.

TABLE 1: Effect of increasing support size on frequency distribution models

MINE REEF HORIZON	HARTEBEESTFONTEIN VAALREEF		ELANDSRAND VCR	
SUPPORT SIZE NUMBER OF VALUES	POINT 72767	1 km x 1 20	POINT 21412	250m x 20
NATURAL LOGS				
VARIANCE	1.598	0.183	1.818	0.124
SKEWNESS	-0.266	-0.346	-0.857	-0.020
KURTOSIS	3.466	2.402	4.554	2.980
3-PAR LOGNORMAL				
BETA (ß) cm.g/t	50.000	50.000	180.000	0.000
VARIANCE	1.155	0.157	0.899	0.124
SKEWNESS	0.262	-0.302	0.180	-0.020
KURTOSIS	2.830	2.400	2.573	2.980
BIAS IN MEAN %	-4.300	-3.200	0.230	0.040

VARIANCES OF THE PARAMETERS OF THE 3-PARAMETER DISTRIBUTION.

The 2-parameter lognormal distribution is a special case of the 3-parameter with the ß parameter equal to zero. The parameters to be analysed for the latter are therefore the mean, logarithmic variance and ß. The mean is the critical parameter to be estimated and is very dependent on the other two parameters. A study (Krige, 1993) of the variability has just been completed by the main author on a very comprehensive scale involving some 30 mines throughout the Witwatersrand basin, seven major reef horizons, some 300 distributions and nearly a million individual underground and borehole values. As the logarithmic variance and ß are strongly negatively correlated, the study was simplified by concentrating on the logarithmic variance (σ_o^2) corresponding to a 2-parameter distribution that has the same untransformed variance as the relevant 3-parameter distribution. This is closely linked with the well-known relative variance as follows:

Relative variance $= \text{Exp}(\sigma_o^2) - 1$

The logarithmic variances (σ_o^2) were correlated for each reef horizon with the corresponding sizes of the population areas and Figure 2 represents an example of this correlation for underground chip samples on the Basal reef in the Orange Free State, with the lognormal de-Wijsian variance/size of area regression model fitted.

From this model the following estimates can be obtained:

(i) the corresponding de-Wijsian semivariogram,

(ii) the indicated average log variance of any support size area within any size of population area,

(iii) the indicated distribution and variance of the variances around the regression line.

The variograms were all scaled to a population variance or sill of unity. The standardised variograms for chip and borehole sample values so obtained for the seven reef horizons are shown in Figure 3 and show a close agreement for all the main reef horizons with the two exceptions for which the data available was not quite adequate.

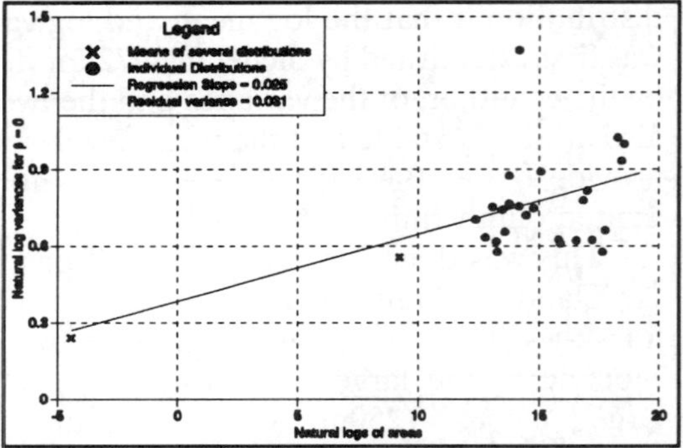

Figure 2 Regression of Logvariance on log area: Basal Reef Chips - cm.g/t-OFS

The significance of the variance analyses is that in any grade valuation based on limited data eg. 10 boreholes, on a new deep level mine, the average log variance for that reef horizon on all or on nearby mines could be a much closer Bayesian type estimate than that based on the limited data (Krige et al ,1990). Furthermore, an even better estimate of the log variance could be a regressed estimate in the form of a weighted average of these two estimates. Such a variance estimate could be combined with a similar estimate of the mean of the log values via the 3-parameter lognormal model to yield a substantially improved grade estimate. This was demonstrated recently (Krige and Bonsu, 1992 a & b) via a practical borehole simulation by repeatedly drawing small sets of values from each of the 20 blocks of 1 km x 1 km in the actual large data base on the Hartebeestfontein mine. The observed actual error variances were obtained on correlation of the estimates based on the small sample sets with actuals available for the same 1 km x 1 km areas (i.e. the averages of some 3000 values within each area). The actual logarithmic

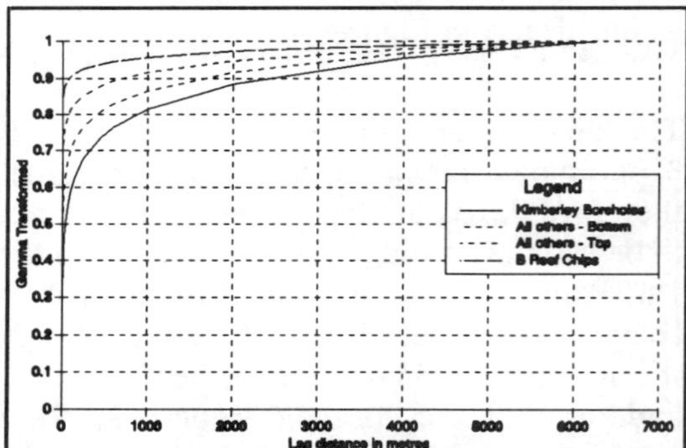

Figure 3 Standardised Semivariograms for 7 Reefs in the Witwatersrand Basin

error variances observed for the orthodox arithmetic mean, Sichel-'t' and the regressed estimate as suggested, were 0.18, 0.16 and 0.08 respectively.

THE COMPOUND LOGNORMAL

This distribution has 4-parameters (Sichel, 1992), the parameters being the log mean, log variance, log skewness and log kurtosis. An inherent characteristic of this distribution it that the log mean and log variance are linearly related. This relation was first established by Sichel in 1972 for diamond deposits (Sichel, 1972). Analyses, by the coauthor, of the variance and the two other parameters have shown that these also are fairly stable and the mean levels indicated for the region or more local area provide better Bayesian type estimates for a new area than that indicated by the limited data within such an area.

This was done for the VCR Reef on the Elandsrand mine on the same basis as for the Hartebeestfontein mine by subdividing the large data base into 20 large blocks (250m x 250m) and comparing estimates based (i) on the regional mean and (ii) on sets of 10 values drawn from each block with the 'actual' parameter values indicated by all the available values in these blocks. This is illustrated in Figures 4 to 6.

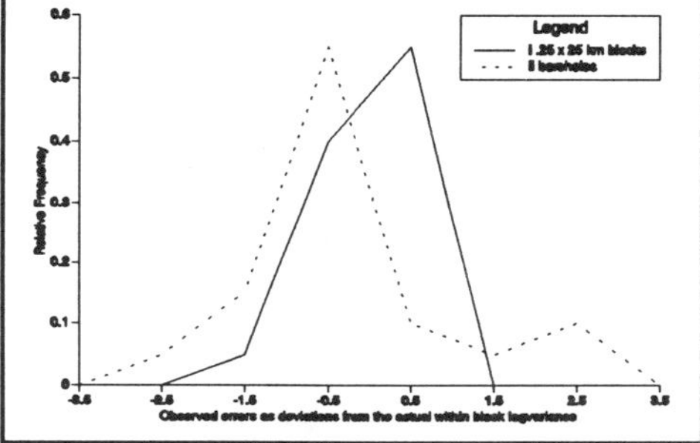

Figure 4 Observed errors of logvariances estimated for blocks on (i) the actual regional mean log variance and (ii)10 boreholes within each block–EGM

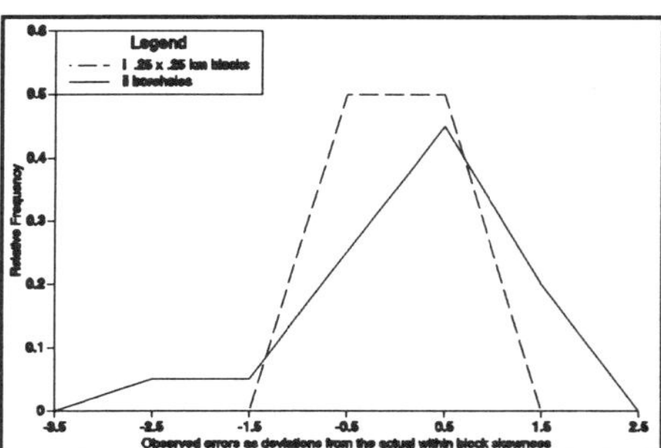

Figure 5 Observed errors of skewness estimated for blocks on (i)the actual regional mean skewness and (ii)10 boreholes within each block-EGM

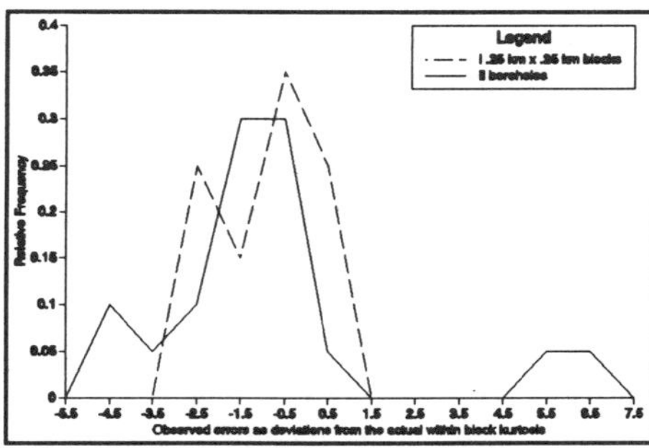

Figure 6 Observed errors of kurtosis estimated for blocks on (i) the actual regional mean kurtosis and (ii) 10 boreholes within each block - EGM

Improved grade estimates as above are therefore also practical for the compound lognormal. This approach has already been successfully used in grade estimates of new mine sections being developed on the VCR reef on the Elandsrand Mine.

SPATIAL PATTERNS FOR PARAMETERS

As the grade variable shows strong spatial patterns for large block supports it is obviously worthwhile also testing the other parameters for spatial patterns. This was done recently (Krige, 1992 a&b, 1993) for the variance parameter and indicated a strong spatial pattern. Such a pattern can be used productively to do macro-kriging of this parameter and thus to further improve the regressed variance estimates mentioned above as well as the block grade estimates. The error variance of the latter for the Hartebeestfontein mine was further improved on this basis from 0,08 on regression to 0,05 on macro-kriging (Krige, 1992 b).

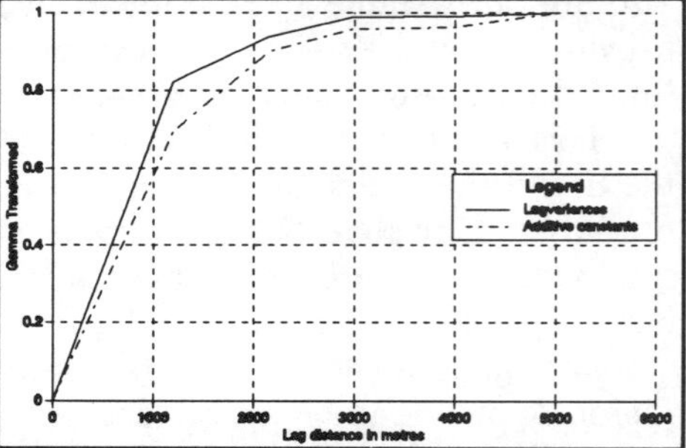

Figure 7 Standardised semivariograms for logvariances and ß values. Hartebeestfontein 1km x 1km blocks

A more recent analysis by Assibey-Bonsu(personal) indicated that the ß parameters for the 20 areas of, 1 km x 1 km areas analysed, also exhibit a spatial pattern. Macro-kriging of this parameter will be done as well, and this should further improve the grade estimates. The standardised semivariograms for the variances and ß values shown in Figure 7 confirm the presence of spatial patterns for both these parameters; the histogram comparisons of the error distributions of the parameters are shown in Figures 8 and 9. From Figures 8 and 9 and also 4 to 6 it is clear that where parameter estimates are based on limited data the regional actual values for these parameters can often provide significantly better estimates and that these can be improved further by regression or macrokriging.

A similar study is proceeding for the case of the compound lognormal on the Elandsrand mine.

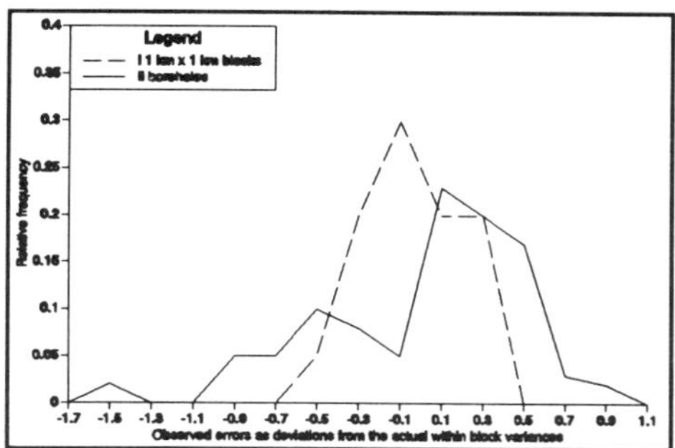

Figure 8 Observed errors of logvariances estimated for blocks on (i) the actual regional mean logvariance and (ii) 9 boreholes within each block - Hartebeestfontein Mine

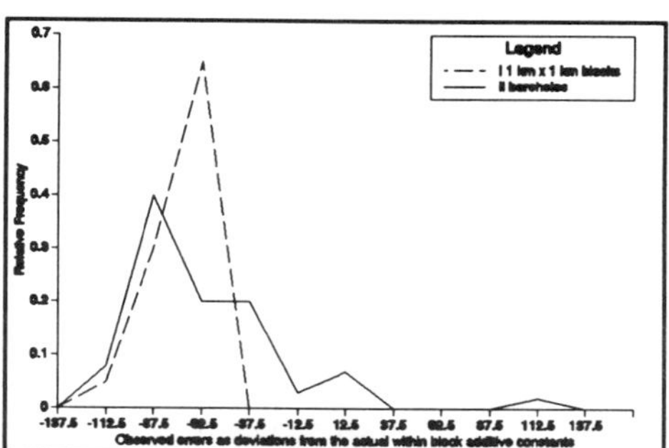

Figure 9 Observed errors of the ß values for blocks based on (i) the actual regional mean ß and (ii) 9 boreholes within each block - Hartebeestfontein Mine

CONCLUSIONS

It has been shown clearly that even after more than 40 years of geostatistical developments and applications in the South African goldfields the valuable information in the available massive data bases for gold grades has not yet been exhausted. Grade valuations of ore blocks and larger virgin areas should be approached via a broader macro perspective and the application of Bayesian type estimates.

ACKNOWLEDGEMENTS

The assistance received from the Anglo American, De Beers Ore Evaluation Department and from Dr Assibey-Bonsu is appreciated.

REFERENCES

KRIGE, D.G. (1978). Lognormal de-Wijsian Geostatistics for Ore Evaluation. Johannesburg, SAIMM Monograph Series, Geostatistics 1,(2nd Edition 1981).

DAVID, M. (1977). Geostatistical Ore Reserve Valuation. Elsevier, Amsterdam.

KRIGE, D.G. (1960). On the departure of ore value distributions from the lognormal model in South African gold mines. J. S. Afr. Inst. Min. Metal., **61**, 231-244. See also discussions ibid., **61**, 333-338 (1961), **61**, 401 (1961), and **62**, 63-64 (1961).

SICHEL, H.S.(1972). Statistical valuation of diamondiferous deposits. APCOM 72, Johannesburg, 1972, pp. 17-24.

SICHEL, H.S., KLEINGELD, W.J., ASSIBEY-BONSU, W. (1992). Comparative study of three distributional models: 3-Parameter lognormal, Compound lognormal and Log-generalised inverse Gaussian distributions. J. S. Afr. Inst. Min. Metal., Vol. 92, No.3, April 1992.

KRIGE, D.G., KLEINGELD, W.J., and OOSTERVELD, M.M.(1990). A priori parameter distribution patterns for gold in the Witwatersrand basin to be applied in borehole valuations of potential new mines using Bayesian Geostatistical techniques. APCOM 90, Berlin, 1990, pp.715-726.

KRIGE, D.G. (1993). Gold grade variability and spatial structures in Witwatersrand reefs; the implications for geostatistical ore valuations. SAIMM, Vol. 93, No. 3, March 1993.

KRIGE, D.G., ASSIBEY-BONSU, W. (1992 a). New developments in borehole valuations of new gold mines and undeveloped sections of existing mines. J. S. Afr. Inst. Min. Metal., Vol. 92, No.3, March 1992.

KRIGE, D.G., ASSIBEY-BONSU, W. (1992 b). The relative rating of macro kriging and other techniques as applied to borehole valuations of new gold mines. Application of Computers and Operations Research in the Mineral Industry (APCOM 92), Ed. Y.C. Kim, Port City Press, Baltimore, Maryland.

MULTIPLE INDICATOR KRIGING OF PRECIOUS METALS AT SILBAK PREMIER MINE, BRITISH COLUMBIA

M. S. Nowak and A. J. Sinclair
Dept. of Geological Siences
The Univ. of British Columbia
Vancouver, B. C. V6T 1Z4

and

A. Randall
Westmin Resources Ltd.
Stewart B. C.

Diamond drill data obtained prior to the present production phase form the main basis of this mineral inventory study of the Silbak Premier deposit, with emphasis on post-1980 data; limited use of blasthole assays from open pit operations during 1990-91 was essential to our structural analysis. Within a mineralized zone the downhole grade distribution is highly erratic in that single high assays can be contiguous with low grade assays. Thus, a high nugget effect is to be expected. Multiple indicator kriging was selected to take into account the presence of multiple populations for both gold and silver.

Traditional autocorrelation functions based on raw assay data for both Au and Ag from Silbak Premier mine were too erratic to model with confidence. Anisotropic, nested, indicator (modified) correlogram models were developed for both Au and Ag, separately for the Main and West zones, using diamond drill hole assays for long range spacings and blast hole assays for short range autocorrelation. A back analysis shows that the anisotropy ratios could have been estimated reasonably well from the geometric form of old stopes from an earlier mining period.

Bench by bench mineral inventory estimates (average grade, tonnage and quantity of metal) by multiple indicator kriging for a large test panel show the method to be well suited to reserve estimations based on exploration data. Multiple indicator kriging results compare favourably with results of other methods and blasthole data.

INTRODUCTION

Silbak Premier polymetallic, epithermal deposit near Stewart, British Columbia, is a bonanza type deposit with a lengthy history of production. The first production phase ended in 1952 and the deposit was dormant for nearly 3 decades. In the 1980's Westmin Resources Ltd. undertook an extensive exploration program and the deposit eventually was put into production in 1989. During the early part of this period of renewed production, substantial discrepancies were found between reserve estimates based on exploration data and actual production. This led to a series of traditional, inhouse, retrospective reserve estimates utilizing both exploration drill data and production blasthole assays. Even with the addition of blasthole information the resulting estimates did not compare as favourably with production as had been hoped. For these reasons Westmin personnel suggested a parallel study by two of the authors (MSN and AJS) in cooperation with the mine staff.

Estimation problems are particularly complex at Silbak Premier because of

R. Dimitrakopoulos (ed.), Geostatistics for the Next Century, 55–63.

(i) "erratic" distribution of values within mineralized zones,
(ii) the presence of 3 distinctly different types of ore to be estimated (massive, stringer, stope fill)
(iii) the presence of different structural orientations of the main ore variety (massive).
(iv) difficulty in recognizing ore type (especially stope fill) during production.
(v) Pre-1980 data are not to the same grid as more recent exploration data and there is some uncertainty regarding locations of the earlier data relative to more recent data.

Data

A total of 41,754 diamond drill core assays (from two drilling phases, pre-1980 and post-1980) and 27,094 blasthole assays (from the production period 1989-1992) were supplied as computer files by personnel of Westmin Resources Ltd. and thoroughly edited for errors and ambiguities in cooperation with mine staff. Diamond drill hole assays represent either in situ rock or large slumped blocks ofstope-fill material. In addition to Au and Ag, some samples were analyzed for Cu, Pb and Zn although these latter three metals did not form part of our study.

Each sample had assigned a rock code and, where applicable, a zone code. The zone codes were assigned by mine staff after careful analysis of the crossections and the level plans within the open pit area. These codes are as follows: M1--a zone of reasonable geological continuity, M2--a zone of short range continuity known as "stringer" mineralization, MS--mineralized stope fill (slumped wallrock), HW--hangingwall zone, and FW--footwall zone. Both HW and FW categories represent waste.

This study is largely restricted to post-1980 drill core data (20,787 assays) because location information for older drill data relative to newer data are uncertain.

General Geology

Silbak Premier deposits are crudely tabular and occur in westerly dipping andesitic fragmental and flow rocks of the Jurassic Hazelton Group. Subvolcanic porphyry bodies are spatially associated with ore. Ores are mostly silicified stockwork and breccia (McDonald, 1990) characterized by several stages of veining and brecciation. Ore breccias have a quartz-rich matrix with highly variable concentrations of pyrite, sphalerite and galena that total less than 5 volume percent on average. Polybasite (the most common silver sulphosalt in the deposit) and electrum (principal gold-bearing mineral in the deposit) form isolated patches or aggregates with sulphide minerals. An analysis of the spatial distribution of precious and base metal grades indicates that high Au assays occur preferentially in silicious breccia and in association with base metal sulphides, especially sphalerite. The principal mineralized zones are crudely tabular but ore zones are highly irregular in detail, particularly the so-called fringing "stringer" variety.

STRUCTURAL ANALYSIS

There are pronounced preferred orientations of ore mineral concentrations as Silbak Premier that are likely to result in anisotropism. Stockworks and siliceous breccias, the main ore hosts have strongly preferred orientations; the Main and West zones are crudely tabular over large distances with average strikes of about 060° and 275° respectively. The main occurences of breccias are in ellipsoidal zones. For the West zone the principal direction of anisotropy of these zones plunges 60° to the west, roughly the direction of elongation of old stopes in the area. Autocorrelation was studied along and across the principal directions of these ellipsoidal ore shoots. Raw assay data for samples averaging about 1.5m in length were used rather than composites. Our structural analysis is limited to massive and stringer categories of *in situ* ore. Stope fill material was assumed to lack structure over distances practical for estimation purposes.

No obvious structure could be interpreted in autocorrelation functions using raw data.

Thus, autocorrelation of indicator transforms was evaluated for various cutoff grdes and showed well-defined stucture for both gold and silver. Thus, we undertook a detailed investigation of the use of indicator kriging (cf. Fytas et al, 1990) in estimating mineral inventory where traditional methods had proved inadequate.

In place of the commonly used variogram, a modified correlogram (cf. Nowak, 1991) has been used:

$$Cor(h) = C(h)/(s_{-h} \cdot s_{+h})$$

where Cor(h) is the standard correlogram, C(h) is the covariance and s_{-h} and s_{+h} are standard deviations. The modified correlogram (1 - Cor(h)) increases with increasing lag to a sill of 1.0 and is particularly useful because it has the same general form as the variogram.

Diamond drill data were adequate to establish autocorrelation for lags in excess of 20m; for shorter lags it was necessary to use production blasthole data. For Au data cutoffs were chosen from 0.3 g/t to 2.0 g/t; for Ag data from 7 g/t to 100 g/t. Nested, anisotropic, spherical models were fitted to the experimental modified correlograms (Nowak and Sinclair, 1993). Models for the West zone (e.g. Fig. 1) are for the three principal geologically defined directions (flattened, cigar-shaped ore shoots plunging 60°W). Very similar anisotropic ratios can be obtained by measuring average stope dimensions from the earlier stage of underground production.

LOCAL RESERVE ESTIMATION

Methodology

Structural analysis of stringer zone mineralization did not produce correlograms that could be modeled, probably due to the discontinuous nature of this type and mineralization as seen in underground workings and the sparsity of data. Thus,

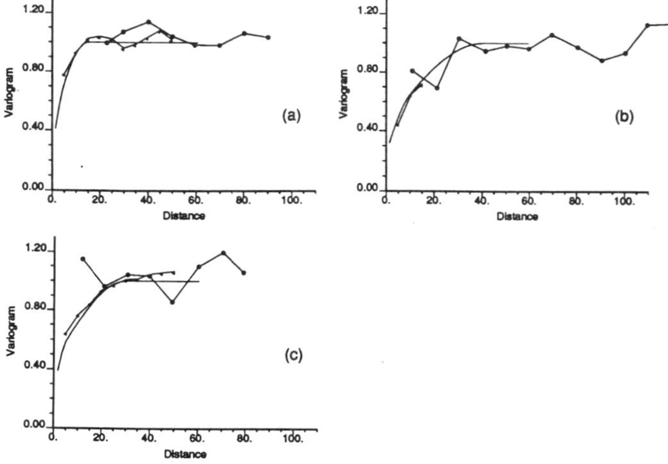

Fig. 1: Examples of indicator, modified correlograms for the West zone, Silbak Premier deposit. Three principal directions of anisotropy for a gold cutoff grade of 0.7 g/t are shown in (a) northerly, (b) plunge 30°E, and (c) plunge 60°W.

estimates of stringer ore will be done by simple averaging of known data within an arbitrarily determined search ellipsoid. Stope fill has occurred by sloughing of *in situ* material from the walls of stopes; any structure that existed prior to sloughing has been destroyed. Thus, grade estimates of stope fill material can be done by a simple averaging of data in a search ellipsoid around the point or block being estimated.

A rectangular mineralized panel comprising 5 levels in a part of the West zone was selected for a detailed comparative study of reserve estimation techniques. A 3-dimensional grid of points (cell = 4m x 4m x 5m) was created in the rectangular panel to approximate a blasthole distribution (67 x 28 x 5 array, i.e. 9380 blasthole sites or grid points) and a bench height of 5 m. For each grid point a local cumulative distribution function will be estimated by multiple indicator kriging. This function is used to calculate, for each grid point, average metal abundance and the probability that the actual value is above a chosen cutoff, information that can be translated directly into tonnage and grade estimates for each of the 5 benches examined.

Each grid node in the rectangular test panel was coded according to ore type by reference to a block model previously coded by mine staff. Of 9380 grid points, 1696 were massive, 154 were stringer and 1378 were stope fill; the remainder were waste. Each grid node was estimated using only neighbouring data (within the search ellipsoid) of the same type (i.e. massive, stringer or stope fill). For estimation purposes we retained all data located in the test panel and within 15 m on the sides and 25m above and below the limiting benches to be estimated. Our 1980's exploration data base consists of 796 assays coded as massive, 100 assays coded as stinger and 273 assays coded as stope fill. In addition, we had available pre-1980 assays as follows: 186 massive, 33 stringer and 226 stope fill assays.

Reserves were estimated for a test panel of 5 benches (555, 560, 565, 570, and 575; bench designation is bench floor elevation in metres above sea level) in two stages:
(i) only post 1980 data were used, and
(ii) using all drill data (pre-1980 and post-1980) in and near the test panel.

The following constraints controlled kriging:
(1) Minimum data required for a point to be kriged was two; maximum was 32.
(2) Maximum search radius along the long axis of anisotropy which plunged 60° in a 270° direction was 50m. This search radius was increased to 75m for stope fill estimates.
(3) horizontal anisotropy ratio, which defines the shape of search ellipsoid, is 0.4 across the structure, and 0.6 along the strike relative to the down-dip length.
(4) Minimum number of informed octants is two.
(5) Ten models of spatial continuity were used for the West zone (Nowak and Sinclair, 1993), although, a small portion of the grid is underlain by the Main zone. For the stope fill estimates models with a very short range of 5m were used, that is almost a pure nugget effect model.

Grid points coded as M1 and stope fill were kriged separately using only data with the corresponding code. Grid nodes coded as stope fill were estimated by kriging procedure although the local cumulative distribution function could have been constructed by simple averaging of indicators inside the search ellipsoid for each particular cutoff.

The 154 grid nodes coded as M2 (stringer) were excluded from the calculations. Based on the global distribution of drill hole assays coded as M2, and on the very short continuity of this ore type, we estimate the probability of exceeding the cutoff is 0.1 for each one of these nodes. This is equivalent to an additional 650 tons of ore on each bench if all of the nodes coded as M2 were estimated, that is, only 2% of expected tonnage on each bench.

The local point cumulative distribution function, F^*, is an output from this type of kriging. It may suffer from order relation problems if $F^*(c_i) > F^*(c_{i+1})$ for $c_i < c_{i+1}$ where c_i and c_{i+1} are the two cutoffs. Order relation errors were corrected

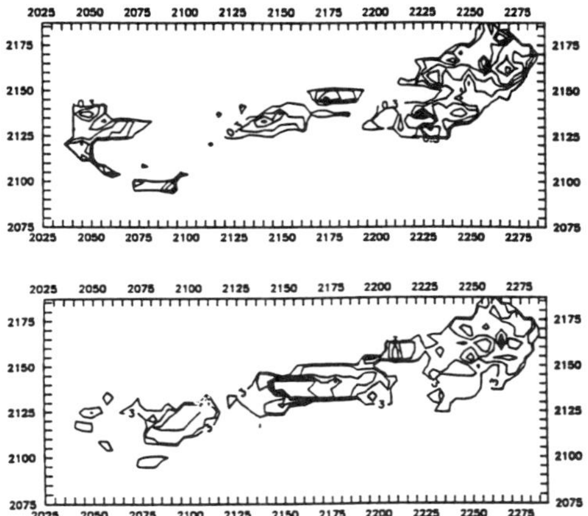

Fig. 2: Comparison of multiple indicator kriging results based on exploration data, with blasthole data. The contoured probability that ore is greater than 0.85 g Au/t, above, is compared with contoured blasthole data below.

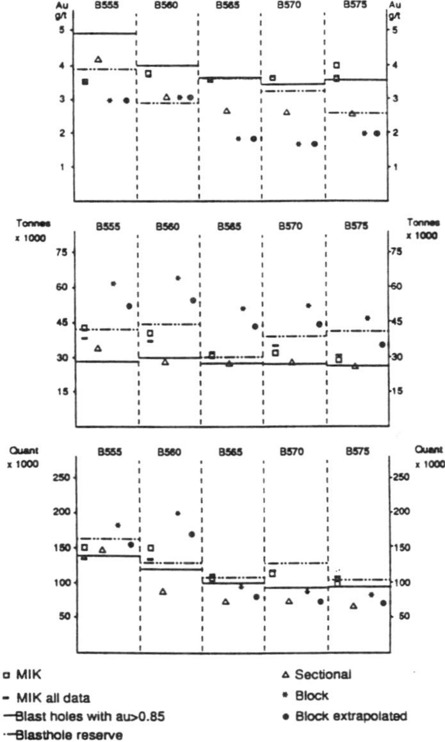

Fig. 3: Bench by bench comparison of gold inventory for the test panel by several methods. See text for explanation of methodologies. Note that MIK is an estimate using only post-1980 exploration data; whereas, "MIK all data" uses earlier exploration data as well.

as follows: values were corrected sequentially upward, sequentially downward, and the mean of the two corrections was retained.

For each of the grid nodes, a probability that the actual grade is above cutoff was retained, and the average grade above this cutoff calculated. Computing actual grade above a cutoff may pose certain problems. F^* is estimated at 10 discrete cutoffs; the declustered global data distribution was used to estimate the behaviour of the function between points. In addition, the following assumptions were made:

(1) the minimum Au value is taken as 0.0 g/t, and the maximum as 12 g/t (approx. 95th percentile of all data larger than 1 g/t Au).

(2) The minimum Ag value is taken as 2.0 g/t, the maximum as 200 g/t (approx. 95th percentile of all Ag data for which Au is larger than 1 g/t).

(3) The cutoff used for our calculation is 0.85 g/t Au, and/or 7 g/t Ag. The Au cutoff is approximately equivalent to the cutoff of 1.02 g/t Au_{eq} (gold equivalent) used for sectional reserves developed at the mine. The cutoff given for Ag is based on the fact that only 7% of Ag assay data is lower than 7 g/t for Au larger than 0.85 g/t.

To determine average grade (of above-cutoff material) on a bench, a weighted average of all calculated point grades belonging to the bench was calculated. Probabilities of exceeding the cutoff were interpreted here as weights. To be able to estimate minable tonnage for each bench, the following interpretation was used:

Table 1: Comparison of results of different estimation methods for Au and Ag for 5 benches of the test panel, West zone, Silbak Premier mine. Estimation methods are as follows: MIK, multiple indicator kriging, 1980 data; MIK1, multiple indicator kriging, all data; BH, contoured blasthole data (including some reserves below cutoff); BH>0.85, blasthole reserves above cutoff; Sect., manual by sections; Block, inverse squared distance block estimates; Blockex, block factored model (see text for details of methodologies). Gr is average grade in g/t; T x 10^3 is metric tonnes of ore; Q x 10^3 is metal content (g) for Au; Q x 10^6 is metal content (g) for Ag.

Bench	MIK			MIK1			BH>0.85			BH			Sect.			Block			Blockex		
	Gr.	T	Q	Gr.	T	Q	Gr.	T	Q	Gr.	T	Q	Gr.	T	Q	Gr.	T	Q	Gr.	T	Q
GOLD																					
555	3.5	42	146	3.5	37	130	4.9	28	137	3.9	42	163	4.1	35	143	2.9	62	180	2.9	52	151
560	3.7	39	146	3.6	36	130	4.0	30	120	2.9	44	128	3.1	29	90	3.1	65	201	3.1	56	174
565	3.5	30	104	3.6	31	111	3.6	27	97	3.6	30	108	2.7	28	76	1.9	52	99	1.9	44	84
570	3.6	31	110	3.6	35	126	3.4	27	92	3.2	39	125	2.6	29	75	1.7	53	90	1.7	45	76
575	3.5	28	98	3.6	30	108	3.5	26	91	2.6	41	107	2.6	27	70	2.0	43	86	2.0	36	72
SILVER																					
555	42	42	1.7	37	37	1.37	43	28	1.2	51	42	2.1	62	35	2.2	46	62	2.8	46	52	2.4
560	44	39	1.7	38	36	1.37	51	30	1.5	45	44	2.0	63	29	1.8	54	65	3.5	54	56	3.0
565	40	30	1.2	38	31	1.18	49	27	1.3	52	30	1.6	63	28	1.8	47	52	2.4	47	44	2.1
570	41	31	1.3	37	35	1.29	51	27	1.4	54	39	2.1	57	29	1.6	39	53	2.0	38	45	1.7
575	40	28	1.1	37	30	1.10	56	26	1.5	45	41	1.8	65	27	1.7	31	43	1.3	31	36	1.1

(1) Each grid node is considered to represent a block of average size (4m x 4m x 5m) on which the node is centered.
(2) The probability of exceeding the cutoff can be interpreted as the portion of a block for which the grade is larger than cutoff e.g. probability 0.5 means that half the block has an average grade larger than cutoff.

The sum of all probabilities assigned to grid nodes on a bench indicates the number of blocks above cutoff. If volume and specific gravity are known for each block, total tonnage for a bench can be calculated. Here, the height of blocks on a bench was taken as the average of all blast holes which returned assay values larger than 0.85 g/t e.g. for bench 555 an average height of 4.72m was determined and was assumed to also apply to silver. Specific gravity used was 2.85 for insitu M1 zone, and 1.5 for stope fill data.

Initial results were obtained using a sample base for which individual samples averaged about 1.5m in length. Because the average bench height is 5m it is necessary to "correct" the local distribution for change of support. A variance correction factor was calculated as a ratio of the variance of 5m composites to the variance of individual assays (circa 1.5m). The indirect lognormal method was used.

RESULTS AND DISCUSSION

Bench by bench estimates of Au and Ag contents of the test panel are listed in Table 1 where they are compared with a variety of other estimates such as sectional, block, extrapolated block model, etc.

Sectional reserves were calculated by mine personnel from bench plans completed at the end of 1990. Individual assays were composited by length to a cutoff of 1.02 g/t gold equivalent based on a silver to gold value ratio of 70:1. The gold equivalent cutoff used is comparable to a Au cutoff of about 0.85 g/t. On benches 555 and 560 Au grades were cut to 13.7 and Ag grades to 685.7 g/t. All reserve blocks had a minimum 5m horizontal width. In situ ore reserve blocks were estimateded separately from caved stope fill blocks. In situ ore blocks had the density factor applied directly to determine tonnage whereas stope fill blocks had a fill factor which ranged from 30-60% applied to determine actual rock volume before applying the same density.

Block model reserves were calculated by mine staff from length-weighted composites on a 5m vertical interval. Using a series of 10m bench plans which showed hole locations, geology, and old stope outlines, 5m x 5m x 5m blocks were assigned to different categories (main, stringer, stope fill). Inverse distance method of estimation was used with the search radius up to 40m in north-south direction, 40m in east-west direction, and 25m vertically. High grades were cut to 17 g/t Au and 340 g/t Ag.

Adjusted extrapolated block model reserves were calculated from block model reserves using empirical, mine-derived factors as follows:

$$\text{Au} = (\text{Block Au})*1.0211$$
$$\text{Ag} = (\text{Block Ag})*1.0039$$
$$\text{Tonnage} = (\text{Block tonnage})*0.8799$$

An in situ grade and tonnage is determined from blast hole data using a polygonal method. All of the above dicussed ore reserve estimation methods ignore dilution arising from the mining process. Blast hole reserves took into account mineability of ore. On each bench, large polygonal areas were selected for mining. All the blast hole grades within a polygon, including some below cutoff, were used to estimate the average grade and tonnage of the polygonal block of ore. The average grade was estimated by a length-weighted, geometric mean of production drillhole assays. Gold grades were cut to 13.7 g/t, and silver to 685 g/t.

An average grade for each bench, for a cutoff 0.85 g/t Au, was calculated as a weighted average of all grades greater than the cutoff. For this estimate Au grades were cut to 34.0 g/t. The choice of the maximum grade allowed does not appear crucial to the averages with one exception; average grade on Bench 555 is very much affected by the cutting limit. The silver average on each bench was calculated as a blasthole length-weighted average of all Ag values for which Au is larger than 0.85 g/t. Silver grades were cut to 685 g/t for blast hole reserve calculations.

Ore tonnage on each bench was estimated using the following assumptions: (i) each blast hole with grades higher than the cutoff represents a volume of block 4m x 4m x h (h = bench height), (ii) each blast hole with an average grade higher than the cutoff is either from M1 zone or is stope fill, (iii) the proportion of blast holes representing M1 mineralization is equal to the proportion of grid nodes coded as M1, (iv) the specific gravity for M1 material is 2.85, and for stope fill is 1.5.

Evaluation

Figures were prepared for each of 5 benches showing the probabilities of exceeding a gold cutoff of 0.85 g/t, as determined by multiple indicator kriging of exploration data. Areas of high probability compare well with the real distribution of high grades indicated from blast holes (Nowak and Sinclair, 1993). Estimated average grades above the cutoff are much less variable, and cover a larger area than do the blast hole grades. This is understandable considering that all estimation methods create smoothed estimates with lower variance than true grades. Only simulation can provide us with a distribution of grades that is both meaningful and comparable in variability to true grades. Generally, the areas of potential mining interest are well indicated from multiple indicator kriging.

Table 1 compares indicator kriging results for the test panel with estimates by other methods. It is apparent that, on average, the multiple indicator kriging method, using exploration drill hole gold assays, returns grades that, on average, give a better indication of reality (approximated by blasthole Au>0.85 g/t) than do the other methods reported here. Similarly, tonnage estimates by multiple indicator kriging are among the best of the methods compared. Block model results and the modification, extrapolated block model results, greatly overestimate the tonnage, especially for bench 555 and 560. Metal quantity of Au on each bench is very well described by the multiple indicator kriging method with block model returning estimates better than sectional reserves.

Average Ag grades calculated by multiple indicator kriging (Table 1) are closer to true grades calculated from blast holes, than any other method. Sectional reserves appear biased and overestimate the average Ag grade on all benches. Note that on three benches average grades from blast hole reserves, which include diluted reserves, are larger than calculated for undiluted resources. This indicates that there are a few blast holes which assayed high in silver although gold content was below the cutoff, hence, they were discarded from the estimate.

CONCLUSIONS

Multiple indicator kriging has been shown to be a viable estimation procedure for mineral inventory at Silbak Premier. In this extremely erratic deposit the method seems better suited to local estimation using only exploration assay information than do the other more traditional methods that were examined. One of the strengths of the multiple indicator kriging method is that it is better at reproducing elements of the local variability shown by the data than are the other methods examined; the principal disadvantages of are complexity, order relation problems and the need for assumptions regarding the mean of data above the uppermost indicator.

ACKNOWLEDGEMENTS

Dr. H. Meade, Vice-President Exploration, Westmin Resources Ltd. proposed this study. Funding was provided by Westmin Resources Ltd. and the Science Council of B. C. We appreciate discussions with R. M. Srivastava and his constructive suggestions to the project. This work has benefitted substantially from extensive discussions with Mr. S. Dykes and Mr. Paul Lhotka, both of Westmin Resources Ltd. GSLIB software (Deutsch and Journel, 1992) was used for many of the calculations. This is contribution #028 of the Mineral Deposit Research Unit, Dept. of Geological Sciences, The University of British Columbia.

REFERENCES

Deutsch, C. V., and A. G. Journel, 1992, GSLIB: geostatistical software library and user's guide; Oxford University Press, New York, 340 p. plus diskettes.

Fytas, K., N. E. Chaouai and M. Lavigne, 1990, Gold deposits estimation using indicator kriging; Can. Min. Metall. Bull. v. 83, no. 934, p. 77-83.

McDonald, D. W. A., 1990, The Silbak Premier silver-gold deposit: a structurally controlled, base metal-rich cordilleran epithermal deposit, Stewart, B.C., unpublished Ph.D thesis, The University of Western Ontario, London, Ont.

Nowak, M. S., 1991, Ore reserve estimation, Silver Queen vein, Owen Lake area, British Columbia; Unpublished M. A. Sc. thesis, The University of British Columbia, Vancouver, B. C. 128 p. plus appendices.

Nowak, M. S., and A. J. Sinclair, 1993, Mineral Inventory of Precious Metal Deposits of British Columbia: Silbak Premier, progress reports; Mineral Deposits Research Unit, Dept. of Geological Sciences, The University of B. C., Vancouver, B. C., 7 reports plus figures and appendices.

MINERAL INVENTORY OF A GOLD-BEARING SKARN, THE NICKEL PLATE MINE, HEDLEY, BRITISH COLUMBIA

A. J. Sinclair and Z. A. Radlowski
The University of British Columbia
Vancouver, B. C.

and

G. F. Raymond
Summerland, B. C.

The South (Bulldog) Pit contains three ore-bearing skarn layers, of which the bottom (Layer Zero), underlain by limestones, is the highest grade (0.164 oz Au/t) and is most extensive physically. The overlying layers (Layers One and Two) have global means of 0.061 oz Au/t and 0.091 oz Au/t. All mineralized layers are separated by barren sills. Surface and underground gold assay data (11,635 samples) for 371 exploration drill holes representing the South Pit and a 250 ft. fringe around the pit, are the basis for this mineral inventory study.

Spatial continuity of the Zero Layer was investigated perpendicular to the mineralized layer and along four directions within the plane of the layer (along strike, down dip, and along both principal diagonal directions). Directional, experimental variograms have been fitted by relative spherical models with $C_0 = 0.2$ and $C = 1.0$. Spatial continuity in the plane of the Zero Layer is isotropic with a range of 15 feet, but across the zone has a range of only 3 feet. Experimental variograms for Layers One and Two are erratic but are consistent with the model for Layer Zero. Crossvalidation of Layer Zero data using ordinary point kriging, polygonal, and inverse squared distance showed that kriging is a marginally better point estimator than are the other two methods.

Reserves were estimated bench by bench and within individual gold-bearing skarn layers separated by barren diorite sills, for a within-pit volume that permitted a direct comparison of results with production and other estimation techniques. Values above 0.5 oz. Au/ton have been shown to have little continuity and have been cut to 0.5 oz. Au/t as determined in a probability graph analysis. Total reserves are estimated to be 652,506 short tons averaging 0.099 oz.Au/short ton, based on a cutoff grade of 0.03 oz. Au/short ton.

Block kriging of exploration drill hole data provides a mineral inventory for the South Pit that more closely approximate production than do estimates by the inverse squared distance method. Moreover, based on exploration data, compare with manual sectional and conditional probability estimates done as a back analysis with parameters derived from production data.

INTRODUCTION

The purpose of this work is to compare production data with a variety of mineral inventory estimates of the South (Bulldog) Pit of Nickel Plate Mine, Hedley, southern British Columbia, owned by Corona Corporation (now Homestake Mining Ltd.). The study is organized into (a) an independent reserve evaluation, and (b) a discussion of production data and various reserve estimates generated through the recent history of

R. Dimitrakopoulos (ed.), Geostatistics for the Next Century, 64–72.

the mine.

The South Pit is centered on layered skarns directly underlain by basal limestone. Skarn mineralization occurs as local, massive sulphide lenses, pods and disseminations in a layered carbonate sequence that strikes northerly and dips 30°W (Fig. 1). Both physical and grade continuities are difficult to establish with certainty because drill holes are, on average, about 70-100 ft apart and locally are more widely spaced (British units are used throughout for consistency with minesite usage).

Three ore-bearing layers separated by barren dykes show spatial continuity throughout much of the study area (Fig. 2). The lowest layer (Layer Zero), in contact with underlying limestones, is best developed and certainly the most continuous with respect to the presence of sulphide mineralization and contained grades, based on the similarity of grade profiles along successive drill holes. The two upper layers (Layer One and Layer Two) have grades continuous over much more limited volumes.

INDEPENDENT MINERAL INVENTORY, NICKEL PLATE MINE (by Z. A. Radlowski and A. J. Sinclair)

Data

Assay and location data used in this study represent 371 drillholes (205 surface and 166 underground) that include all data within a 250 ft fringe around the pit. Geological information has been added to the computer database by digitizing geology from cross-sections supplied by company personnel and assigning a rock type code to each drill hole sample. This was a relatively major undertaking in terms of time, but allows examination of statistical differences and systematic variations in grades as a function of lithology, as well as the ability to isolate of barren dykes as a source of dilution. Three-dimensional coordinates are assigned to the midpoint of each assay. The PC-XPLOR system (Gemcom Ltd.) was used and permitted a variety of graphical output.

Surface drill data consist of 10,320 samples from 205 holes, which are spread over a crude grid which over much of the pit approximates a spacing of 100 ft. The surface drilling subset includes 135 BQ core holes, and 65 rotary holes with reverse circulation plus five percussion holes all 5 1/8" in diameter. Holes are vertical, except for two core holes which both plunge about -60° toward the north. Sample lengths are similar for all drill hole types and despite differences in sample volume the three types of drill data can be treated as of similar support (Radlowski and Sinclair, 1993).

Underground data consist of 1,315 assays from 166 drill holes. The holes are concentrated in mineralized zones and their orientations vary over a wide range. One hundred and fifty three underground holes are of EX core, ten are of AQ core, and three of BQ core. Sample lengths range from 0.2 ft. to 128 ft. but only nine samples are longer than 20 ft.

The histogram for surface drill assays is positively skewed with a mean of 0.022 oz. Au/t. This result is biased because of local concentration of samples in high grade zones. A more meaningful estimate is obtained by declustering. Statistics of declustered surface data are based on an optimum declustering cell of 325 x 325 x 20 ft.^{3}. The arithmetic histograms (Radlowski and Sinclair, 1993) reveal that 87.5% of declustered data have grades below 0.01 oz.Au/t, compared with 77.7% of the raw data. Underground data are comparable.

Probability plots (Radlowski and Sinclair, 1993) indicate two populations to gold assays. Partitioning (cf. Sinclair, 1976) provides a threshold of about 0.5 oz Au/t separating the populations; drill hole profiles show that the high population has very limited continuity.

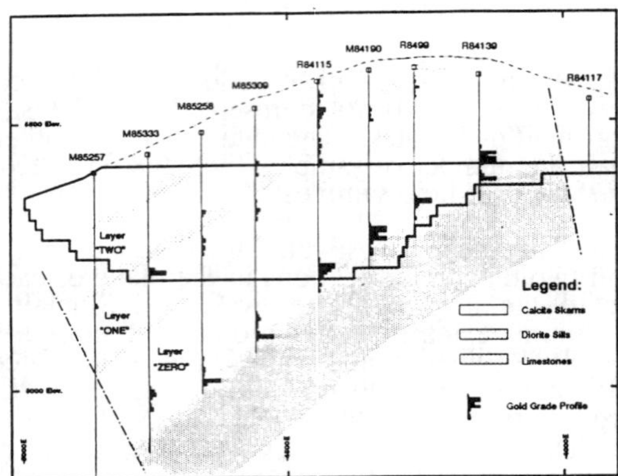

Fig. 1: Vertical E-W section (1000N) through centre of South Pit. Labelled vertical lines (e.g. M85309) are exploration drill holes projected on the plane of the section. Basal limestone and diorite sills are shown as shaded patterns. Relative gold grades in zones of skarn (blank) are indicated by black bars extending outward from drillholes. Ultimate base of South Pit is shown; this, plus an upper horizontal limit extending from ddh M85257, define the volume for which comparative reserve estimates are provided.

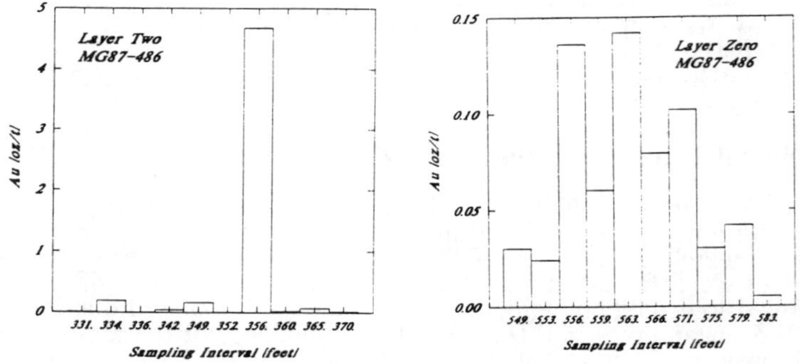

Fig. 2: Examples of gold continuity for (a) extreme, high grades, and (b) low to moderate grades, South Pit, Nickel Plate mine.

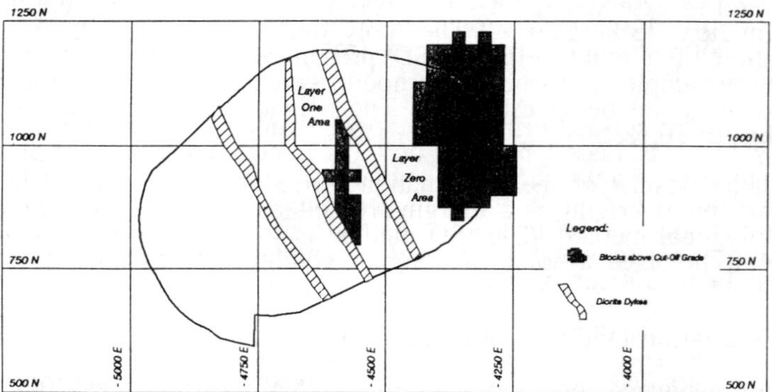

Fig. 3: South Pit outline for 5280 bench showing areas of blocks and fractional blocks estimated by ordinary kriging to have grades $\geq$ 0.03 oz. Au/t (shaded). Projections of barren diorite dykes are crosshatched.

Variogram Modelling

All three gold-bearing skarn layers are parallel to bedding of the enclosing rocks (Fig. 1). The major mineralized layer (Zero) is represented by 429 samples of surface and 210 samples of underground drill holes. The data sets of the other two layers (One and Two) contain about half the number of samples (365 and 358 records, respectively) of Layer Zero, all from surface drill hole samples.

Patterns of spatial continuity have been investigated for raw data (avg. sample length of 6.1 ft) in two principal and two diagonal directions in the plane of each ore zone and in a direction perpendicular to the layering (Radlowski and Sinclair, 1993). Relative variograms were used to minimize the potential problem of non-uniform support (cf. Isaaks and Srivastava, 1988). Experimental variograms were calculated for a fixed angular tolerance of ± 10 degrees and a bandwidth (radius of a search cylinder) of 25 feet. Variograms were determined for the whole subset of each mineralized layer, as well as for separate batches of surface drill hole and underground drill hole data for Layer Zero.

Layer Zero anisotropic, relative spherical model is:

$$\gamma\,(h) \;=\; 0.2 + 1.0\,(3h/2a - h^3/2a^3) \quad a = 15 \text{ ft.}$$

in the plane of the mineralized layer, and

$$\gamma\,(h) \;=\; 0.2 + 1.0\,(3h/2a - h^3/2a^3) \quad a = 3 \text{ ft.}$$

across the mineralized layer. Experimental variograms for Layers One and Two are consistent with this model.

Crossvalidation of Point Estimation Methods

Crossvalidation of surface drill hole assay data from the South Pit area was done by ordinary point kriging, polygonal and inverse squared distance estimation techniques. To decide on a search volume for kriging, the crossvalidation procedure was run for data from Layer Zero for different search ellipsoid volumes (i.e. different lengths of axes) and using 1 to 8 data points per search octant. An unbiased distribution of errors of estimates (i.e. mean error = 0.0 oz Au/t) was obtained for the radii configuration of the search ellipsoid of 70 ft x 70 ft x 14 ft (Radlowski and Sinclair, 1993), coincident in direction with the principal axes of the variogram model.

Polygonal and inverse squared distance (ISD) methods have been applied as alternative point estimation techniques. The polygonal estimate is a nearest neighbour approximation of a point. In the ISD procedure, a point estimate is calculated as the average of samples in its neighbourhood, weighted by the reciprocal of the squared distance to the point being estimated. The neighbourhood is defined by an arbitrary search radius of 70 feet.

Crossvalidation results of these estimation procedures, summarized in Table 1, show that ordinary point kriging is a marginally better estimator than ISD and much better than the polygonal method. The ISD method is substantially better than the polygonal approach but involves a somewhat higher conditonal bias than does ordinary point kriging.

Reserve Estimation-Ordinary Block Kriging

Reserves are estimated for a block size 25 x 25 x 20 ft^3 in the 5200-5420 ft range of elevations, that is, eleven benches, each 20 feet in height (Fig. 1). Calculations are based on analytical data for surface and underground drill holes drilled prior to

Table 1: Comparison of crossvalidation results for Layer Zero data by ordinary kriging, polygonal (nearest neighbour) and inverse squared distance (ISD) methods. Based on 639 estimates using a search ellipsoid with radii 70 ft. x 70 ft. x 14 ft.

Statistics		Data	Estimates			Error on Estimation		
Measure of	Parameter		Kriging	Polygons	ISD	Kriging	Polygons	ISD
Centre	Mean	.145	.145	.170	.157	0	.025	.012
Spread	Standard Deviation	.292	.160	.323	.197	.273	.353	.271
Shape	Coefficent of Variation	2.015	1.103	1.897	1.260	N/A	N/A	N/A
	Skewness	++	++	++	++	-	+	-
Location	Minimum	.001	.001	.001	.001	-2.186	-1.805	-1.829
	Lower Quartile	.016	.043	.016	.045	-.016	-.032	-.009
	Median	.040	.101	.040	.085	+.027	0	+.031
	Upper Quartile	.120	.182	.142	.183	+.107	+.047	+.088
	Maximum	2.844	1.222	2.040	1.592	+.784	+2.000	+.860

Table 2: Ordinary block kriging estimates of geological reserves above a 0.03 oz. Au/t cutoff for the 5280 bench. Results indicate the sensitivity of reserve estimation to change in location of the origin of the block array and the effect of cutting extremely high grades to 0.5 oz. Au/t.

Estimation Conditions			Geological Reserves			
			cut-off grade = 0 oz.Au/ton		cut-off grade = 0.03 oz.Au/ton	
Origin of Block Array x	Origin of Block Array y	Nature of Data	Blocks *	Average Grade oz.Au/ton	Blocks *	Average Grade oz.Au/ton
-5137.5	462.5	Raw data	267	0.065	83	0.197
-5137.5	462.5	High values cut to 0.5 oz/ton	266	0.035	63	0.134
-5150.0	450.0	High values cut to 0.5 oz/ton	265	0.036	64	0.134

* One block is equivalent to (25 x 25 x 20) / 11.2 = 1,116.7 tons

Table 3: Bench by bench comparisons of production and reserve estimates by various methods for South Pit, Nickel Plate mine.

Bench	Production			Ordinary Block Kriging of Exploration Data			Inverse Square Dist. of Exploration Data			Point Kriging of Blast Hole Data			Manual Sectioning of exploration data			Conditional Probability of exploration data		
	Tonnage	Grade	Gold	Tonnage	Grade	Gold	Tonnage	Grade	Gold	Tonnage	Grade	Gold	Tonnage	Grade	Gold	Tonnage	Grade	Gold
5400	60,300	.109	6,573	67,121	.098	6,563	76,200	.087	6,629	81,800	.093	7,607	65,700	.087	5,716	71,600	.080	5,728
5380	52,900	.090	4,761	65,971	.060	3,932	74,500	.077	5,737	73,900	.073	5,395	46,900	.081	3,799	70,500	.063	4,442
5360	51,000	.086	4,386	51,886	.073	3,787	65,900	.088	5,799	64,000	.075	4,800	62,700	.076	4,765	61,700	.063	3,887
5340	44,800	.101	4,525	49,108	.093	4,555	69,300	.121	8,385	64,700	.089	5,758	66,400	.097	6,441	69,800	.079	5,514
5320	76,000	.092	6,992	59,286	.075	4,446	69,900	.145	10,136	85,900	.096	8,246	51,900	.103	5,346	84,300	.093	7,840
5300	55,600	.122	6,783	74,452	.121	9,048	68,700	.197	13,534	68,500	.112	7,672	63,000	.112	7,056	89,000	.101	8,989
5280	72,900	.112	8,165	71,797	.120	8,637	65,700	.204	13,403	70,300	.088	6,186	52,300	.112	5,858	65,700	.103	6,767
5260	57,400	.096	5,510	56,531	.132	7,472	59,800	.215	12,857	56,600	.096	5,434	47,300	.100	4,730	54,100	.107	5,789
5240	50,800	.103	5,232	52,143	.116	6,039	56,000	.181	10,136	40,900	.116	4,744	47,100	.094	4,427	51,700	.103	5,325
5220	62,200	.081	5,038	58,415	.105	6,115	50,600	.156	7,894	N/A	N/A	N/A	34,600	.103	3,564	65,700	.087	5,716
5200	43,500	.072	3,132	45,796	.082	3,718	37,900	.130	4,927	N/A	N/A	N/A	36,400	.085	3,094	52,900	.092	4,867
Total:	627,400	.097	61,098	652,506	.099	64,372	694,500	.143	99,436	N/A	N/A	N/A	574,300	.095	54,795	737,000	.088	64,863
Benches 5240-5400	521,700	.101	52,927	548,295	.099	54,539	606,000	.143	86,615	606,600	.092	55,843	503,300	.096	48,137	618,400	.088	54,280

Note: Tonnage in short tons, Grade in oz. Au/short tons, Gold in ounces

production. Pit outlines, necessary for comparing our reserve estimates with previous estimates and production, were supplied for midbench elevations as sets of digitized points.

Reserves are estimated separately for each bench and then summarized for the entire open pit, based on a cut-off grade of 0.03 oz Au/st. The density factor is assumed constant at 11.2 cubic feet per short ton, as used at the mine site.

Those areas on each bench in which various ore layers were to be estimated were approximated by polygons delimited by projections of contacts of unmineralized sills separating the ore layers and by pit limits. Only areas within these polygons were used in estimation of reported reserves although blocks were estimated in a much larger volume that extends outward from the pit limits. This procedure avoids the smearing effect so common at margins of mineralized zones and, in this respect, optimizes tonnage estimates.

The effects of extremely high, isolated assay values on reserve estimates is a matter of great concern. An examination of grade profiles along drill holes shows that the highest grades generally are not continuous beyond a single sample. In contrast, grades lower than 0.5 oz. Au/t have a substantially greater continuity (Fig. 2). There are several ways of dealing with multiple grade populations (cf. Champigny and Sinclair, 1984). For reasons of simplicity and speed we have elected the conservative procedure of cutting grades to the 0.5 oz. Au/ton threshold documented previously In this way we hope to avoid a potential problem of overextending localized high values.

Before calculating reserve estimates for the entire pit we checked the impact of (i) a change in location of the grid origin, and (ii) the effect of cutting high grades to 0.5 oz. Au/t, on the reserve estimation. This evaluation was performed using Layer Zero data to estimate resources for the 5280 bench as follows: (i) mineral inventory based on raw assay data set and an origin of the block array at mine coordinates x = -5137.5, y = 462.5, z = 5290; (ii) mineral inventory based on data with high values cut to the grade at the 95th percentile (0.5 oz. Au/t) with the same origin for the block array as in (i) above; and (iii) mineral inventory based on cut data as in (ii) above, and the origin for the block array shifted by one-half the block dimension in both x and y directions. The results of these three estimates, calculated for cutoff grades of 0.0 and 0.03 oz. Au/t (Table 2), represent geological reserves that extend well beyond the existing pit. Cutting high grades to a value corresponding to the 95th percentile of the assay data has a tremendous impact on the average grade and a lesser effect on tonnage (number of blocks). In contrast, the change of grid origin does not result in a significant impact on either average grade, tons or total contained metal.

Block kriging estimates were calculated using the variogram model and optimal search ellipsoid described previously (cf. Radlowski and Sinclair, 1993). The block size we estimated, 25 ft x 25 ft x 20 ft, approximating mining selectivity with large loaders mining 20 foot benches, was partly controlled by existing east-west cross-sections 25 feet apart and a mining bench height of 20 feet. The final 3-dimensional block model consists of 56 blocks (easterly) by 36 blocks (northerly) by 11 blocks vertically.

The fraction of each kriged block to be included in reserves was calculated using TechBase software (MineSoft Ltd.) and digitized control polygons described previously. Fractional values vary from zero for blocks outside the defined polygonal areas, to one for blocks totally included within a polygonal area. An example of block classification by kriging (benches 5280 and 5380) is shown in Fig. 3 where pit outlines are seen to cut through blocks estimated to be above cutoff grade.

Results

A complete listing of individual block grade estimates is given in Radlowski and

Sinclair (1993); here, their results are summarized (Table 3). The average waste/ore ratio for the pit (benches 5200 to 5400 inclusive) is 9.1 (cf. Radlowski and Sinclair, ibid).

DISCUSSION OF ORE RESERVE ESTIMATES (by G. F. Raymond)

Nickel Plate gold mine is one of many brought into production during the 1980's that failed to live up to expectations because of a gross overstatement of ore reserves. The contribution of Nickel Plate exploration and production data to the UBC Mineral Deposit Research Unit by Corona Corporation (now Homestake Mining Ltd.) is an important and unselfish step toward helping other mines avoid similar errors in the future. To simulate a feasibilityestimate, the UBC team were given only th3 exploration data for their study. Production data were provided later for comparisons. They did, however, possess the posterior knowledge that there had been ore reserve problems at the mine.

The South Pit data set was chosen for the following reasons:

(1) Although skarn mineralization is extremely erratic, exploration drillholes (DH) are on a fairly uniform grid and sampling and assays are consistent and of reasonably good quality. Comparisons to nearby blasthole (BH) assays indicate that both assay sets are approximately globally unbiased.

(2) The South Pit has been mined out, so both BH assays and reconciled production are available. Mining was generally on 20 foot benches. Grade control involved mining to ore limits contoured from BH kriging point estimates with some resorting of marginal material by visual estimation and grab sampling. The exception is that about 40% of the upper 3 benches was sorted by the earlier method which gave lower tonnages and higher grades. Production is based on mill figures factored to mining estimates on a monthly basis. For the total one year mill period represented, of which about 70% of production was South Pit ore, BH kriging estimates agreed within 5% of milled ore.

(3) Besides the feasibility inverse squared distance estimate done by outside consultants, two subsequent estimates were done by mine staff, a geostatistical conditional probability estimate and a manual sectional estimate.

The feasibility inverse squared distance estimate used 10 foot composites and a small search radius, 50-70 feet along bedding and 7.5 feet across bedding, to produce estimates on a 10 x 25 x 10 foot grid. Although no geological limits were used, strike and dip were varied to follow local bedding. No cutting of high grades was done.

The manual sectional estimate was done by Chief Geologist, W. Wilkinson, based on procedures recommended by consultant H. Bird. The method involved projecting ore grades from 5 foot composites along bedding on section halfway between drill holes to a maximum of 50 ft. Isolated intersections were given smaller influence. Ore grades were cut to 0.5 oz/t (the 95th percentile of ore grades). Projected ore beds were diluted to 20 ft benches by adding waste at 0.02 oz Au/t and then factored to mining experience by increasing ore tonnate by 30% at a grade of 0.01 oz/t. Since experience factors were derived from this South Pit data set, good global agreement has been forced.

The conditional probability method as applied at Nickel Plate has been used in a number of mines by this author (eg. Raymond and Armstrong, 1988) and is still the primary reserve estimation procedure used at Nickel Plate. In place of cutting or factoring, a large number of samples are averaged in kriging to produce a conditionally unbiased estimate that is much smoother than mining selectivity. This estimate is then de-smoothed a determined amount by the conditional probability calculation. The following description applies to the method as it is used property-wide.

First, ordinary point kriging is done on a 25 x 25 x 20 foot grid using the nearest (corrected for anisotropy) ten 20-foot composites within a search radius of up to 150 feet in-bedding and 30 feet cross-bedding. A single relative variogram model is used property-wide with sperical model parameters Co = 2.5, C = 2.5, a = 125 feet in the bedding plane and 25 feet across bedding. To avoid extrapolation beyond one-half the drill hole spacing high kriging variance estimates are discarded. Estimates and composites from the limestone basement are also discarded. A minor cut to 0.5 oz. Au/t on 20 foot composites is still done. This reduces local overestimations around a few high grade intersections but has only a minor effect on overall reserves. An indicator approachto the high grade range was studied extensively but did not give better results. For each kriging estimate probability of ore above cutoff and expected ore grade are calculated based on kriged grade, exploration minus BH kriging variances, and assuming a lognormal distribution of estimation errors. Reserves are tallied by summing tonnage around each grid point multiplied by ore probability.

Comparisons amongst all of the exploration estimates versus actual mining are summarized in Table 4.

Table 4: Percent Difference Between Production and Estimates

Estimation Method	Tonnage	Grade	Gold
Feasibility Inverse Squared Distance	-10%	-32%	-39%
Manual Sectional	+9%	+2%	+12%
Cond'l Probability	-15%	+11%	+2%
Radlowski and Sinclair	+4%	+2%	+5%

The feasibility estimate was a disaster! Contrary to the assumptions in this estimate, mineralization was not continuous. Instead, ore boundaries appeared in mining as an insepeparable mixture of ore and waste BH grades. Attempts at more selective mining using closer BH spacing and smaller bench heights did little to improve mined grades.

Because both the manual sectional and conditional probability estimates were derived from back analysis of at least part of this data their good agreement is expected. In fact, as indicated by the large bench by bench variability, exact agreement with production for this one data set is not necessarily the best estimate. Longer term comparisons indicated that the conditional probability grade estimate was about right but that ore tonnage tended to be over-stated by about 10%.

Sinclair and Radlowski's block kriging estimate is very similar to the feasibility estimate in that both used a small composite size and realtively small search radius. The major difference is their 0.5 oz./ton cut which reduced their estimate by 24% on tons and 32% on grade, resulting in very close agreement to production. Although it did not greatly influence South Pit estimates, their approach of limiting estimates by geology would be important in other Nickel Plate orebodies.

Good agreement in total reserves can always be achieved by choosing the right factors. Bench by bench comparisons to production are an indication of how well each method performed locally. Unexpectedly, all of the estimation methods gave the same magnitude of errors (about ±25% on metal) for individual benches. This was true even of the manual sectional method, which is essentially polygonal. Interestingly, conditional probability gave much better predictions of the BH kriging than the other methods (what it was designed for) but this improvement was negated by the large bench by bench differences between reported production and BH kriging.

CONCLUSIONS

1. For this data set, provided the right factors are chosen, cutting and dilution methods

(both manual and geostatistical) give estimates that are about as good globally and locally as the geostatistical method of conditional probability that relies on smoothing. However, as seen from the Nickel Plate feasibility estimates, choosing the wrong factors can be disastrous. The correct choice by Sinclair and Radlowski (this study) is interesting because (i)it was based only on a knowledge of exploration data, (ii) it used an novel method for identifying the lower threshold of the high, low continuity population, and (iii) the reserve estimation procedure prevented overestimation of tonnage by extending high and medium grades to waste blocks.

2. The conditional probability method minimizes the need for cutting and relies on geostatistical calculation to obtain the estimate. Mining selectivity is allowed for in the BH kriging variance. Geology is used to a more limited extent, resulting in estimates that are quicker to obtain. The advantage is that all parameters are measured; the disadvantage is that a great deal of experience is required to correctly calculate the parameters, especially the variogram.

3. In orebodies as difficult to estimate as Nickel Plate it is worthwhile to apply two or more substantively different methods of reserve estimation and reconcile the results.

ACKNOWLEDGEMENTS

This project was undertaken with the permission of Mr. J. Lovering, Manager, Nickel Plate mine. Extensive cooperation and assistance of W. Wilkinson, Chief Geologist, is very much appreciated. Dr. P. Grimley and N. Champigny provided critical comments on early parts of the study. Funding was provided by the Science Council of B. C. This paper is contribution #027 of the Mineral Deposit Research Unit, Dept. of Geological Sciences, The Univ. of British Columbia, Vancouver, Canada.

REFERENCES

Champigny, N., and A. J. Sinclair, 1984, A geostatistical study of the Cinola gold deposit, Queen Charlotte Islands, B. C.; Western Miner, October pp. 52-66.

Isaaks, E. H., and R. M. Srivastava, 1989, An introduction to applied geostatistics; Oxford University Press, 531 p.

Radlowski, Z. A., and A. J. Sinclair, 1993, Mineral Inventory of Precious metal deposits of British Columbia: Nickel Plate mine progress reports; Mineral Deposits Research Unit, Department of Geological Sciences, The University of British Columbia, 6 reports plus figures and appendices.

Raymond, G. F., and W. P. Armstrong, 1988, Sample bias and conditional probability ore reserve estimation at Valley; Can. Inst. Min. Metall. Bull. v. 81, pp. 128-136.

Sinclair, A.J., 1976, Applications of probability graphs in mineral exploration; The Assoc. Expl. Geochem. spec. Vol. No. 4, 95.p.

INFORMATION MEASURES, INTEGRATION AND GEOSTATISTICAL ANALYSIS

JOINT TEMPORAL-SPATIAL MODELING OF CONCENTRATIONS OF HAZARDOUS POLLUTANTS IN URBAN AIR

BRUCE E. BUXTON and ALAN D. PATE
Battelle Memorial Institute
505 King Avenue
Columbus, Ohio 43201
USA

This paper describes a geostatistical data analysis approach for assessing temporal and spatial variability in the concentrations of hazardous pollutants in urban air. The objective of this analysis is to determine how reliably concentrations measured at one or more times and locations can be used to estimate concentrations at other times and locations. The approach used is that of classical three-dimensional ordinary kriging with a joint temporal-spatial semivariogram model containing an isotropic spherical component, and zonal cosine and linear components in the temporal direction. The method is illustrated using data from the 1990 Atlanta Ozone Precursor Study, an air quality monitoring study conducted by the U.S. Environmental Protection Agency's Atmospheric Research and Exposure Assessment Laboratory (AREAL).

BACKGROUND AND INTRODUCTION

The potential health risk to people caused by airborne chemicals is evaluated using exposure assessment methods. Exposure assessment describes the magnitude and duration of people's exposure to chemicals. There have been many recent studies to assess the concentrations of hazardous air pollutants (HAPs) in the atmosphere of urban environments in response to concern over public exposure to these pollutants. In spite of the studies conducted to date, many questions still exist over the spatial and temporal variability of the concentrations of HAPs, and hence the actual exposures which may occur. Some exposures studies have shown that personal exposures to many pollutants bear little relationship to outdoor measurements from fixed-location sampling. Improved efforts to estimate human exposure from fixed-site monitoring have involved the collection of concentration data at many of the microenvironments (e.g., home, office, car) that an individual or population passes through over the course of a day. The activity profiles of the individual or population combined with the concentration data allow estimates to be made of integrated exposures over a day. However, an assessment of the variability of concentrations within each microenvironment and between similar

R. Dimitrakopoulos (ed.), Geostatistics for the Next Century, 75–87.

microenvironments, such as neighborhood to neighborhood, is crucial to both the design and validation of exposure modeling studies which use data from fixed sites.

In order to assess personal exposures to hazardous air pollutants, it is important to determine ambient concentrations of these pollutants as a function of location and time of day. Furthermore, the accuracy of exposure estimates is often linked to the spatial and temporal resolution of the concentration measurements. For example, for most pollutants, exposure estimates that are based on measurements made at a single urban location are generally less reliable than exposure estimates made with measurements from several locations within the urban area. The reliability with which the true pollutant concentration at a fixed time and location can be estimated from measurements made at one or more neighboring times and locations primarily depends on two important parameters, the variability and degree of temporal and spatial correlation in the measurements. The implications of these parameters on air monitoring network design are quite significant. In order to obtain personal exposure estimates of a specified precision, the number of monitoring sites and times required increases as the variability in pollutant concentration measurements increases, or as the correlation in those measurements decreases.

APPROACH

In recent years air pollution modeling and joint temporal-spatial estimation problems have been addressed in some of the atmospheric and statistical literature. For example, Switzer (1981, 1986, 1989) was one of the earliest statistical researchers to jointly consider temporal and spatial variations in modeling air pollution data. Le and Petkan (1985) followed up Switzer's work by trying to apply his model to rainfall pH measurements collected over six years in the northeastern U.S. In addition, Bilonick (1983) and Guertin and Villeneuve (1989) discuss statistical models for data related to acid precipitation. In 1991, Buxton and Pate proposed a geostatistical model for measurements collected during the Atlanta Ozone Precursor Monitoring Study; however, this model dealt primarily with spatial, rather than joint temporal-spatial, variations.

Objectives

The overall objective of our work with AREAL is to utilize ambient air quality measurements and statistical analysis techniques to characterize temporal and spatial variations in the concentrations of hazardous air pollutants in urban areas, and to identify major sources contributing hazardous air pollutants to urban environments. Four questions of primary concern to AREAL are as follows: (1) Are the concentrations of air toxics consistent across neighborhoods and distances of 0.5 to 10 kilometers (i.e., are measurements at a particular sampling site representative of a larger community)?; (2) Are the ambient concentrations at any particular site so highly variable throughout the day that exposures are also highly variable, and therefore, dependent upon the specific time-activity profiles of exposed individuals?; (3) What do the variations of air toxics in time and space imply about the identity, nature and distribution of the sources?; and (4) What factors exacerbate or mitigate exposures to air toxics by humans as they go about their daily activities?

This paper presents a geostatistical approach primarily intended to address the first of these four questions. In particular, the objective of our geostatistical analysis is:

> To determine, for a variety of hazardous air pollutants, how reliably ambient concentrations measured at one or more times and locations in an urban area can be used to estimate the concentration at other neighboring times and locations.

An important part of this objective is to provide AREAL with a general statistical approach which can be used to quantitatively compare alternative strategies for air quality monitoring. This method should allow AREAL to optimize monitoring network designs by evaluating the marginal costs and benefits of different numbers and locations of monitoring sites, different sampling frequencies (e.g., daily versus weekly monitoring), and different sample integration times (e.g., hourly versus 3-hour sample integration.

Geostatistical Semivariogram Model

In many ways, the geostatistical approach proposed for meeting this objective is quite simple. The required estimation reliability is assessed with a point kriging variance. However, in this case the kriging involves joint temporal-spatial estimation, which in turn requires a joint temporal-spatial semivariogram model. Use of such a model is analogous to a normal three-dimensional kriging situation in space (i.e., coordinates are easting, northing, and vertical elevation, all measured in meters), except that in the joint temporal-spatial case locations are measured in easting and northing, which are measured in meters, and time, which is measured in hours. This requires that one re-interpret normal distances, measured in meters, to be generalized distances, measured in meter-hours.

As with any kriging analysis, the key statistical tool required is the semivariogram function $\gamma(\underline{h})$ (Journel and Huijbregts, 1981), which in this joint temporal-spatial situation is defined as:

$$\gamma(\underline{h}) = 1/2*E[Z(\underline{x}) - Z(\underline{x}+\underline{h})]^2 \qquad \text{(Expression 1)}$$

where $Z(\underline{x})$ is the pollutant concentration measured at location $\underline{x}$ (which is defined by an easting, northing, and time coordinate), and $\underline{h}$ is the vector of separation between locations $\underline{x}$ and $\underline{x}+\underline{h}$. Note that the vector $\underline{h}$ is defined by a shift in location of space and of time, and its length, r, can be interpreted as a generalized distance measured in meter-hours. For example, a vector defined by a shift of 3 kilometers in the easting direction, 4 kilometers in the northing direction, and 1 hour in time would represent a generalized distance of 5.1 kilometer-hours.

A common characteristic of urban air pollution concentrations, particularly in the summer months, is a pattern of strongly periodic diurnal variation. In this pattern, daily temperature inversions and other meteorological and atmospheric chemical processes lead to widely different pollutant concentrations at night as compared with during the day. For example, temperature inversions can cause the concentrations of primary pollutants, for example, benzene to systematically increase at night and decrease during the day. The resulting temporal-spatial variations in the pollutant concentrations can be geostatistically modeled with the following three-dimensional semivariogram function:

$$\gamma(\underline{h}) = C_o + C_1 * \mathrm{Sph}[r,a] + C_2 * t + C_3 * \mathrm{Cos}[t,24] \qquad \text{(Expression 2)}$$

C_o	= isotropic nugget variance, in ppb^2
C_1	= sill variance for spherical structure, in ppb^2
Sph[r,a]	= isotropic spherical structure with range "a"
	$= 1.5*(r/a) - 0.5*(r/a)^3 \qquad r \leq a$
	$= 1 \qquad r > a$
r	= length of vector $\underline{h}$, expressed as generalized distance in kilometer-hours
C_2	= slope of zonal linear structure in the time direction, in ppb^2/hour
t	= time shift of the vector $\underline{h}$
C_3	= sill variance for cosine structure
	= one-half the amplitude of the cosine structure, in ppb^2
Cos[t,24]	= zonal cosine structure in time direction with period of 24 hours
	$= 1-\mathrm{Cos}[2\pi*(t/24)]$

An example of this semivariogram model is presented in Figure 1. Some interesting features to note in the model are as follows:

- For separation vectors within the spatial plane, that is, vectors $\underline{h}$ with a shift in easting and/or northing but no shift in time, the semivariogram model simply amounts to an isotropic nugget variance plus an isotropic spherical structure.

- For separation vectors in the temporal direction, that is, vectors $\underline{h}$ with a shift in time but no shift in easting or northing, the semivariogram model includes the isotropic nugget variance, isotropic spherical structure, as well as the full effects of the zonal linear and cosine structures.

- For separation vectors in all other intermediate directions, that is, vectors h with simultaneous shifts in both the spatial plane and the temporal direction, the semivariogram model includes the isotropic nugget variance, isotropic spherical structure, and a partial effect of the zonal linear and cosine structures which increases for separation vectors more closely aligned with the temporal direction.

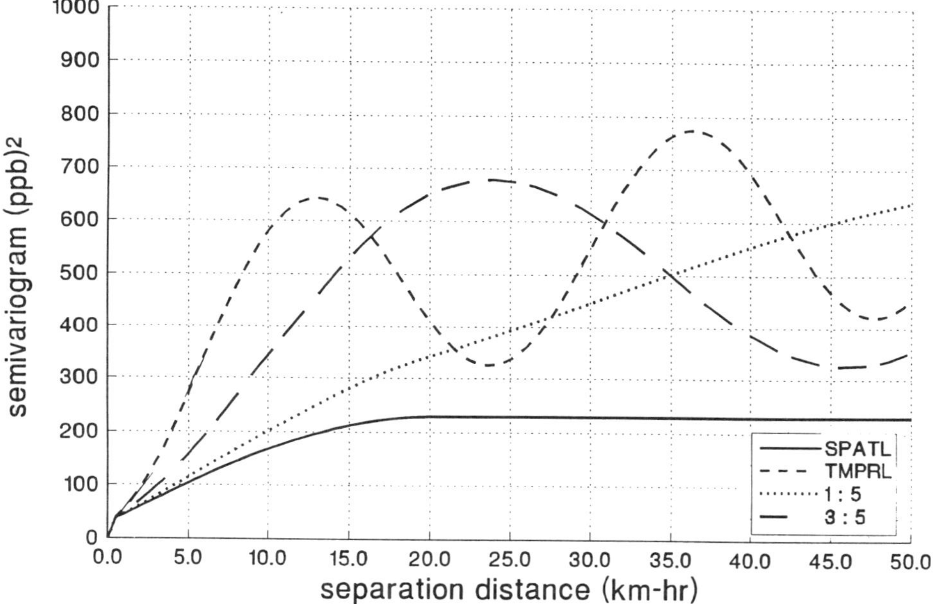

Figure 1. Example of Temporal-Spatial Semivariogram Model. (C_o=30, C_1=200, C_2=4, C_3=200). Four directions are shown, within the spatial plane (SPATL), in the temporal direction (TMPRL), time shift of 1 hour for every 5 km. of spatial shift (1:5), and time shift of 3 hours for every 5 km. of spatial shift (3:5).

Note that this model, which is a combination of a three-dimensional isotropic spherical structure, a one-dimensional zonal linear structure, and a one-dimensional zonal cosine structure, is positive definite since it is a combination of three separate positive definite models (Journel and Huijbregts, 1981).

Kriging Analysis Method

The data analysis objective stated in Section 2.1 can be interpreted as a temporal-spatial interpolation (i.e., estimation) of pollutant concentrations at unsampled times and locations in an urban area based on pollutant concentrations measured at other fixed times and locations in the urban area. For example, the pollutant concentration at 7:00 a.m. on August 2nd might be estimated for all locations across an urban area, based on measured pollutant concentrations monitored hourly at three fixed monitoring stations from 5:00 a.m. on August 1st through 7:00 a.m. on August 2nd (i.e., 27 hourly measurements taken at each of three sites). In contrast, the pollutant concentration at 7:00 a.m. on August 2nd might be estimated for all locations across an urban area, based only on the measured pollutant concentrations monitored at the same three fixed monitoring stations at 7:00 a.m. on August 2nd (i.e., one hourly measurement taken at each of the three sites). Clearly, the first set of estimates, based on 81 measurements, should be significantly more reliable than the second set of estimates, based on 3 measurements.

Ordinary kriging is a linear geostatistical estimation method which uses the temporal-spatial semivariogram function to (1) determine the optimal weighting of the measured pollutant concentrations to be used for the required estimates, and (2) calculate the estimation precision associated with the estimates (Journel and Huijbregts, 1981). As such, kriging can be used as a planning tool to quantitatively evaluate, in terms of estimation precision, the benefits associated with increased numbers of monitoring stations and/or monitoring times.

For the purposes of this study, the utility of the kriging method for designing hazardous air pollutant monitoring strategies was illustrated by: (1) calculating experimental semivariograms and fitting semivariogram models for several different air pollutants; (2) performing an ordinary kriging analysis with several different temporal and spatial data configurations; and (3) comparing the different data configurations using contour maps of relative estimation precision, which is defined as follows:

$$RP = \frac{100*1.96*\sigma_k}{m}$$

where RP = precision, relative to the mean pollutant concentration, associated with a given temporal-spatial data configuration,

σ_k = kriging standard deviation, which changes for each temporal-spatial data configuration,

m = mean pollutant concentration,

1.96 = standard normal value associated with two-sided 95% confidence intervals, and

100 = factor used to express RP as a percentage.

It should be noted that although the semivariogram model (Expression 2) which was used for kriging is positive definite, there may still be a problem with singular kriging matrices if the data searching strategy is not carefully planned. In particular, use of the cosine semivariogram structure requires that all temporal data used for kriging must be closer in time than one "period" (i.e., 24 hours) apart.

ILLUSTRATIVE RESULTS

In this section, the geostatistical approach is illustrated using data for two air pollutants monitored in the AREAL 1990 Atlanta Ozone Precursor Study. The example species are ozone and benzene. Ozone is a secondary pollutant produced by chemical reactions in the atmosphere, while benzene is a primary pollutant that is emitted directly to the atmosphere. This study monitored a wide variety of air pollutants at six fixed sites spaced across Atlanta from about 5 km. to 50 km. apart. Hourly pollutant concentrations were measured at each site, 24 hours a day for more than six weeks (Purdue, et al., 1992).

Results for Ozone

Ambient concentrations of ozone varied during the study from less than 10 ppb to more than 100 ppb. Ozone concentrations typically peaked late in the afternoon (average 6:00 p.m. concentration was 65.9 ppb), and were lowest early in the morning (average 6:00 a.m. concentration was 7.3 ppb). Four experimental semivariograms for ozone, along with their fitted models, are presented in Figure 2. The semivariogram in the temporal direction (TMPRL) clearly shows the strong diurnal variation in ozone concentrations -- the semivariogram peaks at separation distances of 12 hours and 36 hours, and it is a minimum at a separation distance of 24 hours. It can also be seen that for this pollutant the temporal variations (semivariogram cycling between about 180 ppb^2 and 1620 ppb^2) are a more dominant factor than the spatial variations (spatial semivariogram peaks at about 180 ppb^2).

Illustrative results for the kriging analysis of ozone are presented in Figures 3 through 6. These results vary the number of monitoring sites between either one or three sites, and the number of hours of monitoring at each site as either one hour or 27 hours. All results are expressed relative to a mean ozone concentration of 30 ppb. Figure 3 shows that the precision associated with extrapolating ozone concentrations from a single measured value is reasonably good, for example, the estimation precision at the northern outerbelt of Atlanta is plus-or-minus 110%. This figure assumes that only one measurement is taken each day. Figure 4, on the other hand, assumes that continuous hourly measurements are collected, and the most recent 27 data are used to estimate current ozone concentrations. In this second case the increased temporal data improve the relative estimation precision by about 10% (e.g., relative precision at the northern outerbelt improves to about plus-or-minus 100%).

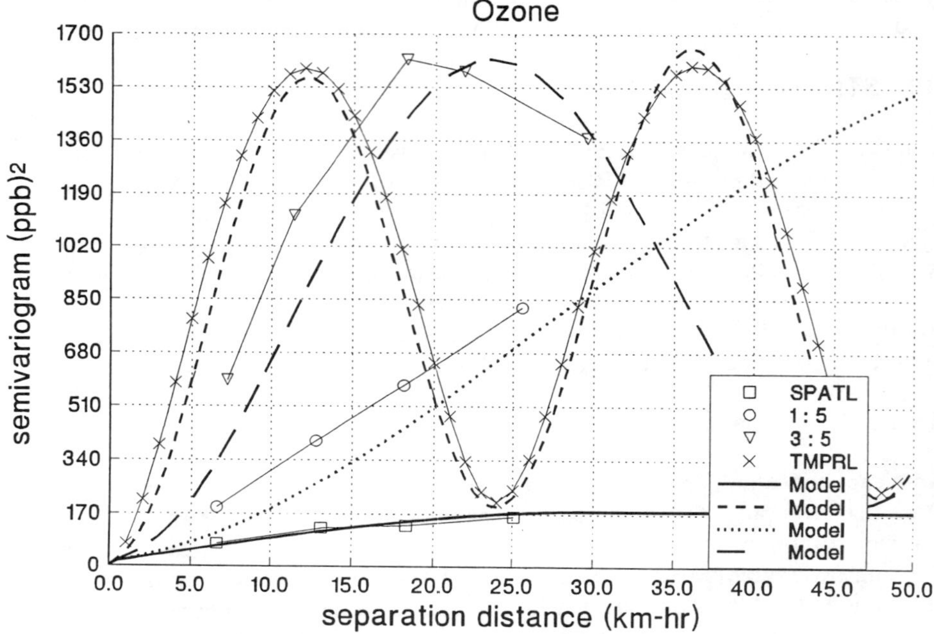

Figure 2. Experimental Semivariograms and Fitted Models for Ozone. ($C_o=10$, $C_1=170$, $C_2=1$, $C_3=720$). Four directions are shown, within the spatial plane (SPATL), in the temporal direction (TMPRL), time shift of 1 hour for every 5 km. of spatial shift (1:5), and time shift of 3 hours for every 5 km. of spatial shift (3:5).

Figures 5 and 6 illustrate the benefits of adding two additional monitoring sites. Figure 5 assumes that only one measurement is taken each day from each of three sites. In this third case the increased spatial data improve the relative precision obtained with one site (Figure 3) by about 20% (e.g., relative precision at the northern outerbelt improves to about plus-or-minus 90%). It is interesting to note that for ozone there is a greater improvement by adding a single measurement at two more sites than by adding continuous hourly monitoring at a single site (Figure 4). Figure 6 completes the illustrative examples for ozone; it assumes that continuous hourly measurements are collected from each of three sites. In this fourth case the increased temporal data improve the relative estimation precision by about 10% over the case where a single measurement is taken at each of three sites (Figure 5).

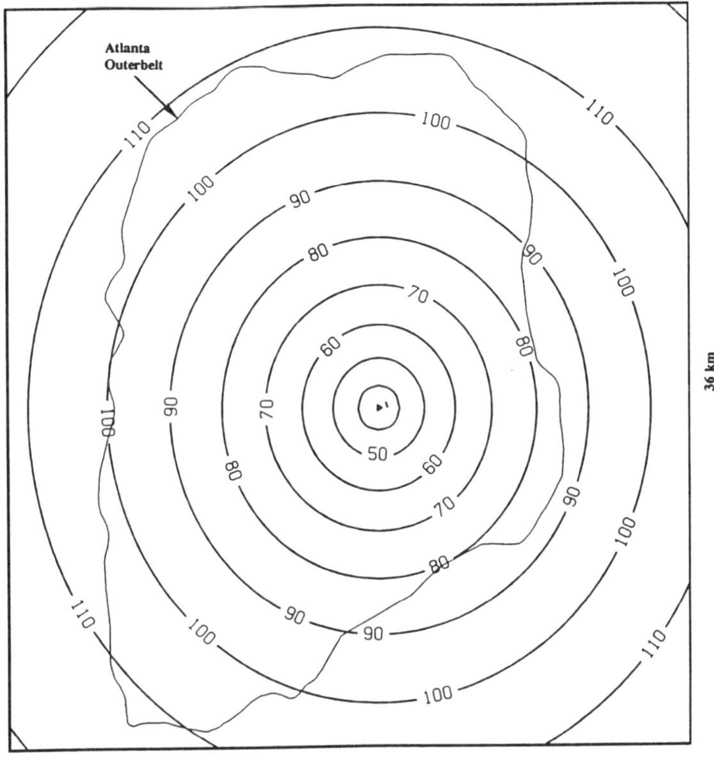

Figure 3. Relative Precision (%) Associated With the Estimation of Ozone at the Current Hour, Using Datum Monitored at One Central Site at the Current Hour.

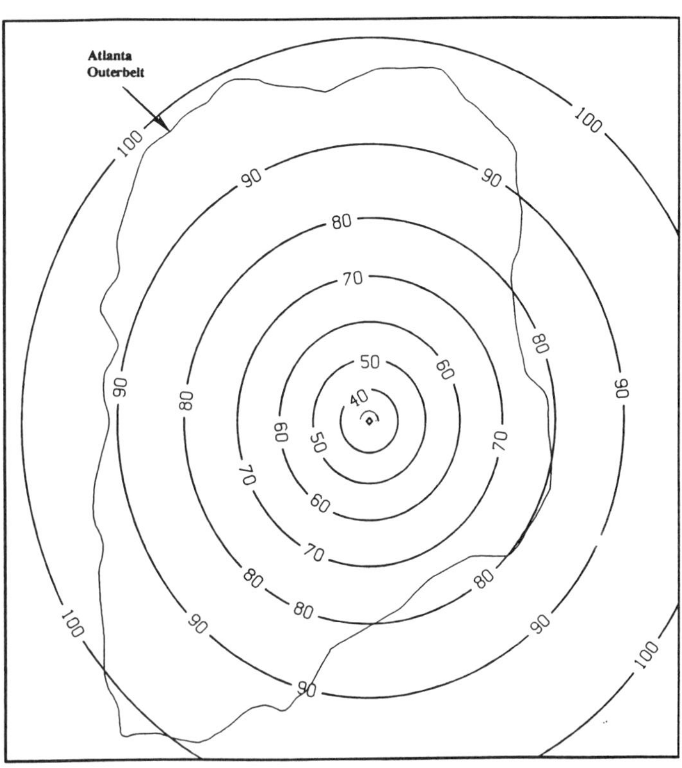

Figure 4. Relative Precision (%) Associated With the Estimation of Ozone at the Current Hour, Using Data Monitored at One Central Site Over the Past 27 Hours.

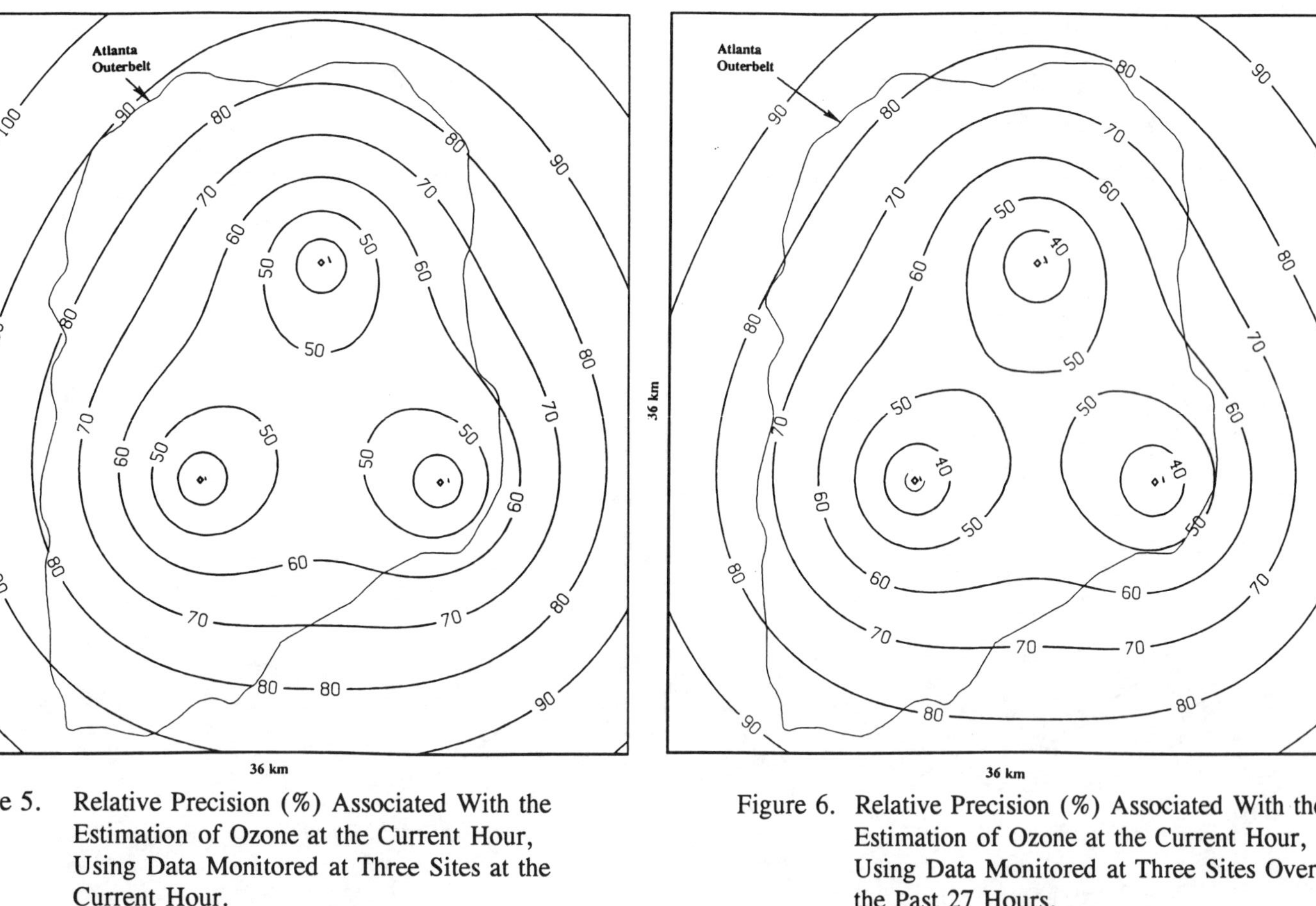

Figure 5. Relative Precision (%) Associated With the Estimation of Ozone at the Current Hour, Using Data Monitored at Three Sites at the Current Hour.

Figure 6. Relative Precision (%) Associated With the Estimation of Ozone at the Current Hour, Using Data Monitored at Three Sites Over the Past 27 Hours.

Results for Benzene

To show the difference in results that can be found for different air pollutants, this section presents a comparative example for benzene. Benzene concentrations in the Atlanta study typically peaked early in the morning (average 2:00 a.m. concentration was 1.2 ppb), and were lowest early in the afternoon (average 2:00 p.m. concentration was 0.4 ppb). Experimental semivariograms for benzene are presented along with their fitted models in Figure 7. As with ozone (Figure 2), the benzene semivariogram in the temporal direction (TMPRL) clearly shows the effect of strong diurnal variation -- the semivariogram shows a strong 24-hour cycle. However, in the case of benzene the temporal variations (cycling between about 0.35 ppb^2 and 0.55 ppb^2) are a less dominant factor in comparison with the spatial variations (spatial semivariogram peaks at about 0.35 ppb^2).

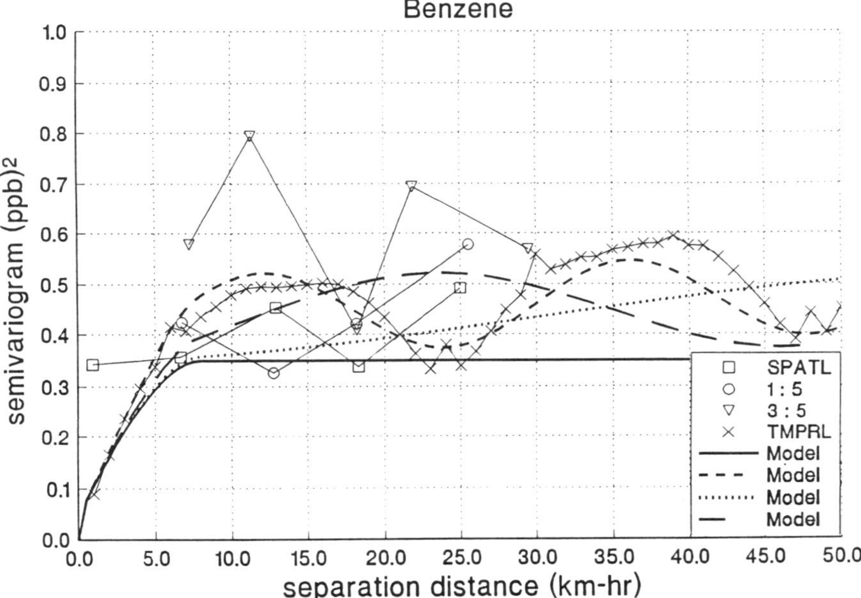

Figure 7. Experimental Semivariograms and Fitted Models for Benzene. (C_o=.05, C_1=.3, C_2=.001, C_3=.08). Four directions are shown, within the spatial plane (SPATL), in the temporal direction (TMPRL), time shift of 1 hour for every 5 km. of spatial shift (1:5), and time shift of 3 hours for every 5 km. of spatial shift (3:5).

A similar set of kriging analyses were performed for both ozone and benzene. Figure 8 presents a single comparative figure for benzene; it shows the relative estimation precision associated with extrapolating benzene concentrations from a single concentration measured at each of three sites. All results in this figure are expressed relative to a mean benzene concentration of 0.8 ppb. In this case the relative precision is not as good as that for ozone (Figure 5). For example, the estimation precision for benzene at the northern outerbelt of Atlanta is plus-or-minus 160% (Figure 8), while the estimation precision for ozone at this same location is plus-or-minus 90% (Figure 5). These results indicate that the variability of benzene concentrations is greater, in a relative sense, than the variability of ozone concentrations.

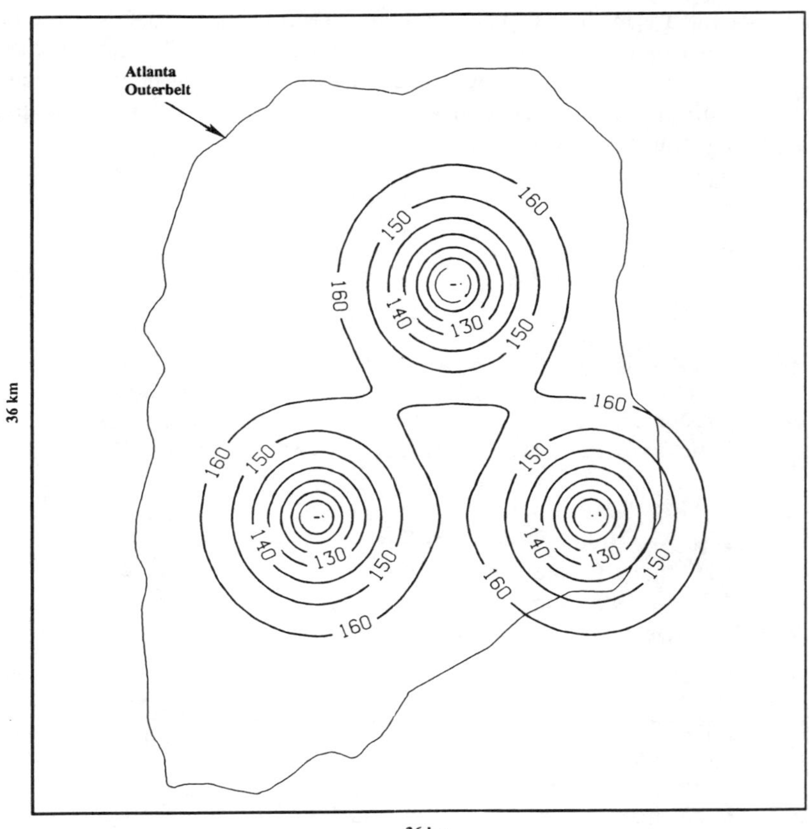

Figure 8. Relative Precision (%) Associated With the Estimation of Benzene at the Current Hour, Using Datum Monitored at Three Sites at the Current Hour.

REFERENCES

Bilonick, Richard A., and Duane G. Nichols, 1983, "Temporal Variations in Acid Precipitation Over New York State -- What the 1965-1979 USGS Data Reveal," Atmospheric Environment, Vol. 17, No. 6, pp. 1063-1072.

Buxton, Bruce, E., and Alan D. Pate, 1991, "Statistical Modeling of Spatial and Temporal Variations for the Atlanta Ozone Precursor Study," presented at the 84th Annual Meeting and Exhibition of the Air and Waste Management Association, Vancouver, British Columbia, Canada, June, 1991, 15 pp.

Guertin, Kateri, and Jean-Pierre Villeneuve, 1989, "Estimation and Mapping of Rank Related Uniform Transforms of Ion Deposition from Acid Precipitation." In Geostatistics Volume 2, edited by M. Armstrong, Dordrecht, the Netherlands: Kluwer Academic Publishers.

Journel, A. G., and Ch.J. Huijbregts, 1981, Mining Geostatistics, reprinted with corrections, Academic Press, London, 600 pp.

Le, D. Nhu, A. John Petkau, 1985, "An Attempted Validation of the Eynon-Switzer Model for the Variability of Rainfall Acidity," SIMS Technical Report No. 90 prepared under support from U.S. Environmental Protection Agency by Department of Statistics, University of British Columbia, September, 1985.

Purdue, Larry J., J. A. Reagan, W. A. Lonnemann, T. C. Lawless, R. J. Drago, G. M. Zalaquet, M. W. Holdren, D. L. Smith, A. D. Pate, B. E. Buxton, and C. W. Spicer, 1992, "Atlanta Ozone Precursor Monitoring Study Data Report," report no. EPA/600/R-92/157, Atmospheric Research and Exposure Assessment Laboratory, U.S. Environmental Protection Agency.

Switzer, Paul, 1981, "Stochastic Processes in the Analysis of Environmental Data," SIMS Technical Report No. 50 prepared under support from U.S. Environmental Protection Agency, Department of Energy, Sloan Foundation, and National Science Foundation, Department of Statistics, Stanford University, July 1981.

Switzer, Paul, 1986, "An Analysis of Hourly Acid Deposition Data," SIMS Technical Report No. 101 prepared under grant from U.S. Environmental Protection Agency, Sloan Foundation and National Science Foundation, Department of Statistics, Stanford University, September, 1986.

Switzer, Paul, 1989, "Non-Stationary Spatial Covariances Estimated from Monitoring Data." In Geostatistics Volume 1, edited by M. Armstrong, Dordrecht, the Netherlands: Kluwer Academic Publishers.

SPATIOTEMPORAL MODELLING: COVARIANCES AND ORDINARY KRIGING SYSTEMS

ROUSSOS DIMITRAKOPOULOS and XIAOCHUN LUO

McGill University, Dept. of Mining and Metallurgical Eng., 3480 University St., Montreal, Qc, CANADA H3A 2A7

This paper presents results on the geostatistical modelling of spatiotemporal data. Permissible space-time covariances and ordinary kriging are considered.

INTRODUCTION

Buxton and Pate (1993) model hazardous air pollutants by extending ordinary kriging in the space-time domain. In support of their *ad hoc* approach, certain results on spatiotemporal modelling are presented here. This is deemed appropriate considering the spatiotemporal nature of a large spectrum of data in earth sciences and engineering. Spatiotemporal phenomena include environmental pollution, attributes of petroleum reservoirs, geohydrologic variables, meteorologic and climatologic parameters, characteristics of renewable resources, etc.

One question that is raised in the approach of Buxton and Pate (1993) is the permissibility of the variogram function they use. A second question relates to singularity problems when solving the space-time kriging system. A last question relates to the physical meaning of spatiotemporal distance measures. Results relevant to these questions are presented next. Both the work by Buxton and Pate as well as ours are based on the definition and properties of spatiotemporal random fields (S/TRF) given by Christakos (1992; 1991). To simplify the terminology, a stationary S/TRF as described here is identical to the space-homogenous, time-stationary S/TRF's of Christakos.

PERMISSIBLE COVARIANCE FUNCTIONS IN SPACE-TIME

This section shows (i) certain permissible spatial covariance models to be also permissible spatiotemporal (s/t) covariances; and (ii) general properties of s/t covariance functions with a general 'zonal anisotropy' representation.

A general relationship between certain spatiotemporal and spatial covariances

Consider Z(s,t) to be a stationary S/TRF in R^nxT with a covariance function,

$$C(s,t;s',t') = C(h,\tau) = C(a^2|h|^2 + b^2\tau^2) \qquad (1)$$

where $|h|^2=\mathbf{h}^T\mathbf{h}$, $\tau=t-t'$ and coefficients $a,b\geq 0$. The spectral density corresponding to eq. 1 can be written

R. Dimitrakopoulos (ed.), Geostatistics for the Next Century, 88–93.

$$S(\boldsymbol{\lambda},\omega) = S(\lambda_1,...,\lambda_n,\omega)$$

$$= (2\pi)^{n+1/2} \int_{-\infty}^{\infty} ... \int_{0}^{\infty} C(r^2) \exp\,[-i(\boldsymbol{\lambda}^T\mathbf{h}+\omega\tau)] \; dh_1 \dots dh_n \, d\tau \qquad (2)$$

where $\boldsymbol{\lambda}^T\mathbf{h}$ denotes the inner product $\lambda_1 h_1 + ... + \lambda_n h_n$, and $r^2 = [a^2(h_1^2 + ... + h_n^2) + b^2\tau^2]$. Note that the covariance in eq. 1 is isotropic in space, but not in both space and time, unless a=b.

If one sets $h'_1=ah_1$, ... , $h'_n=ah_n$, $h'_{n+1}=b\tau$, and $\lambda'_1=\lambda_1/a$, ... ,$\lambda'_n=\lambda_n/a$, $\lambda'_{n+1}=\omega/b$, then

$$r'^2 = h'^2_1 + ... + h'^2_n + h'^2_{n+1} = r^2$$

and

$$\boldsymbol{\lambda}'^T\mathbf{h} = \lambda'_1 h'_1 + ... + \lambda'_n h'_n + \lambda'_{n+1} h'_{n+1} = \boldsymbol{\lambda}^T\mathbf{h} + \omega\tau$$

Consequently,

$$S(\boldsymbol{\lambda},\omega) = S(a\boldsymbol{\lambda}',b\lambda'_{n+1})$$

$$= (2\pi)^{n+1/2}\, 2^{-1}a^{-n}b^{-1} \int_{-\infty}^{\infty} ... \int_{-\infty}^{\infty} C(r'^2) \exp\,[-i(\boldsymbol{\lambda}'^T\mathbf{h})] \; dh'_1 \dots dh'_n \, dh'_{n+1} \qquad (3)$$

If function $C(\mathbf{h}')$ is a non-negative definite function in R^{n+1}, it follows that

$$S(\boldsymbol{\lambda},\omega) = S(a\boldsymbol{\lambda}',b\lambda'_{n+1}) \geq 0, \qquad \text{for all } \boldsymbol{\lambda},\ \omega \qquad (4)$$

Remark 1 Eq. 4 shows that, for the given s/t metric, a non-negative definite covariance model in R^{n+1} is a permissible spatiotemporal covariance model for eq.1.

Remark 2 The reader should be reminded that the above results do not imply the equivalence of distance in time and in space.

Remark 3 Two examples may be noted. A spherical covariance is permissible in R^3 and therefore in R^2xT. The function $\cos(\tau)$ is only permissible in 1D and therefore permissible only for $\mathbf{h}=0$, i.e $\cos(h,\tau)=\cos(\tau)$.

The 'zonal anisotropy' representation of a spatiotemporal covariance

Consider a stationary S/TRF Z(x,y,t) in R^2xT with a covariance function $C(h_x,h_y,\tau)$. Suppose that $C(h_x,h_y,\tau)$ takes the general form

$$C(h_x,h_y,\tau) = C_o + C_1(h_x,h_y,\tau) + C_2(h_x,\tau) + C_3(h_y,\tau) + C_4(h_x,h_y)$$

$$+ C_5(h_x) + C_6(h_y) + C_7(\tau) \qquad (5)$$

where constant $C_o \geq 0$ and C_1, ..., C_7 are permissible covariance functions. Note that purely spatial and temporal covariances are special subcases of s/t covariances for fixed t and s, respectively. The spectral density function of $C(h_x,h_y,\tau)$ in eq. 4 is given by

$$S(\lambda_1,\lambda_2,\omega)= \int_{-\infty}^{\infty}\int_{-\infty}^{\infty}\int_{0}^{\infty} C(h_x,h_y,\tau) \exp (i\lambda_1 h_x+i\lambda_2 h_y+i\omega\tau) \, dh_x \, dh_y \, d\tau$$

$$= S_o\cdot\delta(\lambda_1,\lambda_2,\omega) + S_1(\lambda_1,\lambda_2,\omega) + S_2(\lambda_1,\omega)\cdot\delta(\lambda_2) + S_3(\lambda_2,\omega)\cdot\delta(\lambda_1)$$
$$+ S_4(\lambda_1,\lambda_2)\cdot\delta(\omega) + S_5(\lambda_1)\cdot\delta(\lambda_2,\omega) + S_6(\lambda_2)\cdot\delta(\lambda_1,\omega)$$
$$+ S_7(\omega)\cdot\delta(\lambda_1,\lambda_2) \tag{6}$$

where S_o, ... ,S_7 are the spectral densities of C_o, ... , C_7, respectively, and the delta functions are $\delta() \geq 0$.

Remark 4 Eq. 6 shows that if $C_1(h_x,h_y,\tau)$ is a strictly positive definite function (Berg et al., 1984), then $C(h_x,h_y,\tau)$ is also strictly positive definite. For example, consider that in eq. 5 $C_1(h_x,h_y,\tau) = \text{Sph}[h_x,h_y,\tau; a]$, where Sph[] is a spherical s/t covariance model and a its range. Sph[] is a strictly positive definite function, thus $C(h_x,h_y,\tau)$ is also strictly positive definite no matter what C_o, C_2, ..., C_7 are.

Remark 5 It follows from eq. 6 that if $C_1(h_x,h_y,\tau)$ is only a positive semi-definite function, then $C(h_x,h_y,\tau)$ is also a positive semi-definite function, since the remaining terms are a linear combination of delta functions.

Remark 6 The result from eq. 6 can be generalized and is valid for both R^nxT and R^n.

SINGULARITY PROBLEMS IN SPACE-TIME ORDINARY KRIGING

Prior to the discussion of singularity problems in space-time kriging, it should be noted that the spatiotemporal ordinary kriging system (Christakos, 1992) is found to have a form analogous to that of spatial ordinary kriging. In s/t ordinary kriging all covariance terms of the system are s/t covariances. Implementation differences exist, but are not considered in this paper.

Similarly to the estimation of RF's in the space domain (e.g. Myers and Journel, 1990), kriging S/TRF's with positive semi-definite covariance functions does not guarantee that the s/t kriging matrix is non singular. However, from the representation of a spatiotemporal covariance in eq. 5 and *Remark 4*, it follows that if $C_1(\mathbf{h},\tau)$ is a strictly positive definite function, the absence of singularity problems in the solution of the kriging system is ensured. This becomes apparent if one recalls that the requirements for a positive definite matrix are identical to those of a strict positive definite function.

In the case of semi-positive definite s/t covariances, avoiding certain data configurations are crucial in eliminating singularity problems. The influence of data configurations can be demonstrated on either the general representation of a s/t covariance, or on specific covariance models.

Singularity and 'zonal anisotropy' representations

Singularity problems generated from a 'zonal anisotropy' representation can be demonstrated in the following common case in R^2xT. Consider the case where eq. 5 reduces to

$$C(h_x,h_y,\tau) = C_4(h_x,h_y) + C_7(\tau) \tag{7}$$

that is, the covariance is decomposed into the sum of independent purely spatial and purely temporal components. Then, given four samples at the four corners of a 'rectangle': (x_1,y_1,t_1), (x_1,y_1,t_2), (x_2,y_2,t_1), (x_2,y_2,t_2), the covariance terms are

$$\begin{aligned}
C_{12}(h_x,h_y,\tau) &= C_4(0,0) && + C_7(|t_1-t_2|)\\
C_{13}(h_x,h_y,\tau) &= C_4(|x_1-x_2|,|y_1-y_2|) && + C_7(0)\\
C_{14}(h_x,h_y,\tau) &= C_4(|x_1-x_2|,|y_1-y_2|) && + C_7(|t_1-t_2|)\\
C_{23}(h_x,h_y,\tau) &= C_4(|x_1-x_2|,|y_1-y_2|) && + C_7(|t_1-t_2|)\\
C_{24}(h_x,h_y,\tau) &= C_4(|x_1-x_2|,|y_1-y_2|) && + C_7(0)\\
C_{34}(h_x,h_y,\tau) &= C_4(0,0) && + C_7(|t_1-t_2|)
\end{aligned}$$

Setting $a=C_4(|x_1-x_2|,|y_1-y_2|)$, $b=C_7(|t_1-t_2|)$, $c=C_4(0,0)$, and $d=C_7(0)$ the left hand side of the kriging system becomes

$$\begin{vmatrix}
c+d & b+c & a+d & a+b & 1\\
b+c & c+d & a+b & a+d & 1\\
a+d & a+b & c+d & b+c & 1\\
a+b & a+d & b+c & c+d & 1\\
1 & 1 & 1 & 1 & 0
\end{vmatrix}$$

The sum of rows one and four is the same as the sum of rows two and three, thus the matrix is singular. Furthermore, note that if one adds any number of data to the present configuration the kriging matrix remains singular.

Remark 7 Four data forming a 'rectangular' pattern represent a common case in many types of data sets. For example, two monitoring stations measuring the concentration of pollutants at the same time instant, measurements of reservoir pressure at two wells taken the same day, etc.

Remark 8 Similar results are derived if eq. 7 is rewritten using different combinations of the C_2, ..., C_7 components of eq. 5.

Remark 9 Analogous results are obtained in the general zonal anisotropy in R^3. Furthermore, singularity problems increase as the number of directions increases.

Singularity and covariance models

The singularity problems due to positive semi-definite covariance models are demonstrated using function $\cos(\tau)$ as an example. For $\mathbf{h}=0$, eq. 5 can be reduced to $C(h_x,h_y,\tau)=C(\tau)=\cos(\tau)$. If, say, three data are available at times t_1, t_2, and t_3 separated by distances: $t_1-t_2=2\pi$, $t_2-t_3=\Delta$ and $t_1-t_3=2\pi+\Delta$, then $C(t_1-t_2)=1$, $C(t_2-t_3)=\cos(\Delta)=a$ and $C(t_1-t_3)=\cos(2\pi+\Delta)=a$, thus the left hand side of the kriging system

$$\begin{vmatrix} 1 & 1 & a & 1 \\ 1 & 1 & a & 1 \\ a & a & 1 & 1 \\ 1 & 1 & 1 & 0 \end{vmatrix}$$

is singular. Again, this does not change if any number of additional data are considered.

Remark 11 The covariance function $\cos(\tau)$ will generate a singular kriging matrix, if only one pair of data separated by time distance 2π is present. Note that according to *Remark 4*, if $\cos(\tau)$ is combined with a strictly positive definite function, singularity problems do not occur. For example, consider the covariance $C(h_x,h_y,\tau)=\mathrm{Sph}[h_x,h_y,\tau;a]+\cos(\tau)$, where Sph[] denotes a spherical covariance model and a is its range. Since the spherical covariance is strictly positive definite, singularity problems do not occur.

'NORMALIZED' SPACE-TIME DISTANCE MEASURES

In addition to the notion of spatiotemporal distance, e.g. kilometre-hours, as used by Buxton and Pate (1993), one may consider the idea of unitless s/t distance measures. For a S/TRF Z(s,t) with a s/t metric

$$|(h,\tau)|^2 = |h|^2 + |\tau|^2 \tag{8}$$

one may consider a 'normalized' s/t metric, such as

$$|(h',\tau')|^2 = |h/h_m|^2 + |\tau/\tau_m|^2 \tag{9}$$

Both h_m and τ_m can be derived from general formulae, such as: $g_m=g_{max}-g_{min}$, $g_m=g_{max}$, $g_m=g_{min}$, $g_m=\Sigma_{i=1}^{n}\, dg_i$, $g_m=1/n\ \Sigma_{i=1}^{n} dg_i$, etc., where g is h and τ.

Remark 12 A 'normalized' s/t metric in eq. 9, eliminates the need for physical interpretations of the s/t distance units of the regular s/t domain (Christakos, 1991). However, the implications of this transformation are not fully understood yet. For instance, which of the above equations, or any other, should be used and in which case, or how transformations may relate to the physics of the phenomenon under study, should be further investigated.

CONCLUSIONS

It has been shown that permissible covariance models in R^{n+1} are also permissible covariance models for a specific class of spatiotemporal covariances in R^nxT.

A general 'zonal anisotropy' representation may be used for a spatiotemporal covariance. It has been shown that if a strictly positive definite spatiotemporal covariance is part of this representation, it ensures the uniqueness of the solution of the kriging system. Otherwise, singularity problems are resolved only by avoiding certain data configurations.

It was shown that whenever four data in a data set form a 'rectangle' in time, the simple covariance composed of a space plus a time component will generate a singular kriging matrix. Singularity also occurs in the case a 1D cosine covariance model, when two data points are 2π apart.

From the results presented, it is clear that the approach of Buxton and Pate (1993) as well as that of others (Bilonick, 1985; 1987), although empirically defined, uses both permissible covariance models and avoids singularity problems.

'Normalized' s/t distance measures may be used to provide 'unitless' space time distance measures.

ACKNOWLEDGMENTS

Acknowledgments are in order to George Christakos and Andre Journel for comments and suggestions. Funding was provided from NSERC of Canada grant OPG 0105803 to RD.

REFERENCES

Berg, C., Christensen, G.R., and Ressel, P. (1984) The Theory of Positive Definite and Related Functions, Springer-Verlag, New York, NY.

Bilonick, R.A., (1987) Monthly Hydrogen Ion Deposition Maps for the Northeastern U.S. from July 1982 to September 1984, Consolidation Coal Co., Pittsburgh.

Bilonick, R.A., (1985) "The Space-Time Distribution of Sulfate Deposition in the Northeastern United States", Atmospheric Environment, 19, 1829-1845.

Buxton, B.E. and Pate, A.D. (1993) "Joint Temporal-Spatial Modeling of Concentrations of Hazardous Pollutants in Urban Air", *in this volume*.

Christakos, G. (1992) Random Field Models in Earth Sciences, Academic Press, New York, NY.

Christakos, G., (1991) "On Certain Classes of Spatiotemporal Random Fields with Applications to Space-Time Data Processing", IEEE Transactions on Systems, Man, and Cybernetics, 21, 861-875.

Myers, D. and Journel, A.G. (1990) "Variograms with Zonal Anisotropies and Non Invertible Kriging Systems", Mathematical Geology, 22, 779-785.

GEOSTATISTICS AND DATA INTEGRATION

C. DALY and G.W. VERLY
BP Research Centre, Chertsey Road, Sunbury-on-Thames
Middlesex, TW16 7LN, UK.

Abstract: An important task in the petroleum industry is to integrate various information when modeling reservoir rock heterogeneities. There exist many geostatistical methods which attempt to integrate some aspects of the available information. However no method succeeds in integrating all of it in a practical and satisfactory way. On the other hand, experience has shown that a good geostatistical study conditioned to a limited but relevant amount of information can provide satisfactory results, even from the point of view of the information that has not been incorporated. This gives the possibility of either keeping the results as they are, or improving them using the additional information in a subsequent step.

In this presentation, some examples of such experience are presented. The integrated information consists of geological, geophysical, and/or engineering data. The method used are: sequential indicator simulation with trend, sequential gaussian co-simulation, boolean modeling, and/or simulated annealing. The possibilities, advantages and disadvantages of each method are discussed in the light of the examples. It is then seen how, in some cases, several types of information may be incorporated in a succession of steps.

1. Introduction

An important objective in the petroleum industry is to perform stochastic simulations of reservoir petrophysical properties which respect all available data. These data might include core measurements of permeability and porosity, sonic and density logs, seismic data and well test or production data. We must add to this list subjective geological information usually derived from outcrop analogue studies. Given that all these forms of data are measured on very different supports and are subject to various types of non-correlated or correlated error, it is not surprising that at this point in time, it is still unfeasible to develop a single model which is capable of dealing with all forms of data in a well understood and consistant manner. Some initial steps have been taken in this direction, principally using Monte Carlo techniques (Farmer 1989; Deutsch 1992). However it would appear that more work

R. Dimitrakopoulos (ed.), Geostatistics for the Next Century, 94–107.

is necessary before these techniques are able to incorporate all relevant data in a routine manner.

An alternative to this approach is to prioritize the data and use some of the standard techniques given in the following section to simulate using only the most important data sets. If the data sets which have not been used in this first step are consistant with those that have been used, we often find that our simulation respects the unused data quite well. In these cases we feel that it is possible simply to modify our simulation slightly so that all data is respected. This approach, which generally will not be mathematically rigourous if we decide to stop our modifications to the simulation as soon as our extra data conditions are satisfied, has the advantage of being computer efficient. This consideration is of substantial consequence if we are interested in studies of uncertainty where many realizations are necessary.

2. Review

This section is a brief review of some methods used to generate reservoir heterogeneities. The objective is not so much to describe the methods rather than to point out their limitations in term of data integration.

2.1 Sequential Indicator Simulation (SIS)

This method uses the principle of sequential simulation to generate models of a discrete, e.g. lithotype, or continuous, e.g. permeability, porosity (Journel, 1982; Journel and Alabert, 1988). Sequential simulation consists in selecting one grid point at a time and generating one property value at that location by drawing from its local distribution conditioned to surrounding values, including the simulated ones.

In the case of SIS, the property values are first transformed into a series of 0/1 indicators. Estimates of the local conditional distributions are then obtained by simple kriging or cokriging of the indicators.

The method is simple. It can incorporate several types of data and is convenient for lithotypes. Some problems include consistency of variograms when three or more indicators are considered, an assumption of stationarity, and the impossibility to control shapes when lithotypes are generated. For example, there is no guarantee to have continuous shales.

Usually, SIS only incorporates geological information by means of the variogram and data conditioning. However, since the method is a series of simple kriging, it can handle trends (Langlais, 1993; Figure 1.1). We call this version SISTR, i.e. SIS with trend.

2.2 Sequential Gaussian Simulation (SGS)

Multigaussian theory has been known for a long time (Anderson, 1957) and putting it within the framework of sequential simulation is relatively straightforward (Isaaks, 1990; Verly, 1991, 1992; Gomez-Hernandez and Journel, 1993). In this case, the continuous property values, e.g. permeabilities, are first transformed into normal scores, i.e. values which are normally distributed. If the normal scores are also

1. SIS with trend

2. Cross - stratification

3. Permeabilities; SGS

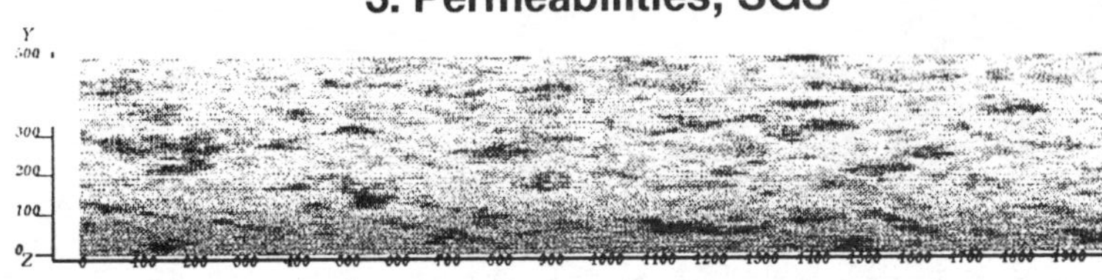

4. Permeabilities; SIS & SGS

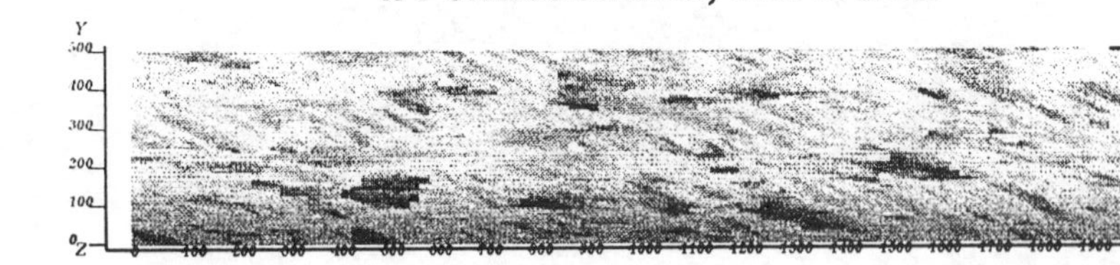

Figure 1 Example of heterogeneity modeling methods. 1) Sequential indicator simulation with trend. 2) Object modeling plus gaussian simulation. 3) Cross-stratification model using sequential indicator and sequential gaussian simulation. 4) "Cross-stratification" model using sequential gaussian simulation alone.

multinormally distributed, any local conditional distribution is normal with mean and variance obtained by simple kriging. The method therefore simulates normal scores. The simulated normal scores are back-transformed to original property values at the end of the simulation.

The method can be extended to several variables and used to incorporate seismic information (Verly, 1992; Figures 2.a, c, and d) and effective permeabilities from well test (Gomez-Hernandez and Journel, 1992). It is then called SGCS for sequential gaussian co-simulation.

SGCS has the advantage of being simple. It also incorporates several types of information, for example geological (e.g. permeability values), seismic (e.g. acoustic impedance), and well test (e.g. effective permeabilities). It does this however in a statistical sense. In other words, there is no guarantee that the physics between the variables is reproduced. A disadvantage is the necessary hypotheses of multinormality. Note that these hypothesis can be relaxed when there are lots of conditioning data (Verly, 1992).

2.3 Object Modeling (OM)

Object Modeling consists in dropping objects within a constant background with various conditioning and stacking rules. The main advantage of OM over SIS is its ability to control shapes and to create models which looks more realistic (Figure 1.2). The disadvantage is that conditioning and stacking rules can get complex. There are also no obvious ways to integrate several types of data, besides a simple trend in the object density.

2.4 Iterative Stochastic Techniques

In the last few years there has been a marked increase in the use of Monte Carlo or iterative stochastic techniques for simulating random functions (Ripley, 1987). Two important techniques are the Gibbs sampler and the Metropolis algorithm. In the former we use Bayesian techniques to obtain the conditional distribution of the process to be simulated and by iterative sampling from this distribution we converge to a realization of the desired stochastic process. In the latter case we simulate arbitrary transitions at each pixel and then accept or reject this transition according to a validation based on the law we wish to simulate. It can be proved that this method also leads to a realization of the desired stochastic process. Some examples of potential applications of these techniques for reservoir simulation and a general formulation of the Gaussian case is given in (Freulon, 1993)

These simulation techniques may be used as part of an optimization procedure in Simulated Annealing wherin an objective function which is to be maximized is represented by a Gibbs distribution. Thus the modes of this distribution correspond to the maxima of the objective function. By sampling from the distribution while the temperature (or variance) is gradually reduced to zero we obtain realizations which converge to modal realizations of the distribution. It can be shown that the particular modal realization obtained is drawn uniformly from the set of all modes of the objective function (Geman and Geman, 1984).

The major problem with these techniques is that convergence may be slow with current computer capabilities thus effectively precluding the possibility of applying

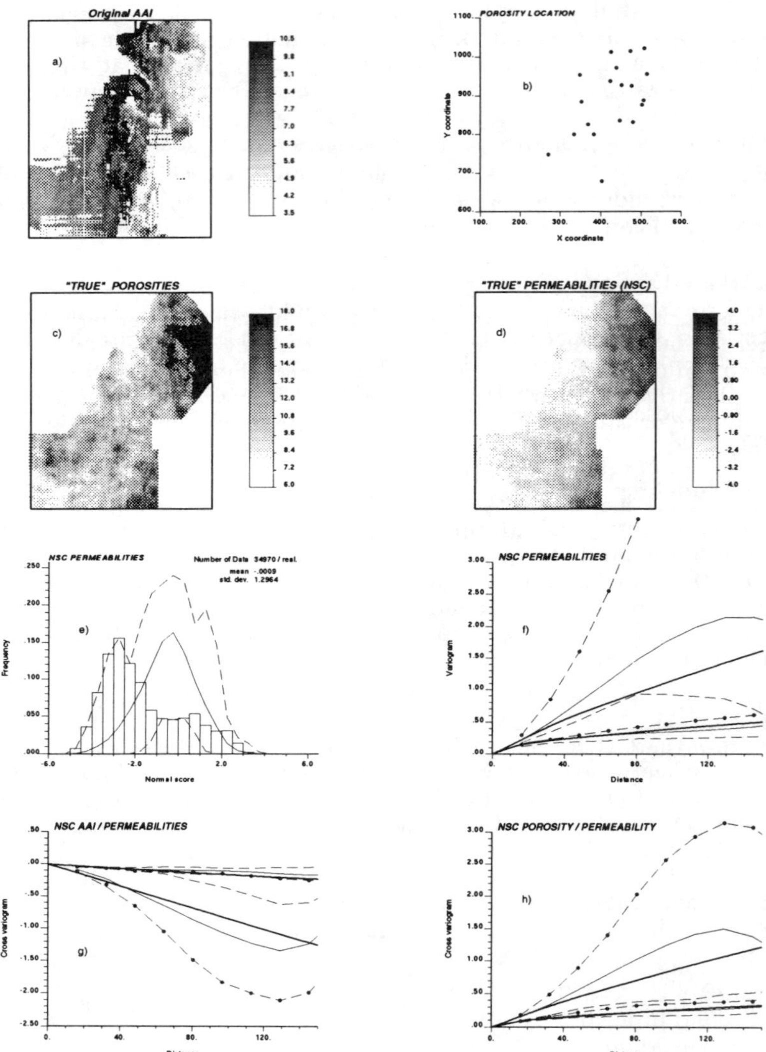

Figure 2 The "reality" model. a) Original absolute acoustic impedance (AAI) map. b) 20 porosity value locations; c & f) First porosity and permeability realizations, chosen as "reality" e to h) Simulated permeability (normal score) distribution, variogram, and cross-variograms with AAI and porosity such that: average simulated distribution/variogram (full lines); mimimum and maximum frequency/variogram values for one realization (dashes); auto-/cross-variogram models (thick lines in f to h); 1st realization ("reality) auto-/cross-variograms (dots in f to h); 1st realization ("reality) distribution (histogram in e).

them to more than a small number of alternative reservoir descriptions. We may use approximate Gibbs sampling techniques or simulated annealing techniques (such as rapid freezing) which converge very quickly. It is true that the approximate techniques will usually bias the result in some sense (e.g. they may not respect the spatial continuity exactly). However if we start from a good initial guess and if the approximate technique is judiciously chosen this bias is usually very small so that we have a quick ad-hoc modifier of our simulation. We will give an example later wherby we adjust some cosimulated images of a North Sea reservoir so that effective permeabilities (supposedly given by well test analysis) are respected.

3. Practical Experience

The previous section was an incomplete list of methods for generating models of heterogeneities but it illustrated a common fact: not all available information is incorporated. Simulated annealing makes no exception. Indeed, one could probably design objective functions that incorporate all available information. But the corresponding algorithms would be very complex and cpu intensive, in other words, impractical.

Experience however has shown that good results can still be obtained if an imperfect method is used, or if all the information has not been incorporated. This happens when the non-incorporated aspects of the information are consistent and partially redundant with those incorporated. Some examples of such experience are now given.

3.1 Ignoring a trend

The objective was to generate porosity models using absolute acoustic impedance (AAI) information from seismic (Verly, 1992; Figures 2.a, b, c). One problem was the strong trend indicated by seismic along one direction (Figure 2.a). The trend was not subtracted prior to the study and SGCS was used to generate the porosity models. The fact that the trend was not explicitly incorporated in the model did not have a negative impact on the results due to the strong conditioning imparted by the seismic.

3.2 Ignoring some geological information

The objective was to study the impact of cross-stratifications on effective permeabilities. Two modeling procedures have been used. The first one is a straight SGS with a permeability variogram covering distances up to several sand units (Figure 1.3). The second procedure is in two steps: 1) SIS for the individual sand units, and 2) SGS for permeabilities within the sand units, with variable dips and correlation lengths (Figure 1.4). The two resulting models look different, but give almost the same single-phase effective permeabilities. Note that an even better model would have been a mixture of OM and SGS as in Figure 1.2, but again this would not have given significantly different effective permeabilities. Single phase effective permeabilities are affected by the tortuosity of the flow, which itself depends on the large scale heterogeneities (Haldorsen at al, 1987). The simpler procedure, i.e. a straight SGS, is able to capture some aspects of the large scale heterogeneities and this was enough to approximately recreate the correct amount of tortuosity. Note that the results could be quite different if two-phase instead of single-phase flow was considered.

3.3 Ignoring production data.

The objective was to generate a reservoir heterogeneity model that incorporated detailed information on those geological features which were determined to have a major impact on fluid flow (Begg et al, 1992). The consistency of the model with production information was then investigated.

The geological model consists of several major facies associations (MFAs), each with its own complex geology on a smaller scale. The method used to numerically generate such a model is a combination of OM, SISTR, SGS, and effective permeability upscaling.

The major geological control on poro-perm (porosity, permeabilities and net:gross) in this case was lithotype. For each MFA, a 2D, generic, lithotype model was generated using OM. These models were gridded on a very fine scale and poro-perm properties assigned using core-plug data. Effective (re-scaled) poro-perm values were then computed within larger 2D blocks. These large blocks have the same size as the blocks used in the reservoir simulator, and are called "reservoir simulator blocks". Variograms of effective horizontal permeabilities were then computed in 2D for each MFA. Correlations between effective horizontal and vertical permeability, porosity and net:gross were also computed. The variogram models were extended from 2D to 3D using geological arguments about likely anisotropy in horizontal continuity. For each MFA, there is therefore a reservoir simulator block permeability variogram and correlation with the other effective properties.

A 3D model of MFAs' architecture, conditioned to the MFAs observed along wells and to a vertical trend, was generated using SISTR. This model is on the large scale: it covers several square miles and the grid cell size is the reservoir simulator block size. Effective permeability within the MFAs architecture were then generated using SGS and the variograms computed earlier.

The final result is a single, but geologically reasonable, model of consistent re-scaled poro-perm that is ready to be used within a reservoir simulator. Very minor ajustments to the description were required in order to achieve an excellent history match to production data from the 60 wells in the model.

3.4 Conclusions

In this section, some examples of models that do not use all available information have been presented. Yet good results were obtained because the features that matters were captured, at least partially. Hence the concept of getting approximate models using methods limited in terms of what they can incorporate, for example OM, SIS, SGS, and then update them with additional information using for example an iterative method. If the models obtained prior to updating are relatively close to the updated ones, it is reasonable to assume that the updating algorithms can be simplified, making them easy to build and not too cpu intensive. Note that this concept of getting a first model and then updating it with simulated annealing has already been used by Deutsch (1992).

4. Case Study

This case study has been designed along the lines of the previous section conclusion. The objective is to generate a series of permeability models incorporating geological, seismic and well test information. A two step procedure is followed: 1) SGCS to generate permeability models incorporating geological and seismic information, and 2) Gibbs sampling to update the models according to well test information.

The original dataset is related to an actual reservoir located in the North Sea. For confidentiality reason, the name and location of the reservoir cannot be disclosed and actual data values have been modified. This dataset consists of more than 30,000 absolute accoustic impedance values (AAI; Figure 2.a) and 20 porosity values measured at well locations (Figure 2.b). This dataset is in 2D and has already been used to generate realizations of a AAI/porosity model (Verly, 1992).

Because no permeability measurements were readily available, it was decided to generate a synthetic dataset called "reality". This "reality" is in fact an extension of the previous model to incorporate permeabilities, i.e. a AAI/porosity/permeability model. From this "reality", permeability samples could be extracted at 18 well locations within the reservoir limits, and effective permeabilities computed around 2 selected wells using the renormalization method (King, 1989). The 18 permeability samples constitute additional geological information, whereas the 2 effective permeabilities consitute the available well test information.

The first step of the modelling procedure, SGCS, was applied to incorporate the geological and seismic information. The second step, an approximate Gibbs sampling, followed to update the models according to the well test information.

The remaining of this section provides additional details on this two step procedure together with a discussion of the obtained results.

4.1 Generating the "reality"

A detailed description of the practice of SGCS to obtain a AAI/porosity model can be found in Verly (1992). This model reproduces:

- the AAI and porosity distributions;
- the AAI and porosity auto- and cross-variograms;
- the 30,000 seismic AAI and 20 well porosity values.

In order to add permeabilities to the model, the following has been considered:

- the permeability distribution is given as $K = e^{Y+5}$, where K is the permeability and Y its simulated normal score;
- the permeability variogram and the two cross-variograms with AAI and porosity are known.

The necessary additional auto- and cross-variograms have been chosen such that:

- the short scale AAI/permeability correlation is null;
- the short scale porosity/permeability is weak;
- the long scale permeability correlations with AAI and porosity are fairly good.

Ten realizations of this AAI/porosity/permeability model have been generated using SGCS. Figures 2.e to 2.h show the histograms and variograms corresponding to the 10 permeability (normal score) realizations, together with the cross-variograms with AAI and porosity. These figures shows that there are significant fluctuations between the realizations. The auto- and cross-variogram models are reproduced on average, at least as far as one can tell from only 10 realizations. Note that an average standard deviation of 1.2 for the simulated normal scores is possible since it depends on the variogram sill (larger than one) and the indirect AAI and porosity conditioning.

Not shown in the figures are the statistics corresponding to the porosity and AAI realizations. There are fluctuations among the porosity realizations, though less severe than among the permeability's. This is due to the 20 conditional porosity values which anchor the realizations. There is almost no fluctuations among the AAI realizations due to the great number of conditioning AAI values.

The first realization has been chosen as the "reality" or "truth". The resulting true porosity and permeability models are shown in Figure 2.c and 2.d. The true AAI model is not shown, but is very similar to the original dataset (Figure 2.a). Indeed, only the missing AAI values have been simulated. The statistics of the true permeability model are shown in Figures 2.e to 2.h. Interestingly enough, these statistics show that the selected "truth" is in fact an extreme realization. It contains a lot of low values, and the corresponding variogram is most variable.

Permeability values have been sampled from the true model at 18 well locations within the reservoir limits. Single phase effective permeabilities have been computed using a renormalization technique (King, 1989) for the two selected well locations.

4.2 Generating models using geological and seismic information

Ten AAI/porosity/permeability realizations have been generated with SGCS. These realizations incorporate the following geological and seismic information:

- 30,000 AAI values;
- 20 porosity values;
- 18 permeability values;
- the variograms, the cross-variograms, and the distributions used to generate the true model.

In fact, the only difference between these realizations and the ones generated as part of the "reality" building exercise are the 18 conditioning permeability values.

The statistics corresponding to the 10 simulated permeability (normal score) realizations are shown in Figures 3.c to 3.f. Two individual realizations are shown in Figures 3.a and 3.b.

The figures show clearly that the simulated statistics are on average much closer to the "true" statistics with the 18 conditioning permeability values than without (Figures 2.c to 2.f). In fact, the true statistics could very well correspond to an 11th (extreme) realization, which is what we would expect.

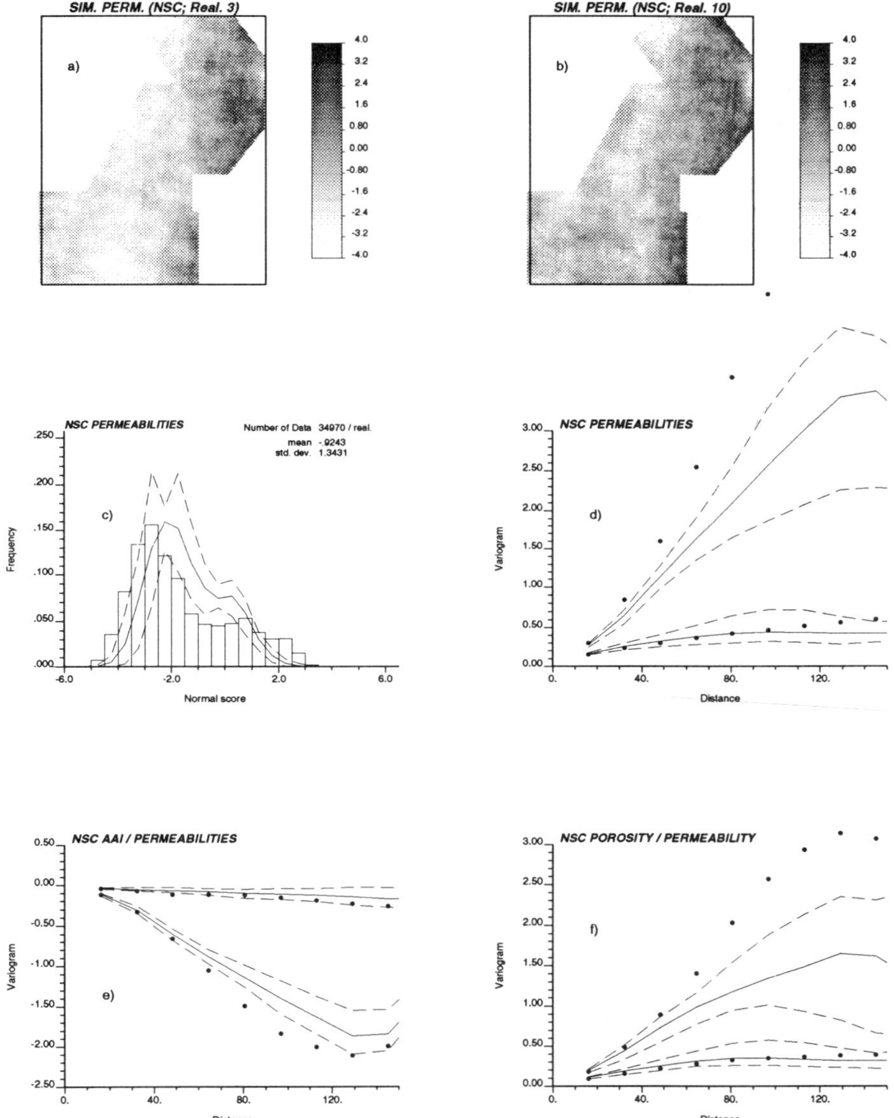

Figure 3 The simulated models. a & b) Two permeability realizations; c to f) Simulated permeability (normal score) distribution, variogram, and cross-variograms with AAI and porosity such that: average simulated distribution/variogram (full lines); mimimum and maximum frequency/variogram values for one realization (dashes); "actual" auto-cross variogram models (dots in d to f); "actual" permeability distribution (histogram in c).

The well test information has not been incorporated at this stage. Table 1 shows the simulated effective permeabilities. The effective permeabilities for the 'true' case are 211.2 and 22.97. Note that the discrepencies between 'true' and the 10 realizations are not very big. This is due to the fact that part of the well test information is contained (or is fairly consistent with) the used geological and seismic information. However we can see that in the first well test all ten realizations give an effective permeability lower than the 'true' effective permeability. This is not due to inadequate modelling. After all, we simulated the 'true' permeability map and used the same model to produce our conditional simulations. In practice, where we would only be confronted with the data, we might be tempted to reject the hypotheses of a Gaussian model, or to change some of the parameters, as we are unable to match the well test directly. From this fairly simple example we see that rarely will we be able to conclude that we have the true model and that assessing the uncertainty in an estimation of say well performance will require analysis of several different types of model. The increased computational load necessitates quick simulation techniques and from a reservoir engineering perspective, quick performance estimators based on these stochastic simulations.

Well 1	Well 2
207.5	23.6
163.9	22.6
125.3	17.9
165.0	11.9
199.8	24.8
150.9	30.5
174.0	20.6
194.6	19.3
161.9	23.7
187.8	30.4

Table 1: Effective permeabilities for the 10 realizations in the neighbourhood of the two wells.

4.3 Updating the models with well test information

We use the renormalization technique within an approximate Gibbs sampler to modify our ten realizations such that the two well test values are matched exactly. In figure 4 we see two of the modified realizations. These two figures (realizations 3 and 10) are extreme cases where a fairly high degree of modification was necessary to match the effective permeability. The modifications are none the less confined to a region 'close to' the zone of influence of the well test, so that for the most part the realizations differ little from the original images (figure 4).

While in principle the updating technique does not guarantee that the variogram will be respected exactly, it does take it into account, and we can see in figure 4d that they are only slighty modified and that the behavior at the origin is not disturbed. The histograms are changed more radically to enable the first well test to be matched.

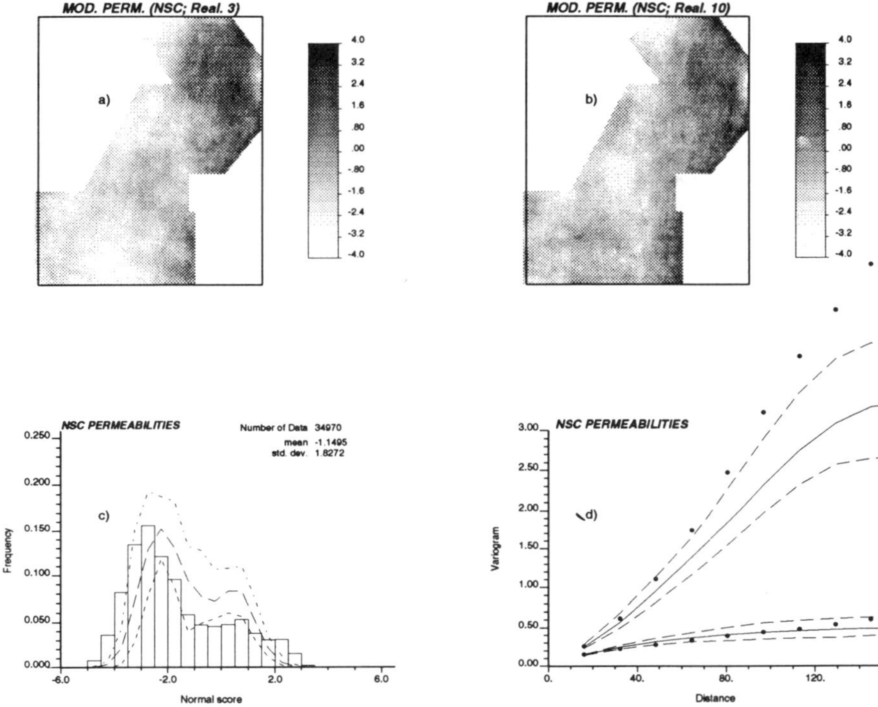

Figure 4 The modified models. a & b) modified versions of the corresponding permeability realizations of figure 3; c & d) Permeability (normal score) distribution and variogram for the modified realizations such that: average simulated distribution/variogram (full lines); mimimum and maximum frequency/variogram values for one realization (dashes); "actual" auto-cross variogram models (dots in d); "actual" permeability distribution (histogram in c).

The updating proceedure takes about 12-20 seconds on a HP750 computer per realization. As such it is useful for studies where many stochastic realizations are needed.

5. Conclusion

In this paper we have looked at several practical examples of data integration for reservoir description. In many cases choosing a simple model incorporating the more important data is sufficient to provide models which satisfy our requirements. This may be because the other data are implicitly respected by the simulation, or simply that they are not relevant for those applications made of our simulations. In other cases, such as the case study in this paper, the extra data information may be included with the aid of a rapid modification of the simulations.

These circumstances, while not completely general, occur regularly in stochastic reservoir description. When this is the case we are capable of obtaining simulations without the complexity and disadvantages (in terms of computer time) of attempting to formulate and sample from a general model. An advantage for practical reservoir modelling is of course in terms of computation time but also in a substantial saving in man-hours. The engineer will spend less time developing and testing a simpler model than a more poorly understood (but theoretically more powerful model). This allows us to consider more realizations, both with fixed parameters and also with variable parameters in our attempt to study the uncertainty within a reservoir.

We have not set out to "solve" the problem of data integration, simply to point out that in certain cases it may be done in a straightforward stepwise manner. The validity of such a technique would have to be tested on a case by case basis and is not fully general. As such the problem of establishing a general methodology for data integration remains a challenging and complex problem.

References

Alabert, F.G. (1987). *Stochastic Imaging of Spatial Distributions Using Hard and Soft Information*, M.Sc. Thesis, Stanford U., 197 pp.

Anderson, T.W. (1958). *An Introduction to Multivariate Statistical Analysis.* Wiley, New York. 374 pp.

Begg, S.H., Gustason, E.R. and Deacon, M.W. (1992). "Characterization of a Fluvial-Dominated Delta: Zone 1 of the Prudhoe Bay Field", SPE 24698, *Ann. Tech. Conf. and Exib. of SPE, Washington, DC, TX*, pp. 351-364.

Deutsh, C. (1993). "Conditioning Reservoir Models to Well Test Information", *Geostatistics Tróia*, Part 1, Ed. Soares, A., Kluwer Academic Publ. , pp. 505-518.

Farmer, C., (1989). "Numerical Rocks", *European Conference on the Mathematics of Oil Recovery.* Ed. P. R. King. Clarendon Press, Oxford.

Freulon, X., (1993) "Conditional simulation of a Gaussian random vector with non-linear and/or noisy observations", *Workshop on Geostatistical Simulations, Fontainebleau 27-28 april 1993* to be published, Kluwer Acad. Publ.

Geman, S. and Geman D. (1984) "Stochastic relaxation, Gibbs distributions and the Bayesian resoration of images.", *IEEE Trans. Pattern Anal. Machine Intell., 6, 721-741.*

Gómez-Hernández, J., and Journel, A.J. (1993). "Joint Sequential Simulation of MultiGaussian Fields", *Geostatistics Tróia 92*, Part 1, Ed. Soares, A., Kluwer Academic Publ. , pp. 85-94.

Haldorsen, H. H., Chang, D. M., and Begg, S.H. (1987). "Discontinuous Vertical Permeability Barriers: A Challenge to Engineers and Geologists" *North Sea Oil and Gas Reservoirs*, Graham and Trotman, London, pp. 127-151.

Haldorsen, H. H., Brand, P. G., and Macdonald, C.J. (1988). "Review of the Stochastic Nature of Reservoir", *Mathematics in Oil Production*, Ed. Edwards S., King P.R., Clarendon Press, Oxford, pp. 109-209.

Isaaks, E.H. (1990). *The Application of Monte Carlo Methods to the Analysis of Spatially Correlated Data*, Ph.D. Thesis, Stanford U., 213 pp.

Journel, A.G. (1982). "Indicator Approach to Spatial Distribution.", *Proc. of the 17th APCOM Int. Symp.*, Eds. Johnson, T.B., and Barnes, R.J, Soc. of Min. Eng. of AIME, Littleton, CO, pp. 793-806.

Journel, A.G., Alabert, F.G. (1988). "Focusing on spatial connectivity of extreme-valued atrributes: stochastic indicator models of reservoir heterogeneities", SPE 18324, *Ann. Tech. Conf. and Exib. of SPE, Houston, TX*. pp. 621–632.

King, P.R., (1989). "The Use of Renormalization for Calculating Effective Permeability" *Transport in Porous Media*, Vol. 4, pp. 37-58.

Langlais, V., and Doyle, J. (1993). "Comparison of Several Methods of Facies Simulation on the Fluvial Gypsy Sandstone of Oklahoma", *Geostatistics Tróia 92*, Part 1, Ed. Soares, A., Kluwer Academic Publ. , pp. 299-310.

Ripley, B. (1987). *Stochastic Simulation.*, John Wiley, N.Y.

Verly, G. (1991). "Sequential Gaussian Simulation: A Monte Carlo Method for Generating Models of Porosity and Permeability". *Special Publication No. 3 of EAPG - Florence 1991 Conference*, Ed. Spencer, A.M., To be published.

Verly, G. (1993). "Sequential Gaussian Co-Simulation: A Simulation Method Integrating Several Types of Information", *Geostatistics Tróia 92*, Part 1, Ed. Soares, A., Kluwer Academic Publ. , pp. 543-554.

Acknowledgements

The authors would like to thank the British Petroleum Co. plc for permission to publish this paper, and John Williams for providing the cross-stratification model.

MODELLING IN THE PRESENCE OF SKEWED DISTRIBUTIONS

CARINA LEMMER
Consultant Applied Earth Sciences
Eagle House, 70 Fox Street
Johannesburg 2001
South Africa

Skewed distributions are usually blessed with exceptionally high outlier values that make log transformations in the modelling process very desirable. This introduces two problems in the determination of a global mean for the relevant deposit: if the distribution is not perfectly lognormal, an unbiased back transform for log estimates is not known analytically; additionally, even if the back transform is known, the high log variances that appear in the exponent make the operation unstable. A general method is presented that addresses these problems without special or subjective treatment, allows determination of variances with the influence of outliers attenuated and proposes an unbiased general back transform. The method is illustrated and evaluated quantitatively against a gold deposit for which the dense data is known.

INTRODUCTION

A complicating factor in the geostatistical modelling of precious metal deposits is often the phenomenon of so-called outliers. These exceptionally high values are important, but their inclusion usually totally obscures the experimental variogram, or leads to overestimation in the kriging process. They typically manifest themselves in the long tails of the skewed, even highly skewed, associated metal value distributions.

For such distributions it proves necessary to consider logarithms of the metal values so that outliers can be included on an equal footing. However, although skewed, metal value distributions are not necessarily perfectly lognormal. The non-lognormality introduces problems of its own, since the back transform is then not known analytically. Even when the transform is known, as in the lognormal case, the variance of the log variable appears in the exponent in the back transform, which makes the results of the back transform very vulnerable to the actual value of the variance used. One has therefore to make sure that the residual influence of the outliers in the log variable is not unduly influencing the variance.

In the following sections these problems are addressed using a modification of the mononodal indicator-grade [MIG] method [1] to render it appropriate for estimation in the presence of skewed distributions.

R. Dimitrakopoulos (ed.), Geostatistics for the Next Century, 108–119.

For the purposes of quantitatively evaluating the MIG method, use was made of dense data from an actual South African gold mine. The data was kindly provided by Gold Fields of South Africa [GFSA], and is not atypical of a shallow tabular precious metal deposit - for instance, a shear zone bound greenstone gold deposit.

The dense data has been regarded as representing the true gold values of the deposit, that would normally not be known. To simulate a typical situation in practice, a sparse data subset was created by extracting gold values from the dense set on a regular "drilling" grid. In addition, a geostatistical cross was "drilled" along the main directions of continuity of the mineralization, that were "known" from geological considerations. To avoid repetition, the gold accumulation values used for this study will henceforth be loosely referred to as "grades".

The paper is structured as follows: after a brief outline of the theoretical background, the evaluation steps in the MIG approach are discussed. Finally the performance of the method, and the premise of variogram proportionality are benchmarked against the "true" properties of the deposit.

THEORETICAL BACKGROUND

It has been previously demonstrated [1] that at a particular grade cutoff, called the mononodal cutoff, the grade and indicator variograms are essentially proportional to each other, and that one can use the latter, which is not affected by outliers, to determine the former. The basic principle underlying the MIG approach relies on the fact that any well-defined density distribution function f(z) associated with a regionalized variable also provides the weight function for a set of orthogonal polynomials [2] $X_n(z)$, say. Their functional forms are <u>completely</u> determined by the moments of the standardized variable $(z-m)/\sigma$

$$r_k = E\{(Z-m)^k/\sigma^k\} \qquad (1)$$

(m and σ^2 are the mean and variance of Z(x) and E{ } signifies the expectation value wherever used). The first three polynomials are $X_0 = 1$, $X_1 = (z - m)/\sigma$ and $X_2 = \text{const.}\{[(z-m)/\sigma]^2 - r_3[(z-m)/\sigma] - 1\}$. The X_n provide a basis [2] for expanding the bivariate density distribution of the variable. This has the consequence of allowing one to display the indicator covariance $C_I(h;z)$ as a series involving products of polynomial covariances $T_{mn}(h) = E\{X_m[Z(x)].X_n[Z(x+h)]\}$ and weight functions $W_n(z)$ for indicator cutoff z. The first terms of this series are

$$C_I(h;z) = \rho_Z(h)\, W_1(z)^2 + T_{22}(z)\, W_2(z)^2 + \ldots \qquad (2)$$

where $\rho_Z(h)$ is the grade correlogram, $C_Z(h)/C_Z(0)$.

The point about this expansion is that the retention of the terms shown, already favoured in deposits with a high nugget effect like gold (and therefore $\rho_Z(h)$ dropping rapidly with increasing lags h), **can be made optimal** by working at the mononodal cutoff, z_n, that is chosen so that the second term is completely suppressed, i.e. z_n is the root of

$$W_2(z_n) = const \times \int_0^{z_n} dz f(z) \left[\left(\frac{z-m}{\sigma}\right)^2 - r_3 \left(\frac{z-m}{\sigma}\right) - 1 \right] = 0 \qquad (3)$$

where f(z) is the density function and (z-m)/σ the standardized variable associated with random variable Z(x), as above. The third moment r_3, or skewness, is defined by eq. (1). This root always exists [1]. The practical consequence of evaluating eq.(2) at the mononodal cutoff is that the grade and indicator covariances become essentially proportional to each other at this particular cutoff. Therefore the associated grade and indicator variograms are also linearly related

$$\gamma_Z(h) = a(z) + b(z)\gamma_I(h;z) \qquad (4)$$

Since one knows that indicator [3] variograms are hardly affected by outliers, it is natural to regard the indicator variogram at the mononodal cutoff as more reliable than the grade variogram - particularly for sparse data - and to determine the parameters of the grade variogram from the above relation, after fitting the indicator variogram. In particular, the sill value of the grade variogram is very unreliably determined in the presence of outliers. Eq.(4) provides a method of estimating this quantity reliably from the indicator sill.

As already remarked, it is essential to first transform the raw grade values Z(x) (gold accumulation in this case) to Y(x) = ln[Z(x)] to minimise the distortion that the outliers introduce, prior to calculating the variograms. This procedure introduces the additional problem of finding a back transform that is (i) unbiased, and (ii) applicable even if the underlying distribution is not perfectly lognormal. The problem is a nonlinear one, requiring as it does the expectation of Z(x) = exp[Y(x)], given the statistics of the Y(x) = ln[Z(x)] variable. However this calculation can be completed by expanding the exponential and then taking the expectation value term by term. One then finds that

$$m = E\{e^{Y(x)}\} = e^{\alpha}\left[1 + \frac{1}{2}\beta^2 + \sum_{k=3}^{\infty} \frac{r_k}{k!}\beta^k\right] \qquad (5)$$

by expanding the exponential. The sum on the right can in principle be evaluated if all the moments $r_k = E\{[(Y-\alpha)/\beta]^k\}$ are known, α and β^2 referring to the mean and variance of Y. For example, the sum can be performed in closed form in the case of the lognormal distribution, where $\bar{r}_{2k} = 2^{[1-k]}[(2k-1)!]/[(k-1)!]$, $\bar{r}_{2k+1} = 0$, so that $m = \exp\{\alpha + ½\beta^2\}$ (refer a text on basic statistics). This fact suggests that in those cases where the underlying distribution is not too far from lognormal, it would be useful to introduce the difference between the actual r_k's of the distribution and the values r_k they would have assumed if the distribution were lognormal. This means adding and subtracting exp[½β^2] in the square bracket in eq.(5) and, remembering that

$$e^{\frac{1}{2}\beta^2} = \sum_{k=0}^{\infty} \frac{\bar{r}_k}{k!}\beta^k$$

it gives the modified expression

$$m = E\{e^{Y(x)}\} = e^{\alpha+\frac{1}{2}\beta^2}\,[1 + e^{-\frac{1}{2}\beta^2}\sum_{k=0}^{\infty}\frac{(r_k-\overline{r}_k)}{k!}\beta^k] \qquad (6)$$

in which the difference between the actual and lognormal moments now appear. The first three terms of the sum are identically zero, since $r_0 = \overline{r}_0 = 1$, $r_1 = \overline{r}_1 = 0$ and $r_2 = \overline{r}_2 = 1$ by construction. If this difference is small, as is usually the case in practice, the expansion can be terminated at the third or fourth moment difference without losing too much accuracy. A similar equation holds for the expectation value of the estimated values Y^*:

$$m^* = E\{e^{Y^*(x)}\} = e^{\alpha^*+\frac{1}{2}\beta^{*2}}\,[1 + e^{-\frac{1}{2}\beta^{*2}}\sum_{k=0}^{\infty}\frac{(r_k^*-\overline{r}_k)}{k!}\beta^{*k}] \qquad (7)$$

Here r_k^* is the estimate of the r_k that depend on the estimates α^* and β^{*2}. By defining $\exp(\eta)$ as the ratio of the right hand side of eq. (6) to that of eq. (7), one has $m^*.\exp(\eta) = m$, i. e. m^* has been rendered unbiased by multiplying it by $\exp(\eta)$. One therefore obtains the following back transform to estimated values Z^* in terms of the original grades, that is by construction unbiased:

$$Z^* = e^{Y^*+\eta} \qquad (8)$$

The unbiasing constant is given by

$$\eta = \frac{1}{2}(\beta^2-\beta^{*2}) + \ln\left\{\frac{1 + e^{-\frac{1}{2}\beta^2}\sum_{k=0}^{\infty}\frac{(r_k-\overline{r}_k)}{k!}\beta^k}{1 + e^{-\frac{1}{2}\beta^{*2}}\sum_{k=0}^{\infty}\frac{(r_k^*-\overline{r}_k)}{k!}\beta^{*k}}\right\} \qquad (9)$$

under the assumption that the estimate of the logs of the grades is unbiased, $\alpha = \alpha^*$. The effect of η is to render the expression for $Z^* = \exp[Y^*+\eta]$ unbiased, i.e. $E\{Z^*(x)\} = E\{\exp[Y^*(x)+\eta]\} = E\{Z(x)\}$.

EVALUATION STEPS

We now implement this approach step by step using the GFSA data.

Description of data

(a) The dense data representing the true values consist of 8185 gold accumulation values spread over a 1000 x 400 metre(m) rectangular area.

(b) The sparse drill hole subset consists of 96 data values on a 100 x 50m sublattice of the rectangle (one drill hole on the corner of each cell, with three drill holes missing), plus a geostatistical cross comprising 56 closely spaced holes (5m apart), aligned with the direction of best grade continuity.

The geostatistical cross serves to determine the variograms over short lags, but will bias the histogram. This means one has to declusterize the data set prior to histogram calculation [4].

Data statistics

The statistics for the dense set of 8185 values and the subset of 96 declusterized values compare as follows (units are arbitrary):

	8185 acc. values	**96 acc. values**
Raw values:		
Mean m:	1158.77	1093.9
Max. value:	44676.0	6640.0
Variance σ^2:	4546064	1755352
Coeff. of var.:	1.84	1.21
Skewness r_3:	6.73	1.70
Kurtosis r_4:	84.53	5.81
Logs of values:		
Mean α:	6.100	6.037
Variance β^2:	2.183	2.829
Std. Dev. β:	1.477	1.682
Skewness r_3:	-0.242	-0.669
Kurtosis r_4:	2.822	2.844
5th ord. r_5:	-1.773	-4.390
6th ord. r_6:	12.784	12.568

Conventional variography

Variograms were calculated for grades and log(grades) for the sparse subset plus cross (96+56=152 values) in the directions of best mineralization continuity and perpendicular to it. These variograms are shown in Figs. 1 to 4. The modelling problem is immediately evident: the variances of the sparse data given above and their logs obviously cannot serve as acceptable sills for these variograms, which in any event do not show discernable structure. (The careful reader will wonder about a possibe proportional effect here:

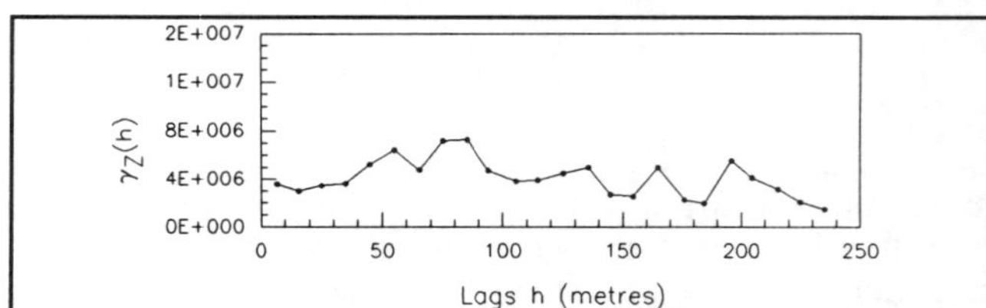

Figure 1. Sparse set grade variogram in direction of best continuity.

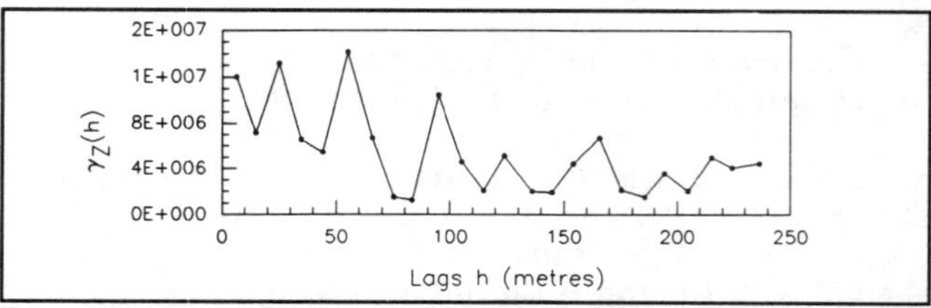

Figure 2. Sparse set grade variogram in direction of poorest continuity.

this was excluded after studying relative variograms).

Mononodal indicator-grade (MIG) variography

(i) *The mononodal cutoff.* For grade distributions that are close to normal, the mononodal cutoff z_n is approximately $z_n = m + \frac{1}{2}\sigma r_3$ [1]. It has been found, however, that in the case of skewed distributions, where the variance - even for logarithms - are affected by outliers, this approximation is not adequate. The mononodal cutoff has to be determined directly by constructing W_2 from eq.(3). It is simple to do so by sorting the 96 log(grade) = y values, calculating the value of $X_2(y) \sim \{[(y-\alpha)/\beta]^2 - r_3.(y-\alpha)/\beta - 1\}$ at each y-value, and accumulating on y. This gives the value of $W_2(y)$. A plot of $W_2(y)$ immediately locates the root $W_2(y_n) = 0$. Fig. 5 shows the $W_2(y)$ function constructed in this fashion for the subset of 96 values and indicates a mononodal cutoff $y_n = 5.42$ in this case. (As an aside a smooth curve results on using the histogram of the y's to construct W_2, but this gives $y_n = 5.33$, which turned out not to be accurate enough, as will be explained presently). Equipped with $y_n = 5.42$ one calculates the declusterized mononodal indicator variance for the 96 data set, and finds the value 0.215. [A reminder: the set of mononodal indicators is defined by: $i(x;y_n) = 1$ if $y(x) \leq y_n$; otherwise $i(x;y_n) = 0$ in this article.]

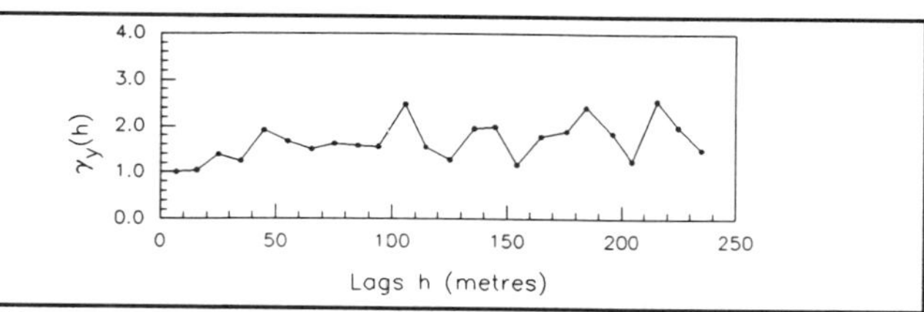

Figure 3. Sparse set log(grade) variogram in direction of best continuity.

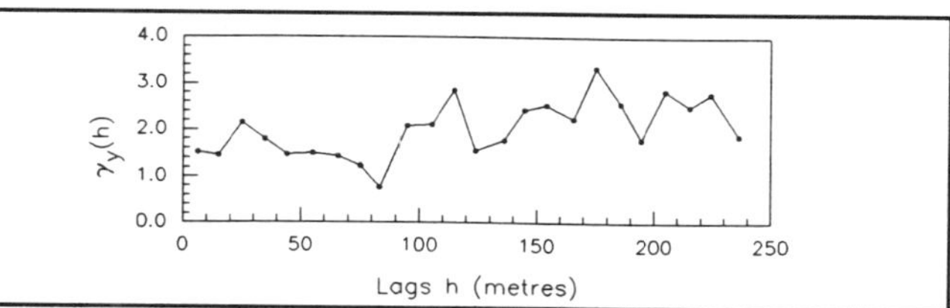

Figure 4. Sparse set log(grade) variogram in direction of poorest continuity.

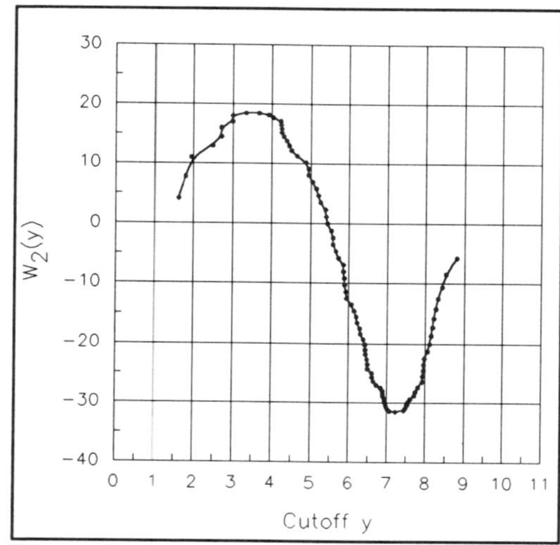

Figure 5. The function $W_2(y)$ for the sparse data set.

(ii) *MIG plot (1st iteration).* We now construct a first grade-indicator plot based on both the grid and the cross drill holes

- 152 values in total. This requires that a pair of grade and (mononodal) indicator variograms be calculated in the direction of best grade continuity. A plot of the grade versus indicator variogram values, that correspond to the same lag distances, is shown in Fig. 6. As a practical measure, variogram values - starting with the shortest lag - should be included in the linear regression fit for obtaining the grade-indicator line, eq.(4), for as far as they seem to line up roughly on the grade-indicator plot. This usually means including those points up to where the variograms reach their sills (one could call these "participating points"). The rest of the points correspond simply to statistical fluctuations beyond the range. However, in the case of the dense data these fluctuations have been dampened, and all points can be used, as we shall discover! The best fit for the grade-indicator line has a slope of 13.190 and an intercept of -0.254. This line has also been added to Fig. 6. One reads off a predicted grade variance of $\beta'^2 = 2.58$ at the declusterized indicator variance value of 0.215. This is at odds with the value $\beta^2 = 2.829$ in the table above, suggesting that outliers have distorted the grade variance; for otherwise these two numbers would agree, - as will be illustrated when we examine the MIG plot for the dense data. The adjustment of the grade variance will, if used to reconstruct W_2, lead to a revised mononodal cutoff. This takes one back to step (i). The importance of the first iteration of the MIG plot is therefore to get the correct number of participating points, which is sensitive to the first mononodal cutoff estimate, and hence the remark above about being as accurate as possible.

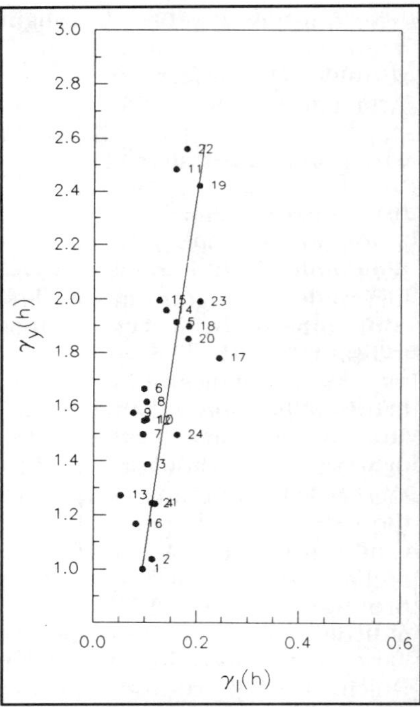

Figure 6. Grade-indicator plot for the sparse data. Mon. cutoff 5.42. Numbers refer to variogram lags. Regression line uses first 5 participating points.

(iii) *Revised mononodal cutoff.* The true mononodal cutoff is characterized by the best line-up of participating points on the grade-indicator plot, which in turn is reflected by minimum experimental deviation from the linear regression fit. A small program was implemented to find this cutoff for which the participating points exhibited minimum deviation from their grade-indicator line. It gave a revised mononodal cutoff of 5.57. It is important to get this correct cutoff for good results. The corresponding declusterized mononodal indicator variance is 0.226.

(iv) *MIG plot (2nd iteration).* A repeat of the grade-indicator plot procedure described under (ii) leads to the results shown in Fig. 7. The corresponding grade-indicator line has a slope and intercept of 13.118 and -1.127, and predicts a grade variance of 1.832 for the indicator variance of 0.226 found above. We have arrived here at a variance that is corroborated by the log(grade) variogram (see Figure 3), reflecting as it does a reliable sill. It is interesting to speculate that this has occurred because the effect of outliers has been attenuated by our procedure. In the following step we will attempt to also upgrade the grade variogram values at the participating lags.

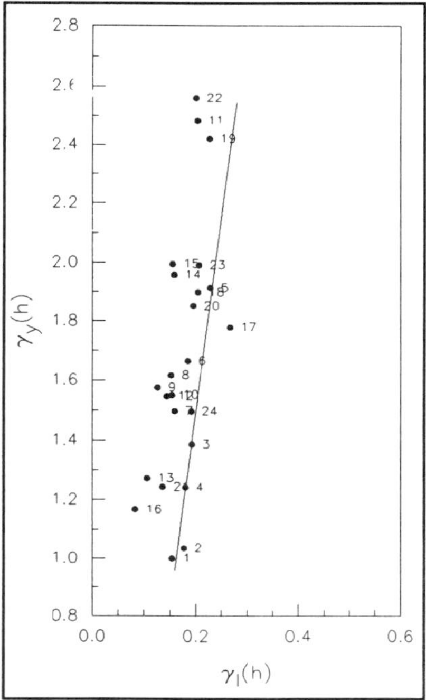

Figure 7. Revised version of Fig. 6 using the new mon. cutoff of 5.57.

(v) *Improvement of the experimental log(grade) variogram in the direction of best continuity*. It is important to start with the log(grade) variogram in the direction of best continuity, since it best defines the grade-indicator line. The premise is that the mononodal indicator variogram should be taken as the more reliable mould over the first few participating lags and that the MIG proportionality should be used to upgrade the log(grade) variogram. What happens beyond the range is not important since a reliable sill is now available. Starting with the shortest lag, and working with progressively longer lags, the procedure is based on the latest grade-indicator plot as follows: the indicator variogram value gives the x coordinate, for which the y coordinate, the log(grade) variogram value, is read off the linear regression line. The new revised log(grade) variogram value usually represents an improvement in the variogram shape, and, having confirmed this, the substitution is made, and the next point is considered. Once the sill is reached the substitution usually no longer represents an improvement, and the procedure is terminated. Fig. 8 gives the log(grade) variogram, in the direction of best continuity, that has been upgraded in this fashion. As we shall show below, the upgrade imparted elements of the detailed shape of the dense data variogram to this sparse data variogram, i.e. two structures became apparent, and could be fitted with spherical models.

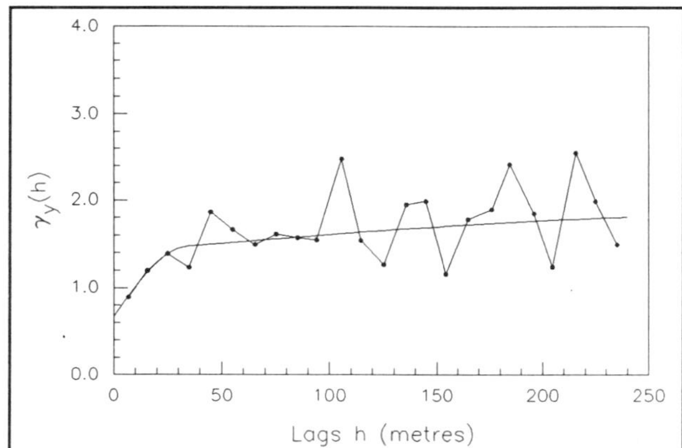

Figure 8. Upgraded log(grade) variogram in the direction of best continuity.

The resulting fit has been added to Fig. 8: a nugget effect of 0.67, a first structure

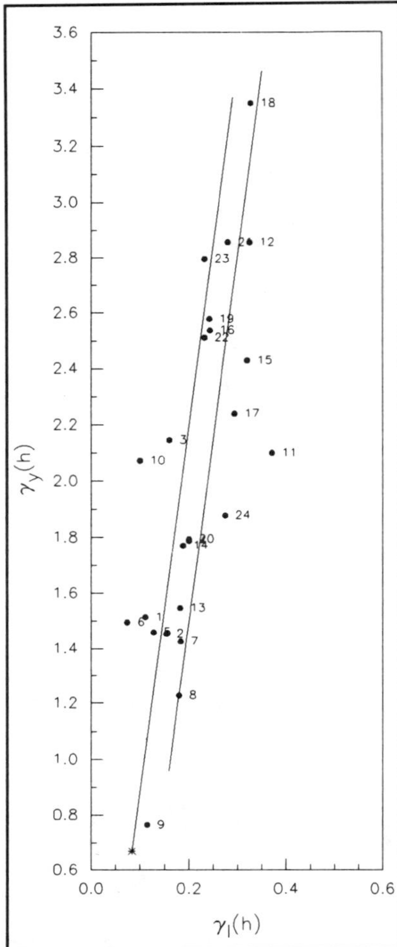

Figure 9. Parallel shift of the grade-indicator line of Fig. 7 to pass through the point (0.084, 0.67), shown as an asterisk.

of sill 0.73 and range 33.5m, and a second structure of sill 0.432 and range 300m.

(vi) *Improvement of the experimental log(grade) variogram in the direction of poorest continuity.* This being the direction of shortest continuity, the variograms, both log(grade) and corresponding mononodal indicator, are usually expected to be more erratic. In this particular case the first few lags of the log(grade) variogram display abnormally low values. The problem was dealt with in the following way. The nugget effect of the model described above for the log(grade) variogram in the direction of best continuity is 0.67, which should be the same for this direction. An experimental nugget effect for the indicator variogram of this direction was inferred from its behaviour at small lags. One finds $\gamma_I(0^+)=0.084$. Thus the linear regression line that will provide the correct nugget effect for the log(grades) when the upgrading procedure described above is used, should pass through (0.084,0.67). However, one should retain the slope (13.118) of the regression line that has already been determined since the covariance proportionality relation, eq. (2), should (ideally) not be direction dependent. The result is the parallel line displaced to the left that is also shown in Fig. 9. The latter line was used to upgrade the first nine points of the log(grade) variogram, shown in Figure 10, with the fitted spherical model: nugget effect 0.67, first structure with sill of 0.73 and range of 20m, and second structure with sill of 0.432 and range of 60m.

Kriging a log(grade) value for each block

The purpose of the kriging exercise was simply to produce unbiased log(grade) estimates for the 100 x 50m blocks, based on the combined variogram model derived. The consideration here was not so much to produce the best possible estimate by including as many drill hole values as possible in a particular kriging system, but rather to have a simple, uniform kriging system for each block, so as to demonstrate unbiasedness before back transformation. Only the four corner drill holes in an ordinary kriging system were used to predict an average log(grade) value for each of the 80 blocks.

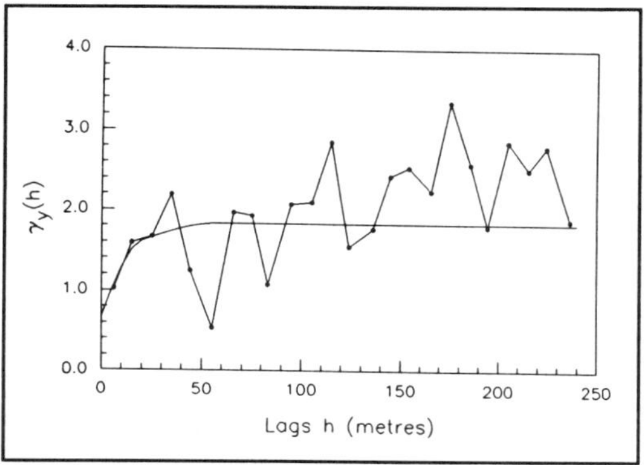

Figure 10. Upgraded log(grade) variogram in the direction of poorest continuity.

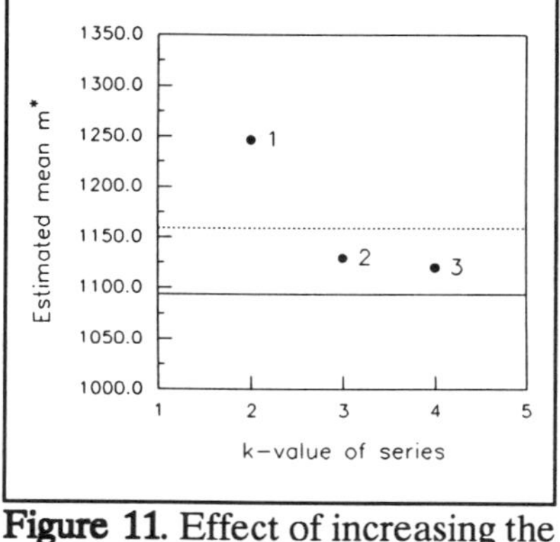

Figure 11. Effect of increasing the number of terms in the series for the unbiasing constant. Solid and dashed lines show the means of 1093.9 and 1158.8 of the untransformed sparse and dense data resp.

Statistics of the kriged log(grade) results

80 log(acc.) block values

Mean α_v^*:	6.049
Variance β_v^{*2}:	0.971
Std. Dev. β_v^*:	0.986
Skewness r_3^*:	-0.381
Kurtosis r_4^*:	2.830
5th ord. r_5^*:	-2.459
6th ord. r_6^*:	11.985

Mean Lagrange param. μ:0.366
Mean kriging var. σ_{Kv}^2: 0.473

A comparison with the drill hole values confirms that the kriging was unbiased and the variance is reduced as is expected. The even standardized moments remained almost the same, and the odd ones were slightly reduced.

Unbiased back transform to grade estimates

The back transform is done according to eqs. (8) and (9). Since the MIG-determined variance value of 1.832 is used here for β^2, the corresponding variance β_v^{*2} to use for the kriged values is determined by [5]

$$\beta_v^{*2} = \overline{C}(v, v) + 2\overline{\mu} - \overline{\sigma_{Kv}^2}$$

which gives 0.524. The standardized moments used for the kriged values are the ones in the table above. [A reminder: for the standard normal distribution $\overline{r}_3 = 0$, $\overline{r}_4 = 3$, $\overline{r}_5 = 0$, $\overline{r}_6 = 15$]. To illustrate the impact of the unbiasing constant, three back transforms were calculated:

(1) *Taking the series up to k=2* [see eq. (9)], i.e. assumption of perfect lognormality. Here the unbiasing constant is zero, and the mean m^* of the back transformed estimates Z^* is 1246.5.

(2) *Taking the series up to k=3*, i.e. including the first non-zero term of the unbiasing constant. The mean m^* is now 1129.6.

(3) *Taking the series up to k=4*, i.e. including the first two non-zero terms of the unbiasing constant. The mean m^* has come down further to 1120.3.

The mean of the 96 raw declusterized drill hole values is 1093.9. The 3rd back transform has come to within 2.4% of this value. Fig. 11 gives a graphical representation of the improvement of m^* with increased terms in the unbiasing constant. Given the variability and non-lognormality of the data, the back transform that takes the series up to k=4 can be considered unbiased.

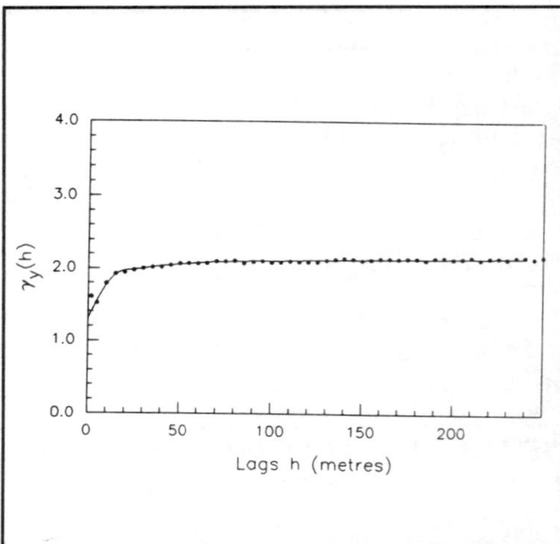

Figure 12. Log(grade) variogram of the dense data in the direction of best continuity. A spherical model fit is also shown.

The "true" properties of the deposit

Conventional variography. With 8185 data values available, conventional variography produces satisfactory variograms - even for a highly skewed distribution. Fig. 12 gives the log(grade) variogram in the direction of best continuity, and the spherical model fitted to it: nugget effect of 1.3, first structure with sill of 0.6 and range of 18m, second structure with sill of 0.16 and range of 80m, and third structure with sill of 0.1 and range of 300m. It is interesting to compare how this model is approximated by the model fitted to the upgraded sparse data variogram.

The mononodal cutoff. In the case of the dense data the approximation $y_n = m + \frac{1}{2}\sigma r_3$ proved to be just as accurate as determining the mononodal cutoff via constructing W_2: both methods gave 5.92. It is interesting to note that there is a specific mononodal cutoff for each data set, and this cutoff of 5.92 did not apply to the sparse set. Not surprisingly, the mononodal indicator variogram in the direction of best continuity strongly resembles the log(grade) variogram - see Fig. 13.

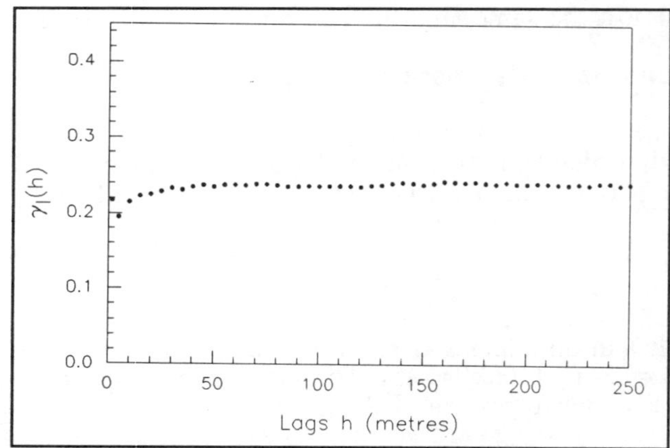

Figure 13. Mononodal indicator variogram for the dense log(grade) values. The mononodal cutoff is 5.92, see text.

The MIG plot. The MIG plot for the two variograms of Figs. 12 and 13 is shown in Fig. 14. The grade-indicator plot, and its linear regression fit, are in this case indistinguishable. In particular, the highest point reflects the two variances accurately. Obtaining such a plot for highly variable real gold data is very reassuring, for it underpins the validity of the MIG method for upgrading variograms.

Unbiasedness of back transform. The mean of the untransformed dense values is 1158.8. This can be regarded as a "true" value for the deposit. The estimated mean value of 1120.3, based on 80 kriged block values, comes to within 3.3% of the true value (see Fig. 11).

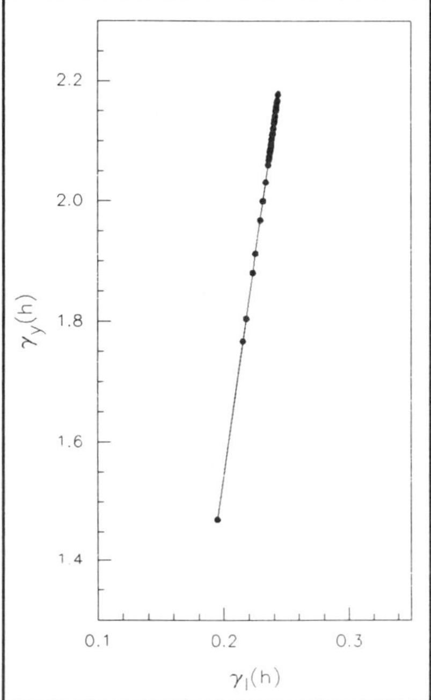

Figure 14. Grade-indicator plot for the dense log(grade) data.

CONCLUSIONS

The last comparison confirms the unbiasedness of the back transform. Furthermore, the MIG method as a whole handled outliers without special or subjective treatment. It is seen as a general method that can be implemented fairly simply by small additions to an ordinary geostatistics package, or the combination of a geostatistics and a spreadsheet package.

ACKNOWLEDGMENTS

I wish to thank Gold Fields of South Africa for providing me with the excellent set of data with which to illustrate the method reported in this paper.

REFERENCES

1. Lemmer, I. C. (1986) "Mononodal indicator variography - part 1: theory, - part 2: application to a computer simulated deposit", Math. Geol., **18**, no. 7, 589-623.

2. Erdélyi, A. et al. (1953) Higher transcendental functions, Vol. III. McGraw-Hill, New York, pp. 396.

3. Journel, A. G. (1983) "Non-parametric estimation of spatial distributions", Math. Geol., **15**, no. 3, 445-468.

4. Deutsch, C. V. and Journel, A. G. (1992) GSLIB: Geostatistical Software Library and User's Guide. Oxford University Press, Oxford.

5. Journel, A. G. and Huijbregts, Ch. J. (1978) Mining Geostatistics. Academic Press, London, pp. 600.

COMMENT ON "MODELLING IN THE PRESENCE OF SKEWED DISTRIBUTIONS" BY C. LEMMER

R. MOHAN SRIVASTAVA
FSS International
800 Millbank
Vancouver, BC
Canada V5Z 3Z4

Since Lemmer first introduced the notion of the mononodal cutoff in the mid-1980's, I have always had a basic problem believing that it really worked. With this most recent contribution, however, I finally have some sense for why it works (and why it would not work in some situations).

The idea behind the mononodal cutoff is that it represents a threshold at which the indicator variogram, $\gamma_I(h; z_c)$, has the same shape as the variogram of the original values, $\gamma_Z(h)$. What has always troubled me, until this paper, is the fact that the mononodal cutoff idea doesn't work for a standard Normal distribution. The mononodal cutoff for a Normal distribution turns out to be the median and the median indicator variogram is not linearly related to the variogram of the original values. It has been known for some time, well before the mononodal cutoff was presented, that the median indicator correlogram for bi-Normally distributed data is related to the correlogram of the original data values through the arcsin:

$$\rho_I(h; M) = \arcsin\left[\frac{\pi}{2}\rho_Z(h)\right]$$

For the particular case of Normally distributed data, the slope of the mononodal indicator variogram at the origin will therefore be steeper than the slope of the variogram of the original values. The fact that the idea would not work, even in the very tractable case of Normally distributed data, led me to conclude that there is something a little too approximate in the way that the series expansion of $\gamma_Z(h)$ is truncated after the $\gamma_I(h; z_c)$ term in the development of the mononodal cutoff.

Lemmer's case study evidence cannot be denied, however; there is definitely something that makes the shape of the indicator variogram at the mononodal cutoff very similar to the shape of the variogram of the original values. I realize now that with skewed distributions, such as lognormal ones, the variogram of the original data values will show a steepness at the origin that is similar to what the mononodal indicator variogram would show. To continue with the example of standard Normal distribution,

R. Dimitrakopoulos (ed.), Geostatistics for the Next Century, 120–121.

if we exponentiate Normally distributed Z values to create lognormally distributed Y values, the variogram of the Y values is related to the variogram of the Z values by an exponential function:

$$\exp[\gamma_Z(\infty) - \gamma_Z(h)] = 1 + \frac{\gamma_Y(\infty) - \gamma_Y(h)}{m^2}$$

The variogram of the lognormal Y values has a steeper slope at the origin than that of the Z values; this is similar to the behaviour we see in the mononodal indicator variogram—its slope at the origin will be steeper than that of $\gamma_Z(h)$.

Lemmer's mononodal cutoff is one of those tools that happens not to work well on Normally distributed data; it will work much better in situations where the original data values are very skewed (a very common case in earth sciences). It would be interesting to have some guide to how the skewness of the data (measured by the coefficient of variation?) influences the degree to which the shape of the indicator variogram at the mononodal cutoff accurately approximates the shape of the variogram of the original values.

REPLY TO "COMMENT ON MODELLING IN THE PRESENCE OF SKEWED DISTRIBUTIONS" BY R. M. SRIVASTAVA

CARINA LEMMER
Consultant Applied Earth Sciences
Eagle House, 70 Fox Street
Johannesburg 2001
South Africa

It should be remembered, as was pointed out in the original publication [1], that the MIG approach is expected to work best in those cases where the deposit displays a high nugget effect. This in turn usually means that the grade distribution is highly skewed.

However, the method is equally applicable to a Normal distribution as long as the nugget effect is substantial. This means $\rho_Z(h)$ stays small and the arcsin function is adequately represented by the linear term of its expansion for small arguments:

$$\arcsin x \approx x + 1/6\, x^3 + \ldots \tag{1}$$

One then recovers the proportionality $\rho_I(h;M) \sim \rho_Z(h)$ that underlies the MIG method. Another way of seeing this is to realize that all the coefficients in the series

$$C_I(h;z) = \rho_Z(h) W_1(z)^2 + T_{22}(z) W_2(z)^2 + \ldots \tag{2}$$

are known and the series may be summed for the case of a Normal bivariate distribution. The result is just the arcsin function again. Only retaining the first term of this series (which MIG does) corresponds to expanding the arcsin to first order as indicated above.

It is not a simple matter to give a general criterion for when MIG will apply in terms of the skewness of the data, as this depends on the magnitude of the neglected terms in eq. (2) above. Were this distribution Normal, with a substantial nugget effect, the ratio of the retained to the neglected term is

R. Dimitrakopoulos (ed.), Geostatistics for the Next Century, 122–123.

$$(\pi^2/24)\ \rho_Z^2(0^+) = 0.4\ \rho_Z^2(0^+) \tag{3}$$

For a nugget effect of half the sill value, i.e. $\rho_Z(0^+) = 0.5$, this amounts to only a 10% correction.

A substantial nugget effect is therefore the best criterion.

Reference

1. Lemmer, I C. (1986) "Mononodal indicator variography - part 1: theory, - part 2: application to a computer simulated deposit", Math. Geol., **18**, no. 7, 589-623.

EXPLORATION OF THE "NUGGET EFFECT"

F.F. PITARD
Francis Pitard Sampling Consultants
14710 Tejon Street
Broomfield, Colorado 80020
U.S.A.

ABSTRACT

Many people believe the "Nugget Effect" is the result of rapid in-situ changes in the concentration of a given constituent of interest, taking place on a very small scale. They are right, however, there is much more to it. Indeed, the "Nugget Effect" is the result of at least seven types of variability:

1. The true in-situ, small-scale, random variability;
2. The variability introduced by Constitution Heterogeneity during all sampling and subsampling stages, which is a function of fragment size and sample or subsample weight;
3. The variability introduced by small-scale Distribution Heterogeneity during all sampling and subsampling stages, which is a function of transient segregation as soon as the material has been broken up;
4. The variability introduced by any deviation from an isotropic module of observation ensuring sampling equiprobability in all relevant dimensions, during all sampling and subsampling stages;
5. The variability introduced by selectivity and poor recovery during all sampling and subsampling stages;
6. The variability introduced by contamination, losses, alteration of physical or chemical properties, and human errors; and
7. The variability introduced by the analytical procedure.

The misunderstanding of all these variability components prevents the effective minimization of the errors they generate. Accordingly, discrepancies between exploration estimates and production realities are likely to occur. This paper intends to be pragmatic in order to set a logical strategy that minimizes the "Nugget Effect", allowing good Geostatistics to proceed smoothly and sucessfully: Failure to do so can lead to errors which are very difficult to correct with statistics.

R. Dimitrakopoulos (ed.), Geostatistics for the Next Century, 124–136.

THE TRUE IN-SITU "NUGGET EFFECT"

As Michel David says, "The "Nugget Effect" is a Chaotic Component. It can be considered as the variance of a totally random component superimposed on the regionalized variable." [1] This random variance is a problem in many gold mines, but experience proves that it can become a problem for many other metals as well, and even for precise and accurate environmental assessments of certain pollutants. More often than not, the question is not to find out if there is a "Nugget Effect" or not, the question is how much "Nugget Effect" (i.e., Vo) there is. A good way to quantify this is by performing very short-range variographic experiments. Then, when Vo is quantified, what is the next logical step?

If Vo is small and acceptable, allowing a good definition of the regionalization, chances are that sampling and subsampling protocols are adequate and under control. If Vo is large and unacceptable, interfering with a good definition of the regionalization, sampling and subsampling protocols must be thoroughly investigated. Furthermore, it is suggested to find out how much Vo can interfere with a good definition of ore boundaries, at a preselected ore grade economical cutoff: This can be done by investigating the proportion L of the mineral of interest that is finely disseminated in the ore near that cutoff. Therefore, we can predict that the value we will find for L may to some extent depend upon the chosen cutoff. The following procedures to estimate L have been developed by C.O. Ingamells and F.F. Pitard[3] and should be the object of more research, which could provide an excellent project to a student.

Collection of Very Small Samples

If a long series of very small samples is taken near the boundaries defined by a selected cutoff, there is a good chance that at least a few samples do not contain any high grade nugget, or geological micro-structure, or veinlet, or blob. In that case, L is the lowest available result.

Search for Discrete Differences Between Samples

If the size of the nuggets or blobs is relatively uniform, L may be found from a series of existing grade determinations (e.g., blasthole results for a given production day), even when the actual number of blobs in all samples is superior to zero. The hypothetical uniform blob size (i.e., average blob size) of the mineral of interest leads to almost discrete differences between successive assays. From C.O. Ingamells [4], each available assay X_i can be expressed as follows:

$$X_i = L + [Z_i s^2 / (X - L)]$$

where s is the standard deviation in a single assay value that can be calculated separately, Z_i is the number of blobs in the ith sample which can be estimated with an histogram, and X is the overall average from all X_i.
After solving for L, and selecting the positive solution only, we obtain the following formula:

$$L = [1/2][X_i + X] + [1/2] \sqrt{[X_i - X]^2 + 4s^2 Z_i}$$

By putting $Z_i = 0, 1, 2, 3$, etc., one obtains a series of possible values for L for each X_i. By comparing these values among all assays, the calculated L which is most nearly the same for each X_i can be found, in some cases by inspection.

Shift of the Mode in Two Series of Results from Two Different Size Samples

This method is the most convenient to use, however it has its limitations because it is based on the assumption that too small samples generating a large "Nugget Effect" give results that are Poisson distributed: It is not always the case. Analysis of two series of samples of different weights Ms_1 and Ms_2 are likely to yield two different skewed distributions, when such Poisson skewness does exist. Ms_2 must be at least ten times larger than Ms_1. If large samples are not available, they can be artificially obtained by compositing neighboring data points, either by clusters or by lines. As sample weight in a series of determinations diminishes, assay results may become more Poisson distributed, and the mean and the mode move farther apart: It is often the case with gold and trace constituents. If the mode value is taken as the most probable result Y of a single determination, two sets of results using samples of weights Ms_1 and Ms_2 will yield two modes Y_1 and Y_2. Then, C.O. Ingamells and F.F. Pitard[3,4] demonstrated that L may be calculated with the following formula:

$$L = \frac{Y_1 Ms_2 [X - Y_2] - Y_2 Ms_1 [X - Y_1]}{Ms_2 [X - Y_2] - Ms_1 [X - Y_1]}$$

The harmonic mean may provide a useful estimate of the modes Y_1 and Y_2 in series of Poisson distributed assay results if histograms are not available.

Discussion About Values of L when Compared to the Local Average Grade X

L can play an important role in ore grade control, yet L does not get the attention it deserves. So, to increase the emphasis, a practical example from cobalt in a lateritic deposit follows this section. If L has a value near the economical cutoff, then nature gave a gift to the ore control engineer: Even a weak sampling protocol can locate the boundary of the minable ore. However, it often happens that L is much below the economical cutoff: The farther L is from this cutoff, the more Vo is likely to have an impact on the precision (i.e., unacceptable Fundamental Error near cutoff) and accuracy (i.e., unacceptable Poisson skewness near cutoff) with which the boundary is located. This leads to important ore dilution. It is not dilution during the mining operation, it is dilution on the maps: There is a difference. This raises the question about the wisdom of high cutoffs too close to the local average X, and too far apart from L, which often leads to the pursuit of an illusory average ore grade the mill will never see, and the devastation of ore reserves. So far, we assumed we were dealing only with a large in-situ "Nugget Effect" Vo', all sources of other errors being minimum. But, if the Sampling Fundamental Errors described later in this paper are not under control and also introduce Poisson skewnesses, then a cutoff faraway from L will most certainly lead to very poor ore grade management: Many assay values are too low, and we have no way of knowing which ones. A few assay values are much too high, and we have no way of knowing which ones. At this stage, we loose the battle, and the company involved is likely to loose millions of dollars in invisible figures.

Example of True In-Situ "Nugget Effect" Affecting Cobalt Assays in a Lateritic Deposit

Prior to a mining test, 12 holes were drilled at unspecified short intervals, and the upper intervals within and immediately below the iron cap were investigated for their cobalt content. The following table shows the cobalt assay values. The spread of the cobalt values goes from 0.01% to 1.07%. But, the mining test clearly showed that if 50-tons samples could have been taken, the true spread should have been from 0.11% to 0.17%. The calculation of L lead to an estimated value of the finely disseminated cobalt around 0.01% or even less. This example clearly demonstrates that the application of any cutoff between 0.01% and 0.14% would not have changed the outcome of the mined out average cobalt content. Though this example is an extreme case, it is not rare to find a similar situation in some gold deposits. Figure 1 summarizes the results: X is the overall average cobalt content, L is the estimated low background content, S'v is the likely spread of the cobalt content content in samples becoming increasingly larger, S is the average standard deviation in a single assay result as a function of sample weight, S' is the standard deviation around the average

Drill depth in meters, and
cobalt assays expressed in parts per 10,000

	1	2	3	4	5	6	7	8
Hole #1	3	28	57	75	39	45	34	25
2	2	2	2	2	2	2	2	3
3	11	20	22	25	10	16	17	12
4	3	3	3	3	2	2	1	2
5	3	12	15	8	7	13	11	4
6	1	1	1	2	2	2	1	2
7	2	2	3	11	107	64	34	14
8	3	3	7	20	16	24	20	24
9	1	2	2	2	3	41	31	46
10	2	2	9	4	4	3	9	8
11	2	2	2	3	5	28	23	33
12	20	11	11	22	21	24	21	20
Average:	4.5	7.3	11.2	14.8	18.2	22.0	17.0	16.1

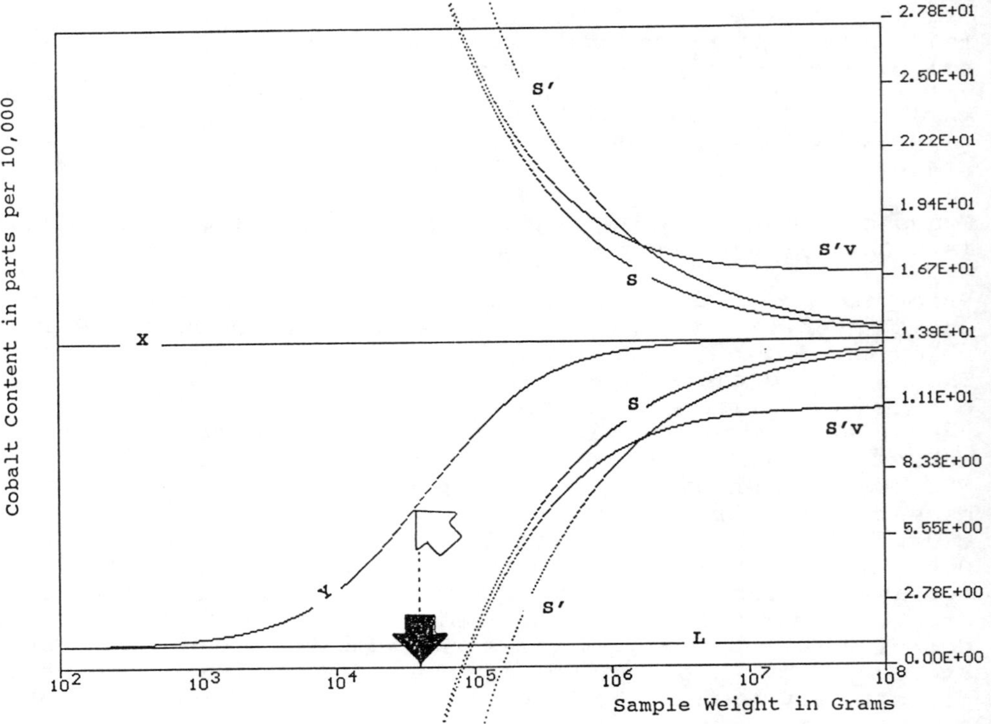

Figure 1. Sampling diagram illustrating the in-situ "Nugget Effect" affecting cobalt assay results in an unspecified lateritic deposit.

of several data points if they were generated by samples introducing no or very little skewness, Y is the most probable result to be expected from one sample as sample weight diminishes. It is important to notice that each data point listed in the above table is generated from a 50-Kg sample, which is pointed out by the dark arrow in figure 1.

Discussion About the True In-Situ "Nugget Effect"

If we call Vo the total "Nugget Effect", and Vo' the true in-situ "Nugget Effect" affecting the original core sample, Vo' can be calculated from Vo when, and only when, all the sampling, subsampling, and analytical errors are under control and second order of magnitude. Vo' is a function of the original sample volume and mass taken in-situ by the drilling machine. Then indeed, the difference Vo - Vo' consists of calculable quantities.

EFFECT OF CONSTITUTION HETEROGENEITY ON Vo

After the core has been broken up, either by a reverse circulation drilling machine or at the laboratory by a jaw crusher, individual fragments are likely to have different contents. Therefore, it is logical to think that the more the differences between individual fragments, the more the sampling challenge: This is intuitive. Constitution heterogeneity at any given stage of comminution is defined as the average difference between individual fragments: The Sampling Theory easily demonstrates that the constitution heterogeneity CH_L of a lot L of broken material can be expressed as follows[2]:

$$CH_L = N_F \sum_i \frac{[a_i - a_L]^2 \; M_i^2}{a_L^2 \; M_L^2}$$

This constitution heterogeneity is responsible for the Fundamental Error FE that affects all sampling and subsampling stages. By duplicating the lot into sublots of different fragment sizes and sublots of different fragment densities, Pierre Gy obtains a simplified version of the above formula which allows him to derive a simpler equation expressing the variance s^2_{FE} of the Fundamental Error, in which **there is nothing empirical** and only legitimate and well addressed approximations.

$$s^2_{FE} = \left[\frac{1}{M_S} - \frac{1}{M_L} \right] Cd^3$$

where C is a characteristic of the constitution heterogeneity **for a given stage of comminution,** which can be estimated by conducting some tests.[2]

For some cases where a quick estimation of s^2_{FE} is needed, Pierre Gy derives a parametric formula for estimating C which must be commented in order to prevent its misuse as has often been the case. C can be expressed as the product of four parameters:

$$C = f\, g\, c\, l$$

where f is a dimensionless fragment shape correcting factor taking into account the fact that fragments are not necessarily cubes (i.e., d^3 is the volume of a cube). g is a dimensionless fragment size distribution correcting factor taking into account the fact that fragments are not necessarily all large fragments (i.e., the definition of d is the square opening of a sieve that will retain no more than 5% oversized fragments). c is a mineralogical factor measuring the heterogeneity generated by fragments of different density, **assuming the mineral of interest is completely liberated**, which is a pessimistic assumption that needs to be corrected by the liberation factor. The mineralogical factor c is estimated with the following formula:

$$C_{max} / fg = c = D_M \frac{[1 - a_L]^2}{a_L} + D_g [1 - a_L]$$

Where C_{max} is the maximum possible heterogeneity when the constituent of interest is completely liberated. D_M is the density of the mineral or constituent of interest, D_g is the average density of the material foreign to the constituent of interest. a_L is the average grade of the constituent of interest expressed as part of one or as a proportion.

l (i.e., little L) is a dimensionless correcting factor taking into account the fact that the constituent of interest is not necessarily liberated: It is called the liberation factor. It is calculated with the following **non-empirical** formula (see reference 2, volume 1, pages 164 and 165):

$$l = \frac{a_{max} - a_L}{1 - a_L}$$

where a_{max} is the maximum possible content reached by fragments of size d in any given stage of comminution, which can easily be determined with an experiment.

Now, sometimes, someone may not be willing to perform such an experiment. So, Pierre Gy proposes a very rough and empirical way to quickly calculate the liberation factor l with the following mineral processing formula:

$$l = \sqrt{d_l / d}$$

where d_l is defined as the size under which it is necesary to crush 95% of the material in order to liberate at least 90% of the constituent of interest. This formula does not belong to Pierre Gy's theory, and it is unfortunate that it has been used as an argument against his academic integrity: He presented it as a quick derivative to prevent the test necessary to estimate a_{max}. Therefore, it must be emphasized again that the only reliable way to estimate the liberation factor is to perform an experiment that can give a reliable estimate of a_{max} for a given stage of comminution.

The empirical approach with d_l usually falls apart when a large proportion of the constituent of interest is **in solution** with the main matrix. In such a case it is just impossible to calculate the liberation size, or the liberation factor in any accurate way by using this empirical approach: It is not a reason to alter the academic integrity of the original formula using a_{max}, with some vague empirical observations.

Of course a_{max} needs to be redefined with every step of comminution. Therefore, the so-called sampling constant C is constant only for a given stage of comminution, and this is clearly stated in **all** Pierre Gy's literature since 1953.[5] Hundreds of experiments around the world have demonstrated that the a_{max} approach is a reliable approach, including cases with coarse gold as explained in reference 6 (volume 2, pages 152 and 153). In nearly all the cases where Pierre Gy's formula lead to excessively large samples, it was clearly because of the misuse of the formula, and the misunderstanding of its underlying principles, and even the misunderstanding of the definition of the parameters like f, g, c, and l.

Going back to the "Nugget Effect", when s^2_{FE} is not optimized for each sampling and subsampling stage, it often becomes a major component of Vo. Indeed, experience at many mines around the world proves that Vo is often artificially introduced because sample and subsample weights are not optimized. A thorough study of the constitution heterogeneity of an ore for a given constituent of interest should be the prerequisite to any feasibility study: Yet, it is rarely done. It is believed that the Geostatistician will repair damages introduced into the data: Maybe yes, maybe not. As long as no bias nor excessive artificial skewness is introduced into the data, the Geostatistician may indeed repair the damage with Kriging techniques. However, if biases and artificial skewnesses are introduced, the Geostatistician should strongly address the limitations of his working model: For him or her, it is self defense. Experience also proves that if s_{FE} (i.e., coefficient of variation) goes beyond an area between $\pm$15 and 25%, it is most likely to introduce an artificial skewness distorting the lognormal model, if such a model applies, and generate an illusion around ore boundaries. If s_{FE} becomes very large, then an illusory patching effect may take place on maps, rending the high grade zones extremely elusive for the ore grade control engineer, the consequences of which can be staggering.

EFFECT OF SMALL SCALE DISTRIBUTION HETEROGENEITY ON Vo

After the sample has been broken up, either by the drilling machine or by the jaw crusher, fragments may segregate because of the omnipresence of gravity in everything we do to that sample. Segregation may take place because fragments have differences in size, density, shape, moisture content, physical or chemical composition, etc... Therefore, we can define small scale distribution heterogeneity as differences between groups of fragments or increments making up the sample or subsamples. This distribution heterogeneity depends on two factors: The constitution heterogeneity and the state of segregation of the material. Distribution heterogeneity is responsible for an additional sampling error affecting each sampling and subsampling stage: The Grouping and Segregation Error GE. The variance s^2_{GE} of the Grouping and Segregation Error can be minimized by acting on three factors:

1. Minimize the variance of the Fundamental Error in sampling protocols, which must be done in all cases.
2. Minimize the grouping factor, by collecting as many increments as practically possible for a given sample weight, with respect to other sampling errors such as DE, EE, and PE which will be defined later in this paper. This is very easy to do, and nearly always successful, yet very few people do this.
3. Minimize the segregation factor, by homogenizing the material before taking a subsample. This is time consuming, expensive, ans provides no guarantee for success, yet many people do this.

EFFECT OF SAMPLING EQUIPROBABILITY IN ALL RELEVANT DIRECTIONS ON Vo

A sample and its subsamples must be probabilistic. However, for them to be accurate, they must be equiprobabilistic. So, any lot (e.g., ore block, sample, subsample) must be scanned by an isotropic module of observation giving a constant sampling probability in all relevant directions. For a three-dimensional lot, the isotropic module of observation is a sphere. For a two-dimensional lot, the module is a cylinder from the top of the lot to the bottom. For a one-dimensional lot, the module is a cross section with parallel plans representing the entire width and entire thickness of the lot which is a stream. For a pie shapped lot, the module is a radial sector. Etc... Any deviation from such an isotropic module of observation introduces a sampling bias, therefore alters the accuracy of sampling: It is called the Increment Delimitation Error DE. Furthermore, because of the transient nature of distribution heterogeneity (i.e., segregation), there is no such thing as a constant bias in sampling. Consequently, DE always inflates Vo. This is a dangerous error, because it is very difficult to quantify or even detect with statistics.

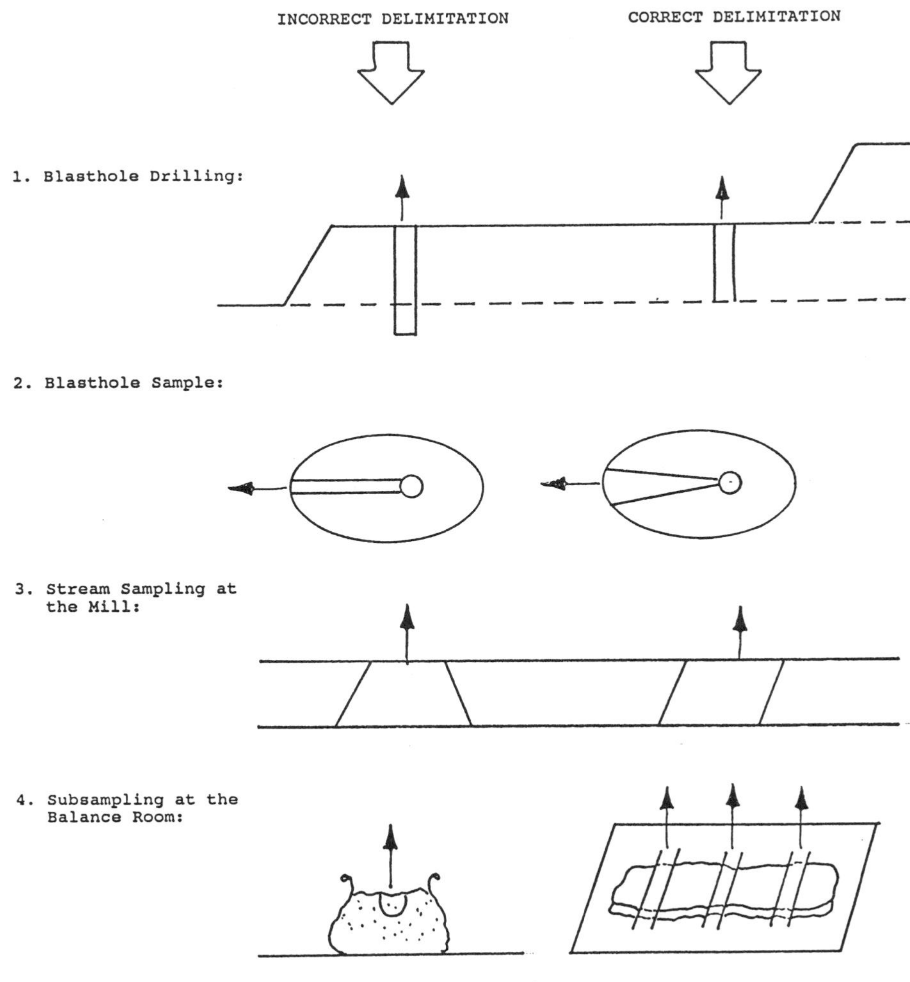

Figure 2. Illustration of a few typical sources of Delimitation bias.

EFFECT OF SAMPLING SELECTIVITY AND POOR RECOVERY ON Vo

Sampling equiprobability must be preserved during the impact of the sampling device with the material to be sampled. In other words, everything that belongs to the isotropic volume of observation must be recovered into the sample or the subsample. The sampling tool is often designed in such a way that some selection takes place: Either too many fine particles are collected, producing a sample with not enough coarse fragments, or vice versa. A poor recovery of the core during drilling is a typical example of this error, which is called the Increment Extraction Error EE. Like DE, EE is responsible for the largest biases encountered in sampling. EE is a dangerous error, because it is very difficult to quantify and even detect with statistics. The only actions that can be taken to minimize DE and EE are preventive actions (see reference 5, chapters 14 and 15).

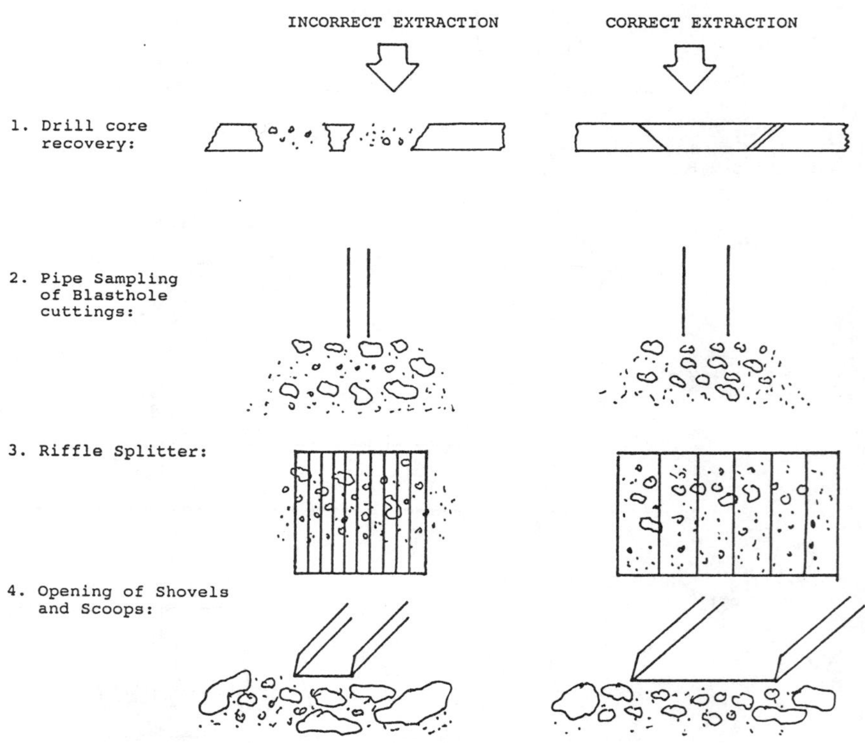

Figure 3. Illustration of a few typical sources of Extraction bias.

EFFECT ON Vo OF CONTAMINATION AND LOSSES DURING SAMPLE PREPARATION

All the sources of errors we saw so far are, one way or another, part of a selecting process taking place at each sampling and subsampling stage. But, there are other errors which are not part of the selective process: They take place between each sampling or subsampling stage. We can indeed contaminate the sample or subsamples with the sampling equipment or the surrounding equipment. We can loose part of the sample as dust, sticky particles remaining in the sampling circuit, smearing on crushers and pulverizers, etc... We can alter the physical or chemical composition of some critical constituents, or introduce human errors. All these errors are referred to as the Preparation Error PE, which is responsible for non-constant biases inflating Vo.

EFFECT ON Vo OF THE ANALYTICAL ERROR

There is no doubt that the Analytical Error AE can inflate the variance Vo. However, only precision problems will. Contrary to sampling biases, relative or absolute analytical biases can be relatively constant: The variogram cannot see these biases, neither directly nor indirectly, or at least identify their effect as such. Therefore, it is of the utmost importance to prevent analytical biases in ore grade control and exploration.

CONCLUSION

The total variance Vo of the "Nugget Effect" is the sum of many components. If n sampling stages (i.e., n = 1,2,3,...,N) are involved between the collection in the field and the last stage at the analytical balance room, Vo can be expressed as follows:

$$Vo = Vo' + s^2_{AE} + \sum_{n=1}^{N} [s^2_{FEn} + s^2_{GEn} + s^2_{DEn} + s^2_{EEn} + s^2_{PEn}]$$

So, if preventive precautions are not taken, the Geostatistician cannot and will not obtain reliable data to work with. As a result, figure 4 illustrates a proposed strategy to be used by Geostatisticians when proceeding with any feasibility study and ore reserves estimation.

REFERENCES

1. David, M., "Geostatistical Ore Reserve Estimation", Developments in Geomathematics 2. Elsevier Scientific Publishing Company, New York 1977.
2. Pitard, F.F., "Pierre Gy's Sampling Theory and Sampling Practice", Volume 1, CRC Press, Inc., Boca Raton, Florida 1989.
3. Ingamells, C.O. and Pitard, F.F., "Applied Geochemical Analysis". Volume 88 in Chemical Analysis Series. A Wiley-Interscience Publication. John Wiley & Sons, 1986.

4. Ingamells, C.O., "Sampling Demonstration", Presented at the IPMI Symposium, San Francisco, March 18-19, 1980.
5. Gy, P.M., "Poids a donner a un echantillon", Congres des Laveries des Mines Metalliques Francaises, Paris, Septembre 1953, et Revue de l'Industrie Minerale, avril 1954, et Revue de l'industrie Minerale, No 38, 1956.
6. Pitard, F.F., "Pierre Gy's Sampling Theory and Sampling Practice", Volume 2, CRC Press, Inc., Boca Raton, Florida 1989.

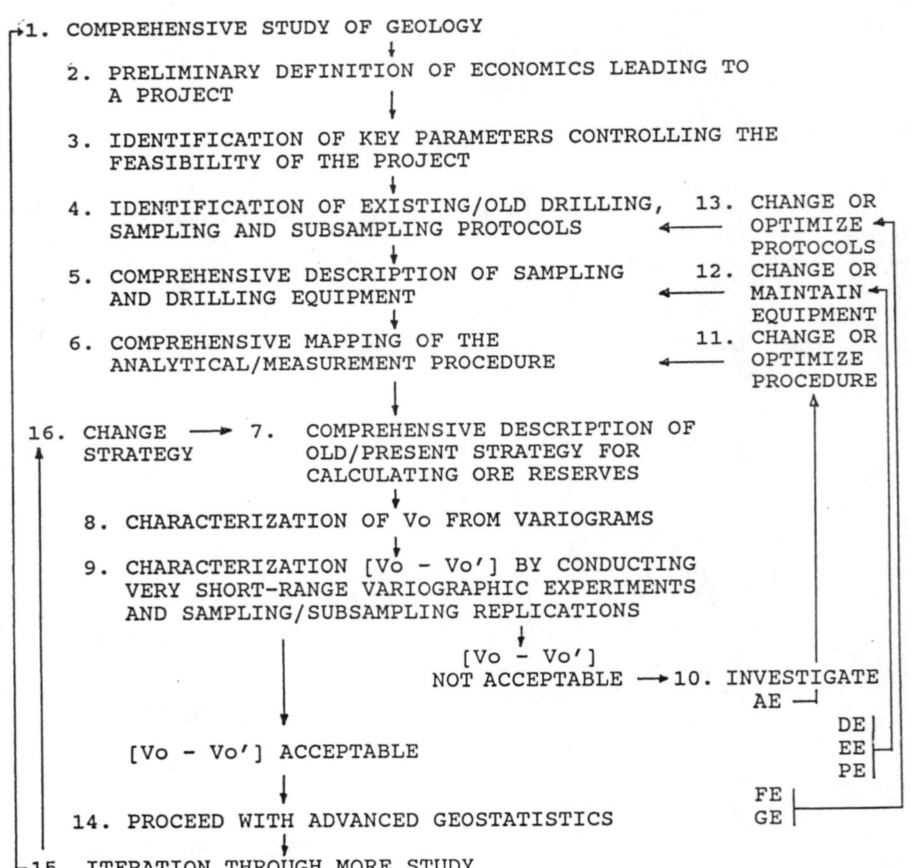

Figure 4. A plan for constant data quality improvement

COMMENTS ON F. PITARD' S "EXPLORATION OF THE NUGGET EFFECT"

D.M. FRANÇOIS-BONGARÇON
Inco Exploration and Technical Services
Copper Cliff, Ontario, P0M 1N0
Canada

In another paper in the same forum, F. Pitard describes the components of the nugget effect from his prospective as a specialist of the theory of sampling of broken ores. In this paper, it is shown that Pitard' s viewpoint and a purely geostatistical approach to these problems can be reconciliated, the latter bringing an added generality to the formula used to predict the reliability of a sample as a function of its mass and of the size of the rock particles. This new viewpoint also clarifies a number of issues related to the proper application of the theory of sampling of broken ores.

INTRODUCTION

In his excellent paper entitled "Exploration of the Nugget Effect", Pitard attracts our attention to a number of sampling issues that are usually ignored by most geostatisticians, although they are an important part of the challenge of accurately predicting reserves. Quite justifiably, Pitard presents his argument through the concept of "Nugget Effect", a term which has become almost proprietary to our geostatistical community, and certainly very familiar to us. Particularly, he points out that the geostatistical nugget effect consists of a "true in-situ" nugget effect (i.e. small-scale intrinsic variability of the grade, or "chaotic component") and of a number of variability components due the various aspects of sampling and sample preparation and assaying procedures.

Here are a few comments in complement and/or qualification to his important remarks, which I hope will help the geostatisticians place Gy and Pitard's theory into the context of daily geostatistical applications. More importantly, I would like to show that if the theory of the sampling of broken ores can hardly be mentioned without direct reference to Gy and Pitard's fundamental contributions, it can neither be fully understood outside of a geostatistical frame of reference. This is probably why, in the first ever practical textbook on geostatistics, David already blew the whistle by devoting a full chapter to "Statistical Problems in Sample Preparation" (David, 1975). His warning, however, has not yet received all the attention it deserves, and we should be thankful to F.Pitard for addressing effective ways of correcting this unfavourable situation.

R. Dimitrakopoulos (ed.), Geostatistics for the Next Century, 137–141.

THE ROLE OF THE TRUE NUGGET EFFECT

As most geostatisticians know, the observed discontinuity at the origin of the experimental variogram, i.e. the experimental nugget effect, reflects short scale variabilities, including all the auto-correlation microstructures (the "true in-situ" nugget effect of the model). Pitard points out that the part of the nugget effect not due to microstructures includes all the sampling and sample preparation variances of the data used in calculating the experimental variogram. The two types of nugget components, the "true" and the "un-true", however, are not as independent as one may think, and we will see that microstructures actually play the main role in the value of sampling variances, especially the variance due to the Fundamental Sampling Error (FSE), and the one due to the Grouping and Segregation Error, which can be shown to be proportional to the latter.

Whether it comes from a mining face or from a piece of drill hole core, once the ore has been broken up and mixed up, the auto-correlation structures that constituted the natural segregation of the ore in place, i.e. the ranges of the macroscopic variogram, are destroyed, sometimes to be replaced by a different type of unwanted segregation. The microstructures at scales smaller than the rock particle sizes, however, remain as they were when the ore was still in place. In most cases, each little rock particle, no matter how small, can still be regarded, under a microscope and in its own right, as a complex, vast field of study as would the entire deposit itself be.

Although the following does not constitute a demonstration, we can invoke the theorem of the standard error of the mean and the geostatistical definition of the variance within a domain to see that the relative variance of the grade of a sample of mass M_S made up of a number of such rock particles drawn at random from a lot of broken ore with total mass M_L (i.e. the variance of the FSE), should be expected to be dimensionally similar to:

$$\sigma_{FSE}^2 = (1/M_S - 1/M_L)\ m\ \sigma_o^2(m) \qquad (1)$$

where m and $\sigma_o^2(m)$ are the mass and grade relative variance of the average particle (what exactly is called here "average" particle would deserve more precisions than space allows, but is not critical to the nature of the result).

Complex, more rigorous demonstrations and more precise expressions of this result in the general case have been documented elsewhere (Gy, 1982) and (Matheron, 1966). Expression (1) can also easily be geostatistically understood as the regularization, over the sample, of the nugget effect variance between particles.

Consequently, the variance of the FSE, as well as the variance of the Grouping and Segregation Errors, which is proportional to it, are proportional to the variance of the grade of some kind of generic particle. This elementary particle variance, in turn, can be interpreted as the variance of the corresponding (very) small support in the lot of ore, and therefore clearly depends on the auto-correlations present within the rock particles, which in turn constitute the bulk of what we earlier called the "true in-situ" nugget effect.

QUANTIFYING THE VARIANCE OF THE FSE

A fundamental consequence of this simple, common sense result, is that quantifying the variance of the FSE in formula (1), as proposed by Gy, amounts to the modelling and quantifying of the relationship between $\sigma_o^2(m)$ and the mass m or, equivalently, a certain, properly defined, "nominal" size d of the rock particles. This, as already stated, amounts to the modelling and quantifying of the true nugget effect as a set of microstructures. The final shape of the formula giving the variance of the FSE as a function of the particle nominal size d, will therefore essentially depend upon this relationship.

Let us then consider the particular case where the true logarithmic nugget effect of the in-situ mineralization can reasonably be adjusted with a single De Wijsian variogram. The adjustment is assumed to hold true over a range of microscopic scales covering the various orders of magnitude of practical particle sizes in the successive comminution stages of the ore to be sampled (this De Wijsian case is believed to be fairly general for precious metal and trace element grades, and quite possibly even for pseudo-normally distributed grades). Then, it can be shown (Francois-Bongarcon, 1991, 1992) that the variance of the particle grades $\sigma_o^2(m)$ is approximately proportional to a term in $1/d^{3\alpha}$ where α is De Wijs' intrinsic logarithmic dispersion coefficient, a parameter that characterizes the variogram of the mineralization and is expected to change in value on a case by case basis. Another author in this symposium, under an assumption of gauge invariance (self-similarity) proposes a model of auto-correlations which more rigorously provides the form of the elementary unit variance as a power function of its inverse support size, with an exponent linked to the form of the variogram (Rose et al., 1993).

In the previous examples, the final formula for σ_{FSE}^2 now appears to be in $d^{3.(1-\alpha)}$ with a variable value of α. As a consequence, the final exponent of d in the formula is, in the most general case, a geostatistical parameter of the ore.

This remark does not invalidate Gy's classical formula written as:

$$\sigma_{FSE}^2 = (1/M_S - 1/M_L)\ f\ g\ c\ l\ d^3 \qquad (2)$$

but only indicates that Gy's "liberation factor", l, actually is a geostatistical parameter that can reasonably expected to be often approximately inversely proportional to some power of d. Its adjustment must therefore rely either on a variographic analysis at nested microscopic scales (usually an impossible or uneconomic feat), or on a series of appropriate sampling experiments.

Written in the classical form:

$$\sigma_{FSE}^2 = (1/M_S - 1/M_L)\ C\ d^3 \qquad (3)$$

however, the formula is not correct if C is (mis-)represented as a "sampling constant", since this quantity C clearly varies with d.

Pitard, conversely, is perfectly correct in defining this quantity C as "a characteristic of the Constitution Heterogeneity **for a given stage of comminution"** (emphasis added), but then it indeed varies with d, and the formula, in this form (3), is unfortunately very misleading for the non-specialist, and very illogical, as the main objective often is to show the quantitative relationship between the variance σ_{FSE}^2 and size d.

A REMARK ON PRACTICAL APPLICATIONS OF GY'S FORMULA

At this stage, it is important to stress a situation of a particular interest to past users of Gy's formula. The application of (3) with C as a constant is often responsible for the determination of abnormally excessive minimum sample masses, a criticism very often heard about Gy's formula.

Strong concerns about the actual exponent of d and the unrealistic sample masses have been clearly voiced in at least one particular instance, by F.Stolze in a discussion following, and included in, one of Gy's early publications (Gy, 1956). Stolze mentioned several examples where significant amounts of practical experimentations had supported a final formula globally in d^2 instead of d^3. The response by Gy, who at that time still regarded C as a constant, consisted in defending formula (3) based on the unit of C, which must remain the one of a specific gravity. Firstly, his argument is not logically correct as it (unnecessarily) takes as a premise that C does not depend on d, and, in any case, it is fully answered by our finding that parameter l in (2), and therefore C in (3), is in effect proportional to the dimensionless quantity $(d_l/d)^{3\alpha}$, where d_l is the liberation size of the ore (Francois-Bongarcon, 1991, 1992), so that the final exponent of d itself can indeed be less than 3 without threatening the overall unit consistency of the formula. In later works (e.g. Gy, 1982), Gy had to introduce qualifying empirical remarks about variations of parameter l with d, consistent with our findings, but unfortunately too particular. Misapplications of a misunderstood formula (3) by most of its users have, however, continued even to the present day. None of these remarks, nor any of those concerns from other people, can, however, be construed as attacks on Gy's academic integrity, that we all keep in the highest respect.

THE LAW OF CONSERVATION OF TOTAL VARIABILITY

Finally, I would like to re-emphasize Pitard's remarks about the totally neglected importance, and the relative ease, of minimizing the impact of Grouping and Segregation Errors by taking a large number of small increments for a given sample weight. This will put this mechanism into proper geostatistical context.

Let us start with the basic but fundamental remark that the regularization of a regionalized variable (the particle grade) over a larger support (the sample) has a drastic variance reduction effect when the variable presents little or no structure (i.e. is nuggety - case a), and only a very slight effect when the variable is strongly auto-correlated (case b). This interesting fact of life is overlooked by most geostatisticians, and usually not emphasized in geostatistical classes and textbooks. It is, however, fundamental when it comes to the sampling theory, and it is the true geostatistical reason why the ore should be broken up as fine as possible and mixed up prior to any sampling operation, to ensure the destruction of large scale auto-correlations, and by the same token a smaller regularized sample variance. Indeed, by means of comminution of the ore, we have the ability to move the state of the lot anywhere from unfavourable case (b) above, towards limit case (a) with minimum sample variance. Conversely, at each stage, distribution homogeneity of the lot is unstable, and the ore tends to re-segregate, which moves the state of auto-correlation of the lot some distance back toward case (b), with a new macroscopic variogram of the broken ore in place reflecting this segregation phenomenon and the corresponding, higher, sample variance. The total of the material, however, remains the same during these transformations. Where does the gained/lost sample variability go ?

What actually happens to the variability is easiest to see when thinking of all the possible samples of a given size as groups in an analysis of variance. The total variability between say, punctual (i.e. particle) grades is preserved. But as the homogeneity of the lot changes, there is a conversion of parts of the between-samples variance into within-sample variance. In a very segregated case indeed, the variance between samples is large (unfavourable), and the variance between particles within a given sample is very small (sub-constant grade within each sample). On the other hand, for a fully homogenized lot, the variance between samples is minimal and equal to the variance of the FSE (a most desirable case), and the particle grade variance within the sample is largest.

In practice, in the presence of a segregated lot, taking a sample of a given size in one single batch therefore clearly places the sampler in an unfavourable, high sample variance case, close to case (b). On the other hand, it is difficult and costly to properly homogenize the lot. But if instead a number of increments (i.e. much smaller individual sub-samples) are taken from various parts of the lot with equal probabilities, and then put together to form the main sample, a large variability between these increments has been achieved (they are small and taken in different parts of a segregated lot), and that variability is then concentrated within the final sample, thus achieving a minimization of the between-samples variance (i.e. the variance of the FSE).

CONCLUSION

Pitard has taken us on a journey towards a better handling of sampling situations, through a discipline that has earned a reputation of being esoteric and, to too many, conceptually unclear. As of recent times, this discipline is being updated, clarified and, in some instances, made more general, thanks to works by Pitard and a few others. Now is the best time for the professionals of the mining industry to climb on board the train, take advantage of Pitard's guidance, and for geostatisticians at large to take the lead in the future advances of a field which, in final analysis, belongs to their kin (Francois-Bongarcon, 1992).

C.D.Rose is gratefully acknowledged for his constructive remarks as a reviewer of this paper, as well as for his enlightening contribution to the subject.

BIBLIOGRAPHICAL REFERENCES

David, M. 1977. Geostatistical Ore Reserve Estimation. Elsevier. New York. 364.

Francois-Bongarcon, D. 1991. "Geostatistical Determination of Sample Variances in the Sampling of Broken Gold Ores". CIM Bulletin, June 1991. Vol. 84, No. 950, pp. 46-57.

Francois-Bongarcon, D. 1992. "Geostatistical Tools for the Determination of Fundamental Sampling Variances and Minimum Sample Masses". Geostatistics Tróia '92. Kluwer Academic Pub. Doordrecht, The Netherlands. Vol 2: 989-1000.

Francois-Bongarcon, D. 1993. "The Practice of the Sampling Theory of Broken Ores". CIM Bulletin, May 1983. Vol. 86, No. 970, pp. 75-81.

Gy, P.M. 1956. "Poids à Donner à un Echantillon - Abaques d'Echantillonnage". Revue de l'Industrie Minérale, No 38. pp. 52-99.

Gy, P.M. 1982. Sampling of Particulate Materials, Theory and Practice. Elsevier. Amsterdam. 431.

Krige, D.G. 1978. Lognormal De Wijsian Geostatistics for Ore Evaluation. South African Institute of Mining and Metallurgy. 50.

Matheron, G. 1966. "Comparaison Entre les Echantillonnages à Poids et à Effectifs Constants". Revue de l'Industrie Minérale, August 1966. 609-621.

Pitard, F.F. 1993. "Exploration of the Nugget Effect". Geostatistics for the Next Century. Kluwer Academic Pub. Doordrecht, The Netherlands.

Rose, C.D., Srivastava, R.M. 1993. "Information, Measures, Integration and Geostatistical Analysis". Geostatistics for the Next Century. Kluwer Academic Pub. Doordrecht, The Netherlands.

A FRACTAL CORRELATION FUNCTION FOR SAMPLING PROBLEMS

C. D. ROSE
Charles Rose Consultants
P. O. Box 4344
Monroe, LA 71211-4344
U.S.A.

R. M. SRIVASTAVA
FSS International
800 Millbank
Vancouver, BC
Canada V5Z 3Z4

Presumably, no one is going to come in the dark of the night and whisper to us the symmetries Nature has woven into Her tapestry. —— A. Zee in *Fearful Symmetry*

ABSTRACT

This paper discusses a recent theoretical model based on a symmetry postulate regarding the serial correlation of sampling units in 1D. The simple postulate leads to analytical expressions with rich theoretical and practical implications, including a spatial correlation function which explains power law relationships often observed in studies of dispersion variances, and which can be related to both the power model variogram used in geostatistical studies and to the fractal dimension used to describe many self-similar processes. From a practical point of view, several case study examples show that the new correlation function does an excellent job of fitting experimental variograms from a variety of mining, environmental, and forestry applications. An additional feature of practical value is that the function has the same form in both ergodic and non-ergodic applications, with only a change in sign of a single scale parameter distinguishing ergodicity from non-ergodicity. This flexibility is important in problems where the available samples cannot be viewed as points within an infinite and unbounded domain. A qualitative interpretation of the new model is developed and supported with case study examples. An extension to irregularly sampled 2D and 3D cases is developed and demonstrated through simulation-based examples.

APPLYING SYMMETRY TO CONSTRUCT THE MODEL

For a more detailed discussion of the mechanics of the theory, the reader is referred to an earlier paper by Rose, (1992). Here, we shall offer some background on the thoughts leading to a symmetry postulate adopted as the basis for the theory.

Often, the serial correlation or variogram model used is a phenomenological model, chosen or constructed from a set of given models to "explain" the structure of the spatial distribution of a finite set of observations using some best fit technique. The choice of model is generally influenced by tradition as well as the geostatistician's personal training, experience, and viewpoint of the world. One might prefer the alternative of choosing a model using at least in part, *a priori* theoretical or qualitative reasoning. Also, it would

R. Dimitrakopoulos (ed.), Geostatistics for the Next Century, 142–155.

prove helpful if parameters of the model were recognizable as regular parameters of mathematical statistics, possibly subject to alternative techniques of estimation as a validation procedure or for refinements of the sample statistics.

At first blush, it does not seem possible to construct a useful general correlation or variogram model from base principles without reference to actual past or current sample data. For encouragement, however, one may consider how the basic models of mathematical physics can be constructed. One approach entails, for example, making many experimental observations and after repeated measurements of, say, kinetic and potential energies both before and after experimental events, and having on each occasion taken into account the experimental error, eventually concluding that energy is conserved. Only after numerous such experiments conducted under highly varied conditions would the ultimate phenomenological model expressed by the appropriate equations be accepted as a "law." Alternatively, a much simpler and more satisfying approach is to recognize and use the power of Emmy Noether's theorem (see Sudbery, 1986, and Zee, 1986). Noether found that in the classical mechanics of a system with a finite number of degrees of freedom, under certain transformations which leave the Lagrangian invariant there are associated constants of motion. For example, with the simple assumption that physical laws do not change with time, Noether's theorem leads directly, without need of laborious experiment, to the conclusion that energy is conserved.

It turns out that momentum of a physical system is conserved if physical laws are invariant under a translation in space, and conservation of angular momentum follows from a rotational invariance. Note that every symmetry results in a conserved quantity.

At the heart of the very useful concept of symmetry lies the question, does physical reality appear different as perceived by different observers with different viewpoints? Will the observer of next year see a different set of physical laws than the observer of last year (related to conservation of energy)? Do two observers located one kilometer apart conclude there is a common set of physical laws (conservation of momentum)? Does the observer whose head is rotated ninety degrees see evidence of the same physical laws that we, unrotated, see (conservation of angular momentum)? Einstein used the notion of symmetry to develop the special theory of relativity. Relativistic invariance means that two observers in relative motion at constant velocity must arrive at the same physical laws, notwithstanding the fact that they differ in their measurements of various physical quantities. If the physical laws observed by the two were not the same, then Nature would be distinguishing between the two observers.

Consider now a hypothetical one-dimensional infinite domain partitioned into an infinite number of contiguous units (supports) of equal but arbitrary finite size.[1] For a characteristic measurable on each support, one may write,

$$\rho_{\infty,1} = 2(\sigma^2_{\infty,2} / \sigma^2_{\infty,1}) - 1, \qquad [1]$$

[1] Extension to additional dimensions is straightforward.

where $\rho_{\infty,1}$ is the serial correlation of units of size 1 whose centroids are separated by distance 1 unit (adjacent units), $\sigma^2_{\infty,1}$ is the dispersion variance over the infinite domain of segments of size 1 unit, and $\sigma^2_{\infty,2}$ is the dispersion variance over the infinite domain of segments of size 2 contiguous units. Now, let $\rho_{N,u}$ be the spatial correlation between units separated at distance u units within a domain of *N* units. Without reference to any actual measurements made on the support units, we know as a matter of simple algebra that regardless of the support (segment size):

If there is no spatial dependence, $\rho_{\infty,u} = 0$ [2]

If there is no spatial dependence, for arbitrary cluster size N units, the expected value of the serial correlation of single units within clusters is given by

$$< \rho_{N,u} > = \rho_{\infty,u} - 1/(N-1) = -1/(N-1) \quad [3]$$

For any degree of spatial dependence, $-1 < \rho_{\infty,u} < 1$ [4]

With these considerations, and taking into account that serial correlation does not carry the baggage of units of spatial dimension, for a natural stochastic population of hypothetically infinite size or a population at least sufficiently large that adjustments for finiteness need not be considered, it is difficult to see how or why the value of serial correlation structure should depend upon a metric chosen by the observer.[2] The choice of metric or choice of segment size, perhaps being the distance from the King's nose to his fingertips, seems irrelevant. Nature invented neither the yard nor the meter, and an infinite spatial domain is infinite notwithstanding the unit of measure of distance.

To conclude otherwise would accept that Nature would be distinguishing between observers inspecting the population on different scales. An observer with a unique scale could have a preferential point of view, which, for a natural stochastic system, seems ludicrous if the domain is of infinite size. Thus, the theory illustrated here is based simply on accepting, for certain natural populations in an infinite spatial domain, the premise of "no unique yardstick." More formally, letting Z_t denote the value of interest for segment t in consecutively numbered segments, one may state,

Postulate: *The serial correlation of Z_t and Z_{t+u} is gauge invariant in the infinite domain.*[3]

We state without proof the belief that the conserved quantity associated with the symmetry is intrinsic stationarity, i.e., $< Z_{t+u} - Z_t >= 0$, and the quantity $<(Z_{t+u} - Z_t)^2 >= 2\gamma(u)$ is conserved (independent of t).

---

[2] It is perhaps not always evident, but with spatial dependence, there is more to the matter of population "size" than merely the number of segments included in the population.

[3] For more on gauge symmetry, see the Appendix.

There are two points that should be made presently lest the reader find us merely proceeding from credulity. First, it is realized there is a goodly stretch from the application of Noether's theorem in symmetries related to the integral of the Lagrangian of a system (the action), to the postulate of a symmetry of the serial correlation in populations of interest to geostatisticians. To this we note that the discussion of Noether's theorem is intended primarily to illustrate the power of the general concept of symmetry when approaching a problem from a theoretical point of view. Furthermore, that useful symmetries are restricted to the action is further contravened by Einstein's applications of symmetry in developing both the special and general theories of relativity.

The second point is the recognition that application of the model will likely be limited for the most part to "natural" populations where the domain of interest is very large in comparison to the support, with high total information capacity, and with support values mainly unaffected by the determinism of outside intelligence. The idea here is that natural populations will tend toward behaving like the model as the spatial domain increases in size relative to the support and the total information capacity increases. In view of these stated limitations, the good fit of the model to experimental data taken from a number of sources is of continuing surprise.

THE CORRELATION FUNCTION AND THE SEMIVARIOGRAM

Using only the assumptions of the model, i.e., that the population is second order stationary and the symmetry postulate holds over the population of interest, one finds (see Rose, 1992),

$$\rho_{N,u} = [N^b/(N^b-1)](1/2)\Delta_u^2(u-1)^{2-b} - 1/(N^b-1), \qquad [5]$$

where $\rho_{N,u}$ denotes the serial correlation between non-overlapping segments with centroids separated in space by a distance of u segments; N is the number of segments in the finite population of interest; b is a scaling parameter in the region $-2 < b \leq 2$ $(\neq 0)$; and Δ_u is the difference operator defined by,

$$\Delta_x f(x) = f(x+1) - f(x) \qquad [6]$$

The corresponding semivariogram is expressed as

$$\gamma'(u) = \sigma_{N,1}^2(1-\rho_{N,u}) = \sigma_{N,1}^2[N^b/(N^b-1)][1-(1/2)\Delta_u^2(u-1)^{2-b}] \qquad [7]$$

Eq. 7 is often not a useful semivariogram, however, in practice, for it neglects the (always real) variance of the errors of measurement of the support values. As a matter of experimental design, one should take steps to assure measurement errors are free of autocorrelation. If measurement errors are uncorrelated with support values and one has been successful in assuring they are not serially correlated, a useful semivariogram may be written as

$$\gamma(u) = \sigma_m^2 + \gamma'(u) = \sigma_m^2 + \sigma_{N,1}^2[N^b/(N^b-1)][1-(1/2)\Delta_u^2(u-1)^{2-b}], \qquad [8]$$

where σ_m^2 is the variance of the measurement errors.

For $b < 1$, in the limit as u approaches zero, $\gamma(u) = \sigma_m^2$. Thus, the value of the semivariogram at the intercept, the "nugget variance," is only the measurement variance.

An excellent approximation to Eq. 8 for $u > 2$ is arrived at by replacing the second order difference by the second derivative with respect to u. This gives,

$$\gamma(u) = \sigma_m^2 + \sigma_{N,1}^2 [N^b / (N^b - 1)][1 - (1-b)(1-b/2)/u^b] \quad [9]$$

COMPARISON WITH TRADITIONAL VARIOGRAM MODELS

Unfortunately, the size of support upon which measurement has been made is often not reported in geostatistical studies, with the assumption that the support of measurement may be considered a point source. Even if there is relatively weak spatial continuity, the effective size of the support is sufficiently large so that it cannot be treated as zero. In addition, if the size of the support is ignored, one cannot directly identify the roles of the measurement variance and dispersion variance in the model. Nevertheless, if the support size is unreported and the lag is expressed in meters or other convenient metric, letting $u = vh$, where v is the conversion factor, from Eq. 9 we have,

$$\gamma(h) = \theta_1 - \theta_2 / h^b, \quad [10]$$

where,

$$\theta_1 = \sigma_m^2 + \sigma_{N,1}^2 [N^b / (N^b - 1)], \quad [11]$$

$$\theta_2 = \sigma_{N,1}^2 [N^b / (N^b - 1)](1-b)(1-b/2)/v^b \quad [12]$$

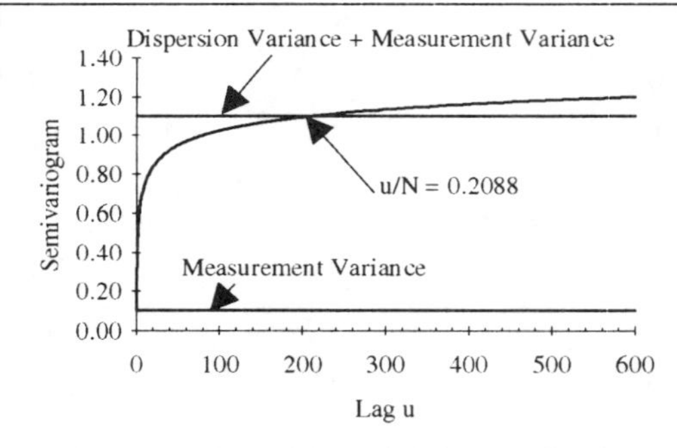

Figure 1. Model semivariogram with N=1000, v =1, $\sigma_{N,1}^2$=1.00, σ_m^2=0.10, and b = + 0.10

If the scaling parameter b is positive, the sequence of population values is ergodic. For practical applications, however, the sequence is often non-ergodic and b is negative. With negative b the value of θ_2 is also negative and the gauge invariant variogram of Eq. 10 takes the form of the classic power model semivariogram with the exception, however, that the coefficient θ_2 depends on b. The asymptotic behavior of $\gamma(h)$ in Eq. 10 as negative values of b approach zero is not the same as that of the classic power model variogram as its positive exponent approaches zero.

Fig. 1 illustrates the shape of an ergodic semivariogram with b = +0.10. Using the same values for the dispersion and measurement variances, Fig. 2 shows a non-ergodic semivariogram with b = -1.50. For $-2 < b < 1$, the model semivariogram intercepts the dispersion variance plus measurement variance line at a ratio u_c / N given by,

$$u_c / N = [(1-b)(1-b/2)]^{1/b} \quad [13]$$

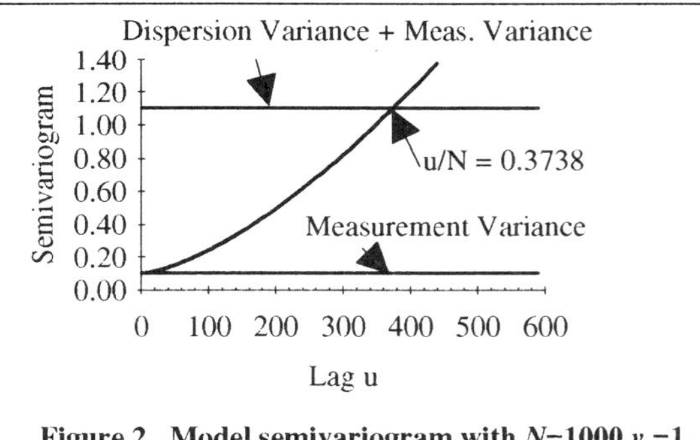

Figure 2. Model semivariogram with N=1000, ν =1, $\sigma^2_{N,1}$=1.00, σ^2_m=0.10, and b =-1.50.

The relationship of Eq. 13 is illustrated in Fig. 3. At b = 1, the semiv .iogram and the dispersion variance plus measurement variance line are the same.

Fig. 4 illustrates the shape of variograms for N = 1000, $\sigma^2_{N,1} = 1.00$, $\sigma^2_m = 0.10$, and various values of b. At b = -1.00 (not shown), the variogram is linear, and between - 1.00 and - 2.00, it is concave upward as a function of u.

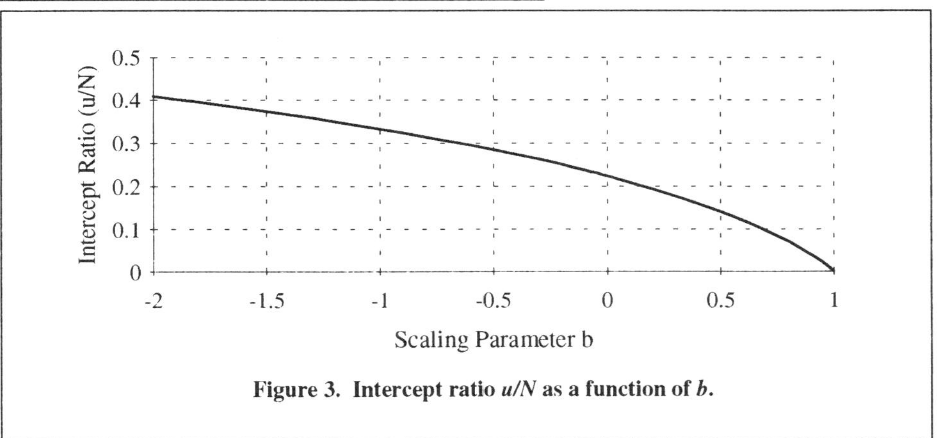

Figure 3. Intercept ratio u/N as a function of b.

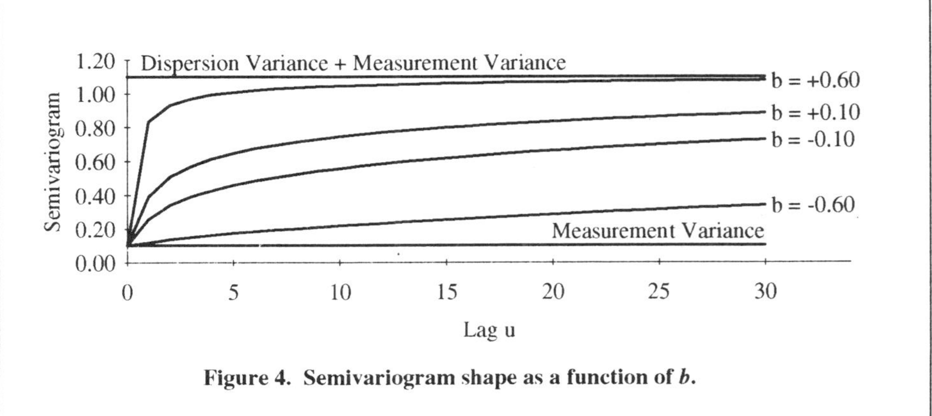

Figure 4. Semivariogram shape as a function of b.

CHANGE OF SUPPORT AND SAMPLING VARIANCES

Letting $\sigma^2_{N,1}$ denote the dispersion variance of information support units (base units) over a region of N total units, the dispersion variance $\sigma^2_{N,k}$ of supports of k contiguous base units is given by,

$$\sigma^2_{N,k} = [(N^b - k^b)/(N^b - 1)]\sigma^2_{N,1} / k^b \qquad [14]$$

The expression for the variance of the mean of systematic samples when the measurement variance is non-negligible is,

$$Var_{sys}(\overline{Z}) = \sigma^2_m / n + (\sigma^2_{N,1} / n)[1 + 2/n \sum_{u=1}^{n-1} (n-u)\rho_{N,ku}, \qquad [15]$$

where n is the number of segments drawn for the sample, and sampling interval $k=N/n$. Eq. 15 may also be written as a function of the semivariogram given by Eq. 10 as,

$$Var_{sys}(\overline{Z}) = \sigma^2_m + \sigma^2_{N,1} - (2/n^2)\sum_{h=1}^{n-1} (n-u)\ \gamma_1(kh), \qquad [16]$$

where the subscript 1 on γ reminds the reader of the information unit support for the variogram.

It should be noted that neither Eq. 15 nor Eq. 16 are model dependent, but do require independence of the measurement errors and knowledge of the measurement and dispersion variances.

CASE STUDY EXAMPLES

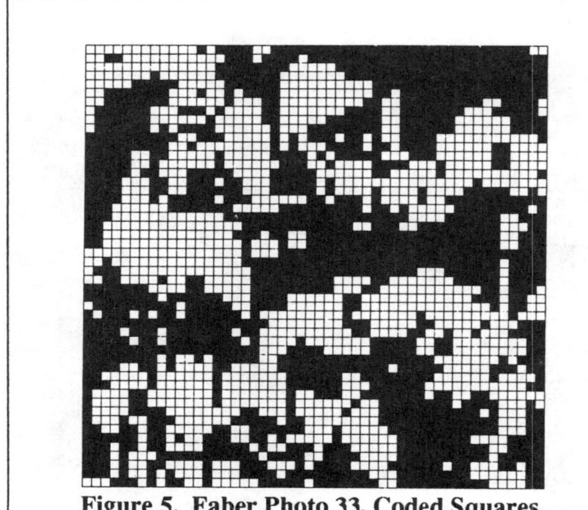

Figure 5. Faber Photo 33, Coded Squares.

Example 1

Faber (1971) published an interpretation of two aerial photographs (photo 33 and photo 86) of the Lake Mickey watershed near Durham, North Carolina. The figure prepared from each photograph has 2500 very small squares in a 50 by 50 square lattice. On the ground, the side of each small square is 4 km. In the figures, an indicator variable is assigned to each of the small squares, with an indicator of "1" if the square covers woodlands, and an indicator of "0" if the square covers something else. Fig. 5 replicates the Faber figure for photo 33, using a black square in place of a "1" and a blank square in place of a "0" for better visibility.

The figure includes a total of 1367 ones, for a proportion of $p = 0.5468$. Thus, N = 50 for each of the two spatial dimensions, and the dispersion variance $\sigma^2_{N,1} = p\,(1-p) = 0.2478$. Fig. 6 shows the fit of the gauge invariant semivariogram model with the assumption that the measurement error variance is negligible. The fitting process entailed selecting the value $b' = 0.265$ which minimizes the sum of squares of the residuals. Note the difference here from the usual variogram modeling process where statistics for two or more parameters require estimating. Fig. 7 shows the dispersion variance of supports of square size 1, 4, 16, and 64 over a trimmed 48 by 48 square area of the figure. The model curve is calculated using Eq. 14, with $b' = 0.265$.

Considering that only a single parameter was fitted, the fit of the model seems quite good. Very similar results were obtained using the Faber photo 86.

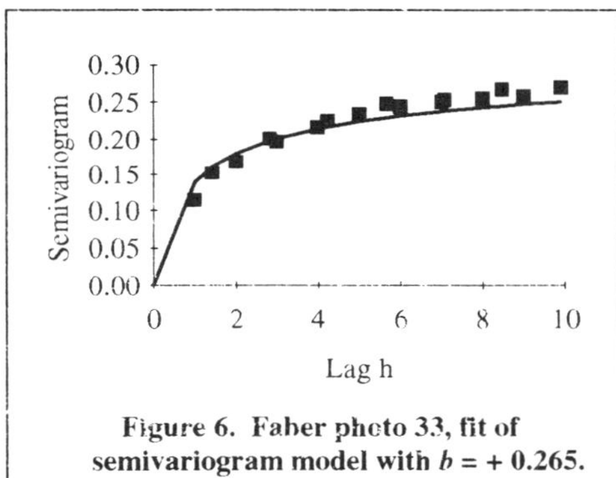

Figure 6. Faber photo 33, fit of semivariogram model with $b = +0.265$.

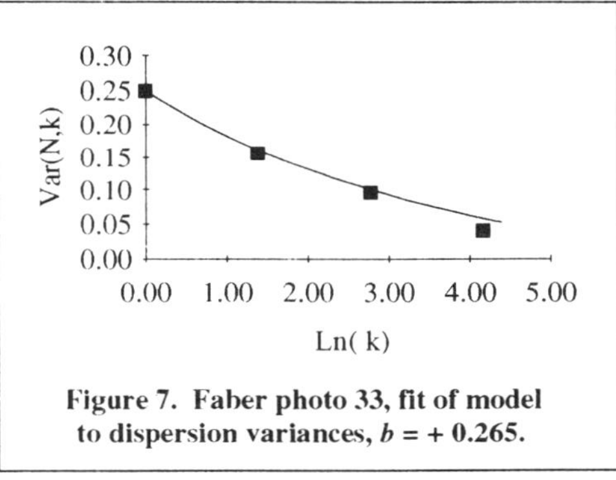

Figure 7. Faber photo 33, fit of model to dispersion variances, $b = +0.265$.

Example 2

Coal for Colstrip electric generating units 3 and 4 located in Colstrip, Montana comes directly from a nearby mine on an overland belt conveyor moving 1550 tons per hour. Samples are collected at the discharge end of the conveyor using a mechanical coal sampling system. The collected and composited sample material is sent at the end of each eight hour shift of operation to a coal laboratory for assay. In the early fall of the year 1991, Rose directed a study on behalf of the supplier of the coal for the purpose of estimating the lot-by-lot variance of the material, the variance of the sampling by the mechanical system, the variance of sample preparation and assay, and biases that might exist against other sampling methods and other laboratories.

As a part of the data collection process, true systematic samples were collected from 90 eight-hour shift lots of coal using a stopped-belt sampling procedure. During each shift, eight stopped belt increments were collected from the conveyor using a divider with dimensions one foot in width and across the full width of the conveyor belt. The increments averaged weighing about 59 pounds each. Timing for the first increment from the lot was selected at random from the 120 one-half minute intervals available within the first hour. The seven successive increments for the lot were collected at one hour intervals. Each stopped belt increment was crushed, subdivided, and sent to a laboratory for assay. One stopped-belt increment for each of the ninety shifts was selected at random for full duplication of the subdividing and assaying process, thereby making information available for an estimate of the variance of increment division and assay (measurement variance). During the test, the mechanical sampling system continued to collect shift samples at the discharge end of the belt. Duplicate assays of a number of these collected sample specimens were also obtained

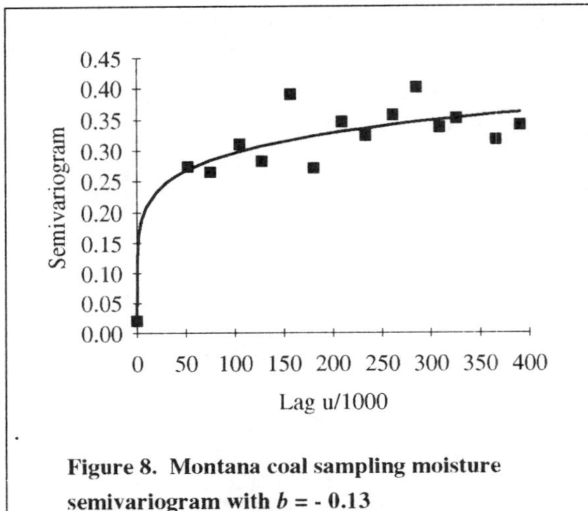

Figure 8. Montana coal sampling moisture semivariogram with b = - 0.13

Fig. 8 shows the semivariogram from the stopped-belt sampling process for coal moisture. The first data point shown at $u = 0$ is the estimate of the measurement variance σ_m^2 (the increment division and analysis variance), which is equal to 0.0200. The second point is at a time distance of one hour, which, in terms of u is,

u = (1550 tons/hr)(2000 lbs./ton)/59 pounds = 52.5×10^3

Data points between intervals of one hour (within restricted ranges) are made available due to shifts being mostly consecutive and using a new random start for each shift. The total number of support segments N within the shift is very large, being about 406,000. The value of $\sigma_{N,1}^2$ was estimated from the fit of the gauge invariant variogram model to be 0.2750. Note that an alternative variogram model, say a linear model, would have given a very different picture of the dispersion and measurement variances, with a high level of unexplained nugget variance. A "two instrument" technique developed by Grubbs (1948, 1982) was used to estimate the lot-by-lot moisture variance of the coal, as well as the variances of sampling of both the stopped-belt and the mechanical sampling procedures. The Grubbs estimate of the moisture variance of the material, lot by lot, is 0.2861. For the stopped-belt sampling variance, collecting the eight increments per lot, the Grubbs estimate is equal to 0.0279. Using the model semivariogram and Eq. 15 (or Eq. 16) of this paper, the estimate of the sampling variance is calculated as 0.0261, thus the two estimates agree quite closely.

It should be noted that the precision of the variance estimates of Grubbs as well as the estimates given by the current variogram model, are affected by the population distribution. A histogram of the Montana coal moisture data looks symmetric and reasonably well approximated by a Gaussian distribution. Transformations might be considered when skewed distributions are encountered.

Example 3

The Naughton electric generating station in Kemmerer, Wyoming is a mine mouth power plant receiving coal from a nearby multiple seam open pit mine. Coal from the lower sulfur seams are sent to Units 1 and 2 of the plant, and coal from the higher sulfur seams sent to Unit 3. Coal is delivered for two to six hours during each of three eight hour shifts daily, with some shifts skipped during the weekends. Each shift's deliveries are generally used to "top off" the plant bunkers with the excess, if any, going to the station stockpile. Handling in the mine is complex: There are thirty or so total seam levels of coal, with some blending (coal mixing) being performed, and a part of the mined coal going to other customers. The mine is the largest open pit coal mine in North America. The Naughton station flue gas stack of generating Unit 1 and the stack of generating Unit 2 are each equipped with continuous emission monitors for measurement of emissions of sulfur dioxide, calculated for regulatory purposes as pounds of sulfur dioxide per million btus of heat input. The variability of measured SO_2 emissions is a function of the variances of measurement errors of the monitor and perhaps combustion variables to some degree, but is believed to be mainly affected by variations in the incoming coal sulfur content.

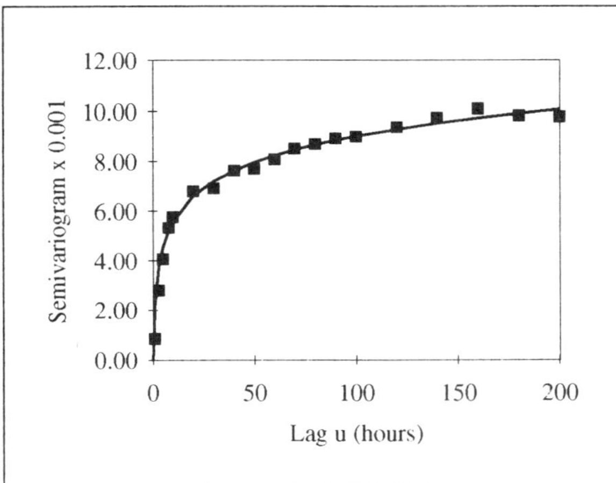

Figure 9. Semivariogram for CEM SO2 measurements, N = 3635, b = - 0 .0085, $\sigma^2_{N,1}$ = 0.01222, and σ^2_m = 0.00000.

Fig. 9 shows the semivariogram for N= 3,635 one-hour average measurements of SO_2 taken by the generating unit 2 CEM during the period August through December, 1991. Measurements made by the CEM were available for all but 79 of the total of 3,635 consecutive hours. The statistics for the parameters b, σ^2_m, and $\sigma^2_{N,1}$ of the semivariogram in Fig. 9 were fitted using all data points shown in the figure with *lag* $u > 8$ hours. The estimate of zero obtained for the measurement variance in the variogram fitting process for this example requires comment. While it is possible the CEM measurement variance is indeed small relative to the dispersion variance, we believe the estimate is likely biased on the low side due to the CEM measurement errors being primarily longer term drift rather than short term instrument fluctuation error. Thus, the assumption in the model of Eq. 10 that measurement errors are free of autocorrelation may not hold in this example.

It is not evident from Fig. 9, but the fit is not very satisfactory for *lag* u of 8 hours or less. Fig. 10 shows an expanded view of the semivariogram at smaller values of u. Note that for $u \leq 8$ hours, the experimental data falls increasingly below the curve of the model as u

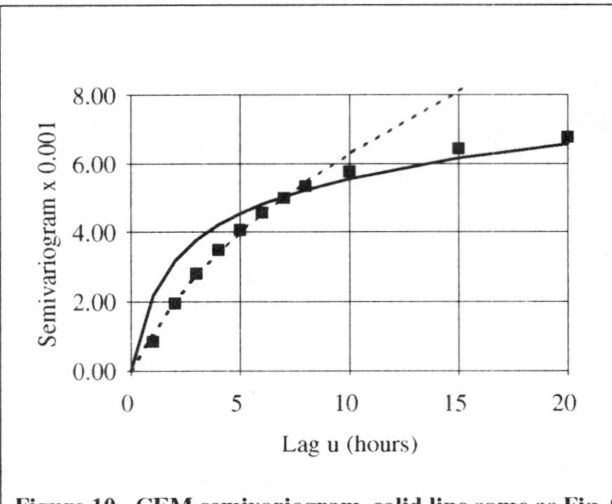

Figure 10. CEM semivariogram, solid line same as Fig. 9, dotted line with $\sigma^2_{N,1} = 0.09123$, *and* $\sigma^2_m = 0.00000$, and $b = -0.543$.

becomes smaller. The dotted line shows the model fit for *lag u* equal to 8 hours and less. The fitted statistics, using the value $N =$ 3635, are $\sigma^2_{N,1} = 0.09123$, $\sigma^2_m = 0.00000$, and $b = -0.543$. The broken symmetry at $u = 8$ hours with a different structure for smaller lags, offers some insight. Recall that the generating plant bunkers are topped off during deliveries in each shift with approximately 8 hours of burn. Thus, generating Unit 2 is being fed material in 8-hour clusters. Due to the numerous mine seams and methods of handling coal in the mine, the cluster of material for the next shift is most often not source-contiguous with the cluster for the current shift. Members of pairs of hourly data values spaced in burn time less than eight hours apart and associated with the same cluster are much more likely to be of similar sulfur content than are members of pairs with the same spacing, but from different clusters. For *lag* $u \leq 8$ hours, the smaller the value of u the greater the likelihood the members of the pair come from the same cluster. This phenomenon, of course, explains the different behavior of the variogram data for small values of u. Unfortunately, cluster boundaries are not identified in the data base, thus one cannot examine the average within-cluster variogram unpolluted with pairs of supports crossing cluster boundaries.

With the thought in mind that such a clustering phenomenon is associated with a broken symmetry, we suggest that such an associated cluster of material may be described as a *clustron*.

USING THE MODEL TO INFER MEASUREMENT VARIANCE

Case studies based on conditional simulation results show that one of the benefits of the gauge invariant model given in Eq. 10 is that it provides a parametric form for the semivariogram that allows inference of the measurement variance.

Fig. 11 shows the results of using sequential simulation to produce a 2D realization whose pattern of spatial continuity is described by the gauge invariant model with $b = 0.25$, no measurement variance and a dispersion variance of 0.083; this realization contains 2500 values on a 50x50 grid. As with the Faber photographs shown earlier, the dimensions of the grid entail that $N = 50$. Fig. 12 shows the location of 56 samples whose values include a measurement variance of 0.010. With the available samples containing some measurement error, one expects that the sample semivariogram will show a nugget effect. The

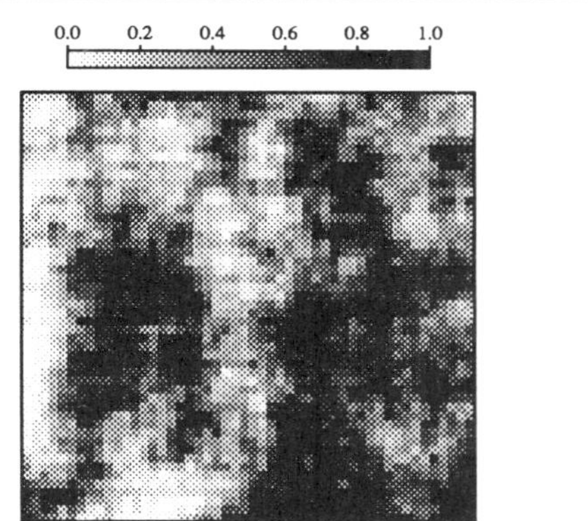

Figure 11. A realization of the gauge invariant model with b=0.25, no measurement variance and a dispersion variance of 0.083.

sparse sampling will also contribute to the visual appearance of a nugget effect since there are too few closely-spaced sample pairs from which to calculate experimental variogram values for very short distances. In Fig. 13 the circles show the experimental semivariogram calculated from the samples shown in Fig. 12. As expected, a visual extrapolation of these experimental values back to the y-axis would lead to the conclusion that the relative nugget effect is more than 50% of the sill; if all of this nugget effect is attributed to measurement error, one would conclude that the measurement variance was greater than the dispersion variance. The solid line in Fig. 13 shows the theoretical model that one would obtain with exhaustive sample information: a gauge invariant variogram model with $b = 0.25$, a dispersion variance of 0.083 and a measurement variance of 0.010. The dashed line shows the results a least-squares fit of the three parameters of the gauge invariant model to the available experimental semivariogram data; for $N = 50$ the best fit is accomplished with $b = 0.285$, a dispersion variance of 0.073 and a measurement variance of 0.020. Though the measurement variance inferred from the gauge invariant model is larger than the theoretical measurement variance, the use of the gauge invariant model still leads to a far better approximation of the nugget effect than would be provided by the use of more traditional models, such as the spherical or exponential variogram model.

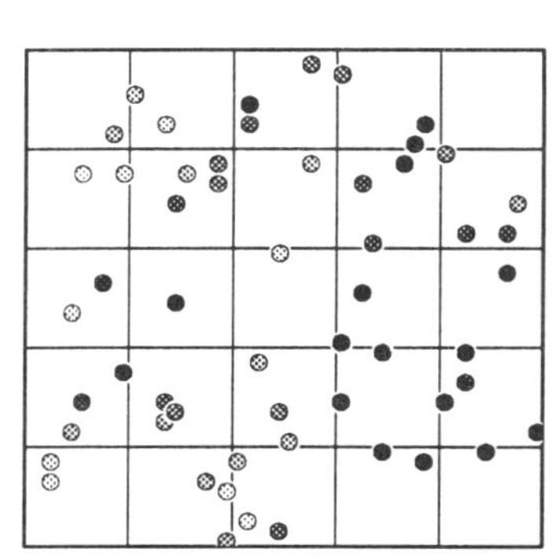

Figure 12. 56 randomly selected sample locations at which the realization in Figure 11 is sampled with a measurement variance of 0.010.

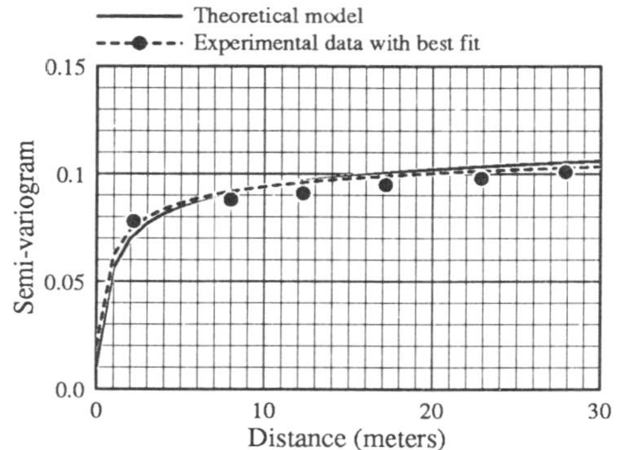

Figure 13. Theoretical semivariogram of the sample values in Figure 12 compared to the experimental semivariogram and its best fit.

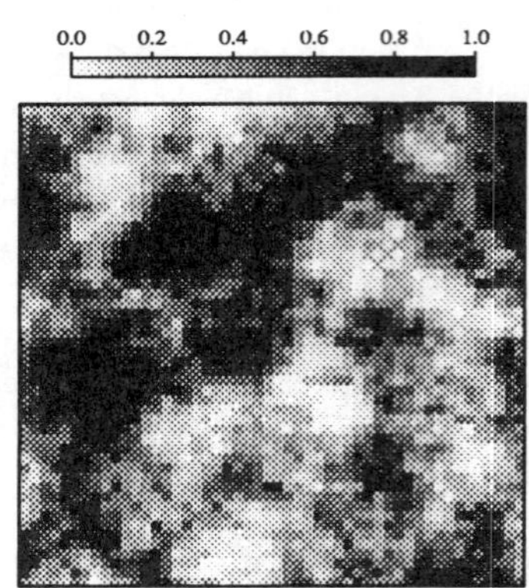

Figure 14. A realization of a spherical variogram model with a sill 0f 0.083, no nugget effect and a range of 10 metres.

The success of the gauge invariant model in predicting the measurement variance from such a sparse sampling is due to the fact that the symmetry postulate underlying Eq. 10 entails a fractal self-similarity. The curvature of the variogram model for short distances can be deduced from the curvature at larger distances.

Though the previous example shows that the gauge invariant model can lead to a good prediction of the measurement variance, the success of this particular example is not surprising since the assumed parametric form of the underlying variogram happens to be exactly correct; the realization in Fig. 11 was created using the same type of variogram model that was later assumed for fitting the experimental variogram in Fig. 13. As a more challenging test of the ability of the gauge invariant model to predict the measurement variance, one could repeat the previous exercise with an exhaustive realization whose underlying variogram is not the gauge invariant model. Fig. 14 shows a realization of a spherical variogram model with a range of 10 meters, and Fig. 15 shows a random sampling of this realization. As in the previous example, the dispersion variance is 0.083, there is no nugget effect for the exhaustive realization and the sample values contain a measurement error whose variance is 0.010. Fig. 16 shows the experimental variogram, its best fit using the gauge invariant model and the theoretically correct underlying variogram model. What

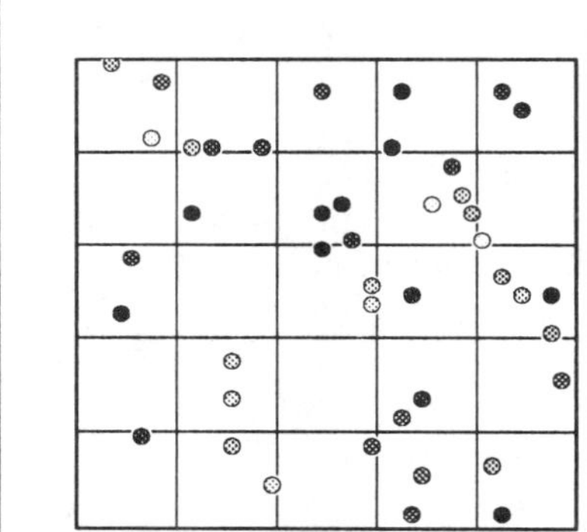

Figure 15. 43 randomly selected sample locations at which the realization in Figure 14 is sampled with a measurement variance of 0.010.

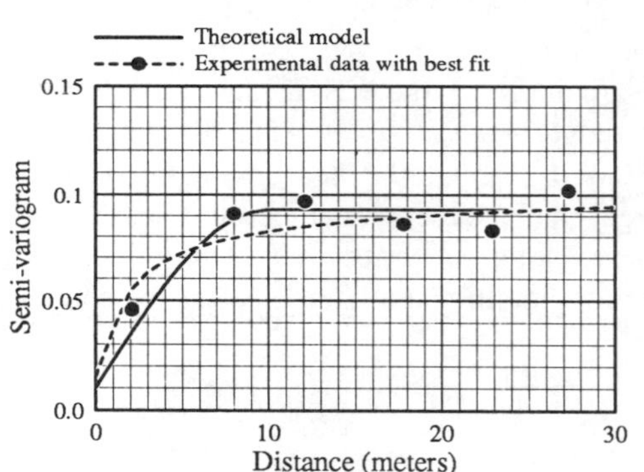

Figure 16. Theoretical semivariogram of the sample values in Figure 15 compared to the experimental semivariogram and its best fit.

is particularly intriguing about this example is that the gauge invariant model does a fine job of predicting the measurement variance even though the exhaustive realization has a different variogram. This may be due, in part, to the way in which experimental variograms are calculated. With few closely spaced pairs of data, it is necessary to group the data into lags within which an arithmetic average is used to estimate the value of the semivariogram. With the semivariogram commonly having a concave downward curvature, the arithmetic averaging in the first few lags inevitably leads to a tendency for a linear extrapolation back to the y-axis to overestimate the true nugget effect. Even when the underlying variogram is not a gauge invariant model, it seems that the use of the gauge invariant model can produce robust estimates of the measurement variance.

BIBLIOGRAPHY

Faber, J. A. J. (1971), "Precision of sampling by dots for proportions of land use classes," *Institute of Statistics Mimeo Series*, No. 773, Raleigh, NC.

Grubbs, F. E. (1948), "On estimating precision of measuring instruments and product variability," *Journal of the American Statistical Association*, 43(2), p. 243.

Grubbs, F. E. (1982), "An introduction to some precision and accuracy of measurement problems," *Journal of Testing and Evaluation*, 10(4), p. 143.

Rose, C. D. (1992), "A fractal model for sampling coal, ores, and other natural populations," *Journal of Coal Quality*, Vol. 11, No.1-2, p. 6.

Sudbery, A. (1986), *Quantum Mechanics and the Particles of Nature*, Cambridge University Press, Cambridge, Chap. 7.

Zee, A. (1986), *Fearful Symmetry: The Search for Beauty in Modern Physics*, Macmillan, N.Y., Chap. 8.

APPENDIX

In *Fearful Symmetry*, A. Zee writes regarding gauge symmetry,

"[Hermann] Weyl named his symmetry gauge symmetry. The term "gauge" comes from low Latin *gaugia*, referring to the standard size of cases, and this sense is retained in such modern usage as "railroad gauge" and "gauged skirt." Curiously, the word entered the permanent vocabulary of physics only because Weyl made a serious but justifiable mistake. We now know that the symmetry responsible for electric charge conservation is described by transformations involving the quantum probability amplitude. Weyl was working before the advent of quantum physics, so he, like everyone else, never dreamed of probability amplitudes. Instead, inspired by the geometric flavor of Einstein's work, Weyl proposed a transformation in which one changes the physical distance between spacetime points. Weyl was reminded of the distance, or gauge, between two rails--hence the name for his symmetry. He showed Einstein his theory, but they were both deeply disappointed that it failed to describe electromagnetism. When the quantum era began, Weyl's theory was quickly repaired. Meanwhile, the term gauge symmetry, although a misnomer, remained. (Incidentally, physicists still do not know whether Weyl's original symmetry is relevant to the world.)"

IMAGE COMPRESSION AND KRIGING

E. A. Yfantis and M. Au
Computer Graphics and Image Processing Laboratory
Computer Science Department
University of Nevada, Las Vegas, 89154

F. S. Makri
Computer Science Department
University of Nevada, Las Vegas, 89154
and Department of Mathematics
University of Patras, Greece

In this research paper we use a kriging-like iterative method with changing interpolation neighborhoods and we also use the data estimated in the previous iteration in order to estimate the data of the present iteration. Our method is used for image compression and it constitutes a lossy compression algorithm. Due to multicolinearity neighboring pixels carry mostly the same information. Therefore if we reduce the amount of information by appropriate sampling of the image then the space needed to store this information is smaller than the space required by the original image. The semivariogram of the image is estimated. Based on this we estimate the zone of influence. If r is the zone of influence of the semivariogram function of the picture, then we sample the picture using a square sampling design with side length equal to $a \times r$, where a is a factor less than one. A kriging estimator is used to estimate the pixels not included in the sample. Under the assumptions of the process being wide sense stationary this kriging estimator is best linear unbiased. The algorithm is lossy, and tests using the image of Lenna are presented. Comparison of this algorithm with JPEG show that our algorithm has graceful degradation. By graceful degradation we mean that drastic reduction on the information used to produce the image results to recognizable images whose features fade away gradually.

INTRODUCTION

The storage required for storing an image in todays high-resolution computers, could be more than one megabyte. This is costly not only in storage space but also in time needed to transfer the information from the disk to memory or from the memory to disk. Image compression reduces the amount of storage needed in memory or permanent storage, as well as the transfer time, since less information needs to be transferred.

Image compression algorithms can be classified in two categories: The first category is referred to as losseles compression, or bit-preserving, or reversible compression[1]. In this one the image reconstructed after compression is numerically identical to the original image on a pixel-by-pixel basis.

R. Dimitrakopoulos (ed.), Geostatistics for the Next Century, 156–161.

The second category is referred to as lossy compression, or irreversible compression. In this one the reconstructed image contains degradations relative to the original image. An excellent review of image compression is given by Rabbani[1]. The algorithm presented here is an extension of kriging[2-5], in the sense that it utilizes the relationship between neighboring pixels as is expressed by the semivariogram and its zone of influence that describes the distance needed to exist between two pixels to be independent or uncorrelated. The text information is stored differently in the computer buffer than the graphics information. Thus for every character in the video buffer we need two bytes, one showing the ASCII code corresponding to the character and next to it another showing the attribute associated to the character. If a text screen therefore consists of 43 rows each having 80 characters then $2 \times 80 \times 43$ bytes are needed to store a screenfull of text. For the graphics mode, if 16 colors are used per pixel then 4 bits are needed per pixel. So for a screen with 640×480 pixels, $(640 \times 480 \times 4)/8 = 153600$ bytes are needed. Manipulating this large amount of data for animation becomes very expensive in terms of both time and space. Computer graphics pictures such as fractals, ray tracing, etc. take a lot of time to calculate, and a lot of space to store in the permanent storage device. They also take a lot of time to store and retrieve information. Compression saves both storage space and retrieval time. An important measure related to data compression is the compression ratio[1,6,7], which is defined to be equal to the ratio of the length of the original data string divided by the length of the compressed data string. This ratio is the degree of data reduction obtained as a result of the compression process. From this definition it is clear that the higher the ratio, the smaller the resulting compressed data. In addition to the cost savings derived from the reduction in storage space data compression may reduce the overall execution time of a program which manipulates a large group of data. Certainly the number of instructions to be executed increases, due to the added instructions to compress and uncompress the data. However, the frequency of disk storage access decreases and the overall time spent by the computer decreases, since it takes more time to access the disk than to execute a set of instructions. Our algorithm was compared to the JPEG lossy compression algorithm, which is available as a utility in the X-window system. Preliminary results show our algorithm produces recognizable images with relatively small amount of information, when JPEG no longer produces recognizable images. Our algorithm was used on the image Lenna and produced recognizable image at compression ratios 63.57 to 1, and 249.26 to 1, while JPEG produced a very distorted picture at 70.24 to 1 ratio, requiring 3499 bytes for storage.

KRIGING AND LOSSY COMPRESSION

We consider a network of points satisfying[2-5] the intrinsic hypothesis within a domain $F \subseteq R^2$. Let X_0 be a point for which we want to estimate the unknown value of the random process based on a network of points, $X_i, i = 1, ..., k$, within the zone of influence. Let the unknown true value at X_0 be $Z(X_0)$, then an estimate of $Z(X_0)$ is given by $Z^*(X_0) = \sum_{i=1}^{n} a_i Z(X_i)$. The weights a_i can be found by solving the system of equations

$$B * A = C \tag{1}$$

where the matrix B is an $(n+1) \times (n+1)$ square matrix and has zeros in the main diagonal, ones in the last row and last column[2], except for the corresponding main diagonal element which is zero, and $\gamma(x_i - x_j)$ in the ith row and jth column. The matrix A is a column vector with $n+1$ rows, the first n containing the unknown parameters $a's$ and the last one the Lagrange multiplier λ. The matrix C is a column vector with the ith row being $\gamma(x_i - x_0)$, for $1 \leq i \leq n$ and the $n+1$ row being 1. For every pixel in the image to be compressed there corresponds an (R,G,B) tuple. When the gray scale is used then each one of the three components of the scale has the same value. Based on the known common (R,G,B) value we calculate the semivariogram, and hence the zone of influence of the underlying random process. If r denotes the zone of influence, then we have shown[4] that if the sampling distance is less than $\frac{r}{2}$, the slope of the mean square error is reduced and gradually becomes less than one and close to zero which implies that decreasing the sampling distance and, therefore, increasing the number of samples does not increase the accuracy of the image at a fixed rate. The points selected initially constitute a network of points on square grids with side length $h \leq \lfloor \frac{r}{\sqrt{10}} \rfloor$. Based on this network of points we first calculate the centerpoints of the squares using a radius of $\frac{\sqrt{10}h}{2}$, where h is the sampling distance. The images shown here were based on a cross like design consisting of 12 points used to estimate the center of the cross. From equation 1, and if the space lags in the semivariograms are much less than the zone of influence, as we choose them here, then the semivariogram can be approximated by a linear equation. Thus $\gamma(h) \simeq mh + b$, where m is the slope in the linear approximation and b is the y-intercept or nugget effect. Based on this approximation and equation 1, we obtain

$$(mhB_1 + D) * A = C_1 \tag{2}$$

where: D is an 13×13 matrix of constants with zero main diagonal and the last row and column all zeros. The array A is a column vector with the last entry being the Lagrange multiplier and the first n entries being the unknown parameters $a's$. The matrix B_1 is a 13×13 matrix of constants with zero main diagonal, and all entries of the last row and last column equal to $\frac{1}{mh}$ exept for the $(13,13)$ entry which is zero. The solution to the above system of equations does not depend on m, h, or b, and it is

$$a_1 = a_2 = a_3 = a_6 = a_7 = a_{10} = a_{11} = a_{12} = -0.0167, a_4 = a_5 = a_8 = a_9 = 0.2834 \tag{3}$$

Two variations of our algorithm were used. In the first one, the original image was sampled using a square grid of length k pixels, subsequently the center of each square was estimated by averaging the 4 vertices of the square, then the centers of each one of the squares formed by the two vertices of a side of a previous square and the two adjacent centers, to that side just estimated. In the second one, twelve points were used to estimate the center of the square. After estimating the centers of the crosses, we use these centers to form smaller crosses, which we calculate the centers of by using the 12, neighboring points that they form a cross with center the point to be estimated. The original image of Lenna required 245775 bytes. Using our cross variation of our algorithm (Figures 2 - 5) with compression ratio 63.57 to 1 the compressed image requires only 3866 bytes and the characteristics of Lenna are well visible. Notice that her eyes are still recognizable. However when the JPEG algorithm was used at the ratio 70.24 to 1, which required 3499 bytes to store the image (Figure 8), the characteristics of Lenna were distorted to the point that the produced image is of lower quality than the image of Lenna produced by our algorithm when a compression ratio of 249.26 to 1 was used (Figure 5).

CONCLUSION

A lossy algorithm based on kriging was introduced. Unlike kriging though the radius of the neighborhood used to estimate the points of the image reduces with every iteration. Therefore while kriging is a low pass filter, our algorithm is a band pass filter. The radius of the points used in the initial estimation is smaller than the zone of influence of the process and the semivariogram within that radius is strictly increasing and approximates a straight line. Theoretical results and preliminary experimental results show that our method is capable of producing an image resembling the original image with a relatively small amount of information and therefore high compression ratio. The absolute distance and the mean square error were used as measures of goodness of image reproduction, or comparison between the original image and the compressed image. Our algorithm was compared with JPEG, which is a lossy compression algorithm available in the X-window environment. When high compression ratios are desired, our algorithm produces recognizable images at a higher degree than JPEG.

Acknowledgment: The authors would like to thank the Chief editor Dr. R. Dimitrakopoulos for his encouragement and the reviewers for their suggestions and time they expended to read our paper. This research was supported by the DOE, Las Vegas Lab. The authors would like to thank Drs R. Jacobson, and F. Miller of DRI, Mr. John Hall, and Mr. Fred Penrod, for supporting this project.

Figure 1
Original image. 245775 bytes are required for its storage.

Figure 2
Compressed image where the lag=2. The estimation is based on 4 points as shown in Fig. 2. 61446 bytes are required for its storage, and achieves a compression ratio of 3.99.

Figure 3
Compressed image where the lag=4. The estimation is based on 4 points. 15386 bytes are required for its storage, and achieves a compression ratio of 15.97.

Figure 4
Compressed image where the lag=8. The estimation is based on 4 points. 3866 bytes are required for its storage, and achieves a compressed ratio of 63.57.

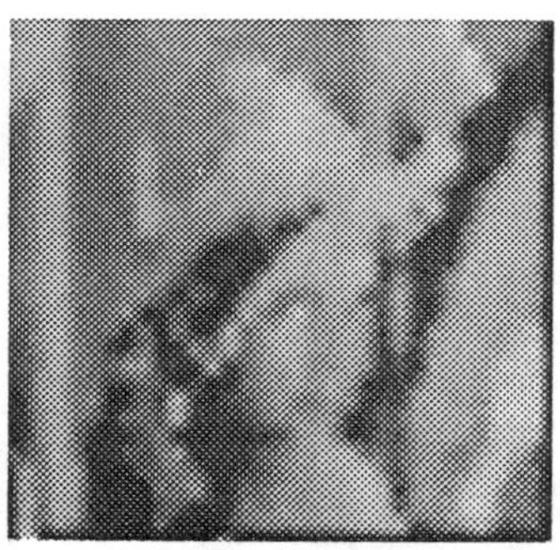

Figure 5
Compressed image where the lag=16. The estimation is based on 4 points. 986 bytes are required for its storage, and achieves a compression ratio of 249.26.

Figure 6
JPEG compression at a quality of 90%. The image takes 62362 bytes to store and has a compression ratio of 3.94.

Figure 7
JPEG compression at a quality of 33%. The image takes 15439 bytes to store and has a compression ratio of 15.92.

Figure 8
JPEG compression at a quality of 1%. The image takes 3499 bytes to store and has a compression ratio of 70.24.

REFERENCES

1. M. Rabbani, "Selected papers on Image Coding and Compression", Milestone Series Vol. 48, pp. xi-xxii, SPIE Optical Engineering Press, Bellingham, WA (1991)
2. A. G. Journel, and G. T. Huijbregts, "Mining Geostatistics," Academic Press , London, 1988.
3. M. David, Geostatistical ore reserve estimation," Elsevier, Amsterdam, 1977.
4. E. A. Yfantis, G. T. Flatman, and J. V. Behar, "Efficiency of kriging estimation for square, triangular and hexagonal grids," Journal of Mathematical Geology, v. 19, no. 3. pp. 183-205, 1987.
5. E. A. Yfantis, and G. T. Flatman, "Sampling Nonstationary Autocorrelated Data," Comput. and Geos., Vol. 14, no. 5, pp. 667-687, 1988.
6. E. A. Yfantis, S. Baker, and G. M. Gallitano, "An Image Compression Algorithm and its Implementation," Finite Fields, Coding Theory, and Advances in Communications and Computing, Lecture Notes in Pure and Applied Mathematics vol. 141, 1992, pp. 417-427, Marcel Dekker, Inc., N.Y.
7. E. A. Yfantis, M. Au, and G. Miel, "An efficient Image Compression Algorithm for Computer Animated Images", Journal of Electronic Imaging 1(4) pp. 381-387, Oct. 1992.

Comments on 'Image Compression and Kriging' by Yfantis, et al.

Pierre Delfiner
7, rue Guynemer
94240 L'hay les Roses
France

The paper on image compression and kriging presented by Yfantis, Au, and Makri, is entertaining and stimulating. The results it shows are interesting and intriguing enough to create a desire to understand more about image compression and see if and how geostatistical techniques could contribute. The following comments are made in this positive spirit.

1. Decimation and aliasing

To compress the image the authors simply use *decimation,* i.e. subsampling to a sparser grid. But it is well know from Signal Processing theory (e.g. Oppenheim and Schafer, 1989) that straightforward downsampling is not a good procedure because dividing the sampling rate by a factor k reduces the observable bandwidth of the signal by the same factor k. If the original signal contains frequencies higher than this reduced bandwidth these frequencies will be folded down into the observable interval, a phenomenon known as *aliasing.* The standard technique to eliminate, or attenuate, aliasing is to use *prefiltering*, that is, apply a low-pass filter to the data before subsampling.

Before proceeding further it is important to pause for a while and discuss the rationale underlying anti-alias filtering. The foundation is the powerful Sampling Theorem (Nyquist, Shannon) which establishes the *equivalence* between a continuous-time signal and its sequence of discrete-time samples, and therefore the ability to do digital signal processing on continuous-time signals without loss of information. But this equivalence requires the signal to be bandlimited and the sampling frequency to be at least double that of the highest frequency present in the signal. Prefiltering purges the studied signal from its higher frequencies so that no further distortion due to sampling occurs. But notice that by doing so we are changing the problem, we are substituting a simpler signal for the one we wanted to analyze in the first place. Is that acceptable? For many signal processing applications it is. For example, as noted by Oppenheim and Schafer ' in processing speech signals only the low frequency band up to about 3-4 kHz is required for intelligibility even

R. Dimitrakopoulos (ed.), Geostatistics for the Next Century, 162–167.

though the speech signal may have significant frequency content in the 4-20 kHz range. Also, even if the signal is naturally bandlimited, wideband additive noise may fill in the higher-frequency range, and as a result of sampling, these noise components would be aliased into the low-frequency band.' The assumption here is that the prefiltered signal is the *interesting part* of the signal.

What are the implications of the above on the image compression problem? Certainly we are interested in reconstructing the original image rather than a filtered version of it. But this is a naive view. The real question is rather: can we do a better job at reconstructing the *original* image if we filter it first than if we don't? The following calculations indicate that the answer is yes. It is preferable to restore the low frequency content of the image well than deal poorly with its high frequencies.

The calculations that follow are only indicative. They are done in one dimension with a linear variogram $\gamma(h) = \varpi h$. Consider N equispaced points with a sampling interval of 1.

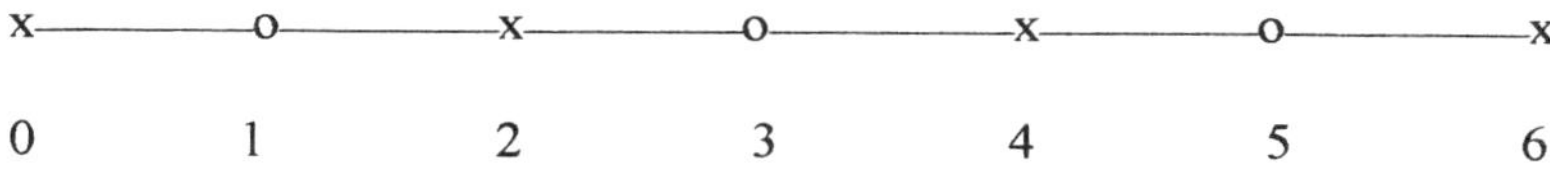

Suppose that we want to keep every other point Z_1, Z_3, Z_5, ... but instead of just dropping Z_0, Z_2, Z_4, ... we form a linear filter involving the value to be kept and its two adjacent neighbors, e.g.

$$\hat{Z}_1 = a\ Z_1 + (1-a)(Z_0 + Z_2)/2$$
$$\hat{Z}_3 = a\ Z_3 + (1-a)(Z_2 + Z_4)/2$$

Then we reconstruct the odd terms such as Z_2 by kriging from its two nearest neighbors. By symmetry this is simply

$$\hat{Z}_2 = (\hat{Z}_1 + \hat{Z}_3)/2$$

Our problem is to determine the weight a so as to optimize the reconstruction of the original sequence Z_0, Z_1, Z_2, ... To this end we must first define an optimization criterion. A natural one is the expected mean square error

$$D^2 = \frac{1}{N} E \sum_i (\hat{Z}_i - Z_i)^2$$

Notice that this criterion involves *all* points and not only the reconstructed ones. For simplicity we will neglect edge effects here so that, approximately

$$D^2 = \frac{1}{2}\left[E(\hat{Z}_1 - Z_1) + E(\hat{Z}_2 - Z_2)^2\right]$$

From the following expression of the errors

$$\hat{Z}_1 - Z_1 = \left[(1\text{-}a)/2\right]\left[(Z_0 - Z_1) + (Z_2 - Z_1)\right]$$
$$\hat{Z}_2 - Z_2 = (a/2)\left[(Z_1 - Z_2) + (Z_3 - Z_2)\right] + \left[(1\text{-}a)/4\right]\left[(Z_0 - Z_2) + (Z_4 - Z_2)\right]$$

and using the orthogonality properties of non-overlapping increments associated with the linear variogram we find

$$E(\hat{Z}_1 - Z_1)^2 = \varpi(1-a)^2$$
$$E(\hat{Z}_2 - Z_2)^2 = \varpi(a^2+1)/2$$

so that

$$D^2 = \varpi(3a^2 \text{-} 4a + 3)/4$$

The weight a_0 that minimizes D^2 is found to be $a_0 = 2/3$. The associated mean square error is ϖ 5/12. The following table (scaled for $\varpi = 1$) show how this compares with other weighting schemes.

Scheme	Weights	Mean Square Error
Optimal filtering	(1/6, 2/3, 1/6)	5/12 = 0.417
Straight decimation	(0,1,0)	1/2 = 0.5
Moving average	(1/3,1/3,1/3)	1/2 = 0.5
'Hanning'	(1/4,1/2,1/4)	7/16 = 0.438

Compared with optimal prefiltering straight decimation increases the reconstruction variance by 20%. It is interesting to notice that straight decimation does as well in this example as a moving average.

The effect is more dramatic in the presence of a nugget effect. Generalizing the above calculations to an arbitrary variogram $\gamma(h)$ we find

$$16D^2 = a^2\left[28\gamma(1) - 12\gamma(2) + 4\gamma(3) - \gamma(4)\right] - 2a\left[14\gamma(1) - 4\gamma(2) + 2\gamma(3) - \gamma(4)\right] + 16\gamma(1) - \gamma(4)$$

$$a_0 = \frac{14\gamma(1) - 4\gamma(2) + 2\gamma(3) - \gamma(4)}{28\gamma(1) - 12\gamma(2) + 4\gamma(3) - \gamma(4)}$$

Consider two cases: (i) a linear variogram with a nugget effect, (ii) a pure nugget effect. The following table shows clearly that decimation results in a higher mean square error than when prefiltering is used.

Variogram	a_0	minimum MSE	decimation MSE	% increase
$\gamma(h)=\delta+h$	19/31	0.96	1.25	30 %
$\gamma(h)=\delta$	11/19	0.539	0.75	39 %

Another generalization is to consider a kriging estimator of Z_2 based on more than two points. Indeed if the variogram of Z is linear it is known that the two adjacent points screen off the influence of all other points. However, after prefiltering the variogram of Z is no longer linear but becomes

$$\gamma_p(0) = 0 \quad \gamma_p(1) = \varpi(2/3 - 5/9) \quad \gamma_p(n) = \varpi(n - 5/9) \quad \text{for } n > 1$$

The kriging estimator of Z_4 based on its four nearest neighbors has the weights (-5/52, 31/52, 31/52, -5/52) i.e. approximately (-0.1, 0.6, 0.6, -0.1). This estimator can be used even if the prefilter has not been optimized for them. Alternatively one could try to optimize the prefilter for these weights but things get more complicated and we will stop here.

To be relevant to image compression the above technique must be extended to 2-D and made recursive so that the compression rate may be adjusted. In 2-D one can consider the 4-point scheme proposed by Yfantis et al. and determine the prefilter that optimizes the reconstruction.

2. Predictive coding

The idea underlying the use of kriging in the context of image compression is to take advantage of the correlation between adjacent pixels to reduce the amount of data required to represent an image. No wonder that an extensive literature exists on similar methods under the generic name of *Predictive Coding*. An excellent account can be found in the tutorial book 'Digital Image Compression Techniques' by Rabbani and Jones (1991) quoted in the references.

If these techniques in essence use Simple Kriging, or Ordinary Kriging, to form a linear estimator from neighboring pixels they place the emphasis on encoding the *differential image* (true - predicted) so as reduce the number of bits to be stored or transmitted. It would have been interesting to visualize and analyze these differential images produced by kriging both to appreciate what is being lost and perhaps how it could be recovered.

Another question discussed relates to the transmission of images as time signals and the necessity for the decoder to use algorithms based on previous pixel values (causal systems). This justifies the wide use of the Markov model with separable covariance function

$$\sigma_{k,l} = m^2 + \sigma^2 \rho_v^{|k|} \rho_h^{|l|}$$

where m and σ^2 are the mean and variance of the image, k and l denote vertical and horizontal displacements, and ρ_v and ρ_h denote vertical and horizontal correlation coefficients, respectively (exponential model).

Rabbani and Jones also report the use of *adaptive prediction* to improve the results at edges where abrupt changes in pixel values occur. In particular there is a method based on switching among a set of predictors based on the most likely direction of the edge. This, and adaptive kriging in general, would be interesting avenues to explore.

3. Optimizing the sampling

An alternative to the decimation method proposed by Yfantis et al. would be to carefully select the points to be retained, rather than selecting them in a regular grid. To illustrate the idea I will refer to a comparison of sampling schemes in the field of topographic mapping (Chilès and Delfiner, 1975). Such application bears resemblance with image processing since a complete reconstruction of topography can be obtained by photogrammetric imaging techniques.

Three sampling patterns were compared: (i) one with data points evenly scattered throughout the studied region; (ii) one with all the high and low points (ridges, peaks, and valleys); (iii) a combination of the above. The first map gave a good reconstruction of the contour lines but the relief did not stand out; the second map gave a poor reconstruction of contour lines but emphasized the relief, even exaggerated it a little; (iii) not surprisingly the third map gave the best results, it was both accurate and interpretable.

Closing the loop with the beginning, all the above suggests that the ideal sampling would be a prefiltered decimated image complemented with selected characteristic points. The implementation details are left for graduate students.

References

Oppenheim, A.V. and Schafer, R.W. (1989). *Discrete-time Signal Processing.* Prentice Hall, Englewood Cliffs, New Jersey.

Rabbani, M. and Jones, P.W. (1991). *Digital Image Compression Techniques*. SPIE Tutorial Text, Vol. TT 7, SPIE press, Bellingham, WA.

Chilès, J.P. and Delfiner, P. (1975). Reconstitution par krigeage de la surface topographique à partir de divers schémas d'échantillonnage photogrammétrique, *Société Française de Photogrammétrie, Bulletin n° 57, 42-50.*

REPLY TO "COMMENTS ON IMAGE COMPRESSION AND KRIGING", by P. Delfiner

E. A. Yfantis
Computer Graphics and Image Processing Laboratory
Computer Science Department
University of Nevada, Las Vegas, 89154

Our reply to Dr. Delfiner's comments are in the same positive spirit as his comments.

1. ANALOG TO DIGITAL IMAGE CONVERSION AND ANTIALIASING

A computer image is digital with resolution the resolution imposed by the hardware. Computer images are generated either by scanning an analog image, or by using a camera as input device, or by drawing an image to a pad, or several other means. Appropriate hardware firmware and software is available today by manufactures for converting an analog image to digital. It is well recognized that such a conversion could create aliasing, a phenomenon whereby lines have a staircase syndrome (here by lines we mean straight lines and other lines), and pixels representing small areas of the original analog picture, have undesirable discontinuities. In order for these undesirable discontinuities which are referred to as aliasing, or the phenomenon of aliasing, to be corrected special mathematical filters are applied. These filters take into consideration the fact that pixels are autocorrelated. The filters replace the intensity, or color and intensity, of the pixel with a new color and intensity. The phenomenon of aliasing is common not only when one converts images from analog to digital by inputing them in the computer by some method, but also by using mathematical methods to draw lines or surfaces in the computer, since our mathematical methods usually have as a domain and range a continuous space, but we use the computer screen, which is limited by the number of pixels available and therefore is very discrete, to represent these continuous phenomena. So aliasing is the staircase phenomenon or the undesirable discontinuities created by mapping continuous images into a discrete space. Antialiasing is a mathematical method used to correct for aliasing. Antialiasing methods usually use the values of neighboring pixels to calculate a new value for a pixel. In the space domain the process of antialiasing results in a convolution which if transformed by a Z-transform, or a Fourier transform it becomes a product. Thus it is easier from both the algebraic and computational point of view to deal with antialiasing in the frequency domain, rather than the space domain. The interesting reader could find this information in any signal processing or computer graphics book published after the mid-eighties. What we like to impress upon our readers here is that the antialiasing is a process of improving a digital image and this process has nothing to do with compression, at least as far as we are concerned and the way we use compression in our lab. We hope that our

R. Dimitrakopoulos (ed.), Geostatistics for the Next Century, 168–170.

reader agrees that if we continue to use antialiasing in the already antialiased image after a few repetitions all the pixels will converge to the same value.

2. COMPRESSION AND SURFACE ESTIMATION

Surface estimation is the process whereby given a set of points in the three dimensional space one is to estimate the surface the points belong to. There is only one surface the points belong to, where the estimates could be many and they depend on the method used and often times on the user of the method and or software. The methods for surface estimation could be classified as stochastic (and these include kriging), deterministic (various numerical methods, splines etc.), hybrid methods (use deterministic methods to reparametrize the domain, and stochastic to estimate the range), chaotic (various fractal methods that calculate high frequency components), and hybrid methods that could employ any combination of chaotic with deterministic and or stochastic.

The problem of lossy compression is in someways a reverse problem to that of surface estimation. Here we have the image but we recognize that there is a lot of redundancy. We restrict our discussion in gray-scale images and one frame, as we did in our short presentation and short paper. Therefore the redundancy is spatial, and the degree of redundancy is expressed by the autocorrelation or semivariogram. Thus high correlation between neighboring pixels of the image implies that they carry similar information. Antialiasing has as a side effect the increase of the autocorrelation. Notice here that since we are dealing with continuous-tone gray-scale images and only one frame, we are not concerned with spectral redundancy (due to the correlation between color planes or spectral bands), or temporal redundancy (due to the correlation between neighboring frames in sequences of images). In all lossy algorithms the idea is to eliminate the redundant information. Philosophically this makes sense but in the application it is difficult to eliminate redundancy without compromising the quality of the image, considering that we would like our compressed image to occupy relatively small computer memory space. One approach is to consider the spectrum of the image and eliminating the high frequencies but this implies greater Δt, therefore decimation. We give a theoretical explanation for this last statement using the one dimensional spectrum for simplicity. It is known that if $[0, F]$ is the frequency domain then $F = \frac{1}{\Delta t}$, or $\Delta t = \frac{1}{F}$. Hence if $\Delta t = 1$, then $F = 1$. Now if we eliminate the high frequencies and F becomes less than one, for example if $F = \frac{1}{3}$, then $\Delta t = 3$. This very last idea was the motive of using kriging as a compression method. The Δt was chosen to be much smaller than the zone of influence. Fluctuations of the values of neighboring pixels correspond to high frequencies. As we stated in our talk the image quality obtained for relatively high compression ratios were better than the JPEG (Joint Photographic Experts Group) which is used as a standard for still picture compression. Kriging pulls information from an omnidirectional neighborhood of pixels or points with known information in order to estimate a point or pixel. Thus it takes into consideration the relationship or dependence of the point with all the points in the neighborhood. If one estimates the unknown pixel or point based on pixels or points along a line then due to the unidirectional nature of the information and also due to the limited amount of information used, the estimate has some undesirable properties. One of them is aliasing of the image. Kriging as a compression algorithm is similar to the autoregressive linear compression algorithms found in the literature. If performed in the space domain, is computationally intensive and therefore very slow. In order to speed up the process computationally we considered a linear semivariogram with sliding neighborhoods. The purpcse of this work was mainly to express our idea or belief that surface estimation and lossy image compression are very related problems. For scientists who share this with us we hope that we can work together towards a unified theory, for the ones they disagree controversy is healthy and necessary for the growth of sciences.

CONDITIONAL SIMULATIONS

THE USE OF NEURAL NETWORKS FOR SPATIAL SIMULATION

P.A. Dowd
Department of Mining and Mineral Engineering
University of Leeds, Leeds LS2 9JT, U.K.

This paper describes current research work in the application of neural networks to conditional spatial simulation. A number of possible approaches are suggested and an example simulation using a feedforward, error back propagation network is shown. The advantages and disadvantages of the neural network approach are discussed and some possible directions for further work are suggested.

INTRODUCTION

In recent years the interest in geostatistical simulation has been rekindled and several alternatives to the original turning bands method have been proposed (Dowd, 1992). Almost all methods proposed to date are algorithmic, ie they consist of a specific set of instructions for the solution of the problem. Neural networks offer two basic, non-algorithmic approaches to conditional simulation. The first, and least controversial, is via the use of suitable networks for optimisation of energy functions or loss functions. The most obvious applications here are neural network implementations of techniques such as the simulated annealing method of conditional simulation (Farmer, 1991, Deutsch and Journel, 1992). The second approach is via the pattern recognition abilities of neural networks and offers the possibility of automatic recognition and reproduction of inherent spatial structures without explicit fitting of variogram models to conditioning data.

One inherent problem with algorithmic approaches to simulation is that a model must be fitted to the experimental variograms which involves many unverifiable assumptions. Often there are too few data to infer a reliable model. The assumed model, with all of its inherent errors, is propagated through the simulation and the reproduction of the model in the simulation is then used as one of the criteria for assessing the simulation. A related problem is the robustness of the simulation to the choice of a model.

Simulated annealing is an example of a non-algorithmic approach although a model variogram must be specified. Neural nets offer the possibility of direct simulation from the conditioning data without the intermediate steps of model fitting. The neural net is taught to recognise the inherent correlation structure encoded within the conditioning data and then reproduce a set of values with this same structure; they also offer the possibility of incorporating constraints directly into simulations and they provide an alternative implementation of the simulated annealing method.

R. Dimitrakopoulos (ed.), Geostatistics for the Next Century, 173–184.

This paper deals with preliminary, experimental approaches and considerable testing and further development work are required to determine whether neural networks can become an acceptable, additional method of conditional simulation. The paper contains more ideas and suggestions than concrete developments and as such is presented in the intended spirit of the forum as a discussion of new ideas and current research.

NEURAL NETWORKS

Neural networks is a term used to describe models which use certain basic structures of the human brain to imitate some of its functions. The models are based on the assumption that the brain's learning units or **neurons** are organised into a hierarchy. Each level in this hierarchy, beginning with the lowest and progressing to the highest, receives inputs which it processes to yield outputs; these outputs then become the inputs to the next layer. The outputs increase in sophistication as they move up the hierarchy. The organised hierarchy of neurons is called a **neural network** or a **neural net**.

To use a computing analogy, the brain is seen as a parallel, distributed processing system in which the basic processing unit is a neuron. These neurons are connected to one another to form a neural network. In computing applications the neurons can be connected in many different ways, with each interconnection defining a different neural network. The three phases in a neural net are : Training, Generalisation and Operation.

Training is the process by which the network "learns" to "recognise" input patterns in the form of a training set of data. Each neural network is accompanied by a set of rules which define how the neural network learns.
Generalisation is the ability of a network to give an acceptable response for inputs which are not members of the training sets.
Operation is the use of the network to perform the tasks for which it was designed.

The first work in the development of neural networks was done in the 1940s by McCulloch and Pitts (1943) although the major advances in the development of practical networks were made in the 1980s; see in particular Hopfield (1982), Hopfield and Tank (1985), Kohonen(1982,1990), Kosko(1988). Much of the work in neural networks has concentrated on pattern recognition (in particular, recognition of handwriting or speech) but there has recently been an explosion of interest in the subject which has led to many new applications including combinatorial optimisation. It is the combination of pattern recognition and optimisation that is of interest in the applications described here.

Neurons

A neuron is a basic cell in the network and operates as a processing unit. It receives any number of inputs and delivers a single output. The neuron consists of an input function, or summation function, the result of which is fed to a transfer function which in turn determines the output. Each input is weighted and summed; this weighted sum is then passed through the transfer function which generates an output value. The basic neuron structure is shown in figure 1.

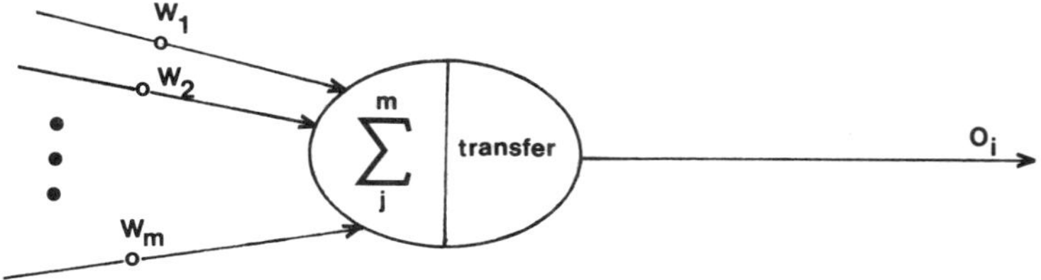

Figure 1 : Basic neuron or processing unit

In general, each input to the network consists of a sequence of components, or measurements, which are usually specified as a vector. For example, if the network is to be trained to recognise the variogram structure inherent in a conditioning data set then each input pattern might consist of two components : the magnitude and orientation of the vector distance between a pair of data; if there are n conditioning data there will be $\binom{n}{2}$ input patterns each consisting of two components. Similarly, each output from the network consists of a vector of components; in the variogram recognition example, the output might only contain one component : the squared difference between the conditioning data values corresponding to the components of the input distance vector.

Suppose that I_{ij} is the jth component of the ith input pattern; O_i is the output of the neuron corresponding to the ith input pattern, w_j is the weight connecting the jth input component to the neuron. The neuron computes the weighted sum of the input pattern values :

$$\sum_j w_j I_{ij}$$

The output from neuron i is then determined by the transfer function :

$$O_i = f\left(\sum_j w_j I_{ij}\right)$$

where f(x) is either a discontinuous step function or a continuously differentiable function defined on an interval (most commonly a hard limiting threshold for the former and a sigmoidal function for the latter). This function determines the state of the neuron.

Training

There are three types of training : supervised, reinforced or unsupervised. In supervised training the network is taught to match training data with corresponding specified outputs. Reinforced training is a form of supervised training in which the precise desired output is replaced by a grading scheme which reports on the performance of the network as it trains. Unsupervised training refers to networks which form their own classifications of the training data; these networks are also called self-organising.

Network architecture

The network architecture is the manner in which the neurons are organised and connected, ie the topology of the network. Neurons can be combined to form layers. Layers can be connected in any number of ways but two general types of connection are distinguished : partial and full. A *fully connected* network is one in which the output from every neuron in one layer is connected to every neuron in the next layer; all other networks are *partially connected.* The strength of a connection between two neurons is given by the connection weight between them. Three types of layer are distinguished : the input layer which accepts the input pattern to the network, the output layer which delivers the network output or response and all intermediate layers or *hidden layers* which represent hidden features of the problem. An example is shown in figure 2.

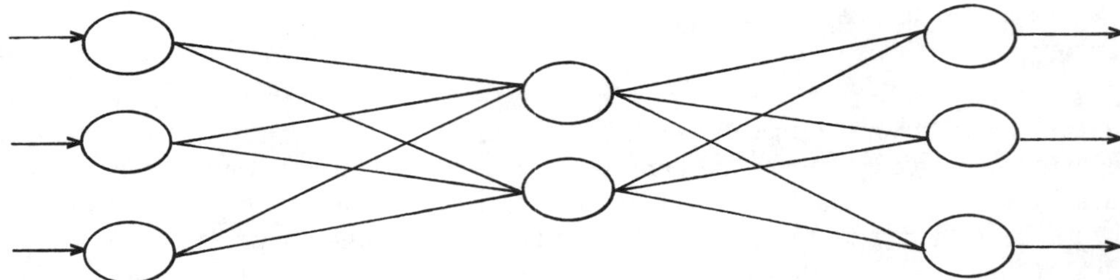

Figure 2 : layers in networks

Feedforward networks are those in which the output from a neuron can only feed forward, ie it cannot become the input for a neuron in the same or any preceding layer. If the network is interpreted as a graph then a feedforward network is one which has no closed paths. Conversely, *feedback* networks are those in which the output from a neuron can be directed back as an input to any neuron in the same or a preceding layer.

Learning rules

Learning rules are based on minimising a criterion function, eg in supervised training networks minimising the difference between training sets and the desired output for each set. There are two common groups of learning rules. The first are derived from differential calculus and are essentially variations of the *gradient descent rule*; the second are derived from statistical mechanics and are essentially forms of simulated annealing.

The most commonly used learning rules for supervised training networks are variations of the so-called *delta rule*. Delta is defined as the difference between the current output of a neuron and the desired output; the connection weights of the network are continuously adjusted so as to reduce the value of delta to a minimum. The basic delta rule is a *least mean squares* learning rule and is one example of the *gradient descent rule* which is an inefficient optimisation method for the non-convex optimisation required in general formulations. Nevertheless it is widely used in neural networks although several alternatives, such as simulated annealing, are used in specific networks.

Suppose that the desired output match for input pattern i is D_i. The mean squared recognition error over all input patterns is :

$$\epsilon = \sum_i [D_i - f(\sum_j w_j I_{ij})]^2$$

Minimising this mean squared error by gradient descent is equivalent to modifying the weights iteratively from an initial value according to the rule :

$$w_j(new) = w_j(old) + \alpha \frac{\delta \epsilon}{\delta w_j}$$

where α controls the rate of modification and is sometimes called the learning rate.

Multi-layered networks use a *generalised delta rule* which is obtained by using the chain rule of calculus to derive a general form of weight modification over the layers of the network. The error is used to adjust the weights which connect the neurons in the input layer to those in the hidden layer(s) and the neurons in the hidden layer(s) to those in the output layer rather than just the weights which connect the neurons in the input layer to those in the output layer. The error is still defined as the least mean squared error.

The mean squared error can also be regarded as an energy function or surface on which a global minimum is sought. Although gradient descent is the most commonly proposed method of finding the minimum point on the energy surface other methods such as simulated annealing are also proposed to overcome the well known difficulties inherent in gradient descent for complex surfaces.

Back-propagation

Back-propagation (Rumelhart et al 1986a, 1986b) networks have a hidden layer and use a generalised form of the delta rule. Error is calculated in the output neurons and then passed back through the hidden layers of the network to the input so that the connection weights on all previous neurons can be altered. For this reason the network is called a back-propagation network. Learning is achieved in three phases :

phase 1 : forward propagation - input training set, proceed through network from layer to layer applying weights, calculate output

phase 2 : calculate total error

phase 3 : back propagation - error is passed back through net causing each connection weight to be modified

Repeat the sequence with the next training set.The network converges when the error is below some specified value.

Learning consists of minimising the squared error on each output for all training sets where the error is defined as a function of the connection weights. Minimisation is usually done by gradient descent but this could be replaced by non-convex optimisation methods. There is no formal proof of convergence for back-propagation networks.

Designing a back propagation network

An accurate mapping is required between the input and the output to ensure that the network will function correctly in both the learning and the operational phases. Given that the particular application fixes the input (eg, spatial location and directed distance between a pair of conditioning data) and the output (eg, value at spatial location and squared difference between corresponding values of data pair) designing the network consists of deciding on the number of hidden layers and the number of neurons in each of these layers. If there are insufficient hidden neurons then in the learning phase the optimising algorithm may not converge or may not yield an accurate mapping. If there are too many hidden neurons there will be too many different solutions and in the operational phase the network will be unable to generalise correctly for new input data. If there is a hidden neuron for each input pattern the network will learn to associate the two. However, this would generally lead to prohibitive computing times and in any case the aim is recognise features in the input rather than respond to each exact input pattern.

There are two approaches to this problem. The first is an *ad hoc* approach which begins by training a network with more hidden layers and neurons than necessary. After training, the activity in each connection is examined and all connections which carry little or no weight are removed; this may entail the removal of entire neurons. The second approach is to generate the architecture automatically. The network begins with no hidden layers; after training a hidden neuron is added and the network is retrained; this procedure is repeated until the performance of the network is optimal, ie the target error has been reached. This latter approach falls into the general area of *genetic algorithms* and specifically Dynamic Neuron Creation (Ash, 1989, Fahlman and Lebiere, 1990). The growth of the network must follow a set of rules so as to guarantee convergence after the creation of a finite number of neurons. The entire network must be retrained after the addition of each neuron. Training is by back propagation and mean squared deviation D(t) between actual and desired output decreases with time t. When no solution can be found the rate of decrease slows dramatically before D(t) has reached the desired error tolerance; a new neuron is then added. The criterion for addition of a new neuron is :

$$\frac{|D(t+\Delta t) - D(t)|}{D(t_o)} < \epsilon$$

where Δt is the interval over which the slope of the error curve is calculated
t_o is the time at which the previous neuron was added
ϵ is slope of the error curve which activates neuron creation

To ensure that at least Δt training steps are taken $t + \Delta t \geq t_o$

Hopfield networks

Hopfield networks are fully interconnected networks : each neuron is connected to every other neuron and there are no obvious input or output layers. The weights on the connections between the neurons are the same in both directions and the Hopfield net is thus said to be a *symmetrically weighted* network. This network uses the principle of dynamic feedback in which the output from each neuron is sent as input to every other neuron; training is achieved when the network becomes stable, ie the output at each neuron is equal to its input. In the simplest form of Hopfield network there is a single layer of neurons and all weights are fixed by storing all input patterns and calculating a correlation matrix between the inputs. As the weights are fixed there is no training phase.

One of Hopfield's contributions was the recognition that the state of the network at any time t can be characterised by an energy function which satisfies the criteria of a Lyapunov function thereby guaranteeing global stability of the network and hence convergence of its dynamic processes. Hopfield (Hopfield and Tank, 1985, Davalo and Nain, 1991) also recognised, via a dual formulation of the problem, that local minimisation of a symmetric quadratic energy function could be implemented by a Hopfield net in which the connection weights are the coefficients of the energy equation. To obtain a global minimisation rather than a local one Ackley et al (1985) proposed the use of hidden units in a Hopfield network together with a simulated annealing (with asynchronous random updating) learning rule; the resulting network is known as a *Boltzmann machine*. A further refinement is the Cauchy machine (Szu, 1987a, 1987b) which uses simulated annealing with random walks and flights to generate new states. Current work at Leeds is aimed at Boltzmann and Cauchy machine implementations of the simulated annealing method of conditional simulation.

The major difference between the Boltzmann machine and the back propagation nets is that the former learns by optimising an energy function whereas the latter uses a generalised delta rule; ie one is statistical and the other is derived from calculus.

APPLICATION

The application has been limited to isotropic simulation in two dimensions. There are n+2 (n = 1, 2 or 3 dimensions) input components in each input pattern : the location of the conditioning data value, the distance between the location of the conditioning data value and the locations of one other conditioning data. There are two components in each training response pattern : the conditioning data value and the value of the experimental variogram corresponding to the input data pair. The network is trained to recognise the relationship between directed distance and experimental squared difference of the corresponding data values as well as the relationship between location and data value and possibly the relationship between location and experimental variogram value.

The advantage of this approach is that there is no need to fit a variogram model and to make the assumptions entailed by the fitting of such a model. Of course there must be sufficient conditioning data to impart the implicit variogram (and other spatial structures) to the simulated values. In essence it is assumed that the variogram model is encoded in the conditioning data. If it is required to impose a specific model, perhaps in cases where there are too few conditioning data at close distances for small scale (eg nugget variance) variogram structures to be encoded, dummy values could be specified. Alternatively, it may be possible to use analogues in the form of previously trained neural nets.

Because of the nature of the transfer functions used inputs for simulation applications must be scaled on the interval [0,1] and output rescaled on the correct interval.

An example

This example uses the data set given in David (1988) page 156*ff*. A total of 77 intersections with a copper vein are recorded over an area of 1800ft x 1300ft. The data values are approximately normal with mean 1.40%Cu and standard deviation 0.28%Cu. An isotropic variogram model with parameters $C_o = 0.035(\%)^2$, $C = 0.040(\%)^2$ and range 600ft, was fitted to the noisy experimental variograms. David used the turning bands method to provide a conditional simulation of 972 values on a 50ft x 50ft grid.

The simulation yielded by an error back-propagation, feedforward network with two hidden layers and 12 nodes in each layer is shown in figure 3. Several dummy values were specified so as to impart variogram behaviour at short distances; without these values the simulation appears to be random but not unlike the structures depicted by the experimental variograms which begs the question of whether the short range structures should have been imposed. Whilst the simulated values conform broadly to the expected pattern they appear to be too smooth in comparison with the conditioning data although this may simply be an artefact of this specific implementation. It is also not clear whether this is a simulation or a particular form of estimation; for example, the network as formulated will not yield multiple simulations from the same conditioning data set.

OTHER POSSIBILITIES

The networks discussed above are largely autoassociative networks and use supervised learning. An autoassociative network stores a pattern by associating it with itself. However, for geostatistical applications, we are more interested in heteroassociative networks, ie networks which store a pattern by associating it with a different pattern. An example of heteroassociation would be an input pattern consisting of spatial location and an output pattern consisting of grade value. Many autoassociative networks can be converted to heteroassociative networks perhaps by adding another layer. Explicitly heteroassociative networks are exemplified by the Adaptive Resonance Theory network. One of the best known networks using unsupervised learning is the Kohonen network.

Kohonen network

The backpropagation network uses supervised learning, ie it depends on a training response for each input in the training set. The Kohonen *self organising map* or *topology preserving map* or *feature map* is an example of an unsupervised learning method (Kohonen, 1982, 1990). The Kohonen net comprises two layers : an input layer and the Kohonen layer. All neurons in the input layer are connected to all neurons in the Kohonen layer; in addition each neuron in the Kohonen layer is connected to all other neurons in the layer that are within a specified neighbourhood and feedback is restricted to these lateral interconnections. There is no separate output layer : all neurons in the Kohonen layer are output neurons. An example of a Kohonen net is shown in figure 4

Every input component is connected to every neuron in the Kohonen layer; each neuron has the same dimensionality as the input vector. If w_{ij} is the weight of the connection between input neuron i and Kohonen neuron j then each neuron j has a vector $\{w_{1j}, w_{2j}, \ldots, w_{nj}\}$. Initially weights are set to small random values. An input vector is presented

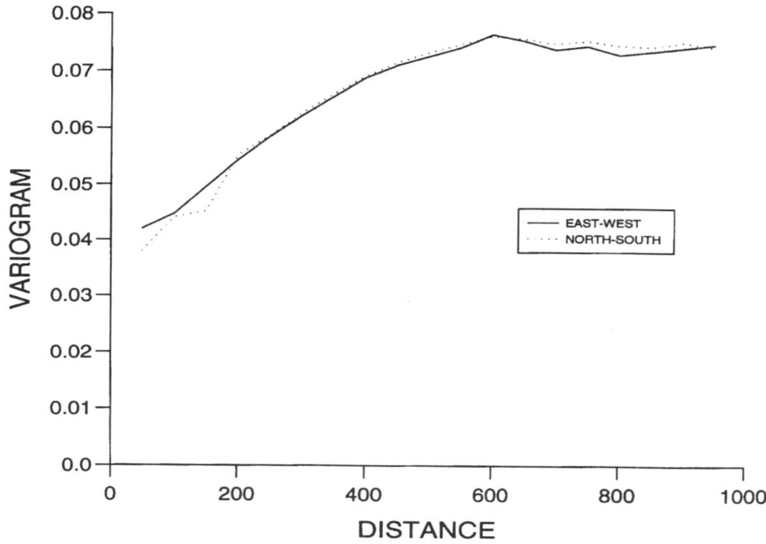

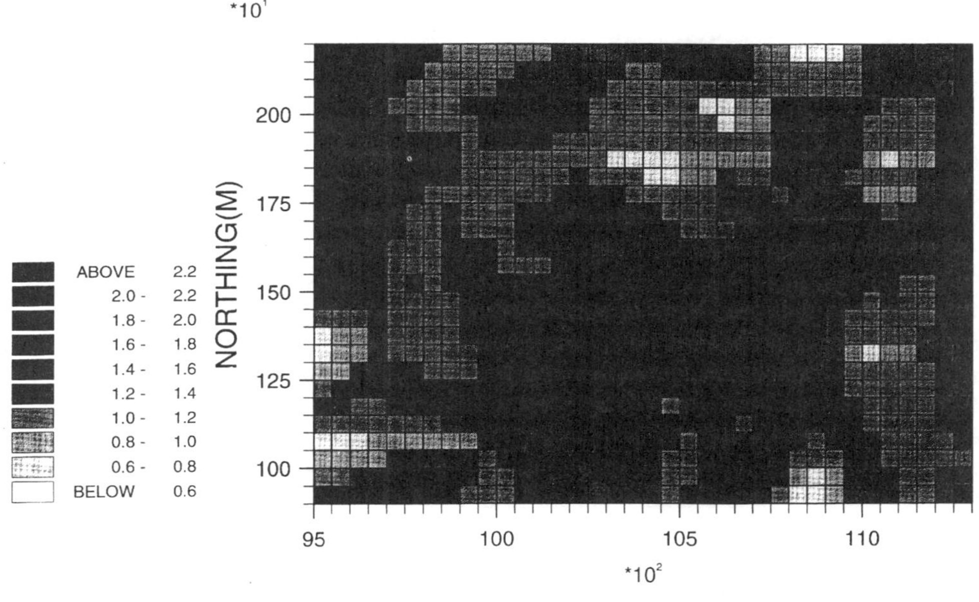

Figure 3 : Simulation results

to the net and the neuron with the weight vector closest (LMS error) to the input vector is chosen. The weight vectors of all neurons within a defined neighbourhood of the chosen neuron are adjusted. Over time the neighbourhood is reduced in equal steps.

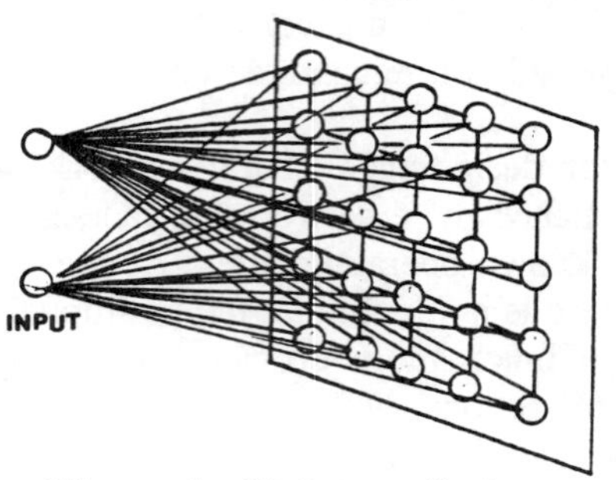

Figure 4 : Kohonen feature map

For spatial simulations the input vector would be the co-ordinates and value of the conditioning data. During the learning phase the relative positions of the neuron vectors are adjusted using the vector distance between them as the adjusting mechanism; this leads to a topographical ordering of the neuron vectors. However, it is unclear at this stage whether we would be simulating or achieving some form of estimation.

Adaptive Resonance Theory

Once multi-layered networks have been trained on a given training set they frequently are unable to adapt to new training data without altering previously set weights and thereby destroying the earlier learning. This can be extremely wasteful in terms of computing time. Adaptive Resonance Theory (ART) (Grossberg, 1988) uses a self-organising network which, once trained, can accept new training data without any detriment to previous learning. This is achieved by distinguishing between two distinct learning states: one in which modifications to the network can be made and one in which they cannot.

CONCLUSIONS : SUCCESSES AND PROBLEMS

Neural networks offer the possibility of a non-algorithmic approach to conditional simulation. Although a function is used to determine the output of each neuron in a network there is no specific set of instructions for the solution of a problem expressed as a neural network and the method is thus non-algorithmic. Unlike the existing algorithmic methods the neural network approach raises many questions most of which we are at present unable to answer. For example :

> How can the performance of the method be measured when there is no input explicit model against which the simulation results can be assessed ?
>
> What is the random component of the method which will allow multiple simulations from the same initial data set ?

How do we know when a solution has been reached ? Convergence to a minimum cannot be guaranteed especially with techniques such as gradient descent.

How do we know whether a solution is correct ? Convergence to a solution when an input pattern is presented is no guarantee that the solution is correct.

There are no general rules for designing a neural network and no obvious way of predicting outcomes or performances of a network. In fact, networks are often viewed as 'black boxes' which use unknown operating rules. At present networks can only be verified by trial and error. There is no doubt that the traditional algorithmic methods of simulation currently provide a much more efficient approach. However, the neural network approach has certain attractions such as :

- possible avoidance of the assumption of a particular variogram model and the potential ability to reproduce **all** spatial structures inherent in the conditioning data without explicitly recognising and modelling them
- inclusion of constraints within the simulation
- alternative methods of implementing simulated annealing; neural networks are very well suited to problems of large combinatorial optimisation (in particular, the Hopfield and Kohonen networks)

To date we have tested fully connected feedforward networks with back propagation of errors; lower or higher connectivity have not been investigated. Results show promise but progress is slow and more work remains before we have a general, practical simulator.

Computing time required for training and operating networks may be prohibitive for large simulations. This is compounded by the inability of some networks to accept new training data without detriment to previous learning. Neural networks are ideally suited to parallel processing and their implementation by simulating parallelism on sequential machines has an obvious limit in terms of processing time for any realistically sized problem.

Many of the computing and design problems can be overcome by the use of specialist languages (see, for example Korn, 1991). In addition there are now several software packages available for neural network design and operation; for example, NeuralWorks Professional II marketed by NeuralWare Inc. Most published source codes, available in several introductory texts (see for example, Müller and Reinhardt) are written in C.

Wu and Zhou (1993) present a method of estimating grade values as a function of spatial location using a feedforward, error back-propagation network. On the data set used the results are plausible and for one cut-off are similar to those obtained from kriging. However, in this application and those described in this paper, the network will "learn" a particular pattern of association, ie it will ultimately use, or assume a model which may appear to work well on a training set but there is no guarantee that it can be generalised.

REFERENCES

Ackley, D.H., Hinton, G.E. and Sejnowski, T.J. (1985). A learning algorithm for Boltzmann machines. *Cognitive Science*, **9**, pp 147-169.

Ash, T. (1989) Dynamic node creation in back-propagation networks. ICS Report 8901, San Diego.

David, M. (1988) Handbook of applied geostatistical ore reserve estimation. Developments in Geomathematics No. 6, Elsevier, Netherlands, 216p.

Davalo, E. and Naim, P. (1991) Neural networks. MacMillan Press, London. 145pp

Deutsch, C. and Journel, A.G. (1992) GSLIB : Geostatistical software library and user's guide. Oxford University Press, New York.

Dowd, P.A. (1992) A review of recent developments in geostatistics. *Computers and Geosciences*, **17**, no. 10, pp 1481-1500.

Fahlman, S.E. and Lebiere, C. (1990) The cascade-correlation learning architecture. in Touretetzky, D. (ed.) *Advances in neural information processing systems*, **2**, Morgan Kaufmann, San Mateo, USA. pp 524-532.

Farmer, C. (1991) Numerical rocks. in Fayers, J. and King, P. (eds.) The mathematical generation of reservoir geology. Oxford University Press, New York.

Grossberg, S. (1988) Neural networks and natural intelligence. MIT Bradford Press.

Hopfield, J.J. (1982) Neural networks and physical systems with emergent collective computational abilities. *Proc. National Academy of Sciences, USA*, **79**, pp 2554-2558.

Hopfield, J. J. and Tank, D.W. (1985) Neural computation of decisions in optimisation problems. *Biological Cybernetics*, **52**, pp 141-152.

Kohonen, T. (1978) Associative memory - system theoretical approach. Springer, Berlin.

Kohonen, T. (1982) Self-organised formation of topologically correct feature maps. *Biological Cybernetics*, **43**, pp 59-69.

Kohonen, T. (1990) Self-organisation and associative memory. Springer-Verlag. Berlin.

Korn, G.A. (1991) Neural network experiments on personal computers and workstations. MIT Press, Bradford Book. 248pp.

Kosko, B. (1988) Bidirectional Associative Memories. *IEEE Transactions on Systems, Man and Cybernetics*, **18**, no. 1.

McCulloch, W.S. and Pitts, W. (1943) A logical calculus of the ideas immanent in nervous activity. *Bulletin of Mathematical Biophysics*, **5**, pp 115-133.

Müller, B. and Reinhardt, J. (1991) Neural networks : an introduction. Springer, Berlin.

Rumelhart, D.E., Hinton, G.E. and Williams, R.J. (1986a) Learning representations by back-propagating errors. *Nature* **323**, p 533.

Rumelhart, D.E., Hinton, G.E. and Williams, R.J. (1986b) Learning internal representations by error propagation. in McClelland, J.L. and Rumelhart, D.E. (eds.) (1986) *Parallel distributed processing : explorations in the microstructures of cognition.* MIT Press. vol 1, p318.

Szu, H. (1987a) Fast simulated annealing. in Denker, J. (ed.) *Neural networks for computing. AIP Conference.* **15**, pp 420-425, Snow Bird publications, USA.

Szu, H. (1987b) Non-convex optimisation. *SPIE* **968**, pp59-65.

Wu, X. and Zhou, Y. (1993) Reserve estimation using neural network techniques. *Computers and Geosciences*, **19**, 4, pp 567 - 576.

JOINT CONDITIONAL SIMULATIONS AND FLOW MODELING

ALLAN GUTJAHR, BRYAN BULLARD and SEAN HATCH
New Mexico Institute of Mining and Technology
Department of Mathematics
Socorro, New Mexico 87801
USA

Joint conditional simulations of two related stochastic fields is of importance in many cases. This is especially the case in groundwater hydrology where permeability and head, for example are related via a partial differential equation.

In this paper we show results for travel times from a joint conditioning procedure that combines both analytical and numerical models. A linearized model is used to develop spectral connections between transmissivity and head fields and the conditional simulation procedure is implemented by using a Fast Fourier Transform field generator. Comparisons with some exact results and test cases are presented.

INTRODUCTION

In many applications of geostatistics it is important to examine the relationship between two or more random fields. For example, the cross-correlation between permeability and retardation or absorption may significantly affect flow and transport in a given region. Similarly the cross-correlations between geochemical element abundances may be useful for prediction of ore reserves. To predict uncertainty of element abundances one could use simulations that incorporate these relationships. In addition to unconditional simulations, conditional simulations are even more useful since they include the measured data. In the current geostatistical literature there has been very little work on such joint conditioning.

While there are several algorithms for generating stationary random fields with prescribed covariance or variogram structure, procedures for joint generation are less well developed. One method is to express each random field as a linear combination of a common set of independent random fields [1] though this leads to a rather restricted set of possible interactions. For a review of spatial

R. Dimitrakopoulos (ed.), Geostatistics for the Next Century, 185–196.

simulation methods see Luster [2]. One could also use matrix decomposition but in that case questions about the validity of possible cross-covariances become crucial. Simply stated it is not easy to check whether a given function is a valid cross-covariance function.

We have developed several algorithms based on spectral representations which can be used for joint generation and conditioning. In the first part of the paper a general spectral algorithm is proposed and some simple examples are presented. In the second part of the paper a more specific model with conditioning is presented and applied. The final section discusses the extension to the general spectral conditioning algorithm.

GENERAL SPECTRAL ALGORITHMS

The method described was explored previously [3,4] but the algorithm given here allows for more general transfer relations. Recall that for jointly second order stationary fields, $V_1(\underline{x})$ and $V_2(\underline{x})$, with covariance structure

$$cov\left[V_j(\underline{x}+\underline{\xi})\ ,\ V_k(\underline{x})\right] = C_{jk}(\underline{\xi}) \tag{1}$$

the auto and cross spectral densities are

$$S_{jk}(\underline{u}) = \int_{-\infty}^{\infty} e^{-i2\pi\underline{u}\cdot\underline{\xi}} C_{jk}(\underline{\xi})\ d\underline{\xi} = \alpha_{jk}(\underline{u}) - i\,\beta_{jk}(\underline{u}) \tag{2}$$

Where $\alpha_{jk}(\underline{u})$ is the co-spectrum and $\beta_{jk}(\underline{u})$ is the quadrature spectrum.

Related quantities that we use are, respectively, the coherency squared and phase

$$\rho^2_{jk}(\underline{u}) = |(S_{jk}(\underline{u}))|^2 \ /\ (S_{jj}(\underline{u})S_{kk}(\underline{u})) \tag{3a}$$

$$\theta_{jk}(\underline{u}) = \arctan\left[-\beta_{jk}(\underline{u})\ /\ \alpha_{jk}(\underline{u})\right] \tag{3b}$$

We start with the spectral representation for two fields [5]. The four spectral components at frequency $\underline{u}$ which consist of real and imaginary parts are treated as a single vector, $[dZ_{R1}(\underline{u}), dZ_{I1}(\underline{u}), dZ_{R2}(\underline{u}), dZ_{I2}(\underline{u})]'$ which has a mean of $\underline{0}'$ and covariance matrix

$$\Gamma = \begin{bmatrix} S_{11}(\underline{u})d\underline{u}/2 & 0 & \alpha_{12}(\underline{u})\ d\underline{u}/2 & -\beta_{12}(\underline{u})\ d\underline{u}/2 \\ 0 & S_{11}(\underline{u})\ d\underline{u}/2 & \beta_{12}(\underline{u})d\underline{u}/2 & \alpha_{12}(\underline{u})d\underline{u}/2 \\ \alpha_{12}(\underline{u})d\underline{u}/2 & \beta_{12}(\underline{u})d\underline{u}/2 & S_{22}(\underline{u})d\underline{u}/2 & 0 \\ -\beta_{12}(\underline{u})d\underline{u}/2 & \alpha_{12}(\underline{u})d\underline{u}/2 & 0 & S_{22}(\underline{u})d\underline{u}/2 \end{bmatrix} \tag{4}$$

Across frequencies all terms are uncorrelated. In the approach here multivariate normality is assumed and thus the conditional distribution of $[dZ_{R2}(\underline{u}), dZ_{I2}(\underline{u})]'$ given $[dZ_{R1}(\underline{u}), dZ_{I1}(\underline{u})]'$ is again multivariate normal [6] with mean

$$\Gamma_{21}\ \Gamma_{11}^{-1}\ [dZ_{R1}\ (\underline{u}),\ dZ_{I1}\ (\underline{u})\ ]' \tag{5a}$$

and covariance matrix

$$\Gamma_{22} - \Gamma_{21}\ \Gamma_{11}^{-1}\ \Gamma_{12} \tag{5b}$$

where

$$\Gamma_{jj} = \begin{bmatrix} S_{jj}(u)d\underline{u}/2 & 0 \\ 0 & S_{jj}(\underline{u})/2 \end{bmatrix} \tag{6a}$$

$$\Gamma_{12} = \begin{bmatrix} \alpha_{12}(\underline{u})d\underline{u}/2 & -\beta_{12}(\underline{u})d\underline{u}/2 \\ \beta_{12}(\underline{u})d\underline{u}/2 & \alpha_{12}(\underline{u})d\underline{u}/2 \end{bmatrix} \tag{6b}$$

and $\underline{\Gamma}_{21} = \underline{\Gamma}'_{12}$.

The diagonal elements of 5(b) are $S_{22}(\underline{u})[1-\rho_{12}^2(\underline{u})]d\underline{u}/2$ while the off-diagonal elements are zero. Namely, given the set of dZ's for the first field, the dZ's for the second field are independent random variables with a variance depending on its spectra and coherence and means depending on the first set of dZ's. This suggests the following algorithm:

1) Generate the $dZ_1(\underline{u})$ components as for field 1.
2) Generate $dZ_2(\underline{u})$ components given the $dZ_1(\underline{u})$ values with the indicated mean and variance relations.
3) Transform the dZ's using a FFT.

In this procedure the coherence and phase enter directly. A covariance approach would require certain conditions on the cross-covariances. In general, it is not easy to decide whether a proposed function can be cross-covariance. In this approach we start with the coherence and phase which are easier to check for validity and have a physical interpretation.

Figure 1

Cross-covariances for Two Examples
No Phase Difference
Exponential covariance and coherence
Solid; observed : Dashed; theoretical

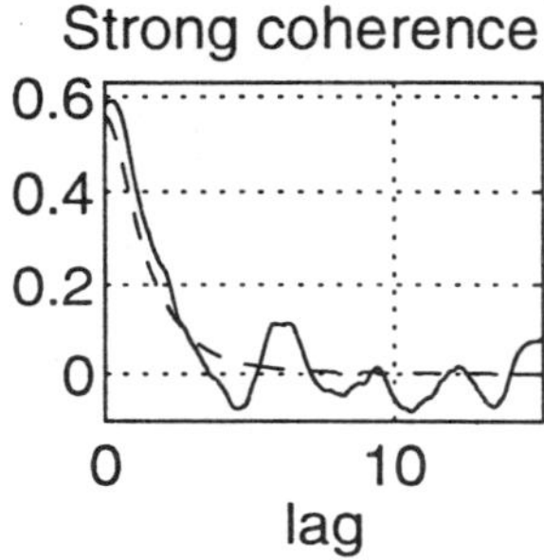

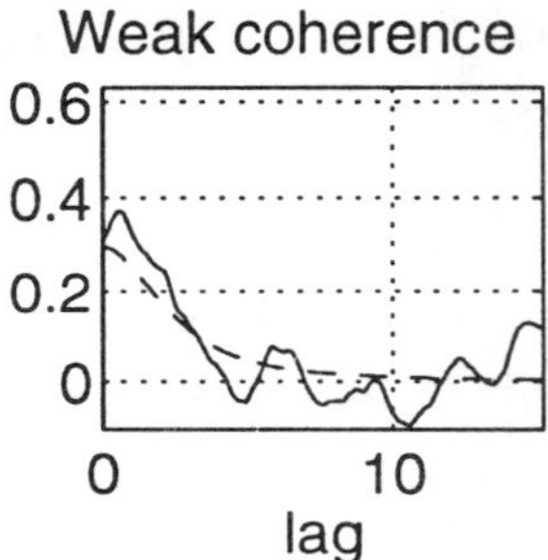

Figure 1 shows some simple one-dimensional examples. In those examples the auto-covariances are exponential, the phase is 0 and the coherency is another exponential. As the parameter of the coherency exponential increases the coherence and cross-covariance between the two series decreases. The theoretical cross-covariances shown were obtained by using the Fast Fourier Transform to invert the cross-spectral densities. Note the agreement between the theoretical and observed cross-covariance is very good. We are currently exploring the algorithm for other coherence and phase functions and developing higher dimensional versions of the algorithm.

MODELS FOR FLOW AND CONDITIONING

In this section we examine a more specific case of conditioning for two fields related by a differential equation.

The equation for steady saturated flow in two dimensions is

$$\nabla \cdot [T(\underline{x}) \nabla \phi(\underline{x})] = 0 \tag{7}$$

where T(x) is the transmissivity, Φ(x) is the head, and where appropriate boundary conditions are also imposed [7]. Given T(x) and the boundary conditions a common procedure is to solve for the head by using numerical methods. In general there is no explicit solution available for this problem.

Our interest will focus on the case where ln[T(x)] is modeled as a second-order stationary stochastic or random process.
Specifically,

$$\ln[T(\underline{x})] = F(\underline{x}) + f(\underline{x}) \tag{8}$$
$$\text{where } E[f(\underline{x})] = 0, \quad F(\underline{x}) = E\,[lnT(\underline{x})].$$

$$\text{cov}\,[f(\underline{x}+\underline{\xi}), f(\underline{x})] = C_{ff}(\underline{\xi}) \tag{9}$$

independent of x.

One approach to analyzing the stochastic version of equation (7) is to linearize the equation, assume the perturbations are small and impose a boundary condition in terms of the gradient of Φ(x) [8]. In the analysis here we allow a linear trend in F(x) (see Gelhar [9] for a simple version of this).

Thus,

$$\phi(\underline{x}) = H(\underline{x}) + h(\underline{x}) \tag{10a}$$

$$\ln\, T(\underline{x}) = A_0 + \underline{B} \cdot \underline{x} + f(\underline{x}) \tag{10b}$$

where f and h have mean 0,

$$H(\underline{x}) = E[\phi(\underline{x})],\ A_0 + \underline{B} \cdot \underline{x} = E[\ln\, T(\underline{x})] \text{ and } \nabla H = -\underline{J}.$$

Then substituting into (7), taking expected values and subtracting,

$$\underline{B} \cdot \nabla h(\underline{x}) + \nabla^2 h(\underline{x}) = \underline{J} \cdot \nabla f(\underline{x}) \tag{11}$$

if products of perturbations (f and h) are dropped.

If both f($\underline{x}$) and h($\underline{x}$) are second order stationary they have the spectral representations discussed in the Appendix. Substituting the integrals and using uniqueness one obtains the spectral connection

$$(i2\pi\underline{u}\cdot\underline{B})\, dZ_h(\underline{u}) - u^2\, dZ_h(\underline{u}) = i2\pi\underline{J}\cdot\underline{u}\; dZ_f(\underline{u}) \qquad \textbf{(12a)}$$

or

$$dZ_h(\underline{u}) = \frac{i2\pi\underline{J}\cdot\underline{u}}{-(2\pi u)^2 + i2\pi\underline{B}\cdot\underline{u}}\, dZ_f(\underline{u}) \qquad \textbf{(12b)}$$

where $u^2 = u_1^2 + u_2^2$.

The spectral connection (12) yields the spectral density of h and the cross-spectral density of f and h:

$$S_{hh}(\underline{u}) = \frac{(\underline{J}\cdot\underline{u})^2}{(2\pi u^2)^2 + (\underline{B}\cdot\underline{u})^2}\, S_{ff}(\underline{u}) \qquad \textbf{(13a)}$$

$$S_{hf}(\underline{u}) = \frac{i\underline{J}\cdot\underline{u}}{-2\pi u^2 + i\underline{B}\cdot\underline{u}}\, S_{ff}(\underline{u}) \qquad \textbf{(13b)}$$

The next phase of the problem involves conditioning the fields to the data. Namely, fields are to be generated consistent with observations of transmissivity, T($\underline{x}_{1i}$), i=1...L_1 and head $\Phi(\underline{x}_{2i})$, i=1...L_2. The objective is to reconstruct fields f($\underline{x}$) and h($\underline{x}$) that agree (within the limits of measurement error) with the observations and that satisfy the relations implied by equation (7). Note that the reconstructed fields will not be unique because the data is generally sparse. Simulation realizations of these conditioned fields can be used to track particles through the system and to generate arrival time distributions at exit locations.To generate these fields numerically we use a Fast Fourier Transform procedure as described in the Appendix [10,11].

Travel times are calculated using the convective velocity

$$\underline{v} = \frac{-T(\underline{x})\nabla\phi(\underline{x})}{nb} \qquad \textbf{(14)}$$

where n is the porosity and b is the thickness of the aquifer. A particle tracking algorithm is used to generate the arrivals at boundaries. In the examples, the

Mizell A [8] spectral density is used for the f fields and $\underline{B} = \underline{0}$:

$$S_{ff}(\underline{u}) = \frac{2\sigma_f^2 \alpha^2 u^2}{\pi[(2\pi u)^2 + \alpha^2]^3} \qquad (15)$$

One difficulty with the spectral conditioning approach used initially and described in the next section, became apparent when larger input variances were used. In that case the fields did not match the heads unless a very fine discretization was used. To circumvent that difficulty the linearity property inherent in the perturbation approach was used to develop a space domain conditioning procedure. In addition the linearized solutions resulted in continuity violations and hence the conditioned solutions for T were used in conjunction with a multigrid solver to get a non-linearized solution.

Several examples are presented to illustrate the techniques developed. In all cases a domain 6.4 by 6.4 is considered with $\underline{J} = (-.1, 0)$ and $\alpha = 1$. Tracking is done to the boundary for travel time curves. A comparison of runs with different grid sizes indicated a size 32 by 32 was sufficient to capture the required details. Here a 64 by 64 grid is used.

Figure 2 shows a comparison of the arrival time curves for the unconditioned problem, and several versions of conditioning. For 30 heads and 30 transmissivity values the non-linearized multi-grid solution is also shown. The data are spread uniformly throughout the domain. The efforts of conditioning on travel times is especially important for transmissivity where a substantial variance reduction is observed.

An interesting related set of figures shows the travel paths for unconditioned (Figure [3a]), 30 heads (Figure [3b]), 30 transmissivities (Figure [3c]) and 30 head and transmissivities (Figure [3d]). Note that the transmissivities would appear to control the arrival times, while the heads control the output location. The results do, however, depend on data location.

EXTENSIONS

The spectral representation theorem offers an additional interesting possibility for doing joint conditioning in the general case where we have observations on both fields. The approach outlined here assumes that the means are known though we expect it is possible to extend this to the unknown mean case. In addition, we assume that the processes are jointly gaussian though some relaxation of that restriction is possible.

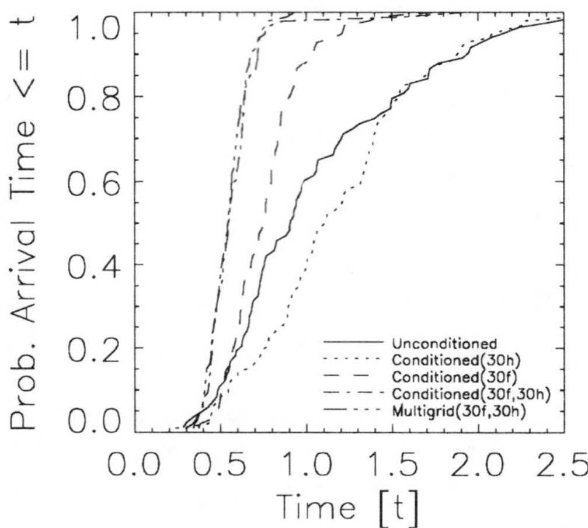

We assume that we have observations $V(\underline{x}_{ki})$, $i=1...L_k$, $k=1,2$ and we wish to carry out a conditional simulation that honors these data points and that has the proper auto and cross-covariance structure. To do this we use the spectral representation theorem and note that the data vectors and the spectral components (namely the dZ_k's) have a joint normal distribution. While this may be a degenerate multivariate normal [6], the conditional distributions discussed below are non-degenerate.

If

$$V_k(\underline{x}) = \sum_{\underline{m}=-\underline{M}}^{\underline{M}-\underline{1}} \exp\{i2\pi(\underline{m}+\underline{1/2})\cdot\underline{x}\Delta u\}\ dZ_k[(\underline{m}+\underline{1/2})\Delta u]$$

is the spectral representation for field k (see the Appendix), then

$$cov[dZ_j[(\underline{m}+\underline{1/2})\Delta u],\ V_k(\underline{x_a})] =$$

$$\exp\{-i2\pi(\underline{m}+\underline{1/2})\cdot\underline{x_a}\Delta u\}\ S_{jk}[(\underline{m}+\underline{1/2})\Delta u](\Delta u)^2$$

Here the covariance operator is applied term by term to the real and imaginary parts separately. In addition one can get the covariances between the data points directly. Consequently one can construct the conditional covariance matrices and hence generate the conditional simulations. Note that the same

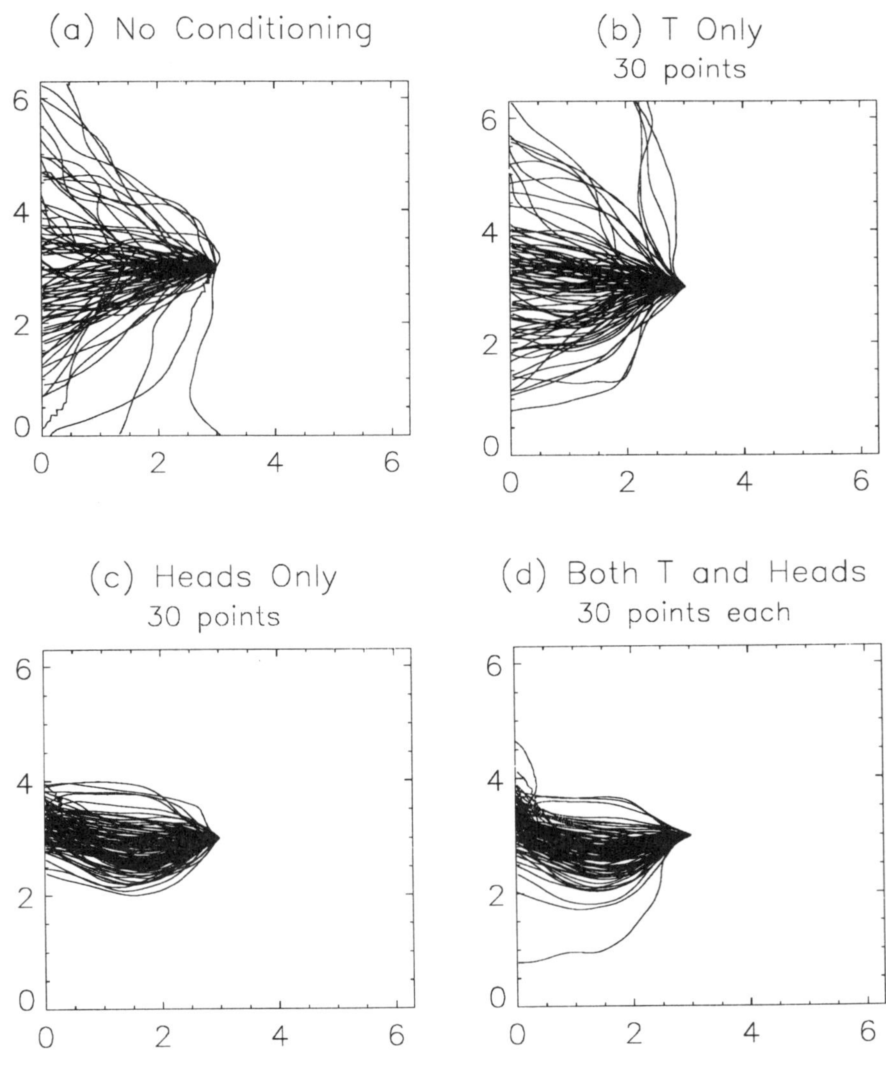
Figure 3
Paths for Various Conditioning Sets
6.4 x 6.4 Domain, Correlation Length = 1.00
(a) No Conditioning
(b) T Only
30 points
(c) Heads Only
30 points
(d) Both T and Heads
30 points each
0
2
4
6

covariance structure can be induced by co-kriging in the spectral domain based on spatial data.

The algorithms presented here are very fast and flexible. They can accommodate many different types of covariances and cross-covariances, can be extended to higher dimensions and can be used to study more than two fields.

ACKNOWLEDGEMENT

The research in this paper was financed in part by the U.S. Department of Energy through the New Mexico Waste-Management Education and Research Consortium (WERC).

APPENDIX

Generation of Random Fields via the Fast Fourier Transform

The basic theorems used in much of the analyses are the spectral representation theorems for mean zero, second-order stationary processes [11,12]. The vector version is stated here but the theorem also applies to single processes.

Spectral Representation Theorem: Let $V_k(\underline{x})$, $k=1,2$, be two second-order stationary mean zero random processes with covariances $C_{jk}(\underline{\xi})$ and associated spectral densities $S_{jk}(\underline{u})$. Then there exist unique (with probability one) complex processes $Z_k(\underline{u})$ such that

$$V_k(\underline{x}) = \int_{-\infty}^{\infty} e^{i2\pi \underline{u}\cdot\underline{x}}\, dZ_k(\underline{u}) \tag{A.1}$$

In addition,

$$E\left[dZ_j(\underline{u}_1)dZ_k(\underline{u}_2)\right] = \begin{cases} S_{jk}(\underline{u})d\underline{u} \text{ if } \underline{u} = \underline{u}_1 = \underline{u}_2 \\ 0 \text{ otherwise} \end{cases} \tag{A.2}$$

$$E\left[dZ_k(\underline{u})\right] = 0 \tag{A.3}$$

Note here, as throughout the paper, the integrals and sums are double sums and integrals:

$$\int_{-\infty}^{\infty} g(\underline{u})\, d\underline{u} = \int_{-\infty}^{\infty}\int_{-\infty}^{\infty} g(u_1,u_2)\, du_1\, du_2$$

$$\sum_{\underline{m}=\underline{0}}^{\underline{M}} g(\underline{m}) = \sum_{m_1=0}^{M}\sum_{m_2=0}^{M} g(m_1,m_2)$$

For the multiple processes case one can rewrite the representation in terms of real quantities

$$dZ_k(\underline{u}) = dZ_{Rk}(\underline{u}) + i\,dZ_{Ik}(\underline{u}) \tag{A.4}$$

To generate a realization for one of the fields ,we start with the integral (A.1) of the spectral representation theorem. We approximate this by truncating at $\pm M \cdot \Delta u$ and discretizing the integral at $(\underline{m}+\underline{1/2})\Delta u$;

$$V_k(\underline{x}) \cong \sum_{\underline{m}=-\underline{M}}^{\underline{M-1}} \exp\{i2\pi(\underline{m}+\underline{1/2})\cdot \underline{x}\Delta u\}\; dZ_k[(\underline{m}+\underline{1/2})\Delta u] \tag{A.5}$$

Here, for the 2-d case, $\underline{m}=(m_1, m_2)$, $\underline{1}=(1,1)$, and $\underline{M}=(M,M)$.

We next take $\underline{x}=\underline{n}\Delta x$, with $\underline{n}=(n_1,n_2)$ and $\Delta x \Delta u=1/2M$, so

$$V(\underline{x}) \simeq W_{2M}^{(n_1+n_2)/2} \sum_{\underline{m}=-\underline{M}}^{\underline{M-1}} W_{2M}^{\underline{m}\cdot\underline{n}}\, dZ_v[(\underline{m}+\underline{1/2})\Delta u] \tag{A.6}$$

where $W_{2M} = \exp\{i2\pi/2M\}$.

The dZ_k's are constructed so that $dZ_k[-(\underline{m}+\underline{1/2})\,\Delta u] = dZ_k^*[(\underline{m}+\underline{1/2})\,\Delta u]$,

$\underline{m}=\underline{0},...,\underline{M}$--1. This procedure will ensure that $V_k(\underline{x})$ is real [10]. The algorithm proceeds as follows:

(i) Generate independent normal random variables with mean 0 and variances 1/2, and form

$$dZ_k[(\underline{m}+\underline{1/2})\,\Delta u] = [\alpha_{\underline{m}}+i\beta_{\underline{m}}] \quad \{S_{kk}[(\underline{m}+\underline{1/2})\,\Delta u]\}^{1/2}$$

for m_1 = -M...M-1, m_2 = 0,1....M-1.

(ii) Set $dZ_k[m_1+1/2)\Delta u,(-m_2-1/2)\Delta]$ equal to

$$dZ_k^*[(m_1+1/2)\,\Delta u,\ (m_2+1/2)\,\Delta u] \quad \text{for } m_2 = 0,\ 1,\ ...M\text{-}1.$$

(iii) Use the FFT to evaluate the required sums and to obtain the field $V_k(\underline{x})$.

REFERENCES

1. A.G. Journel and Ch.I. Huijbregts, Mining Geostatistics, Academic Press, San Diego, (1978).
2. G. Luster, Raw Materials for Portland cement: Applications of Conditional Simulation of Coregionalization, Ph.D. Dissertation, Stanford Univ., Palo Alto Calif., (1985).
3. M.J.L. Robin, Migration of Reactive Solutes in Three-dimensional Heterogeneous Porous Media, Ph.D. Thesis, University of Waterloo, Ontario, Canada.
4. M.J.L. Robin, A.L. Gutjahr, E.A. Sudicky and J.L. Wilson, Water Resources Research, (In press).
5. M. Rosenblatt, Random Processes, Oxford University Press, (1962).
6. T.W. Anderson, Multivariate Statistical Analysis, John Wiley and Sons, (1984).
7. G .De Marsily, Quantitative Hydrogeology, Academic Press, Orlando, Fla., (1986).
8. S.A. Mizell, A.L. Gutjahr, and L.W. Gelhar, Water Resources Research, **19**, 1853 (1982).
9. L.W. Gelhar, Water Resources Research, **22**, 135S, (1986).
10. A.L. Gutjahr, Fast Fourier Transforms for Random Field Generation , N.M. Tech Project Report, (1989).
11. L.Borgman, M.Taheri,and R. Hagan, in Geostatistics for Natural Resources Characterization, Part I, p517, G. Verly et al. eds., Reidel Publ. Co. (1984)
12. L.H. Koopmans, The Spectral Analysis of Time Series, Academic Press, (1974).

THE METHOD OF PROJECTION ONTO CONVEX SETS FOR THE RECONSTRUCTION OF SUBSURFACE PROPERTY MAPS

A. MALINVERNO and D. J. ROSSI
Schlumberger-Doll Research
Old Quarry Road
Ridgefield, Connecticut 06877, U.S.A.

Obtaining a map of the spatial distribution of some physical property in a subsurface reservoir from sparse data is a fundamental problem in reservoir characterization. This problem can be stated in terms of an inverse problem as the search for a map that satisfies certain constraints, i.e., measurements and information on the spatial statistics of the property. We introduce a method (Projection onto Convex Sets, or POCS) that functions both as an estimation procedure, where the objective is to obtain a unique guess of the unknown spatial distribution given some constraints, and as a simulation procedure, where the objective is to obtain a set of nonunique reconstructions that are consistent with some measurements and have given spatial statistics. Numerical experiments that use point measurements at wellbores and volume integral measurements show that incorporating data from the region of the reservoir between wellbores allows us to properly locate large-scale heterogeneities in the reservoir volume.

INTRODUCTION

Reservoir rocks are notoriously heterogeneous, i.e., have physical properties that are highly variable in space. For example, permeabilities are known to vary by several orders of magnitude over short distances [8]. Detailed measurements of physical properties of reservoirs, however, can be taken only in boreholes, and measurements away from boreholes have a much lower resolution. We can then state a general inverse problem in reservoir characterization, that is, to obtain a map of a reservoir property based only on information that is sparse and/or has low resolution. The objective of the reconstruction is to gain some insight into the spatial distribution of reservoir heterogeneities and to infer the paths taken by migrating fluids. The property f will be taken here as a scalar (e.g., density, porosity, fluid saturation), and the vector $\mathbf{x}$ is a position vector. We will use $f(\mathbf{x})$ to denote the actual spatial distribution, or field, of the property within the reservoir, and $\widehat{f}(\mathbf{x})$ to denote the map that we were able to construct. We will examine only two-dimensional fields $f(\mathbf{x}) = f(x_1, x_2)$, where x_1 is a horizontal coordinate and x_2 the vertical coordinate; all our results can be extended to three dimensions.

We see two approaches to this inverse problem in the existing literature, and we will refer to them as "Estimation" and "Simulation" [3]. In an estimation procedure, the objective is

R. Dimitrakopoulos (ed.), Geostatistics for the Next Century, 197–208.

to obtain a unique "best" guess $\widehat{f}(\mathbf{x})$ constrained by some measurements. One example of estimation is the interpolation of point measurements. Clearly, the interpolation method depends on the spatial structure of the function to be mapped; the geostatistical procedure of kriging provides the best linear unbiased estimate of $f(\mathbf{x})$ for spatial statistics specified by a covariance function [15]. Other examples of estimation are the procedures typically employed in geophysical inversion, where the measurements are integrals of a formation property multiplied by some kernel function [18]. As these inverse problems are typically ill-posed (i.e., a small change in the measured data can cause a large change in the result of the inversion), some degree of damping, or regularization, is needed to reduce instabilities in the solution [20]. Alternatively, a unique solution can be obtained by choosing the reconstruction that has a minimum amount of complexity, i.e., that contains only features absolutely required by the data [5]. As the data on which the reconstruction is based are typically sparse and/or have relatively low resolution, the reconstructed maps are smoothed versions of reality, i.e., contain only the large-scale features of the actual $f(\mathbf{x})$, but do not reproduce its small-scale variation.

In a simulation approach, the objective is to obtain a number of maps $\widehat{f}_r(\mathbf{x})$, where $r = 1, 2, \ldots, R$, and R can be as large as desired. Each of these "realizations" is consistent with some measurements and has statistical properties that should resemble those of $f(\mathbf{x})$. An example of this approach is the practice of constructing "conditional simulations" in geostatistics [15]. These simulations are "conditioned" to reproduce a set of point data (typically borehole measurements in reservoir applications), and have second-moment spatial statistics specified by some mathematical model for the variogram. The model variogram is in turn constrained by the borehole measurements or by observations on outcrops. The differences between different reconstructed maps $\widehat{f}_r(\mathbf{x})$ provide a measure of uncertainty [11]. Simulations conditioned on point data only, however, do not incorporate information from the region of the reservoir between boreholes.

The choice between estimation and simulation depends on the problem at hand [12]. For example, an estimation approach can best determine whether there is a given feature (e.g., a diapir) in the reservoir between wells. On the other hand, if the fine-scale heterogeneity of the reservoir matters, a simulation will provide the best answer. It can be argued that the patterns of fluid flow within a reservoir are mostly controlled by the spatial arrangement of extreme permeability values [16]. Although a simulation procedure will not be able to locate exactly reservoir heterogeneities in space, it will generate maps that are statistically similar to reality, and thus should provide a good basis for fluid flow modeling. Tests in actual reservoirs show that conditional simulations give predictions of reservoir performance that are more accurate than those obtained using smooth interpolations [7].

In general, it would be desirable to compare the results of both estimation and simulation. Estimation tells us what features of the reconstructions are constrained by the measurements, while simulation provides a map that supposedly has the same spatial statistics as reality, and thus lets us examine the uncertainty in the reconstruction. We present here an inverse technique, Projection onto Convex Sets, that allows us to perform both estimation and simulation constrained by the spatial statistics of the field and by measurement data taken at different resolutions. We first describe the technique and then show the results of a numerical experiment where we perform estimation and simulation of a synthetic heterogeneous reservoir given only sparse measurements.

PROJECTION ONTO CONVEX SETS

Our inverse problem is to obtain a two-dimensional map $\widehat{f}(\mathbf{x})$ that is consistent with a set of constraints (measurements and information on the statistics of $f(\mathbf{x})$). Let us write our M measurements, or data, as follows:

$$d_j = \int_S f(\mathbf{x}) g_j(\mathbf{x}) dA \qquad (j = 1, 2, \ldots, M), \tag{1}$$

where $g_j(\mathbf{x})$ is some known kernel function that relates the j-th measurement to the field $f(\mathbf{x})$, S is the region of two-dimensional space where $g_j(\mathbf{x}) \neq 0$, and dA an infinitesimal area. In the simplest case of a point measurement at a location $\mathbf{x}_j$, $g_j(\mathbf{x}) = \delta(\mathbf{x} - \mathbf{x}_j)$, and $d_j = f(\mathbf{x}_j)$. In more complex measurement geometries, $g_j(\mathbf{x})$ may be zero everywhere except on a line between a source and a receiver (as in a seismic traveltime experiment), or $g_j(\mathbf{x})$ may be nonzero on some finite area (as in a deep electromagnetic experiment). Note that in Equation 1 we disregard measurement error.

We also wish to impose some constraints on the spatial statistics of the reconstructed map $\widehat{f}(\mathbf{x})$. To quantify spatial statistics, we can use the covariance $C(\mathbf{s})$, where

$$C(\mathbf{s}) = E\{[f(\mathbf{x}) - \bar{f}\,]\,[f(\mathbf{x} + \mathbf{s}) - \bar{f}\,]\}, \tag{2}$$

$E\{\cdot\}$ is the expectation operator, and $\bar{f}$ is the mean of $f(\mathbf{x})$: $\bar{f} = E\{f(\mathbf{x})\}$. Equivalently, we can use the power spectral density $P(\mathbf{k})$, which is the Fourier transform of the covariance function. The power spectral density is related to the total variance σ^2 of $f(\mathbf{x})$ by

$$\sigma^2 = C(0) = \int_{-\infty}^{\infty} \int_{-\infty}^{\infty} P(\mathbf{k}) d\mathbf{k}. \tag{3}$$

The vector $\mathbf{k} = (k_1, k_2)$ is a wavenumber vector; the component wavenumbers are related to wavelengths λ_1 and λ_2 in the x_1 and x_2 directions respectively by $k_1 = 2\pi/\lambda_1$ and $k_2 = 2\pi/\lambda_2$.

The method of Projection onto Convex Sets (POCS) is a simple iterative procedure to obtain a map $\widehat{f}(\mathbf{x})$ that is consistent with a set of measurements of the type specified in Equation 1 and with a given covariance/power spectral density. Detailed mathematical treatments can be found in [2, 4, 10, 25, 26]; we follow here the introduction by Menke [19] and provide only the foundations of the method and a description of the implementation. Consider again the function $f(\mathbf{x})$, and suppose that it is discretized, so that we have values at N locations. We can then imagine a N-dimensional Euclidean function space where any discretized function $f(\mathbf{x})$ is a point in such a space (if $f(\mathbf{x})$ were continuous, one would need an infinite number of dimensions). We can now translate our constraints for the reconstructed map $\widehat{f}(\mathbf{x})$ into sets in function space, which will be called "constraint sets." For example, we can define the set of all the functions that honor the M measurements defined in Equation 1. The solutions to our problem will be functions that lie at the intersection of all the constraint sets, i.e., functions that simultaneously satisfy all the requirements. Figure 1 is a pictorial illustration of the function space and of the intersection set.

The problem now becomes one of finding a function that is in the intersection of all the constraint sets. Imagine starting from an arbitrary location in function space, indicated

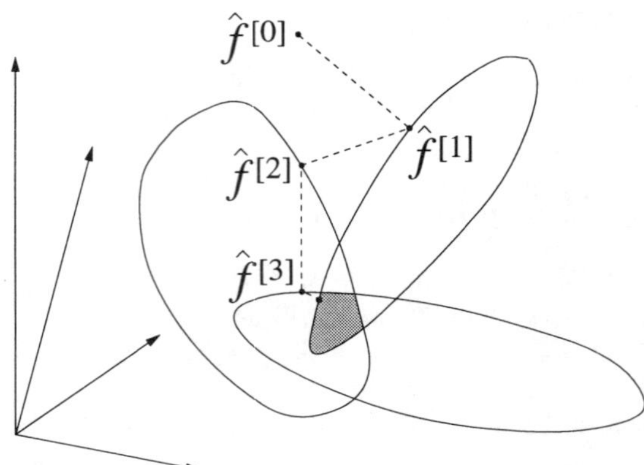

Figure 1. A function can be represented by a point in multidimensional space. Functions that obey some constraints define constraint sets in function space, and a function that obeys all constraints is located in the intersection of all constraint sets (stippled). Modified after Menke [15].

by $\hat{f}^{[0]}$ in Figure 1 (we use the notation $\hat{f}^{[n]}(\mathbf{x})$ for the map $\hat{f}(\mathbf{x})$ at the n-th iteration). If we perform an orthogonal projection of the function onto each of the sets in succession ($\hat{f}^{[1]}$, $\hat{f}^{[2]}$, etc. in Figure 1), we will eventually reach the intersection of all the constraint sets. This is guaranteed to happen if the constraint sets have a simple topology in function space, i.e., are convex. A set of functions is convex if, given any two functions $h_1(\mathbf{x})$ and $h_2(\mathbf{x})$ that belong to the set and any value between 0 and 1 for a scalar β, the function

$$h(\mathbf{x}) = \beta h_1(\mathbf{x}) + (1 - \beta) h_2(\mathbf{x}) \tag{4}$$

also belongs to the set [19]. In geometrical terms, all functions located on a straight segment connecting $h_1(\mathbf{x})$ and $h_2(\mathbf{x})$ in function space must also belong to the set.

Menke [19] lists a number of constraints that define convex sets in function space; examples include the set of all functions that return some measurements d_j as in Equation 1, that have specified local bounds (i.e., $f_l(\mathbf{x}) \leq f(\mathbf{x}) \leq f_u(\mathbf{x})$), local discontinuities (i.e., $f(\mathbf{x}_1) - f(\mathbf{x}_2) = \Delta f(\mathbf{x})$), and local inequalities (i.e., $f(\mathbf{x}_1) \geq f(\mathbf{x}_2)$). Another constraint we wish to impose is on the power spectral density of $\hat{f}(\mathbf{x})$. In this case, it can be shown that all the functions with a power spectral density $P(\mathbf{k})$ that is less than or equal to a desired value $P_0(\mathbf{k})$ belong to a convex constraint set.

The method of POCS can be implemented as follows: start from a guess $\hat{f}^{[0]}$ to the function that is at some location in function space, project the function orthogonally onto each constraint set, and iterate until the function is in the intersection set. In practice, orthogonal projection translates into modifying the function as little as possible to move it onto the boundary of the constraint set. Menke [19] shows that, to project the function $\hat{f}(\mathbf{x})$ to have measurement values equal to d_j as defined in Equation 1 for any form of the kernel function $g_j(\mathbf{x})$, the projection at the n-th iteration is

$$\widehat{f}^{[n]}(\mathbf{x}) = \widehat{f}^{[n-1]}(\mathbf{x}) + \sum_{k=1}^{M}\sum_{l=1}^{M} g_k(\mathbf{x})\,[\mathbf{A}^{-1}]_{kl}\left[d_l - \int_S g_l(\mathbf{x})\widehat{f}^{[n-1]}(\mathbf{x})dA\right], \tag{5}$$

where the elements of the matrix $\mathbf{A}$ are

$$A_{kl} = \int_S g_k(\mathbf{x})g_l(\mathbf{x})dA. \tag{6}$$

To understand the nature of this projection, consider the simple case of point measurements. As noted above, for the j-th point measurement at a location $\mathbf{x}_j$ we have $g_j(\mathbf{x}) = \delta(\mathbf{x} - \mathbf{x}_j)$. Therefore $A_{kl} = \delta_{kl}$, where δ_{kl} is the Kronecker delta (δ_{kl} equals 1 if $k = l$, and is zero otherwise). Substituting these results in Equation 5, we find that for each point measurement d_j at a location $\mathbf{x}_j$ the projection consists of simply setting $\widehat{f}^{[n]}(\mathbf{x}_j) = d_j$.

To project a function to have a spectral density that is less than or equal to a desired value $P_0(\mathbf{k})$ at iteration n, we calculate $\widehat{F}^{[n-1]}(\mathbf{k})$, the Fourier transform of $\widehat{f}^{[n-1]}(\mathbf{x})$. The projection is as follows:

$$\widehat{F}^{[n]}(\mathbf{k}) = c\widehat{F}^{[n-1]}(\mathbf{k}), \tag{7}$$

where

$$c = \begin{cases} \dfrac{\sqrt{P_0(\mathbf{k})}}{|\widehat{F}^{[n-1]}(\mathbf{k})|} & \text{if } |\widehat{F}^{[n-1]}(\mathbf{k})|^2 > P_0(\mathbf{k}); \\ 1 & \text{otherwise.} \end{cases} \tag{8}$$

Note that in the projection operation, only the amplitude of the Fourier transform $\widehat{F}^{[n]}(\mathbf{k})$ is reduced if necessary ($0 < c \leq 1$ is a real number), and the phase is never changed.

The iteration process is continued until $\widehat{f}^{[n]}(\mathbf{x})$ is judged to be in the intersection set. In practice, this translates into continuing the iterations until $\widehat{f}^{[n]}(\mathbf{x})$ fits all the constraints within some small tolerance (e.g., the standard deviation of the measurement error). Unless the intersection set is a single point in function space, the final solution will not be unique, and it will depend on the starting function $\widehat{f}^{[0]}(\mathbf{x})$. In qualitative terms, the final solution will be close in function space to $\widehat{f}^{[0]}$, i.e., it will inherit some of the characteristics of the starting function [22].

A SYNTHETIC FIELD $f(\mathbf{x})$

To demonstrate the method of POCS, we constructed a synthetic two-dimensional field $f(\mathbf{x})$ of a reservoir property, carried out measurements on this field, and attempted to reconstruct it. The synthetic $f(\mathbf{x})$ that we used has an exponential autocovariance function

$$C(r) = \sigma^2 \exp(-r), \tag{9}$$

where r is a dimensionless lag:

$$r = \sqrt{k_h^2 x_1^2 + k_v^2 x_2^2} = k_v\sqrt{\frac{x_1^2}{a^2} + x_2^2}. \tag{10}$$

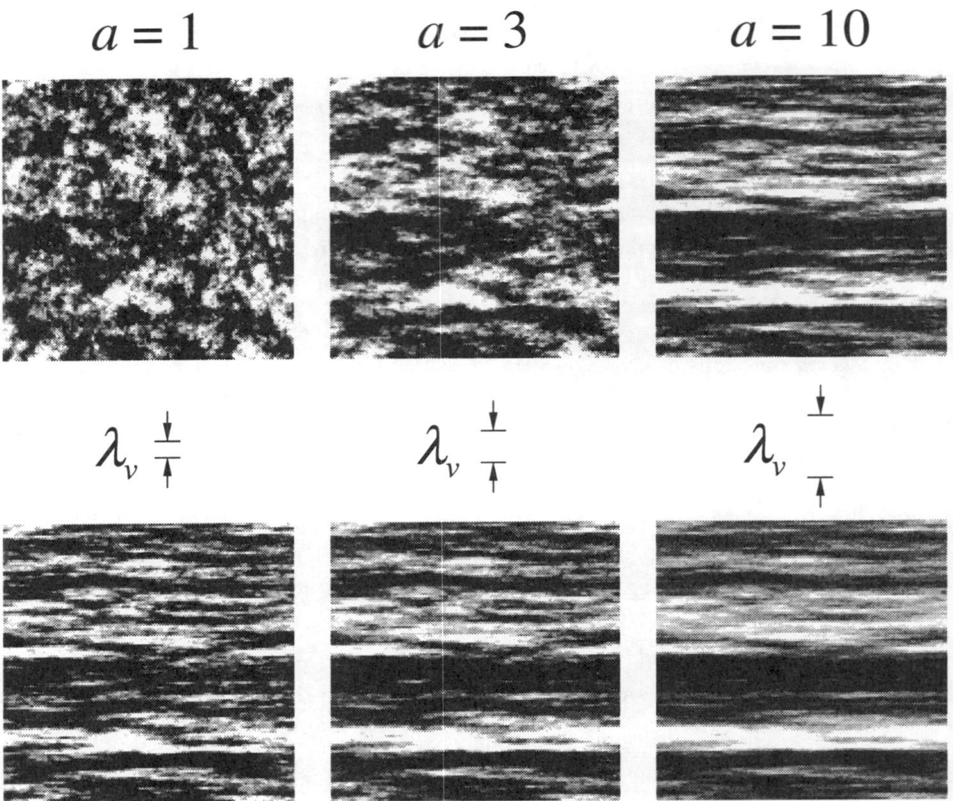

Figure 2. Examples of fields with a covariance as in Equation 9. The top row shows fields with the same properties except for the anisotropy parameter a. The bottom row shows the effect of changing the characteristic vertical length scale λ_v, while a is fixed at 10.

The lag r is dimensionless because k_h and k_v are characteristic wavenumbers, related by the parameter a: $k_h = k_v/a$. If the parameter a differs from unity, the field with the covariance specified by Equation 9 is anisotropic. The parameter a controls the degree of stratification of the two-dimensional $f(\mathbf{x})$; as $a \to \infty$, the field becomes more and more stratified, i.e., all vertical sections tend to become identical.

The two-dimensional power spectral density of a field with the covariance of Equation 9 is

$$P(\mathbf{k}) = P(k_1, k_2) = \frac{a\sigma^2}{2\pi k_v^2 \left(1 + \dfrac{a^2 k_1^2 + k_2^2}{k_v^2}\right)^{3/2}} \tag{11}$$

For wavenumbers $|\mathbf{k}| \gg k_v$ the power spectral density (Equation 11) is proportional to $|\mathbf{k}|^{-3}$. In the terminology of fractal geometry, a function $f(\mathbf{x})$ with a power spectral density

Original Field $f(\mathbf{x})$ Measurements

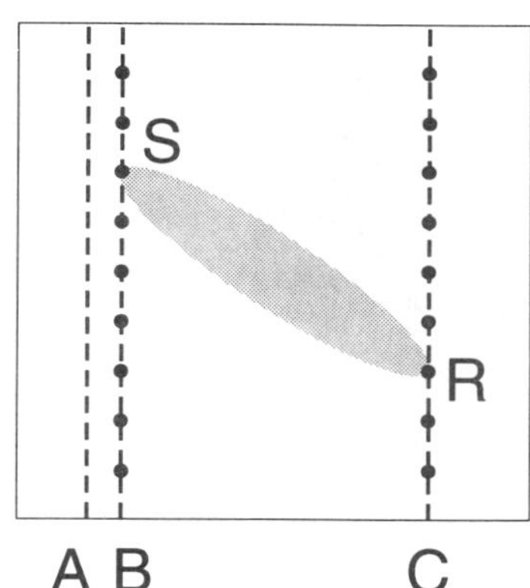

Figure 3. The fictitious field $f(\mathbf{x})$ that we generated for our experiments (left) and the measurements carried out for the reconstructions (right). We took 450 point measurements on the vertical wellbores A, B, and C (dashed lines), and 81 integrals of the field multiplied by a kernel function connecting source and receiver points (dots). The stippled area connecting a source (S) and a receiver (R) shows the region where the kernel function is not zero.

that follows a power law is a self-affine function [17, 24]. In other words, portions of the field observed at different scales are indistinguishable when properly rescaled, and $f(\mathbf{x})$ is said to be scale invariant. Random fields with a covariance as in Equation 9 and a power spectrum as in Equation 11 will be scale invariant for length scales below a characteristic length scale λ_v, which can be defined from the width of the covariance function as [9]

$$\lambda_v = \frac{2\sqrt{2}}{k_v}. \tag{12}$$

The length scale λ_v gives a measure of the typical size of features of the field as sampled in the vertical direction. Figure 2 illustrates a few examples of functions $f(\mathbf{x})$ that have a covariance as in Equation 9, the same variance σ^2, and different values of the parameters a and λ_v. Fields with a covariance as in Equation 9 have been proposed in studies of the spatial distribution of hydrogeologic properties [13, 6], of turbulence [23, 21], and of seafloor topographic roughness [1, 9].

To generate a field $f(\mathbf{x})$ with a power spectral density as in Equation 11, we started from a two-dimensional field of white Gaussian noise, took its Fourier transform, multiplied the value of the transform at each wavenumber by a real number proportional to the square root of the desired power spectral density at that wavenumber, and took an inverse transform. The fictitious field $f(\mathbf{x})$ that we used and the measurements that were carried out on it are illustrated in Figure 3. The field is composed of 150 by 150 samples, has a standard deviation σ of 100 units, a characteristic length scale in the vertical direction λ_v of 14 sample intervals, and an anisotropy coefficient a equal to 10 (i.e., λ_h is 140 sample intervals).

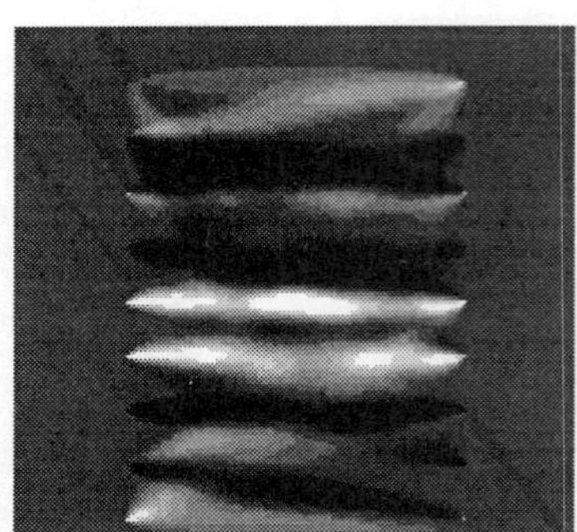

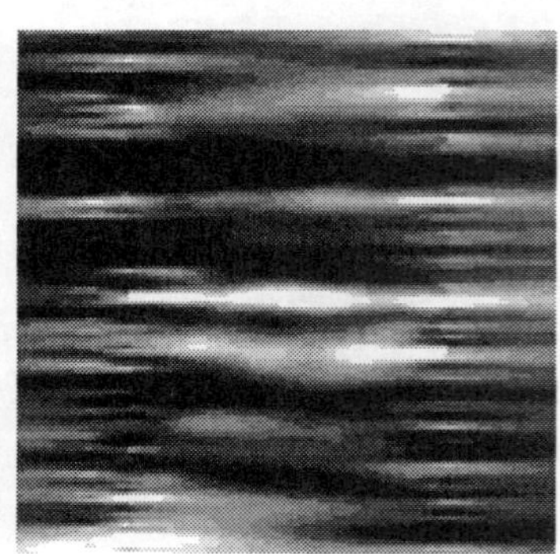

Figure 4. Estimated maps $\widehat{f}(\mathbf{x})$ reconstructed using only 81 cross-well integral measurements and a featureless initial condition (left) and using the 450 point measurements at the wells, the cross-well integral measurements, and an initial condition of a field interpolated by kriging between the point measurements (right). See Figure 3 for the location of wellbores.

RECONSTRUCTING THE SYNTHETIC FIELD

We took two sets of measurements: 450 point measurements of the field along three vertical lines (A, B, and C in Figure 3), simulating a data set from well logs, and 81 integrals of the field multiplied by a kernel function connecting source and receiver points (dots in Figure 3). The synthetic kernel functions are simple two-dimensional cosine bells; Figure 3 shows on the right as a stippled area the region where the kernel function between a source S and a receiver R is not zero. These integral measurements simulate the results of a cross-well experiment. For simplicity, all the measurements are not contaminated by noise.

We then generated a number of reconstructions $\widehat{f}(\mathbf{x})$ using the method of POCS, with different initial conditions $\widehat{f}^{[0]}$ and different constraint sets. We first show the results of experimenting with POCS as an estimation method, i.e., an inversion method to obtain a unique guess $\widehat{f}(\mathbf{x})$ based on some measurements and statistical structure. As noted above, the final reconstruction $\widehat{f}(\mathbf{x})$ obtained by POCS depends on and will be close to the initial condition, i.e., the position in function space of $\widehat{f}^{[0]}$.

We conducted two estimation experiments with different data sets and different initial conditions. In the first example, we used as constraints the 81 cross-well integral measurements and the spectral density of the field, and as initial condition a featureless field of constant value equal to the mean. The left side of Figure 4 shows the result of this procedure. The right side of Figure 4 shows instead the result of our second estimation experiment. In this case, beside adding the 450 point measurements at the three wells to the constraints that had to be satisfied by the solution, we used as an initial condition $\widehat{f}^{[0]}$ a field interpolated by kriging between the point measurements. Adding the point measurements clearly improves the reconstruction (compare Figure 4 to the original field

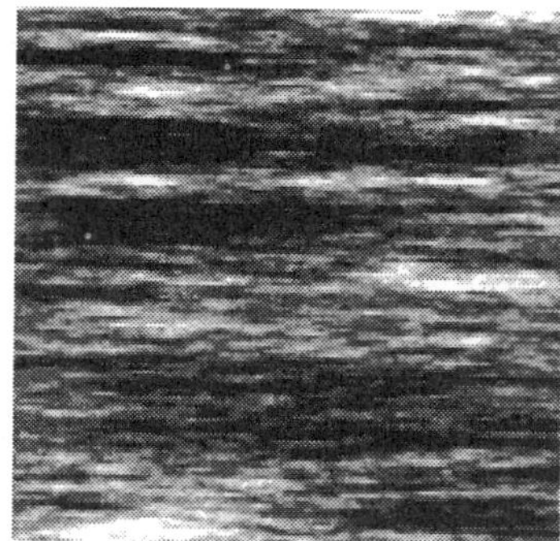

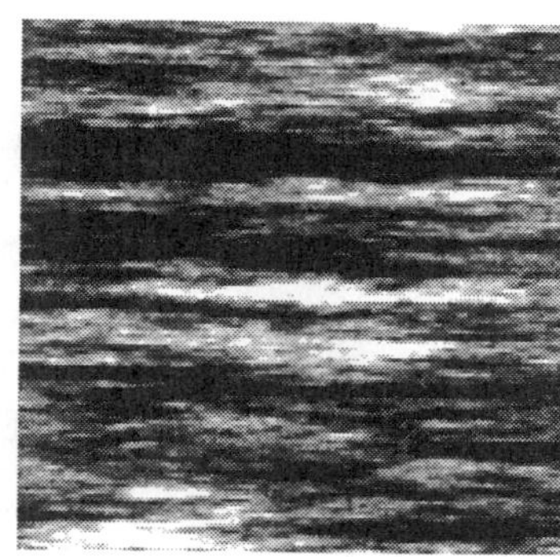

Figure 5. Simulated maps $\widehat{f}(\mathbf{x})$ reconstructed using a spectral constraint and 450 point measurements at three wellbores (left) and using a spectral constraint, 450 point measurements, and 81 cross-well integral measurements (right). See Figure 3 for the location of wellbores.

in Figure 3). The method of POCS thus allows us to easily obtain a solution that honors measurements taken at very different resolutions; note that the reconstructed field on the right of Figure 4 has fine-scale detail near the wellbores, and is smoother in the region between wells.

The spectral density constraint in an estimation procedure plays the role of a regularization factor. By requiring the reconstructed field $\widehat{f}^{[0]}$ to have a spectral density that does not exceed some maximum value, we put a limit to the variations of the field at any spatial frequency. In all our synthetic examples, we assumed that the spectral density of the field was known exactly; in practice we would have to estimate it from the variogram or the covariance of point measurements. Using as a maximum value the spectral density of the field that was inferred from actual measurements, we set a degree of regularization that is not arbitrary, but is controlled by the spatial structure of the property to be reconstructed [14, 20]

The method of POCS can also be used as a simulation procedure. To accomplish this, we used as initial conditions $\widehat{f}_r^{[0]}(\mathbf{x})$ fields interpolated by kriging between the point measurements plus white Gaussian noise of large variance with respect to the variance of the original $f(\mathbf{x})$. This $\widehat{f}_r^{[0]}(\mathbf{x})$ has a power spectral density greater than the threshold value, ensuring that the projection defined in Equation 7 will take place at least once at all wavenumbers, and that the final solution $\widehat{f}_r(\mathbf{x})$ will have a spectral density close to the threshold value. Adding a different field of white noise makes each initial condition $\widehat{f}_r^{[0]}(\mathbf{x})$ different, so that each final map $\widehat{f}_r(\mathbf{x})$ is different. Note that in principle one could start from a completely random initial condition; however, using the interpolator sets the phases of the long-wavelength components of $\widehat{f}_r^{[0]}(\mathbf{x})$ to be somewhat constrained by the point measurements, so that convergence is speeded up. In practice, we obtained final maps $\widehat{f}_r(\mathbf{x})$ that honored all the measurements within 1% of the standard deviation of $f(\mathbf{x})$ and did not

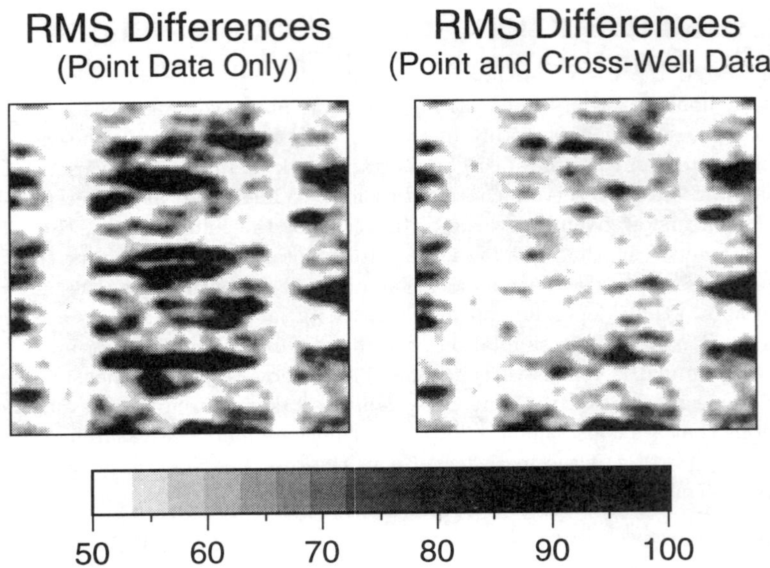

Figure 6. Root-mean square difference between 50 reconstructions and the original $f(\mathbf{x})$. The reconstructions were obtained as in Figure 5 using 450 point measurements at three wellbores (left) and using 450 point measurements and 81 cross-well integral measurements (right). In the darkest areas the error exceeds 100 units, which is the standard deviation of the original $f(\mathbf{x})$.

exceed the threshold power spectral density by more than 1% in 10 to 20 iterations.

We generated two types of simulations. The first set of simulations were constrained by the point measurements only (as in the current geostatistical procedure of "conditional simulation" [15, 16]; Figure 5, left), and the second set by both the point and the cross-well integral measurements (Figure 5, right). A cursory comparison of these simulated fields with the original in Figure 3 shows that, as expected, the simulation that includes data from the region of the field between wells resembles more closely the original.

To quantify how well the simulated reconstructions reproduced the original $f(\mathbf{x})$, we generated 50 reconstructions for both cases and computed a map of the average root-mean-square (RMS) difference between the reconstructions and the original. The results are shown in Figure 6 as images of the RMS difference. For reference, a RMS difference of 100 units means that the reconstructions were, on the average, wrong by as much as the standard deviation of the original $f(\mathbf{x})$. In both cases, the RMS differences are small near the wells, and increase in the region between wells. In this region, the differences are clearly greater if the cross-well integral measurements are not used to constrain the simulation. It is even more important to note that if the cross-well integral measurements are not included areas of large misfits (e.g., where the RMS misfit exceeds 100 units) are found in areas of the original $f(\mathbf{x})$ that contain prominent features (compare Figures 6 and 3).

CONCLUSIONS

We demonstrated the use of an inverse method (POCS) to obtain maps of a reservoir property that honor a variety of constraints, such as point/volume measurements and information on the statistical structure of the property to reconstruct. The method can be easily implemented and provides a rigorous framework to integrate a variety of measurements, a major task in reservoir characterization. While there are other procedures for measurement inversion given a covariance function for the solution (e.g., [14]), the POCS method is attractive in that it allows the implementation of constraints that would be difficult to impose otherwise, such as local bounds and discontinuities [19].

We have shown that, depending on the choice of initial condition, the POCS method can be used as an estimation or a simulation procedure. Our numerical experiments show that the quality of simulated reconstructions that include low-resolution measurements in the reservoir volume in addition to point measurements taken in wellbores is superior to that of simulations that are constrained only by point measurements. Measurements away from wellbores are important if the main features in the reservoir volume need to be properly placed in the reconstruction.

REFERENCES

[1] Bell, T. H., Statistical features of sea-floor topography, *Deep-Sea Res.*, **22**, 883-892, 1975.

[2] Bregman, L. M., The method of successive projection for finding a common point of convex sets, *Dokl. Akad. Nauk. SSSR (English transl.)*, **162**, 688-692, 1965.

[3] Christakos, G. and C. Panagopoulos, Space transformation methods in the representation of geophysical random fields, *IEEE Trans. Geosci. Rem. Sens.*, **30**, 55-70, 1992.

[4] Combettes, P. L., The foundations of set theoretic estimation, *IEEE Proc.*, **81**, 182-208, 1993.

[5] Constable, S. C., R. L. Parker, and C. G. Constable, Occam's inversion: A practical algorithm for generating smooth models from electromagnetic sounding data, *Geophysics*, **52**, 289-300, 1987.

[6] Dimitrakopoulos, R. and A. J. Desbarats, Geostatistical modeling of gridblock permeabilities for 3D reservoir simulators, *SPE Res. Eng.*, **Feb. 1993**, 13-18, 1993.

[7] Emanuel, A. S., G. K. Alameda, R. A. Behrens, and T. A. Hewett, Reservoir performance prediction methods based on fractal geostatistics, paper SPE 16971 presented at the 1987 SPE Annual Technical Conference and Exhibition, Dallas, September 27-30, 1987.

[8] Gelhar, L. W., Stochastic subsurface hydrology from theory to applications, *Water Resour. Res.*, **22**, 135S-145S, 1986.

[9] Goff, J. A. and T. H. Jordan, Stochastic modeling of seafloor morphology: Inversion of Sea Beam data for second-order statistics, *J. Geophys. Res.*, **93**, 13589-13608, 1988.

[10] Gubin, L. G., B. T. Polyak, and E. V. Raik, The method of projections for finding the common point of convex sets, *USSR Comp. Math. and Math. Phys. (English transl.)*, **7**, 6, 1-24, 1967.

[11] Haldorsen, H. H. and E. Damsleth, Stochastic modeling, *J. Petr. Tech.*, **42**, 404-412, 1990.

[12] Hewett, T. A., Fractal distributions of reservoir heterogeneity and their influence on fluid transport, paper SPE 15386 presented at the 1986 SPE Annual Technical Conference and Exhibition, New Orleans, October 5-8, 1986.

[13] Hoeksema, R. J. and P. K. Kitanidis, Analysis of the spatial structure of properties of selected aquifers, *Water Resour. Res.*, **21**, 563-572, 1985.

[14] Jackson, D. J., The use of a priori data to resolve non-uniqueness in linear inversion, *Geophys. J. R. Astron. Soc.*, **57**, 137-157, 1979.

[15] Journel, A. G. and C. J. Huijbregts, *Mining Geostatistics*, 600 pp., Academic Press, New York, 1978.

[16] Journel, A. G. and F. G. Alabert, New method for reservoir mapping, *J. Petr. Tech.*, **42**, 212-218, 1990.

[17] Mandelbrot, B. B. and J. W. Van Ness, Fractional Brownian motions, fractional noises and applications, *SIAM Rev.*, **10**, 422-437, 1968.

[18] Menke, W., *Geophysical Data Analysis: Discrete Inverse Theory*, 260 pp., Academic Press, San Diego, Calif., 1984.

[19] Menke, W., Applications of the POCS inversion method to interpolating topography and other geophysical fields, *Geophys. Res. Letters*, **18**, 435-438, 1991.

[20] Pilkington, M. and J. P. Todoeschuck, Natural smoothness constraints in cross-hole seismic tomography, *Geophys. Prosp.*, **40**, 227-242, 1992.

[21] Tatarski, V. I., *Wave Propagation in a Turbulent Medium*, 285 pp., McGraw-Hill, New York, 1961.

[22] Trussell, H. J. and M. R. Civanlar, The feasible solution in signal restoration, *IEEE Trans. Acoust., Speech, Signal Proc.*, **ASSP-32**, 201-212, 1984.

[23] Von Kármán, T., Progress in the statistical theory of turbulence, *J. Mar. Res.*, **7**, 252-264, 1948.

[24] Voss, R. F., Fractals in nature: from characterization to simulation, in *The Science of Fractal Images*, edited by H. O. Peitgen and D. Saupe, pp. 21-70, Springer-Verlag, New York, 1988.

[25] Youla, D. C., Generalized image restoration by the method of alternating orthogonal projections, *IEEE Trans. Circ. Sys.*, **CAS-25**, 694-702, 1978.

[26] Youla, D. C. and H. Webb, Image restoration by the method of convex projections: Part 1-Theory, *IEEE Trans. Med. Imag.*, **MI-1**, 81-94, 1982.

CONDITIONAL SIMULATION AND THE VALUE OF INFORMATION

Andrew R. Solow
Woods Hole Oceanographic Institution
Woods Hole, MA 02543

Samuel J. Ratick
Clark University
Worcester, MA 01610

One option in decisionmaking under uncertainty is to reduce uncertainty by acquiring information. The decisionmaker will choose to acquire additional information if its value exceeds its cost. This paper describes the use of nested conditional simulation in implementing a Bayesian assessment of the value of information in an explicitly spatial setting. A simple example is given concerning the management of flood damage.

INTRODUCTION

Many decisions must be made in the face of uncertainty. When costly information is available that can reduce uncertainty, the decisionmaker must decide whether or not to acquire it. This is also a decision that must be made in the face of uncertainty, since it must be made prior to the acquisition of the information. The most straightforward approach to assessing the *ex ante* value of information is through Bayesian decision theory (e.g., Berger, 1986). Under this approach, information is acquired if the expected increase in expected net benefits arising from its acquisition exceed the cost of acquisition. Note that the expected change in expected net benefits arising from the acquisition of information is always non-negative. This is due to the possibility that the acquisition of this information will lead to a revised decision.

The purpose of this paper is to describe the use of conditional simulation in implementing this approach in an explicitly spatial setting. This approach grew out of an on-going study of the implications of topographic uncertainty on coastal flood control. The basic problem is the following. A decisionmaker must decide how to respond to the threat of flooding. The damages due to a specified flood event depend on the spatial distribution of elevation within the exposed region. Elevation is measured only at a set of survey locations. The resulting uncertainty about flood damages can be reduced by measuring elevation at additional survey locations, although at some cost. The question facing the decisionmaker is: Should flood control decisions be made under current uncertainty or should additional locations be surveyed?

Spatial considerations enter this problem in two ways. First, not only will no point in the exposed region be inundated if it is above flood stage, but no point *below* flood stage will be inundated if it is *surrounded* by points above flood stage. In other words, flood damges depend on the *joint* distribution of elevation within the exposed region. Second, the specific information whose value is to be assessed is itself spatial.

R. Dimitrakopoulos (ed.), Geostatistics for the Next Century, 209–217.

METHOD

Let $Z = \{Z(t), t \in A\}$ be a Gaussian random field, where t denotes location within some region A, with known mean (which might as well be taken to be 0) and covariance function $\sigma(h)$. For practical purposes, the region A is approximated by a set of N discrete locations, so that Z is an N-variate normal random variable with mean vector 0 and covariance matrix Σ. Partition Z into a set of n observations X and the remaining $N-n >> n$ elements Y. Σ is partitioned conformably as:

$$\Sigma = \begin{matrix} \Sigma_{XX} & \Sigma_{XY} \\ \Sigma_{YX} & \Sigma_{YY} \end{matrix}$$

where $\Sigma_{YX} = \Sigma_{XY}{}^t$.

Suppose that a decisionmaker observes X = x and must select among a set of actions $\{a_1, a_2, \ldots, a_p\}$. The consequences of each action depend on Z. Specifically, let $B_i(z)$ be the net benefits from undertaking action a_i if Z = z. The current expected net benefits for action a_i are:

$$E(B_i(Z) \mid x) = \int B_i(z)\, f(z \mid x)\, dz \qquad (1)$$

where $f(z \mid x)$ is the conditional probability density function (pdf) of Z given X = x. The current optimal action $a^*(x)$ is that with maximal expected net benefits $E^*(x)$.

When net benefits are a complicated function of Z and when N is large, (1) is conveniently estimated by Monte Carlo integration:

$$E(B_i(Z) \mid x) \approx \sum_{j=1}^{K} B_i(z^j{}_{|x}) \,/\, K \qquad (2)$$

where $z^j{}_{|x}$, j = 1, 2, ..., K, are independent realizations simulated from $f(z \mid x)$. The conditional pdf $f(y \mid x)$ of Y given X = x is (N-n)-variate normal with mean vector:

$$\mu_{Y|X} = \Sigma_{YX}\, \Sigma_{XX}^{-1}\, x$$

(i.e., the kriging prediction of Y from x) and covariance matrix:

$$\Sigma_{Y|X} = \Sigma_{YY} - \Sigma_{YX}\, \Sigma_{XX}^{-1}\, \Sigma_{XY}$$

(i.e., the covariance matrix of the corresponding kriging errors). A realization from $f(z \mid x)$ can be generated by, first, simulating a vector from $f(y \mid x)$ and, second, stacking this vector on x. Simulation algorithms are described in Deutsch and Journel (1992).

Suppose, now, that the set of unobserved values Y is partitioned into U and V, where U is a set of m <u>prospective</u> observations. To decide whether to acquire U, the decisionmaker must determine whether the expected increase in expected net benefits from its acquisition exceeds its cost of acquisition C. Suppose that the information is acquired and it happens that U = u. The new expected net benefits for a_i are given by:

$$E(B_i(Z) \mid x, u) = \int B_i(z)\, f(z \mid x, u)\, dz \qquad (3)$$

where $f(z \mid x, u)$ is the conditional pdf of z given X = x and U = u. As before, (3) can be approximated by Monte Carlo integration:

$$E(B_i(Z) \mid x, u) \approx \sum_{j=1}^{K} B_i(z^j_{|x,u}) / K \qquad (4)$$

where $z^j_{|x,u}$, j = 1, 2, ..., K, are independent realizations simulated from $f(z \mid x, u)$. As before, a realization from $f(z \mid x, u)$ can be generated by, first, simulating a vector from $f(v \mid x, u)$ (which is (N-n-m)-variate normal with mean vector given by the kriging predictor of V from x and u and covariance matrix given by the covariance matrix of the corresponding kriging errors) and, second, stacking this vector on x and u.

Once $E(B_i(z) \mid x, u)$ is approximated for i = 1, 2, ..., p, the new optimal action $a^*(x, u)$ (i.e., that with maximal expected net benefits $E^*(x, u)$) can be found. Of course, prior to the acquisition of U, its value is unknown. The current expected value of $E^*(x, U)$ is given by:

$$E(E^*(x, U) \mid x) = \int E^*(x, u) f(u \mid x)\, du \qquad (5)$$

where $f(u \mid x)$ is the conditional pdf of U given x. As before, the integral in (5) can be approximated by:

$$E(E^*(x, U) \mid x) \approx \sum_{j=1}^{K} E^*(x, u^j_{|x}) / K \qquad (6)$$

where $u^j_{|x}$, j = 1, 2, ..., K, are independent realizations simulated from $f(u \mid x)$ (which is m-variate normal with mean vector given by the kriging prediction of U from x and covariance matrix given by the covariance matrix of the coresponding kriging errors).

In summary, $E(E^*(x, U) \mid x)$ can be approximated using the following nested conditional simulation: generate a realization from $f(u \mid x)$; generate K realizations from $f(z \mid x, u)$ and approximate $E^*(x, u)$; repeat the procedure K times and approximate $E(E^*(x, U) \mid x)$ from (6). Note that the first step does not necessitate a full conditional simulation, but only a conditional simulation of values at the prospective sample locations. The ex ante value of U is measured by the expected increase in expected net benefits arising from its acquisition:

$$V(U) = E(E^*(x, U) \mid x) - E^*(x)$$

It is easy to show that $V(U) \geq 0$. The decisionmaker will choose to acquire U if $V(U) > C$.

EXAMPLE

In this section, a highly stylized example of the application of the approach outlined above to a problem of flood control is presented. This example is intended purely as an illustration of the general approach described in the previous section. Certain practical problems arise in the actual application of this approach -- e.g., data analysis, covariance modelling, etc. As these are common to many studies involving conditional simulation, no purpose is served by raising them here, although they will be discussed in a forthcoming report on the on-going flood control study. There are, in addition, a number of factors that arise in the specific application of this approach to flood control -- e.g., developing damage functions, identifying protective measures, etc. These will also be discussed in a forthcoming report.

In this example, the exposed region consists of a 20-by-20 square. Elevation Z(t) within this region is represented by a realization of a Gaussian random field with mean 0 and spatial covariance function:

$$\sigma(h') = 1 - (h'/2\pi)\,(4-h'^2)^{1/2} - (2/\pi)\,\sin^{-1}(h'/2) \qquad 0 \le h' \le 2$$

where $h' = h/10$. This model was chosen for its ease of simulation. The region is approximated by a set of 400 points on a regular 20-by-20 grid. The indicator:

$$I(t) = \begin{cases} 1 & \text{if } Z(t) < 0 \\ 0 & \text{otherwise} \end{cases}$$

is shown in Figure 1. A total of 234 grid nodes have elevation below 0. Interest centers on the effects of a flood of height 0 occurring along the lefthand-side the region shown in Figure 1.

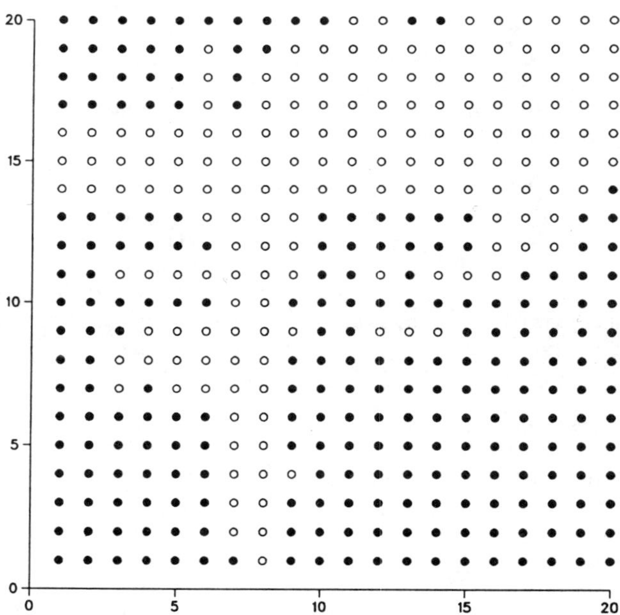

Figure 1. Map of elevation indicator (closed, I(t)=1; open, I(t)=0)

The consequences of such a flood on the region are shown in Figure 2 using the indicator:

$$J(t) = \begin{cases} 1 & \text{if } t \text{ is inundated} \\ 0 & \text{otherwise} \end{cases}$$

A total of 91 grid nodes are inundated.

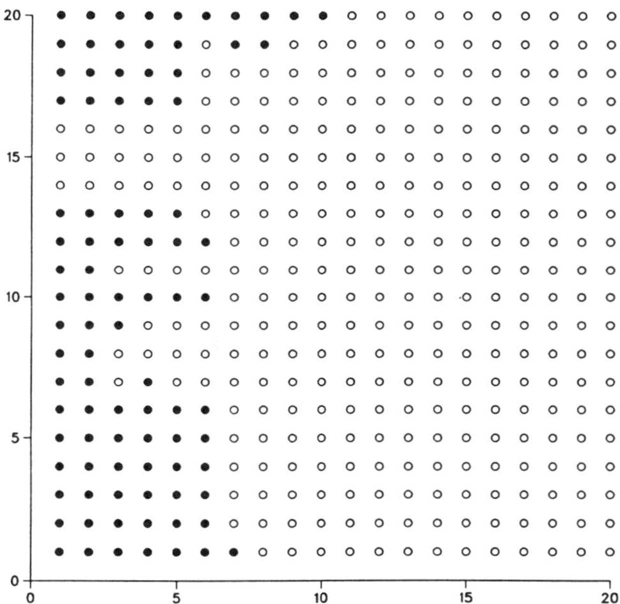

Figure 2. Map of inundation indicator (closed; J(t)=1; open, J(t)=0)

In the face of a certain flood of height 0, the decisionmaker can choose one of two actions: leave the exposed region unprotected (a_1) or construct a dike that will completely protect the exposed region (a_2). The two net benefits functions are:

$$B_1(z) = -k(z)$$

$$B_2(z) = -200$$

where k(z) is the number grid nodes inundated by a flood of height 0 if Z = z and the exposed region is left unprotected.

For the case shown in Figure 1, the optimal decision under perfect knowledge is to leave the region unprotected, incurring net benefits of -91. However, the decisionmaker does not have perfect knowledge. Instead, elevation is known only at the 16 locations shown in Figure 3. The measured elevations at these locations (x) are given in Table 1.

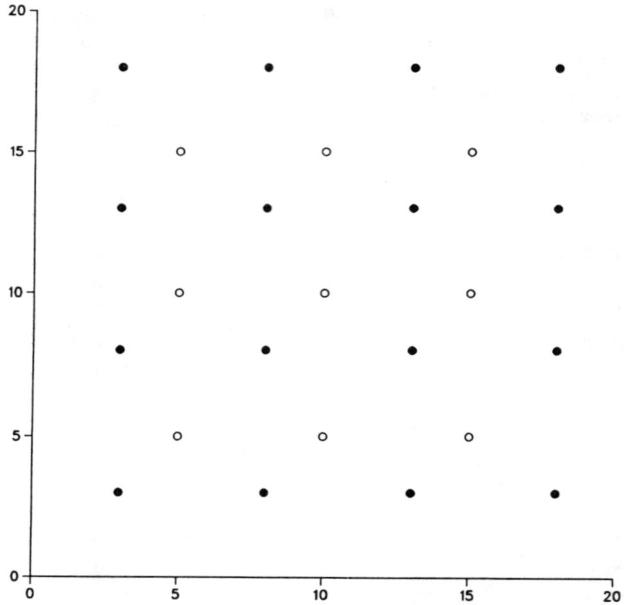

Figure 3. Original (closed) and prospective (open) survey locations

Table 1. Coordinates and elevation of survey locations

coordinate 1	coordinate 2	elevation
3	3	-1.15
3	8	0.34
3	13	-0.37
3	18	-0.60
8	3	0.42
8	8	0.99
8	13	0.30
8	18	0.00
13	3	-0.64
13	8	-0.77
13	13	-0.48
13	18	0.51
18	3	-2.05
18	8	-1.71
18	13	0.41
18	18	1.77

To estimate the expected net benefits from leaving the exposed region unprotected, 100 conditional simulations of Z given X = x were generated and k(z) was found for each simulation. The histogram of these numbers is shown in Figure 4. The average value of k(z) was around 148. Thus, under the current level of uncertainty, the decisionmaker would choose to leave the exposed region unprotected with E*(x) = -148.

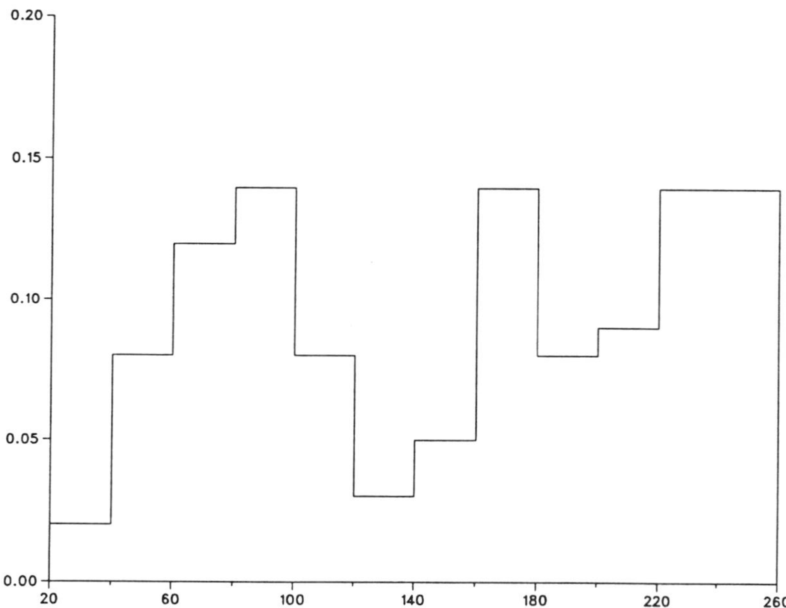

Figure 4. Histogram of $k(z_{|x}^{j})$, j=1, 2, ..., 100

It is clear from Figure 4 that current uncertainty about k(Z) is quite high and the decisionmaker may wish to reduce this uncertainty by measuring elevation at additional survey locations. As a preliminary calculation, it is useful to assess the value of acquiring perfect information. If Z were known, then the optimal action would be:

$$a^*(z) = \begin{cases} a_1 & \text{if } k(z) < 200 \\ a_2 & \text{otherwise} \end{cases}$$

with expected net benefits:

$$E^*(z) = \begin{cases} -k(z) & \text{if } k(z) < 200 \\ -200 & \text{otherwise} \end{cases}$$

The current expected value of E*(z) is:

$$E(E^*(Z) \mid x) = \int E^*(z)\, f(z \mid x)\, dz$$

which is approximated by:

$$E(E^*(Z) \mid x) \approx \sum_{j=1}^{K} E^*(z^j_{|x}) / K$$

where $z^j_{|x}$, j = 1, 2, ..., K, are independent realizations simulated from f(z | x). Figure 4 shows the histogram of $k(z^j_{|x})$ for j = 1, 2, ..., 100, and based on these values E(E*(Z) | x) = -136. This provides an upper bound for E(E*(x, U) | x).

Suppose, now, that the decisionmaker must decide whether to leave the exposed region unprotected based on current information or measure elevation at the 9 additional survey locations also shown in Figure 3. Suppose that the cost of this additional information is 5. The method outlined in the previous section was used to estimate the value of this additional information. Specifically, a total of 20 conditional simulations of Z given x were generated. The value of U was found for each simulation, an additional 20 conditional simulations of Z given x and u were generated, and a*(x, u) and E*(x, u) were found. These values are given in Table 2. Note that in only 1 case was a*(x, u) = a_2. Finally, the average of the values given in Table 2, which is around -146, was taken as an estimate of E(E*(x, U) | x). The estimated value of the additional information is only 2 and does not exceed the cost of acquisition.

Table 2. Values of $E^*(x, u_{|x}^j)$, j=1, 2, ..., 20, generated by nested conditional simulation

j	$E^*(x, u_{\|x}^j)$
1	-142.5
2	-120.0
3	-160.6
4	-132.9
5	-133.6
6	-133.7
7	-156.8
8	-131.3
9	-107.9
10	-151.7
11	-138.6
12	-157.4
13	-135.9
14	-200.0
15	-102.7
16	-190.1
17	-166.6
18	-113.8
19	-161.0
20	-172.1

DISCUSSION

This paper has described the use of nested conditional simulation to assess the value of information in an explicitly spatial setting. It is worth emphasizing that this use of conditional simulation is essentially an algorithm for implementing a longstanding theory. As in other applications of conditional simulation, the real challenge of this approach lies in the details of the stochastic model and the model of the decisionmaking process.

One way in which the example described in the previous section could have been made more realistic is the following. Note that the histogram in Figure 4 appears to be bimodal. This bimodality arises from the presence in some of the conditional simulations of Z given x of a corridor linking the two regions of low elevation along the bottom of Figure 1. When such a corridor is absent, the number of inundated grid nodes is relatively low. When it is present, the number is relatively high. This effect would be easily detected by viewing maps of these conditional simulations. As a consequence, rather than considering the acquisition of the survey locations shown in Figure 3, it would be more sensible to consider the acquisition of survey locations aimed at establishing the existence of this kind of corridor. Thus, the first stage of conditional simulations could be used to optimize the location of prospective additional survey points.

Finally, in the simple example described in the previous section, elevation Z(t) only enters the decisonmaking process through the indicator I(t). In this case, it is not necessary to simulate the full elevation field, but only the indicator field. However, in more realistic applications (e.g., those involving floods of uncertain height), more complete information about elevation would be needed.

ACKNOWLEDGEMENTS

This work was supported by the Economics Project of the NOAA Program for Global Change. The comments of two anonymous reviewers and R.M. Srivastava are gratefully acknowledged.

REFERENCES

Berger, Joseph O. (1986) Statistical Decision Theory and Bayesian Analysis, Springer-Verlag, New York.

Deutsch, Clayton V. and Journel, Andre G. (1992) GSLIB: Geostatistical Software Library and User's Guide, Oxford University Press, Oxford.

COMMENT ON "CONDITIONAL SIMULATION AND THE VALUE OF INFORMATION: A BAYESIAN APPROACH" BY A.R. SOLOW AND S.J. RATICK

R. MOHAN SRIVASTAVA
FSS International
800 Millbank
Vancouver, BC
Canada V5Z 3Z4

Solow and Ratick present an example of the use of conditional simulation for determining whether additional sample information is valuable. When used for this type of risk analysis, geostatistical conditional simulation is simply an adaptation of classical Monte Carlo methods to a spatial setting. There is an implicit assumption with such methods that the outcomes used in the calculations are equiprobable. For example, Solow and Ratick's equation for approximating the expected net benefit is an equally-weighted average of the net benefit calculated on K independent realizations:

$$E(E^*(Z)|x) = \frac{1}{K}\sum_{j=1}^{K} E^*(z^j_{|x})$$

Such a calculation is perfectly reasonable if any one of the K realizations is as likely as any other one. If the realizations are not equiprobable—if the computer code that generates them is not fairly sampling the full space of uncertainty—then the whole approach is compromised. A non-representative set of outcomes will lead to a biased calculation and, possibly, to erroneous decisions.

For most classical applications of the Monte Carlo approach, it is easy to verify that a particular algorithm is producing equiprobable outcomes since the space of uncertainty is usually either a univariate distribution or a set of jointly independent univariate distributions. As long as the algorithm has access to a random number generator that can produce outcomes uniformly distributed between 0 and 1, these can be transformed through the inverse of the cumulative distribution function to values that fairly sample any univariate distribution.

When it comes to spatial simulation, however, it becomes quite awkward to verify that the space of uncertainty is being fairly sampled and that the realizations produced by the computer code are a reasonable basis for risk analysis. The awkwardness arises from the fact that we do not have good analytical definitions of what the space of uncertainty should be, so we have difficulty evaluating whether a particular

R. Dimitrakopoulos (ed.), Geostatistics for the Next Century, 218–219.

set of realizations is a fair sampling of this space. Some geostatisticians, such as Journel, insist that the space of uncertainty cannot be defined for most interesting practical spatial problems. Their view is that *any* algorithm that generates a unique and repeatable outcome from a given random number generator seed can be said to be generating equiprobable outcomes in the sense that their space of uncertainty is simply the set of all possible outcomes that the computer code would generate if it was fed every possible random number seed. While this view has theoretical rigor, it leaves most people uneasy since it undermines our intuitive sense that there is some objective space of uncertainty against which a set of realizations could be judged.

There are now several papers in the technical literature that demonstrate that different algorithms for geostatistical conditional simulation generate different spaces of uncertainty. Deutsch's contribution to these proceedings, for example, shows that the space of uncertainty generated by sequential gaussian simulation is different from that generated by sequential indicator simulation, which is different from that generated by annealing. If conditional simulation is being used only to produce art for the walls of research centers, or to produce pictures that show spatial variability, then the issue of the space of uncertainty is irrelevant. When it is being used for risk analysis, however, this issue becomes critical.

Using Solow and Ratick's flood control example, what sense (if any) could we make from our results if sequential gaussian simulation told us that we should build the dike while simulated annealing told us the contrary? Without an ability to check the realizations against some space of uncertainty that is algorithm-independent, we will never be able to resolve such contradictions. Currently, the only such algorithm-independent space of uncertainty that we know of is the one generated by a multivariate normal distribution. Is it better to cling to the multivariate normal distribution as the only island in a sea of arbitrariness? Or should we boldy set sail with the new and more flexible algorithms, such as annealing, and ignore the problem that we can never tell how badly our boat is leaking?

CHANGE OF SUPPORT AND SCALE ISSUES

FRACTALS, MULTIFRACTALS, AND CHANGE OF SUPPORT

FREDERIK P. AGTERBERG
Geological Survey of Canada
601 Booth Street, Ottawa K1A 0E8
Canada

This paper reviews concepts and methods from two different fields: geostatistics and fractal geometry. Previous fractal interpretations of semivariograms were almost exclusively based on the fractal landscape model. It is demonstrated here that other fractal and multifractal models are useful for the analysis of self-similar sets and random variables in space. Such models can result in exponential and power-law type spatial covariance models which are complementary to the linear and de Wijsian models for the semivariogram of logarithmically transformed data. A simulated multiplicative cascade model is used for illustrating the concepts of self-similarity, multifractal spectrum, and change of the shape of frequency distributions due to change of support.

INTRODUCTION

During the past 15 years there has been a rapid growth of applications of fractal models in many fields of science. Overviews of theory and applications are prodided in Feder (1988) and Falconer (1990). Because of publication of numerous articles, several collections of papers (e.g., Scholz and Mandelbrot, 1989), books (e.g., Korvin, 1992), and software including Roach and Fowler (in press), we are now in a good position to evaluate the usefulness of fractal concepts in the earth sciences.

One of the attractions of fractal models is their simplicity. A distinction is made between fractal dimension D and topological dimension d. In general, when a fractal with $d=1$ is measured with a yardstick of variable length (ϵ), this results in a power-law relationship between the length and the measuring scale. For example, the length $L(\epsilon)$ of a coastline ($d=1$) satisfies $L(\epsilon) = c\ \epsilon^{1-D}$ where c is a constant and the fractal dimension D of the object satisfies $1<D\leq 2$. Such models are easily tested because they reduce to straight lines on log-log paper.

Mandelbrot's (1983) book includes fractal landscapes (computer simulations) which resemble mountainous terrain. The horizontal contours and lines of intersection with vertical planes have the same fractal dimension D for each landscape (D varies between

R. Dimitrakopoulos (ed.), Geostatistics for the Next Century, 223–234.

1.1 and 1.7). In an early application, the thickness of the Sparky sandstone in the Lloydminster area, Alberta, was modelled as a fractal of this type (Agterberg, 1980). Large-scale isopach maps for this sandstone show contours with more details and greater lenghts than those on small-scale maps for the same region. However, the contours on both types of maps have approximately $D = 1.34$, indicating that the fractal model is useful for expressing irregularities and degree of smoothing of contours.

Mandelbrot and Van Ness (1968) showed that the expected value $E(X_i - X_{i+h})^2$ for elevation X_i at point i along a sampling line across the fractal landscape is proportional to h^{4-2D} where h represents distance (or "lag"). Thus a spatial random variable X with a variogram that satisfies a straight line on log-log paper might be representative of a fractal. The fractal landscape also results in a power-law relation for the power spectrum. Burrough (1981) presented 38 examples resulting from a survey of geostatistical literature. Recent applications include the geochemical landscape model proposed by Bölviken et al. (1992). However, the variograms in this paper show well-developed sills not in agreement with the power-law model. Existence of a sill implies existence of finite population mean and variance, whereas the landscape model predicts an infinitely large variance. It is possible to approximate the fractal landscape model by a finite-variance model in which the difference between covariance and variance follows a power-law for small values of h (cf. Falconer, 1990, p.158).

Bruno and Raspa (1989) examined fractal models of surfaces from a geostatistical perspective. They pointed out that commonly used models such as the spherical and exponential models are not compatible with a fractal model predicting a power-law relation for the variogram. On log-log paper these two models result in curves with slopes that are not constant. Conversion of these slopes into fractal dimensions by means of the preceding formula gives $D = 1.5$ at the origin for both models, with a gradual increase to $D = 2$ on the sill. According to the landscape model, the "nugget effect" is fractal with $D = 2$. Bruno and Raspa concluded that fractal models have limited application in geostatistics.

In this paper, fractals are explored from the point of view that assumptions of self-similarity at different scales can provide useful models for the study of the frequency distributions of spatial random variables with change of support. Because of limitations of the simple landscape model, emphasis will be on other fractal and multifractal models which are more flexible in that they may produce covariance models which are equivalent to semivariogram models with a sill. Not all semivariograms can be described in this way but it will be demonstrated that this line of research is more fruitful than thought previously.

The organization of this paper is as follows. First it will be shown that fractal models for logarithmically transformed data often result in models with finite variance for untransformed data. For example, after logarithmic transformation the "signal" in a signal-plus-noise model with exponential covariance function has the characteristics of a fractal with $D = 1.5$ ($d=1$). Next a multiplicative cascade model is discussed which, for given support (block size), produces a de Wijsian semivariogram after logarithmic transformation. The original values for this model have power-law type covariance. This also provides an example of a multifractal which has a spectrum of fractal dimensions.

MODELLING WITH AND WITHOUT THE LOGNORMAL TRANSFORMATION

Suppose that the semivariograms of X and $\log_e X$ are written as $\gamma^*(h)$ and $\gamma(h)$, respectively, with:

$$\gamma^*(h) = \frac{1}{2}E(X_i - X_{i+h})^2; \quad \gamma(h) = \frac{1}{2}E(\log_e X_i - \log_e X_{i+h})^2 \quad (1)$$

If it can be assumed that the mean $EX=m$ and the variance of X (to be written as *var*) are finite, the covariance $cov(h)$ satisfies:

$$cov(h) = E(X_i X_{i+h}) - m^2 = var - \gamma^*(h) \quad (2)$$

This covariance is related to the semivariogram for logarithmically transformed values by means of (cf. Matheron, 1974; Agterberg, 1974, p.339):

$$cov(h) = var \cdot e^{-\gamma(h)} \quad (3)$$

provided that $\log_e X$ is approximately normal (Gaussian) with variance $\sigma^2 >> \gamma(h)$. An example of application of Eq.(3) is as follows.

De Wijs (1951) used assay values from the Pulacayo sphalerite-quartz vein in Bolivia for example. Along a drift 118 channel samples had been obtained at 2.00-meter intervals (see Fig. 1). The massive sulphide vein was on average about 0.50 meter wide but all samples were cut over the anticipated stoping width of 1.30m. Consequently, lower grade wallrock material containing disseminated sphalerite was included in the samples before assaying. These channel samples provide estimates of the zinc concentration in 2m-long blocks measured along the vein in the direction of the vein. As shown by de Wijs, the zinc concentration values are approximately lognormally distributed.

Estimated covariances for the 118 zinc values are shown in Figure 2a. The scale for the covariance is logarithmic. Also shown is the straight line corresponding to a signal-plus-noise model of the type $cov(h)=c\ var\ e^{-ah}$ ($a=0.1892$; $c=0.5157$). As described in Agterberg (1967), the noise can be removed by using a bilateral exponential filter whose coefficients are functions of a and c. The resulting estimates of the signal, also shown in Figure 1, probably are better estimates of average concentration in consecutive 2m-long blocks than the original channel samples.

The exponential covariance model of Figure 2a corresponds to a linear semivariogram

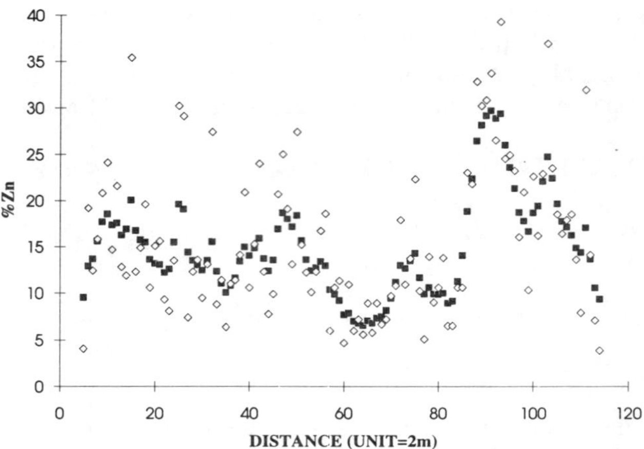

Figure 1. Pulacayo sphalerite vein deposit. The 118 zinc concentration values (open squares) were filtered in order to extract signal (black squares) (after de Wijs, 1951; Agterberg, 1967, 1974).

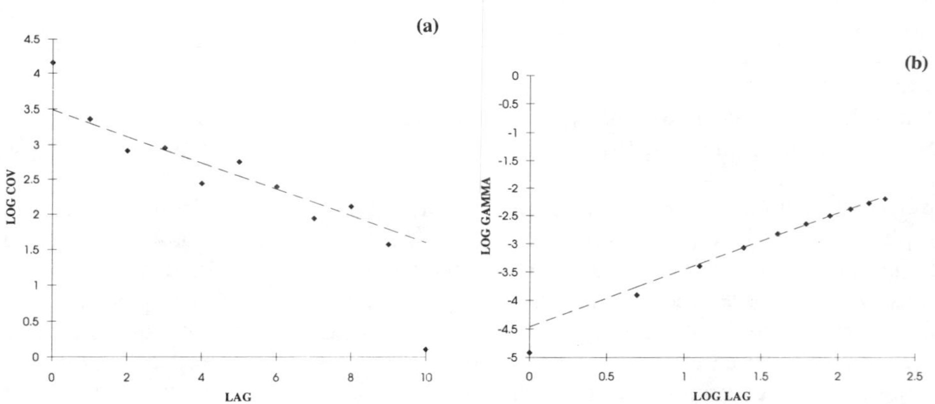

Figure 2. (a) Exponential function (straight line with slope $-a=-0.1892$) fitted to covariance estimates for zinc values of Figure 1. Nearly half of the variance of the zinc values can be attributed to noise (nugget effect). (b) Semivariogram for logarithmically transformed zinc signal values with logarithmic distance scale. Straight line has unit slope. This is in agreement with exponential covariance model for untransformed data.

for logarithmically transformed zinc values. This follows immediately from substitution of $\gamma(h) = ah$ into Eq.(3). Consequently, the semivariogram of logarithmically transformed zinc values plotted on log-log paper should be a straight line with slope equal to one. In Figure 2b it is shown that this model is approximately satisfied. It can be concluded that the logarithmically transformed zinc values behave as a fractal which is a special case of the simple landscape model. From $4\text{-}2D = 1$, it follows that $D=1.5$. This is the Hausdorff dimension of ordinary Brownian motion (cf. Feder, 1988, p.163).

THE MODEL OF DE WIJS

De Wijs (1951) argued as follows. Suppose that an orebody with average metal concentration m is cut into halves of equal volume. The two parts will have different metal concentration values which can be set equal to $(1+d)m$ and $(1\text{-}d)m$, with $d>0$. Let $\eta = (1+d)/(1\text{-}d)$ represent the ratio of the greater concentration value divided by the lesser concentration value. De Wijs assumed that η for two adjacent blocks remains constant regardless of size and shape of the blocks. Thus η would also control ore concentration variability when the two parts are cut into halves, and when the process of cutting is continued during successive stages. The only restriction mentioned by de Wijs is that the process should be stopped so that all blocks remain significantly larger than individual ore mineral grains. Because volumes of blocks are kept equal at each cut, η also represents the expected value of the ratio of amounts of metal in the two blocks.

Figure 3 illustrates the preceding model with $\eta = 7/3$, after $k=8$ stages of cutting. Initially (at stage $k=1$), a hypothetical rod-shaped orebody with total amount of metal arbitrarily set equal to 25,600 (and volume equal to 256,000 so that $m=0.1$) was cut in half and a random number generator was used to determine which half would receive the greater amount of metal ($= (\eta+1)/\eta = 1.43$ times as much as it had before). This random allocation of values with constant ratio was continued during successive cuts. The end product of this multiplicative cascade model is a sequence of $2^8=256$ values for amount of metal ("measure" in Fig. 3). Because the 256 blocks have the same horizontal length scale ($\epsilon = 256{,}000/256 = 1000$), these values are proportional to the concentration values X_i ($i=1,2,\ldots,256$). The average concentration value ($m=0.1$) remained the same during successive stages of cutting. Obviously, $E(X_i) = m$ for any value of i. It is noted that $\eta=7/3$ was selected for the computer simulation experiment of Figure 3, because it applies to turbulent flows. This allows comparison of results to be derived for this example with theoretical and experimental results of Meneveau and Sreenivasan (1987).

The first and last $2^7=128$ values of Figure 3 both can be regarded as an end product obtained after $k=7$ stages of cutting. Likewise, smaller subgroups of values can be interpreted to be the result of fewer stages of cutting. In this sense, the simulated orebody is self-similar on the average. As originally shown by de Wijs (1951), $\log_e X_i$ is a binomial random variable. Consequently, X_i is a logbinomial random variable which cannot be distinguished from a lognormal random variable for large k. If X_k denotes a

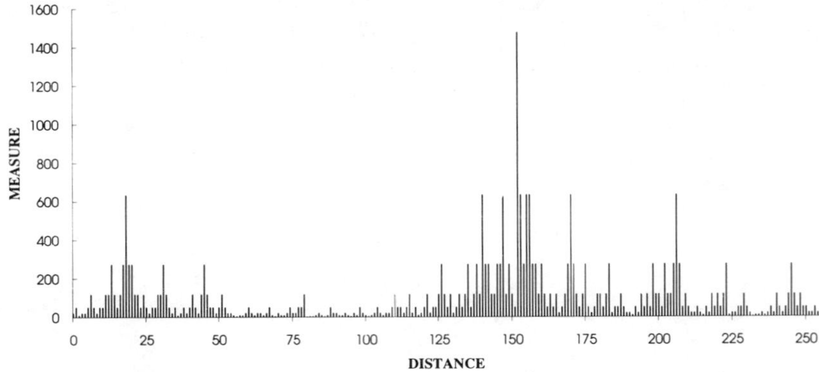

Figure 3. Example of result obtained by model of de Wijs after 8 successive stages of cutting blocks into halves. At each cut, amount of metal in a block was divided into two parts with 70% going into one half and 30% into the other. This ratio (η=7/3) was kept constant but the block with the greater measure was selected at random. This is an example of a binomial multifractal measure.

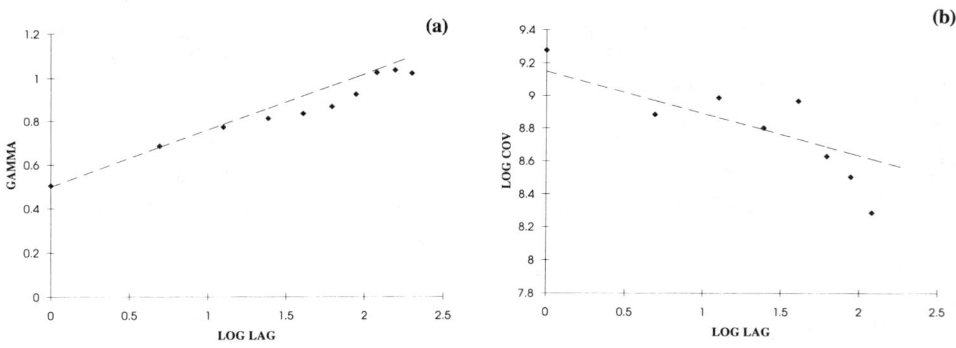

Figure 4. (a) Estimated values of semivariogram $\gamma(h)$ for logarithms of 256 values of Figure 3 plotted against $\log_e h$. Dip of straight line is equal to $\beta = 0.233$ representing theoretical (model of de Wijs) value derived from η=7/3 by means of Eq.(6). (b) Log-log plot for covariance of 256 (untransformed) values of Figure 3 versus distance (lag). Dip of straight line is $-\beta$=-0.233 for theoretical power-law relationship. The corresponding semivariogram (not shown) has a sill. Estimated covariances for lags greater than 8 (not shown) deviate strongly from straight line.

value selected at random from a sequence of concentration values obtained after k stages of cutting, the logarithmic variance satisfies (cf. Agterberg, 1961, Eq.(32), p.157):

$$\sigma^2(\log_e X_k) = \frac{k}{4}[\log_e \eta]^2 \tag{4}$$

Note that this variance tends to infinity as k is increased.

Matheron (1962, p.75) introduced the De Wijsian semivariogram with:

$$\gamma(h) = \beta \cdot \log_e h \tag{5}$$

where β is a constant. He showed that, if two blocks with volumes v and V have similar shapes, then the logarithmic variance of the small blocks (volume v) within the large blocks (volume V) is proportional to $\log_e(V/v)$. Similarity of shape is satisfied for the example of Figure 3 where $V/v=2^8=256$ by assuming that the rod-shaped orebody and its parts can be approximated by straight-line segments. Then the logarithmic variance and β are fully determined by η, and Eq.(4) can be written as:

$$\sigma^2(\log_e X_k) = \beta \cdot k \log_e 2; \quad \beta = \frac{1}{4\log_e 2}[\log_e \eta]^2 \tag{6}$$

(cf. Matheron, 1962, p.308-309). Use of $\eta=7/3$ yields $\beta=0.233$. It can be expected that, after logarithmic transformation, the semivariogram for the values of Figure 3 is logarithmic. This is confirmed experimentally in Figure 4a where estimated values of $\gamma(h)$ are approximately linearly related to $\log_e h$ with $\beta=0.233$.

Substitution of Eq.(5) into Eq.(3) gives:

$$cov(h) = var \cdot h^{-\beta} \tag{7}$$

Consequently, the log-log plot for the covariance of the original zinc values would be a straight line with slope equal to $-\beta=-0.233$. This prediction is confirmed in Figure 4b (for the first eight lags only). As a generalization, suppose that there is a nugget effect γ_0. Then:

$$\gamma(h) = \gamma_0 + \beta \cdot \log_e h \; ; \quad cov(h) = c \cdot var \cdot h^{-\beta} \tag{8}$$

where $c = \exp(-\gamma_0)$. If the de Wijsian semivariogram is used to compute the semivariogram of average values for adjoining blocks (regularization, cf. Journel and Huijbregts, 1978, p. 82), this is approximately equivalent to adding a constant term to the logarithmic semivariogram of Eq. (5), resulting in another power-law relation for $cov(h)$.

Several authors have derived Eq.(7) following a different route. For example, Rose (1992) proposed the following model. If $\rho(h)$ is the autocorrelation function (=covariance divided by variance) for a series of concentration values for adjacent blocks of the same volume and variance $\sigma^2(1)$, then the variance of the average of two successive values satisfies $\sigma^2(2) = \sigma^2(1)\{1+\rho(1)\}/2 = \sigma^2(1)2^{-b}$ where $2^{-b} = \{1+\rho(1)\}/2$. Suppose that, because of self-similarity, the variance of the average of 4 successive values is equal to $\sigma^2(4) = \sigma^2(2)2^{-b} = \sigma^2(1)2^{-2b}$. For the variance of the average of 2^k successive values X_i:

$$\sigma^2(k) = \sigma^2(1)/2^{kb} \tag{9}$$

The corresponding autocorrelation function is:

$$\rho(h) = \frac{1}{2}[(h+1)^{2-b} - 2h^{2-b} + (h-1)^{2-b}] \tag{10}$$

This equation had also been derived by Mandelbrot and Van Ness (1968) for increments in fractional Gaussian noise. Voss (1985) used it to for computer simulation of fractal landscapes. Replacement of the second-order difference for $\rho(h)$ in Eq.(10) by a second derivative yields Eq.(7) with $\beta=b$ and $var=\sigma^2(1)$. This equivalence of results is not surprising: because averaging groups of 2^k successive values such as those shown in Figure 3 is the reverse of the process used to generate these values, the self-similarity implied by Eq.(9) is approximately satisfied.

THE MODEL OF DE WIJS AS A MULTIFRACTAL

Multifractals (cf. Stanley and Meakin, 1988) are spatially intertwined fractals with a continuous spectrum of fractal dimensions. Meneveau and Sreenivasan (1987) presented a multifractal cascade model for the dissipation field in fully developed turbulence. This model provided better results than an earlier fractal model for turbulent flows with a single value of D. Recently many authors including Lovejoy and Schertzer (1990) have developed other types of multifractal models.

Schröder (1991) introduces the concept of multifractals by discussing the model of de Wijs. The only difference between Figure 3 and the patterns more frequently used to illustrate multifractals is in the random choice of which part receives the greater measure at a split. If the greater value always is assigned to the block on the right, the result is a deterministic multifractal as discussed by Feder (1988) and Schröder (1991). As explained before, the random multifractal of Figure 3 with its approximately power-law type covariance was generated by using a random number generator for independence of "left-right" choices from level to level. Each bar can be regarded as the amount of metal $\mu(\epsilon)$ in a cell of size ϵ.

Evertsz and Mandelbrot (1992, p.931) define the "coarse" Lipschitz-Hölder exponent α as:

$$\alpha = \frac{\log_e \mu(\epsilon)}{\log_e \epsilon} \tag{11}$$

The measure of Figure 3 has only 9 possible values. In general, any positive value is possible but values can be grouped by defining classes as in a histogram. The coarse Lipschitz-Hölder exponent serves to label the blocks covering the set supporting a measure. In general, there are $n_\alpha(\epsilon)$ blocks for each α. By addition of measures for successive pairs of adjacent blocks, new measures for larger blocks can be obtained. For example, there are 128 measures averaging 200 for cell size 2000 in the situation of Figure 3.

Suppose that, for given ϵ, there are $n(\epsilon)$ measures μ_i. Then the partition function for moment q is defined as:

$$\chi_q(\epsilon) = \sum_{i=1}^{n(\epsilon)} \mu_i^q \tag{12}$$

Taking subsets of cells with Lipschitz-Hölder exponents between α and $\alpha+d\alpha$, and replacing the measures μ_i of the blocks by ϵ^α (cf. Eq. 11), this becomes:

$$\chi_q(\epsilon) = \int_0^\infty n_\alpha(\epsilon)\,(\epsilon^\alpha)^q\,d\alpha \sim \int_0^\infty \epsilon^{q\alpha-f(\alpha)}\,d\alpha \tag{13}$$

The distribution of α is described by means of a function $f(\alpha)$ defined by assuming that $n_\alpha(\epsilon)$ is proportional to $\epsilon^{-f(\alpha)}$. The value of each integral is determined by α's close to the value for which the exponent $q\alpha$-$f(\alpha)$ is a minimum. Introduction of a function $\tau(q)$ yields:

$$\chi_q(\epsilon) = c\cdot\epsilon^{\tau(q)}; \quad \tau(q) = q\alpha(q) - f(\alpha(q)); \quad \frac{\partial}{\partial q}\tau(q) = \alpha(q) \tag{14}$$

The function $f(\alpha)$ can be interpreted as the negative of the Legendre transform of $\tau(q)$. If S_α represents any subset of blocks with the same coarse Lipschitz-Hölder exponent α, then the multifractal spectrum $f(\alpha)$ provides an estimate of the fractal dimension of S_α.

A log-log plot of partition function versus block size was constructed for Figure 3 by means of Eq.(12). Six different block sizes (ϵ) were used. According to the first part of Eq.(14), the slopes of the lines in this plot are equal to $\tau(q)$. Figure 5a shows 41 estimates of $\tau(q)$. A cubic interpolation spline was used for obtaining a continuous curve passing through all points in Figure 5a. The first derivative of this curve gave α (see Fig. 5b). Figure 5c shows the final multifractal spectrum $f(\alpha) = q\alpha - \tau(q)$.

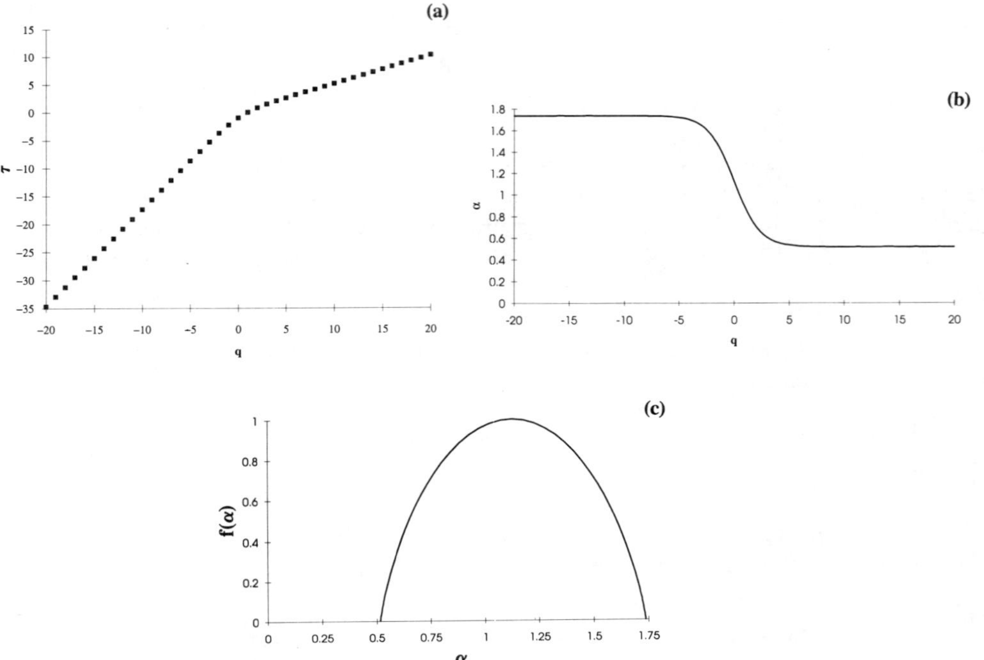

Figure 5. (a) Relationship between $\tau(q)$ and q. (b) First derivative with respect to q of cubic interpolation spline-curve for τ constructed through points shown in (a). The result is α as a continuous function of q. This curve would reduce to a horizontal straight line for a fractal. (c) Multifractal spectrum $f(\alpha)$ ($= q\alpha-\tau$) corresponding to spline-curve for τ and its first derivative (b). Because of choice of η ($=7/3$), Figures 6b and c resemble curves shown by Meneveau and Sreenivasan (1987, Figs. 1 and 2).

If α would be constant, the curves of Figure 5 all would reduce to straight lines representing a fractal. The fact that this does not happen confirms that the measure of Figure 3 is multifractal. For decreasing value of q, $\tau(q)$ approaches a straight line with slope α_{max}, and for increasing q, a straight line with slope α_{min}. It can be shown (cf. Evertsz and Mandelbrot, 1992, p.932) that $\alpha_{max} = \log_2(\eta+1)$, and $\alpha_{min} = \log_2(\eta+1)/\eta$. Consequently, the domain of $f(\alpha)$ ($=\alpha_{max} - \alpha_{min}$) should be equal to $\log_2 \eta = 1.222$ which is also the maximum width of the curve shown in Figure 5c.

The multifractal formalism looks primarily at the frequency distribution of the measure being analyzed as opposed to the spatial distribution structure where covariances and semivariograms are relevant. Any permutation of a multiplicative cascade-generated multifractal measure (including the deterministic multifractal discussed at the beginning of this section) produces exactly the same multifractal spectrum.

ACKNOWLEDGEMENTS

I am grateful for critical review of the first draft of this paper and helpful comments by A.J. Desbarats (Geological Survey of Canada, Ottawa), A.D. Fowler (Department of Geology, University of Ottawa), D.E. Roach (Canadian Centre for Remote Sensing, Ottawa), A. Solow (Woods Hole Oceanographic Center) and an anonymous reviewer. Quiming Cheng (Ottawa-Carleton Geoscience Centre) is thanked for comments as well as assistance in preparation of the diagrams. This paper is Geological Survey of Canada Contribution No. 13593.

REFERENCES

Agterberg, F.P. (1961) "The skew frequency-curve of some ore minerals", Geologie en Mijnbouw 40, 149-162.

Agterberg, F.P. (1967) "Mathematical models in ore evaluation", J. Canadian Operations Research Soc. 5, 144-158.

Agterberg, F.P. (1974) Geomathematics, Elsevier, Amsterdam, 596p.

Agterberg, F.P. (1980) "Mineral resource estimation and statistical exploration", in A.D. Miall (ed.), Facts and Principles of World Petroleum Occurrence, Canadian Soc. Petroleum Geol., Memoir 6, pp. 301-318.

Bölviken, B., Stokke, P.R., Feder, J., and Jössang, T. (1992) "The fractal nature of geochemical landscapes", J. Geochemical Exploration 43, 91-109.

Bruno, R., and Raspa, G. (1988) "Geostatistical characterization of fractal models of surfaces", in M. Armstrong (ed.) Geostatistics, Volume 1, Kluwer Academic Publishers, Dordrecht, pp. 77-89.

Burrough, P.A. (1981) "Fractal dimensions of landscapes and other environmental data", Nature 294, 240-242.

De Wijs, H.J. (1951) "Statistics of ore distribution", Geologie en Mijnbouw 13, 365-375.

Evertsz, C.J.G., and Mandelbrot, B.B. (1992) "Multifractal measures", Appendix B in H.-O. Peitgen, H. Jürgens and D. Saupe, Chaos and Fractals, Springer-Verlag, New York, pp. 922-953.

Falconer, K. (1990) Fractal Geometry, Wiley, New York, 288p.

Feder, J. (1988) Fractals, Plenum, New York, 283p.

Journel, A.G., and Huijbregts, Ch.J. (1978) Mining Geostatistics, Freeman, New York, 600p.

Korvin, G. (1992) Fractal Models in the Earth Sciences, Elsevier, Amsterdam, 408p.

Lovejoy, S., and Schertzer, D. (1990) "Multifractals, universality classes, and satellite and radar measurements of cloud and rain fields", J. Geophys. Research 95 (D3), 2021-2034.

Mandelbrot, B.B. (1983) The Fractal Geometry of Nature (updated and augmented edition), Freeman, New York, 468p.

Mandelbrot, B.B., and Van Ness, J.W. (1968) "Fractional Brownian motions, fractional noises and applications", SIAM Review 10, 422-437.

Matheron, G. (1962) Traité de Géostatistique Appliquée, Mém. Bur. Rech. Géol. Minières, 14, 333p.

Matheron, G. (1974) "Effet Proportionnel et Lognormalité ou: le Retour du Serpent de Mer", Note Géostatistique No. 124, Centre de Morphologie Mathématique, Fontainebleau, 43p.

Meneveau, C., and Sreenivasan, K.R. (1987) "Simple multifractal cascade model for fully developed turbulence", Phys. Review Letters 59, 1424-1427.

Roach, D.E., and Fowler, A.D. (in press) Dimensionality analysis of patterns: fractal measurements, Computers & Geosciences.

Rose, C.D. (1992) "A fractal model for sampling coal, ores, and other natural populations", J. Coal Quality 11, 6-13.

Scholz, C.H., and Mandelbrot, B.B. (1989) Introduction to "Fractals in Geophysics", Pure Applied Geophysics 131, pp. 1-4 (followed by other papers on pp. 5-313).

Schröder, M. (1991) Fractals, Chaos, Power Laws, Freeman, New York, 429p.

Stanley, H.E., and Meakin, P. (1988) Multifractal phenomena in physics and chemistry, Nature 335, 405-409.

Voss, R.F. (1985) "Random fractal forgeries", in R.A. Earnshaw (ed.) Fundamental Algorithms in Computer Graphics, Springer-Verlag, Berlin, pp. 805-835.

CHANGE OF SCALE IN RANDOM MEDIA

Dominique JEULIN
Centre de Géostatistique
Ecole des Mines de Paris
35 rue Saint–Honoré, 77305 FONTAINEBLEAU
France

When considering the physical properties of heterogeneous media and their fluctuations at various scales, it is necessary to introduce appropriate techniques and models. The problem is complex, since in the general case, no linear (or even non linear) rule of combination of the physical properties given on the lower scale can predict the higher scale properties. This problem is encountered in many situations, such as flows in porous media, elastic properties or strength of composite materials. It is solved by appropriate methods, using case models of random media.

The macroscopic (or effective) properties (like an overall permeability) can be estimated from partial knowledge on the microstructure (lower scale) by variational methods. In that case, approximations of average properties of infinite random media are obtained.

The statistical distribution of the properties of finite domains can be obtained theoretically for physical properties sensitive to local heterogeneities, as in strength of materials. Scale effects, like the change in mean properties and variance with the size of specimens can be predicted and compared to experimental data.

Using numerical simulation of random media, the macroscopic properties and their distributions can be estimated from the solution of a discrete version of partial differential equations. This can be obtained by finite element calculations (for instance in the case of the mechanical behavior of materials) and by other techniques, such as lattice gas simulations for flows in porous media. In other cases, approximate methods, based on geodesic propagations on random microstructures, can also provide good estimates of transport or mechanical properties.

INTRODUCTION

To predict the overall behavior of heterogeneous media from its structure is of importance in many fields of applied sciences. In materials science, this is the way

R. Dimitrakopoulos (ed.), Geostatistics for the Next Century, 235–244.

to design and to produce materials with appropriate microstructures. For the simulation of flows in porous media, it is necessary to estimate the permeability at the scale of the discretization used for the calculations. In any case, models are required to solve the problem of change of scale. This includes the emergence of a macroscopic physical law and the estimation of its macroscopic coefficients (namely the "effective properties"). Another point of interest is the scale effect on the fluctuations of the effective properties as a function of the microscopic variability for specimens with finite size.

The first point can be addressed by models of periodic structures which are deterministic, using homogenization techniques [1]. The second point requires a probabilistic approach involving the use of models of random media, as presented below.

In this paper, the following problems will be outlined: estimation of the effective properties of random media, scale effects on the distribution of overall properties (illustrated for the strength of heterogeneous solids) and use of simulations. The text, which is limited to a short introduction to the main topic, is not a general review of this broad domain. It is restricted to a presentation of the techniques, together with their main limitations and expected developments "for the next century". More details and application to various physical problems are given in the references.

EFFECTIVE PROPERTIES OF RANDOM MEDIA

From a macroscopic point of view, the behavior of a physical system can be considered as its response to demand. In the current practice, the physical variables are well identified (for instance the stresses and the deformations in a solid) and are related by constitutive relationships. The physical meaning of the variables and the validity of the constitutive relationships always depend on the scale of observation [2].

It is not always easy to give an answer to the problem of the effective parameters, since it can depend on the boundary conditions. An instructive example was presented by G. Matheron [2] in the case of the composition of permeability of random media. Darcy's law and the macroscopic permeability exist for uniform flows in porous media; for radial flows across two concentric boundaries, the observed macroscopic permeability depends on the geometry, and therefore is not an intrinsic property of the medium.

The methods of calculation of the effective properties for periodic or random structures, also called **homogenization techniques**, are well documented [1−3]. In the case of probabilistic models, only linear constitutive behaviors (like Darcy's law) are fully studied. For simplicity, we will refer here to stresses and deformations in elastic media, related by a proportionality tensor through Hooke's law [4].

To summarize, various approaches can be followed, as illustrated by some major references:

- The difference between the stress field $\sigma(x)$ (or the deformation field $\varepsilon(x)$) of the heterogeneous medium and of a reference homogeneous medium, can be iteratively calculated [4–6].
- A perturbation method develops the difference between the macroscopic and the microscopic coefficient as powers of a small parameter α (using the Schwydler tensor in [2]).
- Particular trial fields are chosen to minimize an energy criterion (J. Willis in [1]). Powerful variational methods, that can be coupled to the above mentioned iterative procedures, give bounds for the effective properties according to the available information on the random microscopic coefficient. New bounds were recently derived by G. Matheron [7].

The common property of these estimators is that they account for correlation functions of the medium up to order n (for n iterations, or for a development up to α^n) and for the Green operator corresponding to a homogeneous medium subject to the same boundary conditions. Theoretical correlation functions for models of random media [2, 8–10] can be used in the calculations, and should be preferred to empirical correlation functions, which are hardly available for high orders (for $n>4$, three dimensional maps of the medium are required). Interesting new results are expected in this direction.

Some empirical approaches derive estimates of the effective properties by averaging the punctual values of the coefficient after an appropriate transformation (such as a power law); this kind of averaging has no theoretical foundation, except for the case of log symmetrical two dimensional media, where the geometric average is exact [2]. Otherwise, no such average exists. Instead, the spatial law of the random coefficient is required in the estimate, at least through the correlation functions.

The main limitations of these methods are the following:

- The above developments diverge for a zero or infinite elastic modulus (pores or rigid inclusions). In these conditions, other techniques are available, like the self–consistent method (A. Zaoui in [1]). It accounts for the interaction between a particle and the equivalent homogeneous medium, and gives good order of magnitude results. In a probabilistic version, each ellipsoidal particle sees a random field (S. Kanaun in [11]).
- These methods cannot be applied to media with very long range correlations, for which it is hard to describe an equivalent homogeneous medium.

Improvements are expected in the following directions:

- Better estimators of effective properties could be obtained with information about the connectivity of the components of heterogeneous media. The presence of continuous paths in each phase can be particularly important in the case of transport properties, where percolation effects usually appear. This is studied in statistical physics for simplified geometrical models, which give interesting qualitative indications on the behavior of heterogeneous media, despite the use of unrealistic descriptions. These kinds of indications are hardly present in correlation functions, at least for stereological reasons, since partial information on connectivity requires third order statistics in the plane and fourth order statistics in the three–dimensional space to estimate the Euler Poincaré characteristic (or connectivity number) [9,10].

• The extension to non–linear constitutive laws (for instance an elastic–plastic behavior) is another challenge. It is difficult in that case to use Greens operators, unless an incremental calculation is made. In that case, it is necessary to follow the local changes of the moduli during the load. This requires sequential models that depend on the initial structure and on the evolution of the load. Without simplified assumptions on the geometry or the behavior, this type of extension seems difficult in the frame of a probabilistic approach. Some progress is being made by variational methods, where only information on volume fractions is available [12–14].

• The study of the fluctuations of macroscopic properties, which is of interest for bounded domains, is still lacking. Only some approximations of the point histogram (or of the point variance) of the effective property in an infinite random medium are available [15–17]. Similarly, approximations of the covariances of the fields $\sigma(x)$ or $\varepsilon(x)$ are given in [11].

FRACTURE STATISTICS MODELS

The fracture of heterogeneous media is sensitive to the presence of heterogeneities. It results from a local weaker critical stress, or from a local stress concentration. In both cases the incidence of local heterogeneities at a small scale is preponderant, contrary to the case of effective properties where large smoothing effects occur during the change of scale. Therefore appropriate methods and models are required. This is the case of other situations where the input involves changes and instabilities in the medium, as for plastic deformations in the case of the mechanical behavior.

To study the fracture of a material under various loading conditions, fracture statistics models that are consistent on various scales are required. They involve random function models with a point support, and appropriate change of supports, as briefly introduced below.

A first step required to develop fracture statistics models is the **choice of local and of global fracture criteria.** A local criterion is sensitive to the fracture initiation, while a global or macroscopic criterion accounts for the fracture of a domain.

Various local criteria can be used: the fracture is initiated at points in the structure where some intrinsic property of the material is reached, as the result of the applied load. Usually, this property is the critical stress $\sigma_c(x)$, or more generally the critical stress intensity factor $K_{Ic}(x)$ for the tensile fracture in linear elastic brittle materials [10,18]. When there is competition between several fracture mechanisms, as for cleavage and intergranular fracture in rocks and in metals, multivariate criteria and multivariate random function models can be used [10,18]. The local fracture energy $\gamma(x)$ corresponding to the creation of a fracture surface is used in [19,20].

The following macroscopic fracture criteria, involving different fracture assumptions, were proposed [10,18]:

- The **weakest link model** is well suited for the brittle fracture of materials; it corresponds to a sudden propagation of a crack after its initiation.
- Models with a **damage threshold** generalize the previous one; they are valid for a fracture with several potential sites for crack initiation.
- Models with a **Griffith crack arrest** criterion compare, for each step of a crack path, the local fracture energy $\gamma(x)$ to the stored energy $G(x)$ due to the deformation of the material.

Formally, the first type of criterion uses a change of support of the information by the operator $\wedge$ (infimum); the second family is connected to a change of support by convolution; finally, the last criterion involves a change of support by the operator $\vee$ (supremum).

In the present approach, an equivalent homogeneous medium with a random critical stress is used. This simplification, which separates the applied field and the critical field, enables us to obtain closed form results without any simulation. This is justified for media with a single component, like polycrystals in metals or in rocks. However, this approach cannot account for small scale stress fluctuations induced by the microstructure when the components have unlike mechanical behaviors. In that case, new models or the use of simulations (as explained later) will be developed. The three mentioned macroscopic fracture criteria give the following models:

i) Fracture statistics models for brittle materials: based on the weakest link assumption, they induce the fracture of a part as soon as for a single point x_0 we have $\sigma(x_0)>\sigma_c(x_0)$. To estimate the probability of fracture with this assumption, it is necessary to know the probability distribution of the minimum of the values $(\sigma_c(x)-\sigma(x))$ over the loaded domain. This can be done for some random structure models when using the deterministic field $\sigma(x)$ seen by an equivalent homogeneous medium [10,18]. For instance, the Boolean random varieties describe structures with different geometrical defects: points or grains, fibers, strata. A particular model of this type gives the well−known Weibull distribution, often used in applications, as a probability of fracture under a homogeneous stress field. For the weakest link models, the size effect, i.e. the decrease of the median strength with the volume of the specimen, depends on the choice of the model and of the statistical properties of the defects (size, shape, critical stress). For a fixed population of defects, the size effect increases in the order: strata, fibers, grains.

ii) Fracture statistics models with a damage threshold. It is possible to generalize the weakest link criterion where a critical volume fraction of the defects with a fracture stress lower than the applied stress can be introduced. Asymptotic results, valid for samples larger than the microstructure, are obtained from the bivariate distribution of the random function $(\sigma_c(x)-\sigma(x))$. For a homogeneous stress field, they predict no size effect for the median stress, and a convergence toward a deterministic behavior.

For defects like points, lines, the critical volume fraction should be replaced by a critical density of defects (for instance in numbers). In the case of a distribution according to a Poisson process, it appears that large specimens are less sensitive to the most severe defects.

Similar models with two fracture modes in competition were built from a mosaic structure simulating a polycrystal [10].

iii) Fracture statistics model with a crack arrest criterion. The fracture probability was calculated for two loading conditions, inducing a stable or an unstable propagation, and for two types of random media (Poisson mosaic and Boolean mosaic). The predicted size effects depend on the tail of the distribution F: for a very slow growth, the fracture probability can decrease with the size of the specimen.

The main advantages of these models are the following:

- They can be easily introduced as a post−processor calculation in a finite element code [21,22].
- They depend on few parameters (2, 3 to 4), and they can be tested from data on various scales: on the macroscopic scale, by means of the experimental distributions obtained from mechanical tests on various specimens geometries; on a microscopic scale, by means of image analysis measurements as illustrated for ceramics in [21,22].
- Exact theoretical results, coherent at different scales, are available.
- Finally, various scaling laws are obtained, according to the chosen fracture criteria, or to the appropriate random structure models, reflecting the situations occurring with experimental data.

CHANGE OF SCALE AND SIMULATIONS

In most situations, the theoretical derivation of change of scale models is intractable, so that it may be of interest to use simulation techniques. The main drawbacks of an approach based on simulations lay in the limited size of simulations due to the high costs of calculations. Difficulties of extrapolation may result from this situation. The literature on this topic is extensive. However, we introduce here some examples based on original methods.

Using **finite element calculations** on simulated random functions, the following mechanical problems could be addressed:

- Evaluation of size effects for the dispersion of the elastic modulus of random media with a finite size [23]. The results of this study can be used when simulations with irregular meshes, or at different scales, are required.
- Combining random function models to simulate the distribution of defects and finite element calculations, it was possible to model the progression of fracture and to estimate its probability for two dimensional composites, which was compared to experimental results [24]: unidirectional carbon/epoxy composites, and a monolayer bi–directional SiC/SiC composite. The key point of the approach concerns the choice of a correct mesh to simulate the distribution of defects. The identification of the parameters of the model is made from mechanical tests at a small scale. The effect of the distribution of the fracture stresses of defects could be deduced from the simulations, underlining the impact of the tail of the distribution that controls the progression of the damage in the material. A multiscale approach is presently being developed for these materials.
- From **finite difference calculations on simulations**, or alternatively from variational methods using Fourier series expansions and simulations on parallelepiped tessellations of space, the composition of Darcy's permeability was solved on bounded domains in [25].
- At a lower scale, flows in porous media can be simulated using the **lattice gas model** [26–27]. This is obtained by interaction of particles moving on a graph (hexagonal for two dimensional simulations). For porous media, the particles are allowed to move on the edges of the graph included inside the pores. At each step of the simulation, the particles can change at random the direction of their velocity in a collision preserving mass and impulse; then they move one step in the direction of their velocity. In the boolean version of the model, an exclusion principle is applied, so that at most a single particle of a given velocity is allowed on every vertex of the graph. This has for consequence the fact that the simulation can be made through iterations of binary operations, without any round–off errors, so that the longtime behavior of the system can be studied. On a larger scale than the grid, the lattice scale obeys the Navier Stokes equations [27]. From a probabilistic point of view, the model simulates random walks of populations in interaction. For simulations on a heterogeneous medium, its properties are taken into consideration for boundary conditions (for instance reflexion of particles on grains of a porous medium). Examples of application concern the estimation of the macroscopic permeability of simulated porous media [28–29], and its changes according to various conditions: fluctuations through different realizations of a random porous medium, effect of the pore volume fraction, of the grain size, and of anisotropy. With the same model, viewed as the simulation of random walks, it is possible to investigate the dispersion in porous media. As compared to classical random walk simulations, it is more suited to the study of the dispersion

in complex flows, since a velocity field coherent with the geometry of the porous medium and with the boundary conditions is obtained. Promising extensions of this model are in progress: the filtering of suspensions in porous media and the formation of aggregated structures [30]; diphasic flows in porous media [31]; flows of liquid gas mixtures and evaporation [32], reaction–diffusion models [33] simulating random spatio–temporal microstructures.

The interest of these models is to introduce physical effects with very few parameters. They are however limited to the description of processes on a very small scale, and therefore cannot be implemented on the scale of the flow around a plane or inside an oil–reservoir, since at present no computer could handle such large data sets. It is promising to use them to find constitutive laws reflecting the behavior on a meso scale and their statistical fluctuations; this can be later introduced in a post processing mode on a larger scale in finite element or in finite difference calculations, exactly as for the above mentioned case of damage progression in random composites [24].

By **geodesic propagations** on graphs with valued edges, simulations of propagations on heterogeneous media can be made. Based on the principle of Fermat, they were developed with efficient algorithms to estimate effective physical properties related to propagation phenomena [34–36]: crack propagation based on a minimal fracture energy criterion [34,36]; diffusion of a product in a composite material [35], sound velocity and fluid invasion in porous media [37]. A good agreement between predicted and measured properties was observed. After an appropriate valuation of the edges of a graph obtained from images of the medium, the method provides an estimation of the macroscopic property on the scale of observation, and also interesting descriptions of the connectivity of random structures, for which various scale effects could be studied [36–37].

CONCLUSION

In this presentation, complementary approaches for solving the change of scale problem in random media were briefly introduced. Based on theoretical calculations or on numerical calculations made on images of the structure or on simulated media, they can be applied to many physical problems, including mechanical properties of materials or flows in porous media. Some progress is expected in various directions: construction of random structure models, development of variational methods coupled with connectivity criteria, modelling of interactions of cracks in fracture processes and generalization of the use of three dimensional simulations with special purpose computers.

REFERENCES

[1] Sanchez Palencia E., Zaoui A. (ed) Homogenization Techniques for Composite Media, Lecture Notes in Physics vol. 272, Springer Verlag, 1987.

[2] Matheron G. Eléments pour une théorie des milieux poreux, Masson, Paris, 1967.
[3] Ericksen J.L., Kinderlehrer D., Kohn R. and Lions J.–L. (Ed.). Homogenization and Effective Moduli of Materials and Media, Springer–Verlag, 1986.
[4] Beran M. J. Statistical Continuum Theories, Wiley, 1968.
[5] Kröner E. Statistical Continuum Mechanics, Springer Verlag, 1971;
[6] Kröner E. in Modelling Small Deformations in Polycrystals, Ch. 8, Statistical Modelling, Elsevier, 1986.
[7] Matheron G. Quelques inégalités pour la perméabilité effective d'un milieu poreux hétérogène. *Cahiers de Géostatistique*, 3, "Compte Rendu des Journées de Géostatistique, 25–26 Mai 1993, Fontainebleau", Ecole des Mines de Paris, 1993, pp. 101–105.
[8] Matheron G. Random sets and integral Geometry, J. Wiley, 1975.
[9] Serra J. Image analysis and Mathematical Morphology, Academic Press, London, 1982.
[10] Jeulin D. Modèles Morphologiques de Structures Aléatoires et de Changement d'Echelle. Thèse de Doctorat d'Etat ès Sciences Physiques, University of Caen, 1991.
[11] Kunin I. A. Elastic Media with Microstructure II, Springer Verlag, 1983.
[12] Talbot D.R.S., Willis J.R. Variational Principles for Inhomogeneous Non–linear Media, IMA Journal of Applied Mathematics, vol. 35, pp39–54, 1985.
[13] Ponte Castaneda P., Willis J.R. On the overall properties of nonlinearly viscous composites, Proc. R. Soc. Lond. A 416, pp. 217–244, 1988.
[14] Willis J.R. On methods for bounding the overall properties of nonlinear composites, J. Mech. Phys. Solids, vol. 39, n°1, pp. 73–86, 1991.
[15] Beran M. J. Fields fluctuations in a two phase random medium, J. Math. Phys., vol 21, (10), Oct. 1980, p. 2583–2585.
[16] Kreher W., Pompe W., Field fluctuations in a heterogeneous elastic material. An information theory approach, J. Mech. Phys. Solids, vol. 33, n°5, pp. 419–445, 1985.
[17] Bobeth M., Diener G., Field fluctuations in multicomponent mixtures, J. Mech. Phys. Solids, vol. 34, n°1, pp. 1–17, 1986.
[18] Jeulin D. Random Functions and Fracture Statistics Models, In A. Soares (ed), Geostatistics Troia '92, Kluwer Academic Publ., Dordrecht 1993 (Quantitative Geology and Geostatistics 5) Vol. 1, pp. 225–236.
[19] Chudnovsky A. and Kunin B. Statistical Fracture Statistics, in M. Mareschal and B.L.Holian (ed), Microscopic Simulations of Complex Hydrodynamic Phenomena, Plenum press, New York, 1992, pp. 345–360.
[20] Jeulin D. Some Crack Propagation Models in Random Media, communication to the Symposium on the Macroscopic Behavior of the Heterogeneous Materials from the Microstructure, ASME, Anaheim, Nov 8–13, 1992. AMD Vo. 147, pp. 161–170.

[21] Berdin C., Baptiste D., Jeulin D., Cailletaud G., Failure models for brittle materials, in J. G. M. van Mier et al. (eds), Fracture Processes in Concrete, Rock and Ceramics, E. & F.N. Spon, London, 1991, pp. 83–92.
[22] Berdin C. Etude expérimentale et numérique de la rupture des matériaux fragiles, Thesis, Ecole des Mines de Paris, May 1993.
[23] Cailletaud G., Jeulin D., Rolland P., Size effect on elastic properties of random composites. Note N– /92/G, Ecole des Mines de Paris.
[24] Baxevanakis C., Boussuge M., Jeulin D., Munier E., Renard J., Simulation of the development of fracture in composite materials with random defects, to appear in Proc. of the International Seminar on Micromechanics of Materials, MECAMAT'93, Fontainebleau, 6–8 July 1993.
[25] Le Loc'h–Lashermes G., Etude de la composition des perméabilités par des méthodes variationnelles, Thesis, Ecole des Mines de Paris, 1987.
[26] Frish U., Hasslacher B. , Pomeau Y., Phys. Rev. Lett. 56, p. 1505, 1986.
[27] Frish U. & al., Complex Systems, 1, p. 649, 1987.
[28] Rothman D. H., Geophysics, 53, p. 509, 1988.
[29] Jeulin D. Flow and Diffusion in Random Porous Media from Lattice Gas Simulations, in T.M.M. Verheggen (Ed), Numerical Methods for the Simulation of Multi–Phase and Complex Flow, Springer–Verlag, pp. 106–123, Berlin 1992.
[30] Brémond R., Jeulin D., Random Media and lattice gas simulations, Project of communication to the "Geostatistical Workshop", Fontainebleau, 27–28 May 1993; Note N–07/93/MM, Ecole des Mines de Paris.
[31] Rothman D. H. Macroscopic laws for immiscible two phase flow in porous media: results from numerical experiments, Journal of Geophysical research, vol. 95, n°B6, pp. 8663–8674, June 1990.
[32] Appert C., Zaleski S., Lattice gas with liquid–gas transition, Phys. Rev. Lett., vol. 64, n°1, pp. 1–4, June 1990.
[33] Dab D., Boon, J.B., Cellular approach to Reaction–Diffusion systems, in Cellular Automata and Modeling of Complex Physical Systems, (eds) Maneville P., Boccara N., Vichniac G. Y., Bidaux R., Springer Proceedings in Physics 46, Springer–verlag, Berlin, 1989.
[34] Jeulin D., On image analysis and micromechanics, Revue Phys. Appl. vol 23, pp. 549–556, 1988.
[35] Jeulin D., Vincent L., Serpe G., Propagation algorithms on graphs for physical applications, J. of Visual Communications and Image Representations, vol. 3, n°2, pp. 161–181, June 1992.
[36] Jeulin D. Damage simulation in heterogeneous materials from geodesic propagations, Engineering computations, vol. 10, pp 81–91, 1993.
[37] Jeulin D., Kurdy M. B., Quelques paramètres géodésiques pour caractériser la connexité de milieux biphasés, Note N–24/89/MM, Ecole des Mines de Paris.

DIRECT CONDITIONAL SIMULATION OF BLOCK GRADES

D. MARCOTTE
Mineral Engineering, École Polytechnique
Case Postale 6079, Succursale A
Montréal, H3C-3A7
Canada

A sequential approach is presented based on the Hermitian model for the change of support and either disjunctive kriging or multiGaussian kriging for the estimation of the conditional distributions. Theoretical block variograms are well reproduced by the simulations. Also, it is shown that taking block grade information into account can modify substantially the results of the simulation.

INTRODUCTION

The interest in simulations comes from the complexity of the problems tackled nowadays by geostatisticians. Short term and medium term mine planning is an example of a problem better solved by simulations than by estimation. Other examples are, the change of support for non additive variables like hydraulic conductivity, or the modelling of petroleum reservoir performance.

In mine planning, it is necessary to predict not only the average grade of a block, or a combination of blocks, but also to forecast the dispersion of true values around the estimated average. A simulation will display the desired variability. The simulation should ideally be conditioned by all the available information, including core samples, mined blocks, stopes, bulk samples of various types. Note that a known 'block' grade can correspond to several spatially separate volumes of different sizes and shapes.

Journel and Huijbregts (1978, p. 511-515) describe three different algorithms for conditional simulation of blocks, none of which can conditioned on data of different support size (core, block or stope). In Gomez-Hernandez (1993), the block-block and block-point covariances were experimentally derived from point simulations. The approach is feasible only if few different block sizes and shapes need be considered in the simulation. In effect, for p different block sizes, a total of $p(p+1)/2$ variograms will have to be modelled consistently. In mining, many different block sizes and shapes can be encountered in a single deposit, thus the need for a change of support model to derive the required covariances analytically is justified.

R. Dimitrakopoulos (ed.), Geostatistics for the Next Century, 245–252.

This note uses the sequential simulation algorithm coupled with disjunctive kriging or multiGaussian kriging with the Hermitian model for the change of support to obtain direct and efficient simulation of blocks conditioned by data defined on various support sizes.

SEQUENTIAL SIMULATION ALGORITHM

The algorithm is based on the definition of conditional distribution. Levy (1937) and Rosenblatt (1952) showed that a multivariate joint distribution can be obtained by a sequence of independent drawings from conditional distributions. This idea was the basis of works by Journel and Alabert (1989), Isaaks (1990) and others. The algorithm starts by defining a random path visiting the nodes to be simulated. The value x_n at the nth visited node is obtained by drawing "u" from a uniform (0,1) distribution such that $F_n(x_n) = u$, where F_n designs the conditional distribution at node n given the n-1 previous nodes. Differences between the various approaches appear in the way the successive conditional distributions are estimated. It could either be obtained as the result of indicator kriging (IK), multiGaussian kriging (MG), or disjunctive kriging (DK), although this last method has not been discussed in the literature for simulation purposes.

The Hermitian model

Matheron (1975a) presented this model for change of support in the frame of disjunctive kriging (DK) (Matheron, 1975b). The DK equations are:

$$\sum_{i=1}^{N} \lambda_k^i \; T_k(x_i, x_j) = T_k(x_i, x_0) \qquad j=1...N; \quad k=1...K \tag{1}$$

where the x_i are locations with informing data and x_0 is the location to be estimated. $T_k(x_i,x_j)$ is defined as $E[\eta_k(x_i),\eta_k(x_j)]$ and the η_k are the normalized Hermite polynomials (note: $T_0=1$ and T_1 is the correlation coefficient between Y and Y_v). The Hermitian model starts with a Gaussian transform of the point support data:

$$Z = \Phi(Y) = \sum_{k=0}^{\infty} C_k \eta_k(Y) \tag{2}$$

where Y is a normal (0,1) random variable, the η_k are the normalized Hermite polynomials, and the C_k coefficients are determined by numerical integration. Similarly, a Gaussian transform of the block grades is defined as:

$$Z_v = \Phi_v(Y_v) = \sum_{k=0}^{\infty} C_k D_{k,v} \eta_k(Y_v) \tag{3}$$

where Y_v is also normal (0,1). To estimate the $D_{k,v}$ coefficients, it is convenient to assume that (Y,Y_v) also follows a bivariate Hermitian distribution:

$$f(y,y_v) = \sum_{k=0}^{\infty} T_k(y,y_v)\ \eta_k(y)\ \eta_k(y_v)\ g(y)\ g(y_v) \tag{4}$$

where 'g' is the normal density. The $D_{k,v}$ coefficients are then obtained by:

$$D^2_{k,v} = \frac{1}{V^2} \int_v \int_v T_k(Y(x), Y(x'))\ dx\ dx' \tag{5}$$

Table 1 shows the T_k coefficients expressions for the point-point, point-block and block-block pairs on the assumption that each pair is Hermitian. In Table 1, v_i, v_j indicates two blocks of possibly different sizes and shapes. Each block can be composed of disjoint sub-blocks. The DK system (1) can be constructed using equation (5) and Table 1.

Table 1: Coefficients used in the DK system (1) for the different pairs involved.

Pair	T_k coefficients
point x_i-point x_j	$T_k(Y_i, Y_j) = \rho^k_{ij}$
point x_i-block v_j	$T_k(Y_i, Y_{v_j}) = \frac{1}{D_{k,v_j}} \frac{1}{V_j} \int_{v_j} T_k(Y_i, Y(x))\ dx$
block v_i-block v_j	$T_k(Y_{v_i}, Y_{v_j}) = \frac{1}{D_{k,v_i} D_{k,v_j}} \frac{1}{V_i V_j} \int_{v_i} \int_{v_j} T_k(Y(x), Y(x'))\ dx dx'$

After the DK system (1) is solved, the DK cumulative function is obtained with:

$$F_{DK}(x_0;y) = G(y) - g(y) \left[\sum_{k=1}^{K} \frac{H_{k-1}(y)}{k!} \sum_{i=1}^{N} \lambda^i_k\ H_k(y_i) \right] \tag{6}$$

where

N is the number of points or blocks in the kriging neighbourhood.
λ are the solutions of (1) (solutions of K different simple kriging systems).
H_k is the non-normalized Hermite polynomial.

Once the cdf is estimated with (6), and corrected for order relation, a simulated value is obtained as follows:

i. a value p is drawn from a uniform(0,1) distribution
ii. equation (6) is solved numerically to find y such that $F_{DK}(x;y) = p$
iii. Z_v is computed using (3) and (5)

The sequential algorithm coupled with DK generate by construction bivariate laws of the form of (4) and the theoretical model-based point and block variograms and cross-variograms are reproduced.

A multiGaussian approach

Once the Gaussian variates corresponding to point and block supports Y and Y_v are determined with the aid of the Hermitian model for change of support, it is tempting to assume a multiGaussian hypothesis in order to simplify the computations. In effect, under a MG hypothesis, only one simple kriging system has to be solved instead of K for the DK approach. However, point and block Gaussian variates can be multiGaussian only if the gaussian transform (2) is linear (Verly, 1984). Nevertheless, the MG approach was used to compare its performance with the more rigorous, but computationally more demanding, DK approach.

SIMULATIONS

The steps to obtain a block conditional simulation based on conditioning points or blocks are:

i. Estimate the C_k coefficients of the Hermite expansion for the point transform function and perform the Gaussian transform.
ii. Compute and model the variogram of the point Gaussian variable.
iii. Estimate the $D_{k,v}$ coefficients, one set for each different block size present in the simulation.
iv. Transform the block grades to Gaussian variable using the $D_{k,v}$ and C_k coefficients.
v. Define a random path to the points or blocks to simulate.
vi. For each point or block in its turn:
    - perform the DK (or MG)
    - draw a probability value
    - find the equivalent Gaussian value
    - find the equivalent original value
    - add the point or block to the conditioning list.

Simulation 1

The point original variate is lognormal with coefficient of variation of 1 and a mean of 5; units are arbitrary. The corresponding normal (0,1) point variate has an exponential variogram with unit sill and practical range 'a'. As the distribution is lognormal, C_k coefficients are determined a priori (see Journel and Huijbregts, 1978, p. 474). An expansion of order 4 (5 coefficients) was used. Along a line, four thousand points are simulated at an interval of a/12, two thousands blocks of size a/6, and one thousand blocks of size a/3 and 2a/3 are simulated at their block size intervals.

Figure 1 shows the theoretical variograms and the experimental variograms of the simulation obtained by DK and MG for the transformed variates for various block sizes. Both methods give virtually identical results in this example and reproduce the theoretical variograms. Thus the computations performed are consistent with the model. Simulated MG and DK values showed correlations in the range .91-.99.

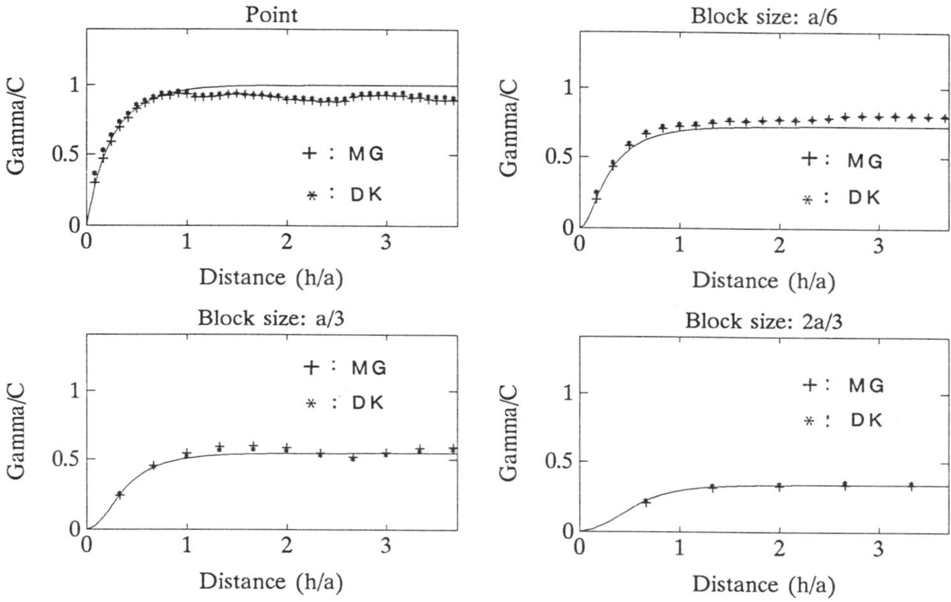

Figure 1: Experimental and theoretical variograms for the transformed variates; Simulation 1. Units are normalized.

Simulation 2

This example established the potential impact on simulation of taking into account known block grades. For this purpose, the configuration of points and blocks shown in Figure 2 is used. The values for points and block A have been obtained with a preliminary simulation in order to insure consistency of the grades with the variogram function. Variogram and statistical parameters are the same than for simulation 1. Point values of Figure 2 tempt to represent respectively rich and poor zones of an hypothetical deposit. For each zone, 500 simulated grades for block B were obtained (by DK) assuming that:

- i. block A is ignored
- ii. block A is relatively rich
- iii. block A is relatively poor

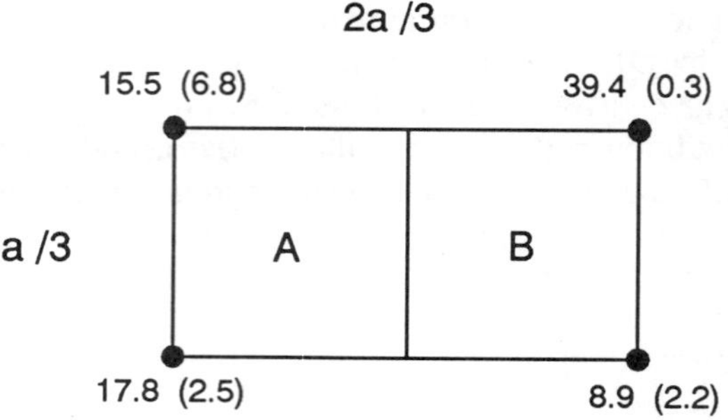

Figure 2: Data configuration for simulation 2, rich zone (poor zone).

Figure 3 shows the histograms of the 500 simulated blocks for each case in the rich zone (left) and in the poor zone (right) of the deposit. In the rich zone, the mean and the shape of the histograms of the simulated block B are modified by the inclusion of known block A. However, the standard deviation is relatively unaffected by the inclusion of block A. In the poor zone, the mean, the standard deviation and the shape of the histograms are all strongly affected by the inclusion, in the conditioning set, of block grade A.

Note that the standard deviations are significantly lower in the poor zone than in the rich zone of the deposit. Using direct kriging variance or approximations based on combinations of linear kriging variances and block dispersion variances (Kim and Baafi, 1984) would give a unique estimate of variability for both zones unless some kind of proportional effect is included (either by using a relative variogram or by performing a nonlinear kriging). Clearly this estimate would be inadequate.

DISCUSSION AND CONCLUSION

The main advantage of the two proposed algorithms is to facilitate the inclusion of conditioning block grade data in the simulation/estimation procedure. Due to the change of support model, the approach is flexible as each conditioning block data can be of different sizes and shapes. This would not be realistically possible with a numerical based approach. Numerous variograms would have to be modelled, with no guarantee concerning their consistency, also, small anisotropies induced by the block sizes, shapes and orientations would not be taken into account.

As shown in simulation 2 above, the inclusion of block data can change substantially the results of simulation/estimation in certain circumstances. In simulation 2, when incorporating information from block A, the standard deviation was between 1/3 to 5 times the standard deviation obtained without block A. Such variability fluctuations could impair the performance of the mine concentrator and could necessitate adjustments in the operation of the mine.

When only point conditioning data are available, the proposed approach still has the advantage of a possible gain of an order of magnitude in computing time and space requirement relatively to the point based approach. On the other hand, the main disadvantage of the method is that it can work only on additive variables. Also, it is a Gaussian method and, in this respect, may not be suitable to model adequately particular fields (Journel, 1993).

Despite the fact that point and block Gaussian variates can not be jointly multinormal, the MG approach provided virtually identical results to the DK approach in the simulations performed. Both reproduced fairly well the theoretical block variograms. Whether this holds for other variograms and statistical parameters remains to be verified.

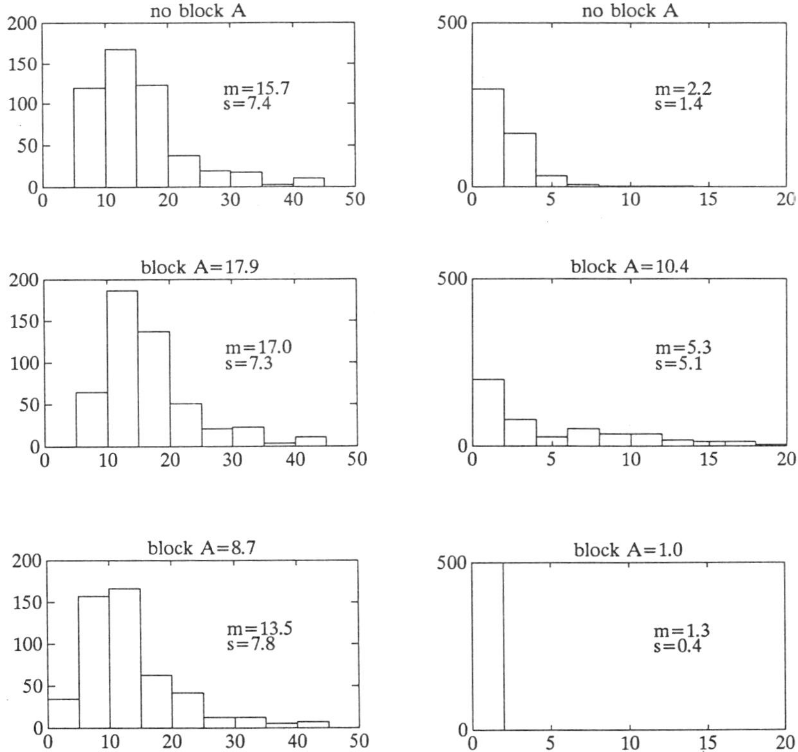

Figure 3: Histograms of the 500 simulated blocks B in the rich zone (left) and in the poor zone (right). Three different scenarios: block A is not used in the simulation (top), block A is relatively rich (middle), block A is relatively poor (bottom).

ACKNOWLEDGMENTS

I am more than indebted to Michel David for all these years that he was my guide in the field of geostatistics. Comments by A.G. Journel were helpful in improving the manuscript.

REFERENCES

Gomez-Hernandez, J. (1993) "Regularization of hydraulic conductivities: a numerical approach", in A. Soares (ed), Geostatistics Troia'92, Kluwer Academic Publishers, Dordrecht, pp. 767-778.

Isaaks, E.H. (1990) "The application of Monte-Carlo methods to the analysis of spatially correlated data", Unpublished PhD thesis, Stanford University, Palo Alto, CA., 213 p.

Journel, A.G. (1993) "Geostatistics: roadblocks and challenges" in A. Soares (ed), Geostatistics Troia'92, Kluwer Academic Publishers, Dordrecht, pp. 213-224.

Journel, A.G. and Huijbregts, Ch. J. (1978) Mining geostatistics, Academic Press, London.

Journel, A.G. and Alabert, F. (1989) "Non-Gaussian data expansion in the earth sciences", Terra Nova, 1, 2, 123-134.

Kim, Y.C. and Baafi, E.Y., (1984) "Combining local kriging variances for short-term mine planning", in G. Verly et al. (eds), Geostatistics for natural resources characterization, NATO-ASI, serie C, v. 122, Reidel, Dordrecht, pp. 185-200.

Levy, P. (1937) Théorie de l'addition des variables aléatoires, Gauthier-Villars, Paris.

Matheron, G. (1975a) "Forecasting block grade distributions: the transfer functions", in M. Guarascio et al. (eds), Advanced geostatistics in the mining industry, NATO-ASI, serie C, v. 24, Reidel, Dordrecht, pp. 237-251.

Matheron, G. (1975b) "A simple substitute for conditional expectation: the disjunctive kriging" in M. Guarascio et al. (eds), Advanced geostatistics in the mining industry, NATO-ASI, serie C, v. 24, Reidel, Dordrecht, pp. 221-236.

Rosenblatt, M. (1952) "Remarks on a multivariate transformation", Annals of mathematical statistics, 23, 3, 470-472.

Verly, G. (1984) "The block distribution given a point multivariate normal distribution", in G. Verly et al. (eds), Geostatistics for natural resources characterization, NATO-ASI, serie C, v. 122, pp. 263-290.

CHANGE OF SUPPORT AND TRANSFORMATIONS

D. E. Myers
Department of Mathematics
University of Arizona
Tucson, AZ 85721 USA
USA

ABSTRACT

The practical and theoretical effects of using non-point support data for estimating variograms or on the kriging equations when estimating spatial averages, i.e., block kriging, are well-known. Under an assumption of lognormality the proportional effect is also well-known. While other transformations are commonly used in statistics only the log and indicator transforms are widely used in geostatistics, the latter has the advantage of generally not requiring an inverse transform.. Additional theoretical and empirical results are presented on the interrelationship between non-point support data, non-linear transformations and variogram estimation, modeling. The non-point support data may incorporate spatial averages or compositing of point support data.

INTRODUCTION

Let Z(x) be a regionalized variable defined in 1, 2 or 3 space and H(Z) a real linear functional, i.e., a mapping of Z into the real numbers. The two most common examples of H are point evaluation, i.e., H(Z) is simply $Z(x_0)$, and spatial averages Z_V, the average value of Z over a volume V. Much of geostatistics has been concerned with one of two problems; estimation of the linear functional H(Z) or estimation of a probability distribution associated with H. For example, let x_0 range over all possible values in a region or let V range over possible congruent volumes within the region. The resulting probability distributions for the point valuation functional and the spatial average functional are of interest in many applications. These problems generalize when transformations are allowed, either on the domain of Z or on the range of Z.

STRUCTURE FUNCTIONS

Linear geostatistics is based on the use of a structure function, i.e., the variogram or

R. Dimitrakopoulos (ed.), Geostatistics for the Next Century, 253–258.

covariance of Z. The relationship between the covariance of the common linear functionals of Z(x) and the structure function for Z(x) is well known. The relationship between the probability distributions is generally not known or at least only under strong assumptions. The relationship between the covariances of non-linear functionals of Z(x) and the structure function of Z(x) is generally not known. Since the class of continuous mappings is very large and contains as a subset the mappings with continuous second derivatives many problems involving non-linear functionals or non-linear estimators can be resolved by generating those functionals or estimators using such transformations or mappings. A number of known results will be described in this general context and then open problems will be presented.

GENERAL ESTIMATION PROBLEMS

Extending the approach in Cressie (1993) define general classes of non-linear functionals

$H_1(Z) = g(Z)$ (1a), $H_2(Z) = g(Z_V)$ (1b), $H_3(Z) = (g(Z))_V$ (1c)

where g(u) is a real valued function defined on a subset of the reals. If in addition g has an inverse then the following is of interest.

$$H_4(Z) = g^{-1}((g(Z))_V) \quad (1d)$$

Note that in general $H_3(Z) \neq H_2(Z)$ and $H_4(Z) \neq Z_V$. In the case of (1a), (1c) one can simply transform the data $Z(x_1),...,Z(x_n)$ into new data $g(Z(x_1)),...,g(Z(x_n))$ and use a linear estimator of the form

$$H^*(Z) = \Sigma_{i=1,n} a_i g(Z(x_i)) \quad (2)$$

Although $g^{-1}g(Z) = Z$, in general $g^{-1}(H^*(Z)) \neq \Sigma_{i=1,n} a_i Z(x_i)$

Examples

Let I(x;z) be the Indicator function associated with Z(x), Journel (1983) That is, I(x;z) = $g_z(Z(x))$ where $g_z(u) = 0$ if u > z and 1 otherwise. As has been pointed out by Cressie (1993), there are two equivalent problems. First, the non-linear estimator can be used to estimate a probability distribution function, or it can be used to estimate values of the non-linear functional obtained by applying the indicator transform to the point evaluation functional. That is, for a certain non-linear functional F we have

$F^*(z) = \Sigma_{i=1,n} a_i g_z(Z(x_i))$ (3) or $g_z(Z(x_0)) = \Sigma_{i=1,n} a_i g_z(Z(x_i))$ (4)

For a volume V centered at the origin, let V_x be the volume rigidly translated to the point x. Z_{Vx}, as is common in the literature, denotes the spatial average over the translated volume. Although $Z(x_1),...,Z(x_n)$ (multivariate) lognormally distributed is equivalent to $LnZ(x_1)$, ...,$LnZ(x_n)$ being multivariate normal, determining the distribution of Ln Z_{Vx} is more difficult than for $(\text{Ln } Z)_{Vx}$. Although the sum of normal random variable is again normal the same is not true for lognormal random variables although it is often considered to be approximately true empirically. Journel(1980) makes this assumption in deriving the bias adjustment for lognormal kriging.

In the case of the indicator transform Cressie (1993) has suggested an alternative. Since it is straightforward to estimate Z_V by a linear combination of the data for Z, why not estimate $g(Z_V)$ as the transformation of a linear combination of the data (not necessarily one of the usual kriging estimators). More explicitly,

$$H\hat{}(Z) = g(\Sigma a_i Z(x_i)) \tag{5}$$

While in general $E\{g(\Sigma a_i Z(x_i))\} \neq g(E\{\Sigma a_i Z(x_i)\})$ unbiasedness of $H\hat{}(Z)$ could be imposed as a constraint. Similar conditions on the variances could be imposed and Cressie minimizes the variance of the error of estimation subject to two constraints and hence uses two Lagrange multipliers. $H\hat{}(Z)$ is essentially the simple block kriging estimator for Z_V BUT an additional constraint has been imposed. The new "kriging" equations require only the covariance function of Z (as well as the point-to-block and block-to-block variances computed from that covariance). Interestingly enough, *these new kriging equations do not depend on the transformation g*. This is because g has been assumed to have continuous second derivatives, i.e., a smoothness condition has been imposed on g. The resulting estimator is viewed as an approximation to $E[g(Z_V) \mid Z(x_1),...,Z(x_n)]$. By analogy with the usual simple kriging estimator one is inferring the conditional distribution of $g(Z_V)$ rather than the distribution of Z_V. Finally it is easy to see from these new kriging equations that the system may be unstable when the block size is too large, i.e., the block-to-block variance is too small.

DISTRIBUTIONS AND TRANSFORMATIONS

Let U be a random variable and g a one-to-one differentiable mapping. Then the probability distribution of g(U) is completely determined by the probability distribution of U. If $U_1,....,U_n$ are jointly distributed random variables and $W_1(U_1,....,U_n),...,W_n(U_1,....,U_n)$ are on-to-one transformations with continuous partial derivatives and whose Jacobian does not vanish then the joint distribution of $W_1,...,W_n$ is determined by the joint distribution of the $U_1,....,U_n$. As a special case let $W_i = g(U_i)$. This is essentially the problem considered above. However in geostatistics the joint distribution function for Z is generally NOT considered known and hence the general result is not useful. Note that even in the case of "nice" transformations and "nice" joint distributions it may be difficult to compute the new joint distribution. The change of variable theorem is not directly applicable for such functions as $g(u) = u^2$ and a multivariate normal distribution. Although the problem of obtaining the distribution of a function of a random variable is a difficult one, computing the first and second moments (from the distribution of the original random variable) is straightforward and the conditions are less restrictive than the general change of variable theorem. This idea was exploited by Matheron (1985).

Because the Gaussian is essentially the only multivariate distribution that is characterized by the bivariate correlations it is the only distribution for Z(x) that allows easy computation of the variogram of g(Z) in terms of the variogram of Z. Note however that the variogram of $(g(Z))_{Vx}$ is easily computable in terms of the variogram of g(Z). That is,

$$\gamma_{(g(Z))V}(h) = 0.5\ \mathrm{Var}\{(g(Z))_{Vx+h} - (g(Z))_{Vx}\} = (1/V^2) \int_V \int_V \gamma_{g(Z)}(x-y)dxdy \qquad (6)$$

Hence if the data is transformed the variogram of g(Z) can be estimated and modeled, from this the variogram for $\gamma_{(g(Z))V}(h)$ is computable.

ALTERNATIVE APPROACHES

When the transformation is sufficiently smooth, i.e., the second derivative is continuous, then the "value" at one point can be approximated by the value at a nearby point. Linear functionals can then be applied to this approximation. Instead of transformations of Z, the regionalized variable can be considered as a non-linear transformation. Matheron (1985) utilizes this approach to approximate Z_{Vx} by Z(x). The approximation is valid at least for small V. Consider x to be a (vector) random variable uniformly distributed over the region of interest. Then Z(x) is obtained as a transformation applied to this (uniform) random variable. Since the distribution of x is known and the transformation is "known" the moments of Z(x) are computable in two ways; one in terms of the transformation and the uniform distribution of x, secondly in terms of the unknown distribution of Z(x). Note that this relationship extends to composite transformations, g(Z(x)) at least for functions g where the moments exist. By combining the idea of approximating Z in terms of its first two derivatives and equivalence of the two methods for computing moments, it is possible to approximate the distribution of g(Z(x)) and of $g(Z_V)$.

REVERSE PROBLEMS

The change of support problem is generally thought of in terms of determining the characteristics of or estimating Z_V given data for Z and some form of structural information for Z (such as the variogram). However there are many problems where the reverse is equally important. Block size is important in mining applications because it is related to selectivity and hence to tonnage. Block size is also important in environmental remediation because it is related to the scale of remediation. The simplest form of remediation for contaminated soil consists of removing soil to some fixed depth over a specified area. In a manner not dis-similar to that of cut-off grades for ore, contaminants usually have toxicity levels. These may be support dependent however and are exposure related. Although the average concentration of a contaminant will generally decrease as the block size increases, exposure potential may not decrease with block size. Although the cost of remediation is related to the total amount of soil removed it is also related to the number of blocks of soil to be removed and there will be a minimum block size related to the equipment to be used. If the distribution of average concentrations, for blocks of a certain size, is known can we infer the distribution for smaller blocks. In particular if none of the blocks have an average contaminant concentration above the toxicity level, is there some assurance that none of the blocks of a fixed smaller size will have average concentrations in excess of the toxicity level? Can average concentrations for small blocks be estimated from large block concentrations. This is a serious problem because in comparison to the assaying of ore samples the laboratory analysis of environmental samples is usually very costly. The use of large blocks corresponds to the use of composited samples.

Because toxicity levels are very low for many contaminants it is not uncommon to have data reported as "non-detects" or "not quantified". In the latter case the substance was detected but at such a low concentration that the results are unreliable. Because there are examples of toxicity levels very close to the detection levels it is not appropriate to consider non-detects as zeros. Note that compositing may actually make this problem worse. The question of how to estimate variograms in the presence of such data has received little attention.

When g is the indicator transform (cut-off value = z) then $(g_z(Z))_V$ can be interpreted as the proportion of (points in) V where the value of $Z(x) \leq z$. Alternatively this is the probability that if a point is chosen at random in V then the value will be $\leq z$. For a collection of disjoint, congruent sub-blocks the variogram quantifies the correlations between these probabilities. Hence if the variogram of $(g_z(Z))_V$ has a very long range (compared to the size of the region of interest) and a small sill then the probability distribution is nearly the distribution of the entire region.

ASPECTS OF COMPOSITING

Let $Z^T = [\, Z(x_1),...,Z(x_n)]$ and A a k x n matrix with non-negative entries satisfying two additional conditions; (1) for any one column at most one row has a non-zero entry, (2) $AU_n = U_p$ where U_n , U_k are column vectors with all entries 1's. A has the effect of compositing the data vector Z. Note that in a multivariate case Z would be an n x p matrix but the compositing matrix would function in the same way. Let $\mathbf{Y} = [\, Y(x_1),...,Y(x_k)]^T = A\mathbf{Z}$ be the composited data vector, $\mathbf{Z} = [Z(x_1),....,Z(x_n)]^T$. Then for any point x_0, the composited kriged estimate for $Z(x_0)$ be given by

$$Z^*(x_0) = \Sigma\lambda^c{}_i Y(x_i) \qquad (7)$$

Although the Y's will not be associated with locations as such the covariance matrix C_Y with entries $C_{ij,Y} = \mathrm{Cov}\{\, Y_i, Y_j\}$, is expressible in terms of the covariance matrix of Z and hence in terms of the variogram of Z. Let $\lambda = [\lambda_1,....,\lambda_p]^T$. Then $C_Y = AC_ZA^T$ and the weights in (7) are found as the solution to

$$AKA^T \lambda + U_p\mu = AK_0$$

$$U_p^T \lambda = 1 \qquad (8)$$

where K is the matrix of variogram entries for Z (between sample locations) and K_0 is the vector of variograms (between the sample locations and the location to be estimated). Note that columns in the coefficient matrix will not coincide with the column on the right hand side even if x_0 is one of the data locations. Hence the values of the components of composited samples can be estimated using the composit sample data.

REFERENCES

Box, G.E.P. and Cox, D.R. (1964) "An Analysis of Transformations", J. Royal Statist. Soc. B26, 211-243

Cressie, N. (1993) "Aggregation in Geostatistical Problems", in Geostatistics Troia '92, A. Soares (ed), Kluwer Academic Publishers, pp25-35

Crozel, D. and David, M. (1985) "Global Estimation Variance:Formulas and Calculation", Math. Geology 17, 785-796

Journel, A.G., (1980) "The Lognormal Approach to Predicting Local Distributions of Selective Mining Grades", Math. Geology 12, 285-303

Journel, A.G., (1983) "Non-parametric Estimation of Spatial Distributions", Math. Geology 15, 445-468

Lantuejoul, Ch. (1988) "On the Importance of Choosing a Change of Support Model for Global Reserves Estimation", Math. Geology 20, 1001-1019

Matheron,G. (1971) "The Theory of Regionalized Variables and Its Applications. Les Cahiers du Centre de Morphologie Mathematique", Fas. 5. ENSMP, Fontainebleau, 212p

Matheron, G. (1985) "Change of Support for Diffusion-Type Random Functions", Math. Geology 17, 137-165

Zhang, R., D.E. Myers and A. Warrick (1990) "Variance as a Function of Sample Support Size", Math. Geology 22, 107-121

ESTIMATING RECOVERABLE RESERVES: IS IT HOPELESS ?

M.E. Rossi and H.M.Parker

Mineral Resources Development, Inc.,
1710 South Amphlett Blvd, Suite 302,
San Mateo, California, USA 94402

This paper examines "established" models that are used for correcting point distributions into block distributions. This is known as the "change of support" effect, and is commonly characterized by a "variance correction" factor. Mining takes place at a very different volume (usually called Selective Mining Unit, or SMU) than the volume of composites.

The correction for change of support is a very difficult task, since very little is known a-priori about the block distribution of interest. Ore grades average arithmetically, so that the mean of the block distribution is the same as the mean of the point distribution. It is also known that the block variance will be smaller than the point variance. However, as we go through the perils and tribulations of some open-pit gold mines in Nevada and Southern California (USA), we will show that these two pieces of information are not enough, and that the problem is not an easy one.

We will discuss the implication of using a "local" distribution (a distribution of grades for each block) and a "global" correction factor to obtain a "local" recoverable block distribution. Systematic overestimation of the grades above cutoff is a potential danger, particularly if the mean grade of the block is used to obtain a corrected distribution. A comparison with the discrete gaussian method for change of support suggests that the latter is more robust when correcting "local" distributions using "global" correction factors. The theoretical inconsistency of the approach warrants further discussion.

R. Dimitrakopoulos (ed.), Geostatistics for the Next Century, 259–276.

The next step in volume-variance correction is, undoubtedly, the use of conditional simulation to obtain "experimental" factors. From a theoretical standpoint, the advantages are obvious. From a practical standpoint, the unwise use of conditional simulations (where geologic factors are ignored, or where an inappropriate model is assumed for the deposit), will hamper the results.

INTRODUCTION

The estimation of recoverable reserves in mining applications is currently based on very limited information about the block distribution of metal grades. There are a wealth of methods and techniques that help estimate "point" distributions, but relatively little research has been done to develop robust methods for estimating block distributions. The information that is available is:

- The "point" distribution itself, with its variogram model;
- A "theoretical" model for the block distribution, obtained applying a correction based on a number of limiting assumptions; and
- Models to correct the "point" variance into the "block" variance based on assumptions about the underlying distributions, eg. affine correction, discrete gaussian correction, and indirect lognormal correction.

The importance of dealing with recoverable reserves was generally recognized very early on in geostatistics, but it was M. David's early work (1977) that pioneered the necessary applied research to demonstrate the significance of estimating recoverable reserves, while Journel and Huijbregts (1978) synthethized the theoretical foundations for the most common methods for change of support.

The first decision to be made is which variogram model to use to determine the variance correction factor. In the case of skewed distributions, there is no straightforward answer. Outlier grades will significantly impact the variance correction factor regardless of which variogram estimator one uses. Thus, a subjective decision is required, hopefully based on economic or risk-related criteria, to decide if and where to trim the original distribution when estimating a variogram. Also, the variogram model used to correct for change of support may not be (and need not be) the same as the variogram model used to interpolate the point distribution.

The variogram models (for the point and the block distributions) are typically defined using all the data available, i.e., they are "global". The models used to correct the distributions are also typically applied to "global" distributions. However, in a mining operation, recoverable reserves need to be predicted on a much smaller scale, i.e.

"locally". In practice, this transition from global to local prediction is done haphazardly, in the hope that the "global" variogram is applicable at a local scale (individual SMUs, for example), and that the model applied to the global "point" distribution (eg., affine correction) is also valid at a local scale.

In addition, the models used are based on a certain SMU size. Grade control panels rarely correspond to the shape and, more importantly, to the size of the "theoretical" SMU, sometimes by a factor of two or three. Further, the use of the models mentioned above accounts only for internal dilution, ignoring altogether contact dilution problems.

Although most of the discussion is centered on correcting distributions based on point support to those based on SMU's, therefore involving a reduction in variance, it is also possible to adjust the distribution of interpolated block grades in the opposite direction. If the variance must be adjusted upward, so that a distribution with the requisite SMU variance results, the models that can be used are either the affine or the indirect lognormal method. It is not possible to use the discrete gaussian method for this purpose, and the affine correction may result in negative quantiles.

Although relatively simple models and procedures are available to correct for change of support, it will be shown that in practice they sometimes provide inaccurate solutions. In assessing recoverable reserves, the need for a high degree of flexibility in the tools used is evident.

THE TRIMMING EFFECT AND THE VARIANCE CORRECTION FACTOR

The variance correction factor (*VCF*) is defined as the ratio of the dispersion variance of the SMU's to the dispersion variance of the samples or composites. Generally, the *VCF* is a value smaller than one, since the dispersion variance of the SMU's is smaller than the composite or "point" distribution. The *VCF* is calculated from the variogram model, and, for a given SMU size, it depends also on the block discretization chosen, since it is based on an estimate of the change in variance from points to SMUs, (dispersion variance), see Journel and Huijbregts (1978), Isaaks and Srivastava (1989). We will also refer to the affine correction factor (*ACF*), which is simply the square root of the *VCF*.

Although the variogram model as a whole is used in calculating the point-within-SMU dispersion variance (required to compute the *VCF*), the most important parameter of the model is its nugget effect. This nugget effect changes significantly when the data used in the calculation are clipped above a certain threshold (trimmed). If a variogram is trimmed, the experimental points are much more easier to model, and the nugget effect can drop significantly. Although it can be argued that the nugget effect observed in experimental variograms can be largely due to sampling and assaying errors, as opposed to natural variability of the grade distribution, the issue is irrelevant at the time of estimating recoverable reserves. Understanding the sources of the observed nugget effect is a separate problem that requires considerable attention and effort.

The larger the nugget effect with respect to the total variance (sill) of the variogram, the smaller the *VCF* will be, i.e., the more severe the change of support will be. To understand this, recall that the variance of the SMU can be found with:

$$\sigma^2_{SMU} = \overline{\gamma_D} - \overline{\gamma_{SMU}}$$

where the first term is the average variogram within the deposit, and the second term is the average variogram value within the SMU. It is reasonable to assume that the average variogram function within the deposit is simply the variance of the sample grades, at least as long as the range of the variogram is much smaller than the dimensions of the deposit (see, for example, Parker, 1980). The SMU is in practice discretized using a regular grid, and its average variogram is approximated using an arithmetic average of all possible combinations of the variogram values within the SMU. Therefore, the *VCF* can be expressed as:

$$VCF = 1 - \frac{\frac{1}{n^2}\sum_{i=1}^{n}\sum_{j=1}^{n}\overline{\gamma}(\boldsymbol{h_{ij}})}{\overline{\gamma_D}}$$

M. David (1977) provides charts to obtain the variance of different volumes (rectangular prisms) as a function of the sill and nugget effect of standard variogram models. These charts are useful to understand the interaction of the variogram parameters in the calculation of the variance of the SMUs.

Figure 1 shows the effect of trimmed data on the *ACF* for a gold deposit in southern California. To place the trimming grades in perspective, Figure 2 shows the effect of trimming on the grade-tonnage curves. At the 0.02 opt economic cutoff grade, there are significant differences in grade, depending on the amount of trimming; there is a 20 percent difference between the average grade for no trimming and a 0.1 opt trim. The 0.1 opt trim affects 4.77 percent of the database. So the question remains in the practitioner's mind: which variogram model (i.e., trimming level) should one use?

The correction factor used to calculate recoverable reserves in this deposit is 0.71, which corresponds to trimming at 0.15 opt. No justification was available for the arbitrary choice. To some extent, it seemed to have been supported by historic data, based on limited blast hole-to-model reconciliation studies. Figure 1 shows that the affine method has been applied beyond its recommended limits. Journel and Huijbregts (1978, p. 471) imply the *ACF* should not be lower than 0.85, and yet in all cases shown in Figure 1 the *ACF* is below this value. Indeed, it is our experience that most gold deposits where the *ACF* is now being used have *ACF's* well below 0.85. As we learned in this particular case, the consequences of such transgressions can be significant. On the other hand, alternative methods that require fitting some type of polynomials and are based on a more general permanence of distribution hypothesis (such as the discrete gaussian method, see Matheron, 1975a, disjunctive kriging, see Matheron 1975b, or multigaussian conditional distributions, see Verly, 1983), were not implemented because of their complexity.

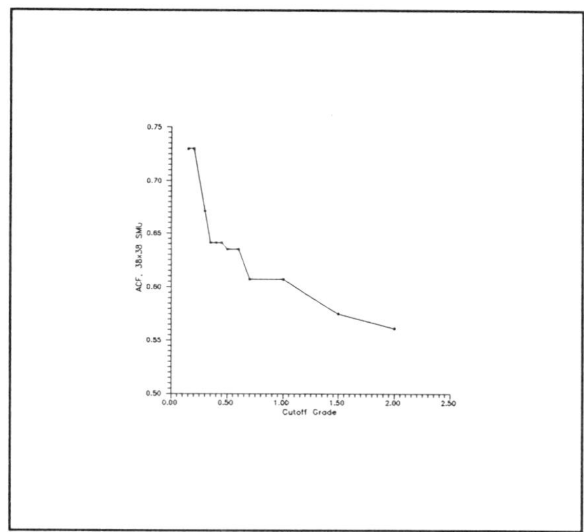

Figure 1: Effect of Trimming on the *ACF*

GLOBAL CHANGE OF SUPPORT

The following illustrates the application of the most common methods for change of support, i.e., affine correction, indirect lognormal correction, and discrete gaussian methods. The basis for the study is a detailed conditional sequential indicator simulation (on a 2 by 2 ft grid) done on one bench of a gold deposit in Northeastern Nevada. The conditioning blast hole data set is laid out approximately on a 16 by 16 ft grid. The area simulated was 500 by 500 ft, for a total of 62,500 nodes. The upper left histogram in Figure 3 is the histogram of the simulated values, i.e., the point distribution.

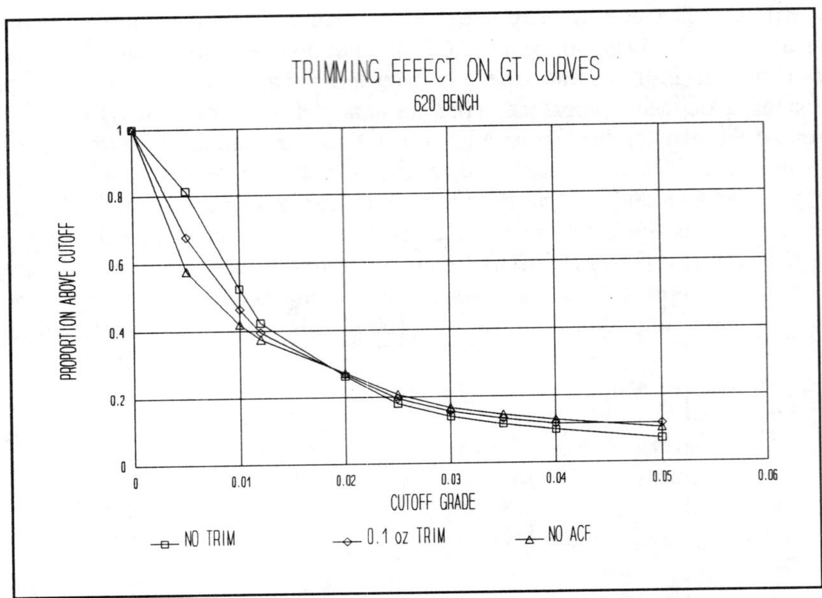

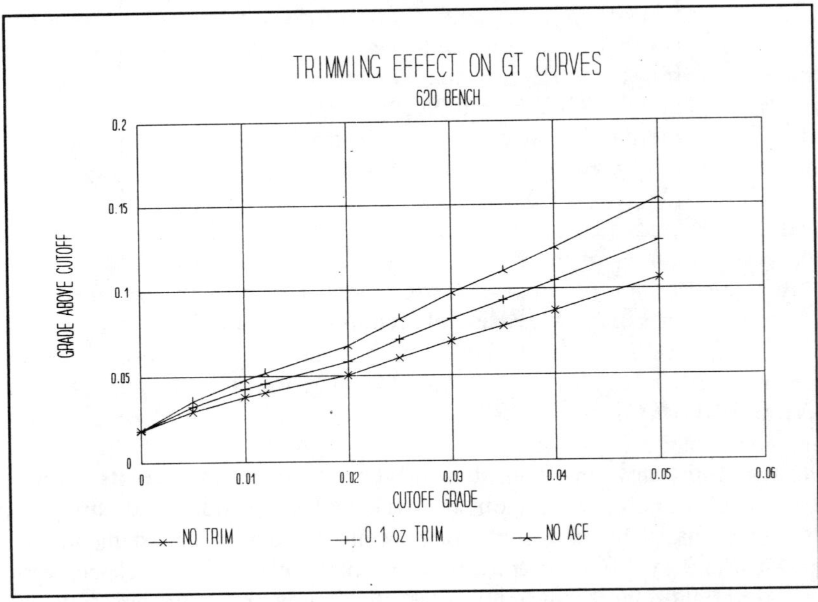

Figure 2: Effect of Trimming on GT Curves, 620 Bench

Three different SMU sizes were used in this study: 8 by 8 ft SMU, 16 by 16 ft SMU (which approximates the mine's "nominal" SMU size, and 32 by 32 ft. The true reduction in variance was obtained by averaging on the 2 by 2 ft grid within each SMU size, and calculating the corresponding histogram and statistics. These histograms are also presented in Figure 3. The corresponding *VCF* factors are:

- For the 8 by 8 ft SMU, $VCF = 0.61$; Thus, $ACF = 0.78$.
- For the 16 by 16 ft SMU, $VCF = 0.53$; Thus, $ACF = 0.73$.
- For the 32 by 32 ft SMU, $VCF = 0.44$; Thus, $ACF = 0.66$.

Recall that the affine correction factor (*ACF*) is the square root of the *VCF*. Using the true *VCF* for each case, three methods for change of support were applied to the point distribution:

- The affine correction method, with overall mean of 0.0339 opt, and *ACF*s derived from the *VCF*s above for each SMU size.
- The indirect lognormal correction method, see Journel and Huijbregts (1978) or Isaaks and Srivastava (1989).
- The discrete gaussian method (Journel and Huijbregts, 1978).

The lognormal shortcut (David et al, 1977) was not considered in this study, because it was developed primarily to deal with porphyry copper-type deposits. In the case of gold deposits, it has been shown that it may produce significant biases for selected cutoffs, see for example Sullivan and Verly (1983). The study was done on the same mine, but for a different orebody, as the one presented here. These biases could not be removed by in-fill drilling.

The results are presented in the form of grade-tonnage (GT) curves. Figure 4 presents the GT curve for the 8 by 8 ft SMU, Figure 5 presents the GT curve for the 16 by 16 ft SMU, and Figure 6 presents the GT curve for the 32 by 32 ft SMU. Table 1 presents a summary of the relative differences for the economic cutoffs of interest. The percentages are relative to the true values. Notwithstanding the use of the true *VCF*, an unknown luxury in the real world, the discrete gaussian method appears to perform better. The *ACF* factor applied in all cases is outside the recommended range of applicability of 30 percent in variance reduction. There are large changes in variance, even for SMU's which are too small to be mined in practice. For example, the 8 by 8 ft SMU presents about a 39 percent reduction. These results are all based on the assumption of perfect knowledge of the *VCF* to be applied. The DG method proves to be more robust with respect to significant changes in variance, perhaps because of its assumption of generalized permanence of the distribution. However, it is not consistently so, since it produces biased results at higher cutoffs. In addition, estimating global recoverable reserves is insufficient for detailed planning.

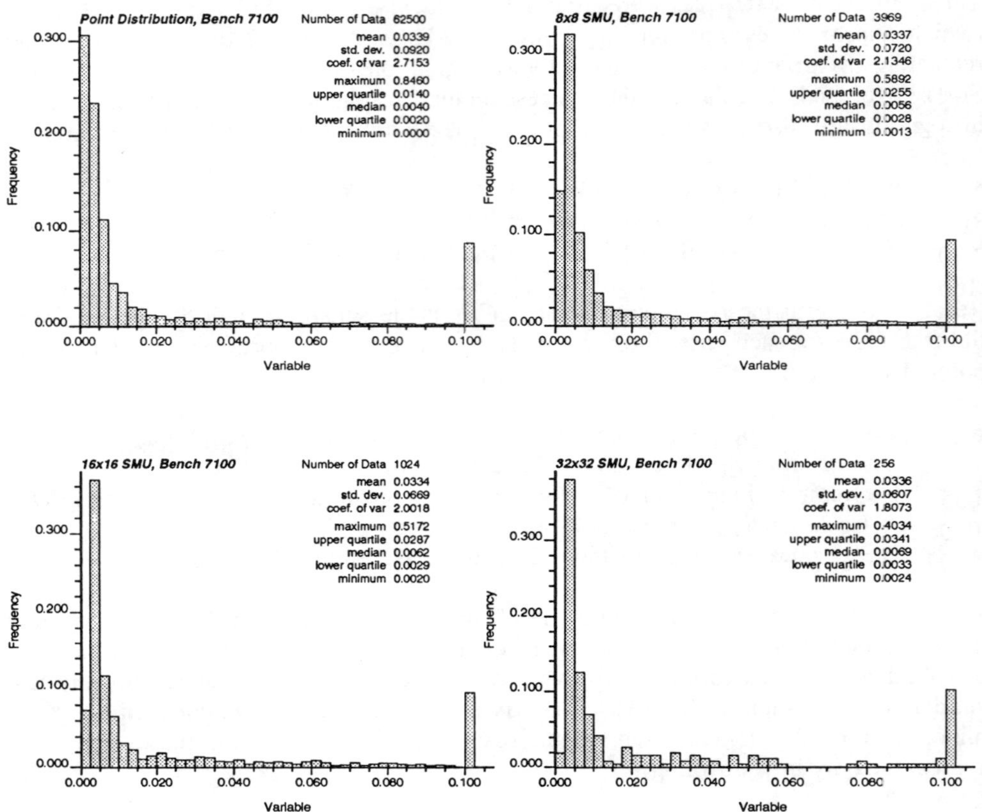

Figure 3: Histograms and Univariate Statistics of the Point Distribution, the 8x8 ft SMU, the 16x16 ft SMU, and the 32x32 ft SMU distributions.

LOCAL CHANGE OF SUPPORT

As mentioned before, the change of support methods commonly used are applicable to global distributions. In real life, however, practitioners require corrections that are accurate at a smaller scale. This is the case, for example, at operating mines, where each phase (which could include a large number of SMUs) requires the prediction of tonnage and grade above cutoff that is going to be recovered. A global distribution is not very useful in this instance. The necessary transition from a global scale (over the whole deposit or orebody) to a local scale (partial benches and mining phases) is difficult at best. An example of an affine correction applied to a block-by-block method is described next.

Table 1: Global Recoverable Reserves Relative Percent Error[1]

8x8 SMU

Cutoff (opt)	Affine Ppn	Affine Grade	Ind. Log. Ppn	Ind. Log. Grade	Dis. Gauss. Ppn	Dis. Gauss. Grade
0.055	22.8	-11.0	11.4	-15.3	-1.3	3.1
0.070	23.5	-12.0	9.8	-14.8	0.0	2.7
0.100	19.4	-10.2	4.3	-12.9	-2.2	4.4
0.200	6.8	-6.9	4.4	-12.6	11.4	-0.1

16x16 SMU

Cutoff (opt)	Affine Ppn	Affine Grade	Ind. Log. Ppn	Ind. Log. Grade	Dis. Gauss. Ppn	Dis. Gauss. Grade
0.055	27.3	-13.1	13.9	-18.9	0.0	2.6
0.070	27.4	-13.9	11.1	-17.9	0.7	2.3
0.100	27.1	-15.2	6.3	-16.7	2.1	1.9
0.200	17.4	-15.0	9.5	-18.9	21.7	-8.4

32x32 SMU

Cutoff (opt)	Affine Ppn	Affine Grade	Ind. Log. Ppn	Ind. Log. Grade	Dis. Gauss. Ppn	Dis. Gauss. Grade
0.055	28.0	-10.9	10.4	-17.8	-5.5	6.8
0.070	36.5	-21.3	17.6	-26.4	7.4	0.0
0.100	34.3	-24.2	11.8	-26.3	9.8	-2.7
0.200	15.4	-16.9	4.4	-22.3	20.5	-9.2

Multiple indicator kriging (MIK) is a method that has become popular in the last few years, and is being applied at a number of gold mines in the western United States. As described in Journel (1986), the MIK method provides, for each block being estimated, a local (posterior) conditional distribution, but importantly on the small support of the input information, generally blast holes or exploration composites. Partly, the appeal of the method is that it can be interpreted as providing the proportion of tonnage and corresponding ore grade for a series of cutoffs, in particular those of economic interest. Earlier, we mentioned the problems associated with using methods (eg., the affine correction) beyond their range of applicability. We also mentioned the difficult decision as to which *VCF* to use. In calculating the

---

[1] Positive implies overestimation, negative implies underestimation.

VCF, should an experimental variogram be trimmed or not? The third key issue that needs to be dealt with is which mean to use when applying the change of support method. Let us analyze the affine method, for example.

The mathematical expression of the correction is:

$$q' = ACF * (q-m) + m$$

where *ACF* is the affine correction factor, q is the quantile of the distribution we want to correct, m is the overall mean of the distribution, and q' is the new quantile. The AC method is based on the idea that the distribution does not change in shape or mean, only the variance changes. Therefore, the variance of the point estimates can be shrunk around a common mean to obtain the distribution of the SMUs estimates.

In the formula presented above, both the *ACF* and m are parameters obtained from the overall distribution. If the method is applied on a block-by-block basis, using a single *ACF*, but a different mean m for each block being corrected, (for example, the E-type estimate of the block conditional distribution), significant biases may occur. The consequence of using the estimated block average to obtain the SMU distribution is likely to produce higher grades above the global mean of the distribution, and lower grades below it. To illustrate, assume an E-type estimate for the block of 0.04 opt, the global mean of the distribution is 0.019 opt, the *ACF* is 0.70, and the grade (quantile) to be corrected is 0.03 opt. Then:

$$q' = 0.70 * (0.03-0.04) + 0.04 = -0.007 + 0.04 = 0.033$$

Using the global mean, all other parameters being equal:

$$q' = 0.70 * (0.03-0.019) + 0.019 = 0.0267$$

If the E-type estimate for the block is below the global mean of the distribution, the results are:

$$q' = 0.70 * (0.03-0.01) + 0.01 = 0.014 + 0.01 = 0.024$$

Using the global mean, the result is 0.0267 opt. In this example, the global declustered mean (0.019 opt) is close to the economic ore grade cutoff (0.020 opt). As is shown above, in the richer portions of the deposit (E-type estimates above 0.019 opt) the application results in an up-grading for a given quantile compared to making the correction using the global mean.

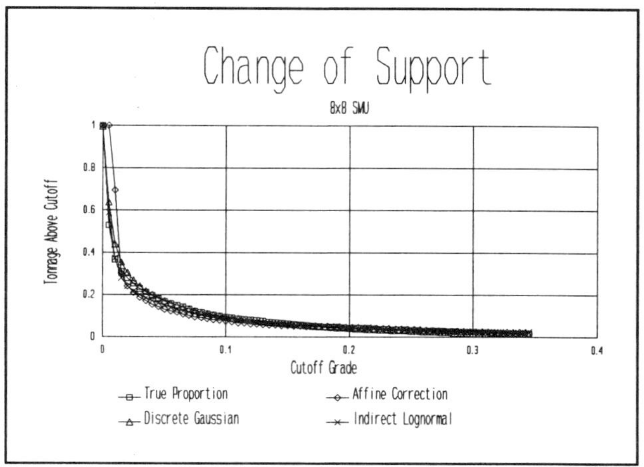

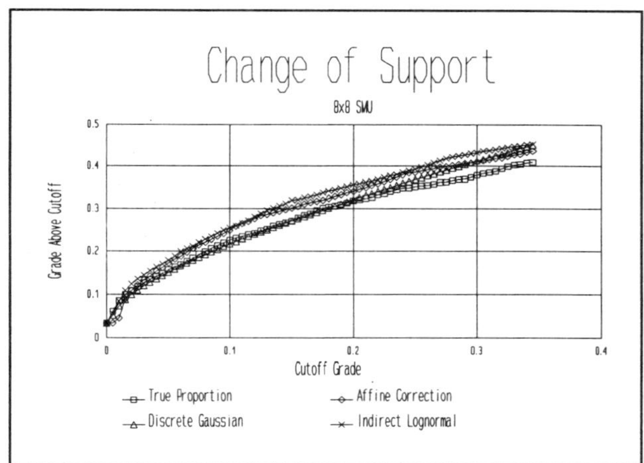

Figure 4: 8x8 ft SMU, Grade-Tonnage Curves.

The converse is true in poorer portions of the deposit (E-type estimates below 0.019 opt); the application results in a down-grading for a given quantile compared to making a correction using the global mean. As most of the ore lies in blocks with E-type estimates above the global mean, application of the *ACF* using an E-type estimate can result in overestimation of the recovered ore grade.

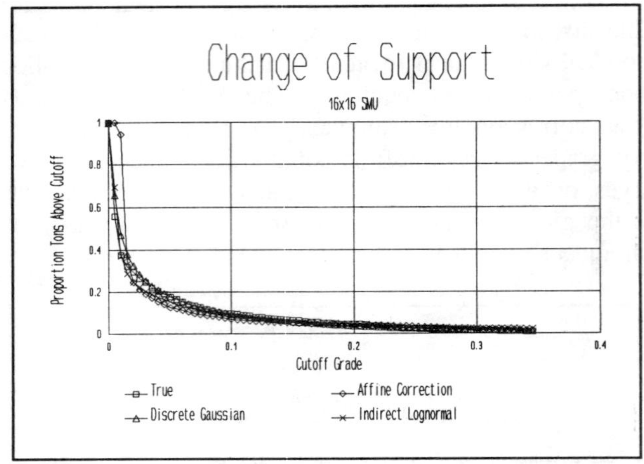

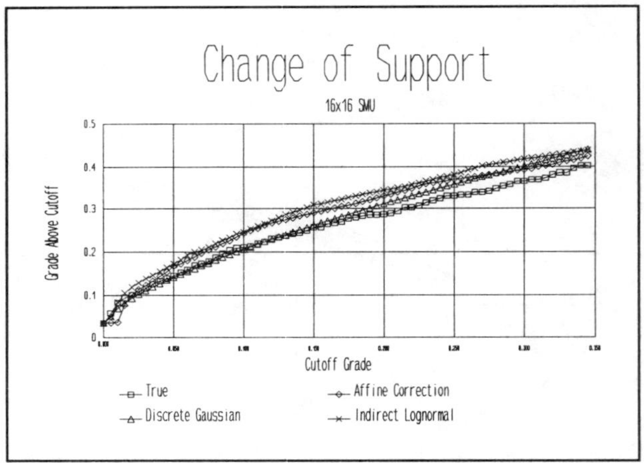

Figure 5: 16x16 ft SMU, Grade-Tonnage Curves.

"Quasi-Local" Change of Support

The same affine correction can be applied by sub-zones, or areas. If there are enough samples to obtain a variogram model, and thus a *VCF*, the correction can be applied on a "local" basis, for example on a bench-by-bench basis.

The comparison is made between the same global *ACF* correction applied to each posterior (MIK) conditional distribution, and using the E-type estimate as the average. The second method takes the declustered mean of the bench composites as the mean of the distribution, and also uses a different *ACF* value for each bench, i.e.,the model now calls for a local (bench) distribution. This implies, naturally, that a variogram model per bench is required. The grade-tonnage curve presented in Figure 7 and the corresponding summary in Table 2 show that the resulting overestimation of grade above cutoff is about 16.7 percent at the 0.02 opt cutoff. The tonnage curves, on the other hand, are much closer, because the 0.02 opt cutoff is approximately the global mean of the distribution. Presumably, the use of bench *ACF's* and bench means would be more accurate.

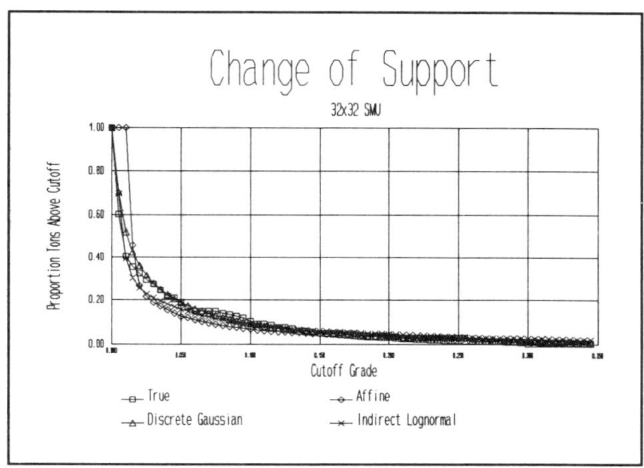

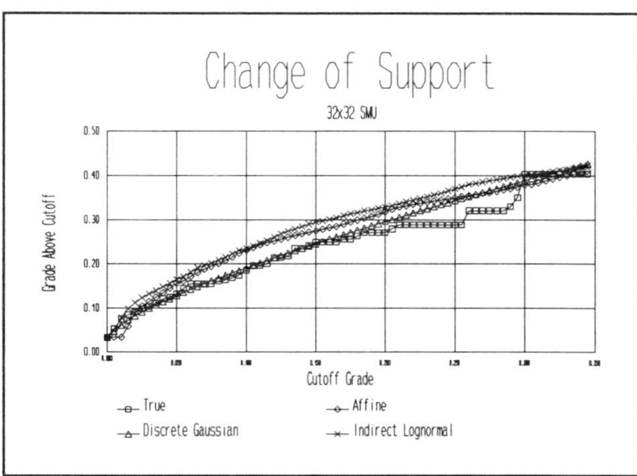

Figure 6: 32x32 ft SMU, Grade-Tonnage Curves.

Table 2: Recoverable Reserves using Local *ACFs*

Cutoff	Global *ACF*[2] Ppn	Global *ACF*[2] Grade	Local *ACF* Ppn	Local *ACF* Grade	Diff. (Global-Local) Ppn	Diff. (Global-Local) Grade
0.012	0.482	0.044	0.499	0.040	-3.4%	10.0%
0.02	0.336	0.056	0.332	0.048	-1.2%	16.7%

In addition to the *ACF* described above, a change of support correction using the discrete gaussian (DG) method was done. It proved to be more robust than the affine method (as seen also in the global change of support). Table 3 presents a summary of the grade-tonnage results obtained using different methods. These curves were obtained using the two different ACFs, plus the DG method. Four different distributions are compared:

- The SMU (target) distribution (from blast hole averages); "True" in Table 3.
- The "experimental" affine, calculated from the ratio of the SMU distribution's standard deviation to the point distribution standard deviation;
- The "theoretical" affine correction, calculated from the variogram models;
- The DG correction.

None of the three change of support methods provides an accurate answer. However, both affine corrections provide unacceptable curves, including the unrealistic projection that 100 percent of the bench is above 0.012 opt. This is to be expected, since the change in variance used violates the basic AC's assumption that the shape of the distribution does not change. The DG model performs better, although overestimating the tonnage above cutoff (and underestimating the grade) for cutoffs lower than 0.025 opt, and underestimating the tons above the 0.025 opt threshold.

Table 3: Recoverable Reserves using Different Methods
Relative Percent Error[3]

Cutoff	True - Ex. Affine Ppn	True - Ex. Affine Grade	True - Theor. Affine Ppn	True - Theor. Affine Grade	True - DG Ppn	True - DG Grade
0.012	-58.9	35.6	-58.9	35.6	-14.0	16.0
0.02	-7.4	22.3	-16.6	27.7	-8.7	14.8
0.03	19.0	11.7	19.0	15.7	5.3	9.4
0.05	43.6	-0.3	48.2	-0.1	23.3	1.9

---

2 "Global" uses block e-type estimates for the mean, "Local" uses bench means.

3 Positive implies overestimation, negative implies underestimation.

Non-traditional Change of Support Methods

Weighted average estimators have the property of producing smoothed distributions of block estimates. The variance of the block distributions may be less than the variance of the SMU's. This is particularly true where sample-block correlation is weak, the nugget effect is low and the SMU size is small. The dispersion variance of kriged estimates is a function of the kriging plan, and particularly of the number of samples used for kriging. The kriging plan can be "tuned" so that the distribution of kriged estimates has the same coefficient of variation as SMU's, thus incorporating the change of support process into the block model building process.

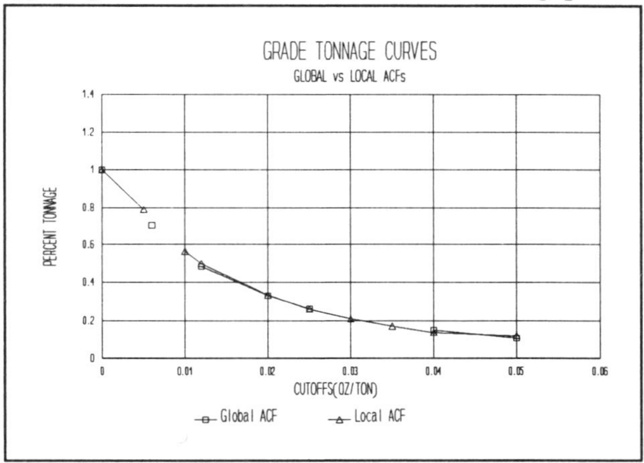

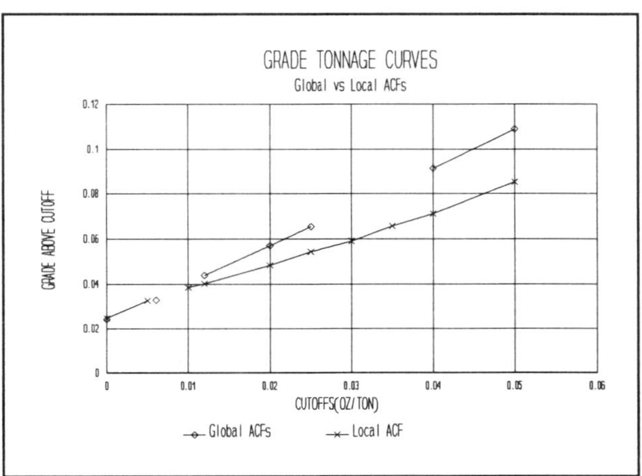

Figure 7: GT Curves Comparing Global vs Local (Bench) affine corrections

Size and Shape of an SMU

An alternative approach is to use conditional simulations to develop the "ground truth" SMU distribution, which is then used as a yardstick. Recoverable reserves can be determined with a conditional simulation on a very small grid, for example using procedures described in Isaaks (1990), and then averaging conditionally simulated grades over various geometries to form SMU's. Grade-tonnage curves for simulated SMU's can then be prepared for mining phases. The grade-tonnage curves can then be used to develop adjustment factors for the model.

In Roditis et al., 1993, these non-traditional approaches to change of support are described in more detail. In summary, it should be noted that the change of support correction is not constant across a deposit and that conditional simulation provides a powerful tool to assess its impact within local areas.

An issue that has not been addressed in geostatistical research is the SMU size and its geometry. All traditional methods for change of support are based on pre-established SMU sizes. Therefore, the dispersion variance and the target block variance are defined. Unfortunately, open pit mining practices are such that rarely those SMU sizes are used in practice. The SMU is a concept that is used, and is very useful, at the mine design and initial mine planning phases of the operation. This includes operation sizing, mine layouts, and equipment selection, hopefully all based on orebody characteristics. However, during production, the blast patterns and grade control panels used usually do not resemble the planned SMU size and geometry. These panels are typically much larger than an SMU, and their shapes are far more irregular, as they follow the local variations and caprices of the orebody.

Since ore/waste selection is done on a panel basis, rather than on an SMU basis, accurate prediction of the original SMUs becomes an irrelevant problem. The issue is rather how to accurately forecast grade and tonnage of the whole panel, and to determine whether it should be sent to the waste dump or to the mill. Therefore, the *VCF* required can vary on a panel by panel basis. In addition, contact dilution can become a major factor. Very little research has been done in the geostatistical community to address contact dilution problems.

CONCLUSIONS

Summarizing, facing a change of support problem requires taking into account:

- The effect of trimming, as it relates to the calculation of the *VCF*;
- The effect of local changes in the distribution, which has a bearing on the mean used to apply the correction;

- The use of block conditional distributions to estimate the in-situ reserves which do not provide a solid foundation to estimate the recoverable reserves, particularly at a local scale. If a change of support correction is necessary, great care must be used to avoid biases in the GT curves.
- Because of the irregularity of the grade control panels, and thus the mixture of sizes of SMU's which are actually mined, doubt is cast on the applicability of complex methods such as the Multigaussian or the Disjunctive Kriging methods to predict recoverable reserves.

RECOMMENDATIONS, OR THE LIGHT AT THE END OF THE TUNNEL

The following general procedure is recommended to solve or avoid altogether the number of problems that have been presented so far, stemming from theoretical and practical caveats of current models for change of support.

- Develop a reserve model using ordinary kriging or a combination of ordinary and indicator kriging which gives grade-tonnage curves for blocks which are approximately the same as those for panels.
- Perform a conditional simulation, hopefully based on blast hole data, on a small grid.
- Sample the conditional simulation, extracting a set of simulated blast holes, on the same grid used by the mine.
- Use some form of kriging to interpolate grades on a very tight grid.
- Assign grades to panels, defining the location of the diglines.
- Compare panel grades and tonnages to reserves which are modelled using exploration data and develop suitable correction factors on a local to global basis.

This general method does not require the use of an SMU, does not require knowing beforehand the variance correction factor, and it simply avoids the problems associated with using theoretical assumptions that may be violated in practice. In addition, it is applicable both at the mine design phase and for grade control. The difference between the two is the amount of information available to condition the initial simulation. Further, the approach also resolves the issue of global and local change of support.

ACKNOWLEDGEMENTS

We would like to acknowledge several mining companies' contribution to this work, and their authorization to publish their data. In particular, Mark Springett, Gold Fields Mining Corporation, Matt Thiel and Ian Douglas, Independence Mining Company, and Chris Carter, Kennecott Corporation, are gratefully acknowledged. The interpretations and opinions expressed herein are solely those of the authors.

REFERENCES

David, M., 1977, *Geostatistical Ore Reserve Estimation*, Elsevier, Amsterdam.

David, M., Dagbert M., Belisle J.M., 1977, *The Practice of Porphyry Copper Deposit Estimation for Grade and Ore-Waste Tonnages Demonstrated by Several Case Studies*, Proceedings, 15th APCOM Symposium, Brisbane, Australia.

Isaaks, E.H., 1990, *The Analysis of Spatially Correlated Data Using Monte Carlo Methods*, PhD. Thesis, Stanford University.

Isaaks, E.H., and Srivastava, R.M., 1989, *An Introduction to Applied Geostatistics*, Oxford University Press, 561 p.

Journel, A.G., and Huijbregts, Ch.J., 1978, *Mining Geostatistics*, Academic P., 600p.

Journel, A.G., 1983, *Non-parametric Estimation of Spatial Distributions*, Math. Geology, Vol. 15, No. 3, pp.445-468.

Matheron, G., 1975a, *Forecasting Block Grade Distributions: The Transfer Functions*, in "Geostat 75", pp. 237-251.

Matheron, G., 1975b, *A Simple Substitute for Conditional Expectation: Disjunctive Kriging*, in "Geostat 75", pp. 221-236.

Parker, H.M., 1980, *The Volume-Variance Relationship: A Useful Tool for Mine Planning*, in Geostatistics, pp 61-91, McGraw-Hill.

Roditis, I., Rossi, M.E., and Parker, H.M., 1993, *Evaluation of Existing Geostatistical Models and New Approaches in Estimating Recoverable Reserves*, to be presented at the XXIV APCOM, October 31- November 3, Montreal, Canada.

Sullivan, J.A., and Verly, G., 1983, *The Estimation of Recoverable Reserves Using Conditional Recovery Functions*, Stanford University, Department of Applied Earth Sciences.

Verly, G., 1983, *Estimation of Spatial Point and Block Distributions: the Multigaussian Model*, PhD. Dissertation, Stanford University, Department of Applied Earth Sciences.

AN ANNEALING PROCEDURE FOR HONOURING CHANGE OF SUPPORT STATISTICS IN CONDITIONAL SIMULATION

R. MOHAN SRIVASTAVA
FSS International
800 Millbank
Vancouver, BC
Canada V5Z 3Z4

ABSTRACT

Simulated annealing offers tremendous flexibility in the kind of information that can be honoured in a conditional simulation. Several researchers have shown that the speed of annealing (and, therefore, its practical utility) is related to whether or not the energy function used by the annealing procedure can be readily updated. If it is readily updateable, and does not have to be recalculated from scratch, annealing can be as fast as other commonly used geostatistical techniques for conditional simulation.

Change of support is a critical concern in all areas of geostatistical application. Through training images and exhaustive data sets, it is possible to build histograms that describe how the change of support will affect a random function. In some applications, such as mining, one has direct access to observations at different levels of support. This information can be fed directly into an annealing procedure so that the resulting realization has the correct behaviour under change of support.

This approach to spatial simulation is computationally fast, since the statistics are easy to update, and pedagogically simple, since the support effect is a concept that is readily understood. Though it does not explicitly incorporate variogram information, it does honour two-point statistics implicitly through change of support statistics.

SIMULATED ANNEALING

Simulated annealing[1,2] has recently attracted considerable interest in the simulation of earth science phenomena because it can accommodate a wide variety of conditioning information. Unlike most geostatistical simulation procedures, which explicitly honour only a few select statistics, such as a histogram and a variogram, simulated annealing can honour nearly any statistic imaginable.

An outline of the algorithm

As a brief introduction to how simulated annealing works, consider the simple problem of producing a realization of a 0/1 indicator variable that honours the information

R. Dimitrakopoulos (ed.), Geostatistics for the Next Century, 277–290.

traditionally used in conditional simulation: the histogram, the variogram and some known values at specific locations. Simulated annealing could be used to tackle this problem as follows:

1. Assign to any grid cell that corresponds to a sample location the correct indicator value. If the sample value at that location has an indicator of 1, then assign a value of 1; if it has an indicator of 0, then assign a value of 0.
2. Assign 0's and 1's randomly to all cells that remain empty after the sample data have been used as conditioning information, with a probability of p_0 of selecting a 0 for any particular cell, and a complementary probability, $p_1 = 1 - p_0$, of selecting a 1. The values of p_0 and p_1 should be chosen to respect the global proportion of 0's and 1's.
3. Select an "energy function", E, that measures the deviation between some statistical measurements taken from a training image and the corresponding measurements taken from the initial grid. The choice of this energy function is addressed later.
4. Select a starting temperature, t. The choice of an initial temperature, the number of swaps that should be tried before decreasing the temperature and the rate at which the temperature decreases are all part of what is commonly referred to as the "cooling schedule" in the annealing literature. Advice on how to select these parameters can be found in the references[1,2].
5. Select at random two cells that are not intersected by samples and swap their indicator values. Calculate the energy of this new grid, E_{new}, and compare it to the energy of the old grid prior to the switch, E_{old}. If the new energy is less than the old one (i.e. if the statistical measurements taken from the new grid are closer to those of the training image) then keep the swap. If the new energy is greater than the old one, then keep the swap with probability p_{accept}:

$$p_{accept} = \exp\left(-\frac{E_{new} - E_{old}}{t}\right)$$

   and undo the swap with the complementary probability $1 - p_{accept}$.
6. Repeat the previous step many times and decrease the temperature t slightly.
7. Repeat the previous two steps until the energy is suitably close to zero.

The final grid that results from this procedure will honour the conditioning data from the available samples since the original values in those conditioned cells are never swapped. The final grid will also honour the global proportion of 0's and 1's since the initial grid (after step 2) had the correct global proportions and the swapping of 0's and 1's from one cell to another never alters the global proportion. At the end of step 2 the grid has two of the statistical properties we require: it honours the histogram (the proportion of 0's and 1's in the case of an indicator variable) and it honours certain known values at specific locations. In terms of its variogram, however, this initial grid is a complete disaster; an experimental variogram on this grid would show it to be a pure nugget effect with no spatial correlation whatsoever.

The key to annealing is its iterative updating of the grid using the energy function as a guide to whether a particular swap is successful or not. For the purposes of our indicator simulation, where we want to honour the indicator variogram, we could choose the energy function to be

$$E = \sum_{i=1}^{n} \left[\gamma_{grid}(h_i) - \gamma_{model}(h_i)\right]^2$$

where $\gamma_{grid}(h_i)$ is the experimental variogram value for lag h_i calculated from our grid and $\gamma_{model}(h_i)$ is the corresponding theoretical value calculated from our variogram model. This energy function summarizes with a single number how far away the variogram of our grid is from the variogram that we would like to honour.

After the indicator values in several cells are swapped, the experimental variogram of the new grid might actually be a little bit closer to the variogram; all that we require is a little bit of luck in stumbling across a set of swaps that happens to impart some spatial continuity to the grid. The acceptance/rejection procedure in step 5 of the algorithm described above guarantees that we always accept progress towards our final goal and is, at the same time, somewhat tolerant of lack of progress. Initially, with the temperature, t, set to some very high value, the probability of accepting lack of progress will be very high. As the iterative algorithm proceeds, we decrease t from time to time and the chance of accepting lack of progress also decreases.

By brute force trial and error, the computer will eventually find a series of swaps that imparts to the grid a better spatial continuity. Ultimately, it is possible to get the value of E to be as close to zero as we could want, at which point the grid of indicators finally has the correct variogram. It is correct in the sense that at every lag that contributes to the evaluation of the energy function, the experimental variogram from the grid matches the theoretical value from the variogram model. For lags that do not explicitly enter into the evaluation of the energy function, there is no guarantee that the variogram is correct.

Advantages of annealing

The single biggest advantage of annealing over other geostatistical procedures for conditional simulation is that we can incorporate into our energy function any statistic that we can dream of. In the indicator example given in the previous section, we made the energy function depend only on the variogram. If, in addition to honouring the variogram, we wanted to make sure that there were a certain number of 2×2 groups of cells that were entirely 1's, we could define the energy function to be

$$E = w_1 \cdot \sum_{i=1}^{n} \left[\gamma_{grid}(h_i) - \gamma_{model}(h_i)\right]^2 + w_2 \cdot \left[N_{2\times2,grid} - N_{2\times2,model}\right]^2$$

where $N_{2\times2,grid}$ is the number of times that we observe in our grid a 2×2 group of cells that is entirely 1's and $N_{2\times2,model}$ is the number of times that we would like to see this happen. Since the differences between the variogram values are not likely to have the same order of magnitude as the differences in the N values that we have added to our energy function, it is common to include weighting factors, w_1 and w_2, that prevent one statistic completely dominating another in the evaluation of E.

In recent years, researchers in flow modelling have successfully incorporated some quite complicated statistics into their conditional simulations, such as information from well test analysis, from tracer tests or from seismic data.

Disadvantages of annealing

The principal practical disadvantage of annealing is that it is a relatively recent development. There is not yet enough accumulated wisdom and experience with annealing and there is little technical literature that provides the details of the tradecraft involved in designing an appropriate energy function and a cooling schedule. Even with existing public domain software, first-time users often find themselves completely unable to produce sensible results. Though the relative maturity of the more established techniques currently counts as a disadvantage for the use of annealing, this imbalance will diminish with time as more case studies get published and more people have a chance to experiment with the flexibility of the annealing approach.

The other notable practical drawback of simulated annealing is that it can be abysmally slow if the energy function is complex. Since the annealing procedure often requires many thousands, if not millions, of swaps to find a grid whose statistics match those of the training image, the only practical applications of annealing involve energy functions that are easy to update. If the energy function cannot be updated after each swap, but must be completely recalculated instead, then the computational effort required by simulated annealing is prohibitive.

The principal theoretical drawback of annealing is the lack of a consistent mathematical framework. The basic algorithm is an optimization procedure, and this raises many questions about what space of uncertainty is being explored by this algorithm and whether it is producing realizations that can be called "equally probable".

CHANGE OF SUPPORT

For nearly 40 years, geostatisticians have wrestled with the problem of change of support[3]. The distribution of values measured on small volumes is not the same as the distribution of the same variable but measured on a larger volume. For additive variables, such as metal grades in mining applications, porosities in flow applications or contaminant concentrations in environmental applications, we typically find that although the mean of the distribution remains the same for any support size, the spread of the distribution tends to decrease as the support size increases and the distribution also tends to become more symmetric.

This support effect has important implications for many geostatistical problems. If ignored (or not recognized), the support effect can cause estimates to be very badly biased. Particularly for applications that deal with truncated statistics, such as ore grades above some cutoff or contaminant concentrations above some remediation threshold, the failure to recognize that large volumes have less variable average values than do small volumes can lead to completely misleading conclusions.

In most practical applications, the available samples are all collected with one support size (typically quite a small one) and information about the distribution of values for larger support sizes is not available. If, however, we have access to a training image or an exhaustive data set that can serve as an analog of the problem we are studying, then it is possible to compile quite detailed information on the support effect. Such exhaustive analogs could include production information from previously mined-out areas in mining applications, outcrop studies in petroleum applications, or the results from a deterministic process simulation.

Annealing with change of support statistics

Most of the recent work on earth science applications of annealing have either used energy functions based on traditional geostatistical statistics, such as the variogram, or have incorporated complicated statistics that reflect flow behaviour. Davis[4], on the other hand, has suggested an opposite approach. Rather than incorporating more and more sophisticated statistics into the energy function, why not use only the most simple statistic: a histogram?

At first glance, Davis' suggestion appears irrelevant because annealing always exactly honours the histogram by virtue of the fact that the grid is initialized with the correct histogram and then the values are simply swapped around the grid. The key to appreciating Davis' idea, however, is to realize that we are only honouring the point support histogram through the initialization and swapping, we are not directly honouring the histogram of any larger support. His idea is to construct the energy function from change of support histograms alone:

$$E = \sum_{i=1}^{n} \sum_{j=1}^{m_i} \left[p_{i,j,grid} - p_{i,j,model} \right]^2$$

where $p_{i,j}$ is the proportion of values in the j-th class of the histogram for the i-th support configuration.

This energy function requires us to select the n support configurations whose histograms we want to honour and, for each of these, to decide how many classes, m_i, we want to use. The choice of these support configurations is the critical piece of tradecraft for this new approach to simulation. If directional anisotropies exist, certain support configurations may be more advantageous than others.

The calculation of the $p_{i,j}$ values can either be done by direct counting or by resorting to a parametric model. In most applications, it is likely that we would mix the two types of calculation—we might calculate $p_{i,j,model}$ from a parametric model and $p_{i,j,grid}$ from a direct counting on our grid.

There are two main advantages to this type of spatial simulation procedure: it is rapid and it is easy to understand. Its speed comes from the fact that the energy function is composed entirely of histogram information that is easily updated. When the values of two cells are swapped, the change of support statistics will be affected only in their immediate vicinity. With very few calculations, it is possible to figure out how the number of observations in each class of the histogram will be affected.

The second advantage, the ease of understanding, is important to anyone who has struggled to teach the nuts and bolts of conditional simulation. Many people who are quite curious about the concept of conditional simulation are never actually able to apply it because they end up confused by the details; if the semivariogram doesn't get them, the kriging likely will. With annealing to change of support statistics there are two key concepts that need to be taught. The first is the principle of annealing, the second is the idea of the support effect.

Though most people are initially skeptical of the annealing algorithm, this stems not from lack of understanding (they actually understand it very well); what they are dubious about is whether a sensible realization will evolve from brute force trial and error. Once they have seen through practice that it does, in fact, work, the simplicity of the algorithm becomes very appealing.

The support effect is equally easy to appreciate. Most earth scientists have a strong instinct for the fact that quantities measured on small volumes will be more variable than those measured on larger volumes. Once these two ideas, annealing and support effect, are understood, a newcomer to spatial simulation has mastered all of the important details.

An interesting spinoff of the procedure is that it honours two-point statistics, such as the variogram, even though they do not explicitly enter into the calculation of the energy. This is due to the fact that the effect of change of support on a spatial distribution is a reflection of the spatial continuity of the variable. If there is tremendous spatial continuity, then the reduction in variance and the increase in symmetry will be slow as we go from small volumes to larger ones; if there is virtually no spatial continuity, then the reduction in variance and the increase in symmetry will be rapid. So even though users need know nothing about variograms, and need not struggle through variogram analysis and modelling, their realizations will honour the variogram implicit in the change of support statistics they have chosen.

MODELLING FRACTURE NETWORKS

As the critical role of fractures has become more apparent in fluid flow and contaminant transport studies, the characterization of fracture networks has received considerable attention in a wide variety of applications: nuclear waste repository design,[5,6] groundwater studies[7] and petroleum reservoir modelling[8,9]. The problem of modelling fracture networks offers an ideal opportunity to test the idea of annealing to change of support statistics.

The traditional approach

The most common approach to the stochastic modelling or fractures involves the use of statistics that describe the orientation and size distributions of fracture planes, usually modelled as elliptical disks[10]. Once the distribution of the orientations, lengths and widths of these elliptical disks have been inferred, a "boolean" or "marked point" process is used to generate a collection of disks that honour these distributions. Typically, a random number generator is used to select a location for the center of a disk, and to select an orientation, a length and a width from the appropriate distributions. Disks continue to be randomly located, oriented and sized until the total number of simulated fractures reaches a user-specified threshold.

If sample data from wells reveal the location and orientation of specific fractures, this simple procedure can be modified so that simulated disks intersect the wells at the correct depths and with the correct orientation. Further enhancements to the basic procedure allow the orientation and size distributions to be locally customized to accommodate regions where the fracture density is thought to be higher or lower than average, or where the orientation or size of the fractures is thought to be different from the global distribution. The technique has also been extended to incorporate the theory of percolation[7,11,12].

Limitations of the traditional approach

This disk-based approach has gained widespread acceptance due to its conceptual simplicity and its computational convenience. Despite its popularity, however, there are significant limitations to the usefulness of this approach.

One limitation stems from the fact that the size distribution of fractures can rarely be inferred from field data. Boreholes and wells are too small to see fractures greater than a few centimetres. Even though observations from underground workings and outcrop studies may extend the scale of observation to tens of metres, the largest fractures are typically truncated by the edges of the observable region, making the statistical inference for these most critical large fractures very difficult.

This difficulty with the inference of the size distribution and volume density of the large fractures has led some researchers[5] to adopt a fractal assumption in which the sizes and densities of the smaller visible fractures are extrapolated using fractal scaling laws. Other researchers[9] have assumed specific distributional models for the fracture length distribution—lognormal, exponential and Pareto being the most popular—and used the information from the smaller fractures to select parameters for the distribution; once the parameters of the distribution have been selected, the behaviour of the tails can easily be predicted.

A second shortcoming of this approach based on disks is that the statistical parameters it honours, fracture orientation and size distributions, may not be the most relevant characteristics to be honouring if flow simulation is the final goal. Figure 1 shows two fracture networks that have the same distribution of fracture lengths and orientations. As shown in Figure 2, the fracture network in Figure 1a contains a connected cluster of fractures that completely span the region; the one shown in Figure 1b does not. From the point of view of contaminant transport, the existence of such a continuous path through the fractures is of critical importance. The simulation of fracture networks by randomly located disks does not directly take into account information about fracture spacing, fracture intersections or the size of connected clusters. Other spatial statistics such as these may have more impact on the flow behaviour of the fracture network than do the size and orientation distributions.

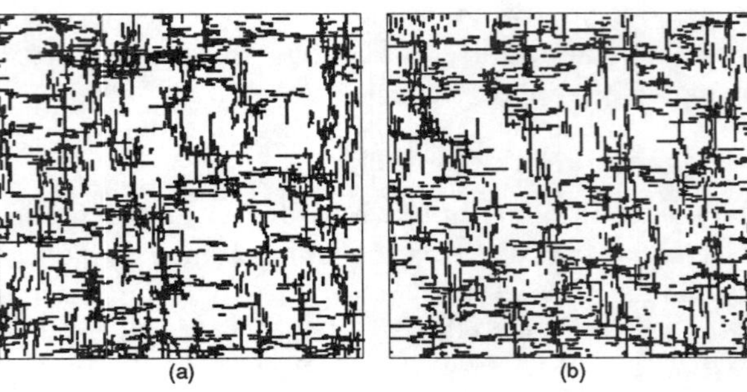

Figure 1. An example of two fracture networks with the same length and orientation distributions but different cluster characteristics.

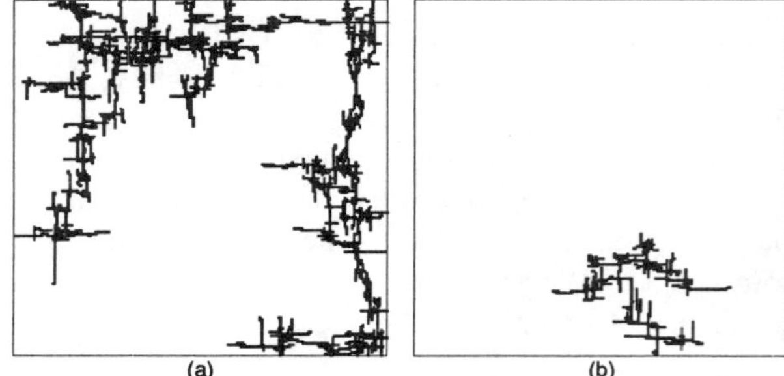

Figure 2. The largest connected clusters for each of the two examples shown in Figure 1.

The third major shortcoming of this disk-based approach is that the fractures that are being modelled are, in reality, not elliptical disks; it is doubtful that they are even planar. The decision to model fractures as thin disks is simply a convenient approximation. A comparison of actual field data with simulations based on disks suggests that this convenient approximation may not always be a good one. Figure 3 shows

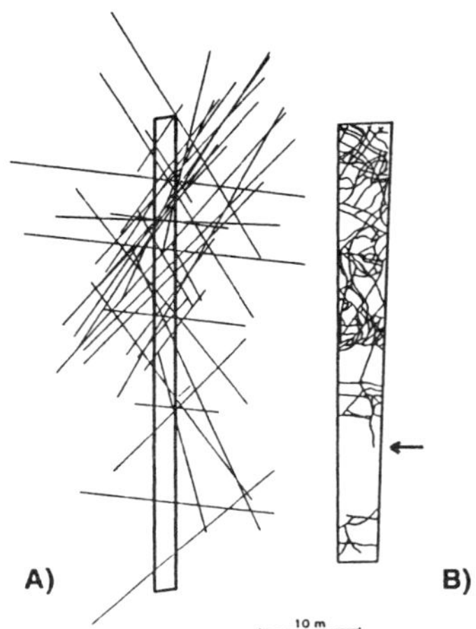

Figure 3. An example of actual fractures from a chalk formation (right) and a corresponding realization from a boolean simulation procedure (left) [from Stolum et al. (1991)].

an example of the actual fractures from a chalk quarry along with the corresponding simulation based on treating the fractures as disks. The flow characteristics of the actual and simulated fracture networks are quite different.

The fact that the fractures being modelled are not really disks at all further exacerbates the problem of inference. Even with abundant outcrop data and underground observations, it is usually quite awkward to define where a particular fracture starts, where it stops and what its resulting length should be. With the field data shown in Figure 3, for example, where is the end of the fracture that starts near the arrow on the right? Does it continue for more than 10 metres in the vertical direction? Or does it terminate in a few metres? Or does it disappear off the right of the visible section?

Fracture density

Rather than treat individual fractures as separate geometric objects, entire clusters of fractures could be treated as objects. The focus of the fracture characterization could then be shifted from the level of the individual fractures to the level of clusters. Rather than focusing on point observations, we could deal instead with variables that represent volume averages, such as the fracture density, F, which can be calculated from borehole or well data as follows:[6]

$$F = \frac{d}{z} \sum_{\theta=0}^{90} \frac{f(\theta)}{\sin(\theta)}$$

where z is the thickness or length of the interval of core; d is the diameter of the cylindrical volume that F pertains to; θ is the inclination of the fracture; and f is the number of fractures measured in the interval for each inclination. Similar formulas can be used to calculate fracture densities from outcrop data and from underground observations.

The fracture density will change from one interval to the next, so it is appropriate to think of it as a variable that depends on spatial location. For this reason, we will refer to the fracture density as $F(x)$, the x being used to denote spatial location. Statistical summaries, such as histograms and variograms of $F(x)$ can be obtained from the analysis of $F(x)$ in available boreholes and in outcrop studies or underground workings. The determination of the direction of maximum continuity can be guided by regional tectonic considerations. Once histograms, variograms and conditioning data have been assembled, geostatistical simulation can be used to create several possible renditions of the fracture density in the subsurface.

HONOURING MULTIPOINT STATISTICS

Fracture density statistics are ideal for the annealing procedure proposed by Davis. The effect of support on the distribution of fracture density contains a lot of information about the spatial arrangement of the fractures, much more information than can possibly be contained in the parameters of the traditional boolean approach.

At the "point" level of support, where a "point" corresponds to a cell in the grid, the fracture density is always 0 or 1. One could construct a histogram of this binary variable and include it as information in the energy function. As pointed out earlier, however, there is no need to include the proportion of 0's and 1's in the energy function since these global statistics are honoured by virtue of the way the grid is initialized and the way that cells are swapped.

For support volumes larger than a single cell, the fracture density can take on more than two values. For example, if we take a 2D grid of the fracture indicators and examine areas that are 2×3 the number of fractured cells in this volume could be any integer value from 0 to 6. The histogram showing the frequency of each of these seven outcomes is not honoured by the initialization and swapping mechanism; if we wanted our final grid to honour the fracture density statistics for 2×3 support volumes from a training image, we would have to include in our energy function a calculation of the difference between the histogram of the fracture density from the training image and the corresponding histogram from our simulated grid.

Since the histograms of fracture densities can be rapidly updated, it is possible to include many histograms for many different support volumes in the energy function. When such an energy function reaches zero, the final simulated grid will have fracture density statistics that match those of the training image at many different levels of support. By honouring changes in the fracture density histograms as the support volume changes, the final grid will also honour some multipoint statistics that are difficult to capture in conventional geostatistical algorithms.

Figure 4 shows an example of some results from this annealing procedure. Figure 4a shows the training image, a 2D area taken from a physical model of crack propagation[13] that has been discretized into 40,000 cells on a 200×200 grid. Figure 4b shows one possible realization that was created by the simulated annealing procedure outlined above. No conditioning data were used in the creation of this realization, so the resulting realization is not intended to match the training image at any particular location, but is intended only to mimic the spatial arrangement of 0's (white) and 1's (black). The energy function used by the simulated annealing proce-

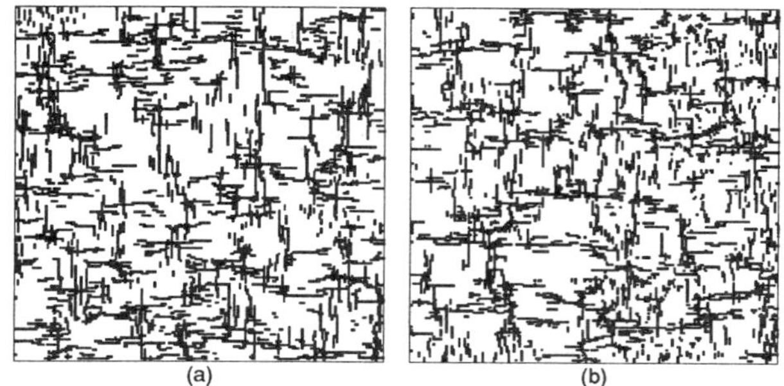

Figure 4. A training image and one possible outcome from annealing to change of support histograms.

dure consisted solely of histograms of fracture density statistics for different support levels.

A total of 20 different support configurations were incorporated into the energy function. For some of these 20 support configurations, Figure 5 compares the histograms from the training image to the corresponding histograms of the simulated realization. Several attempts to use only those support configurations that lined up with the obvious N-S/E-W anisotropies did not produce satisfactory results, even though all of the histograms were matched very well. The inclusion of the square support configurations was the key to a visually acceptable result. Though this is but one initial experiment with the technique, this experience suggests that the selection of appropriate support configurations may not

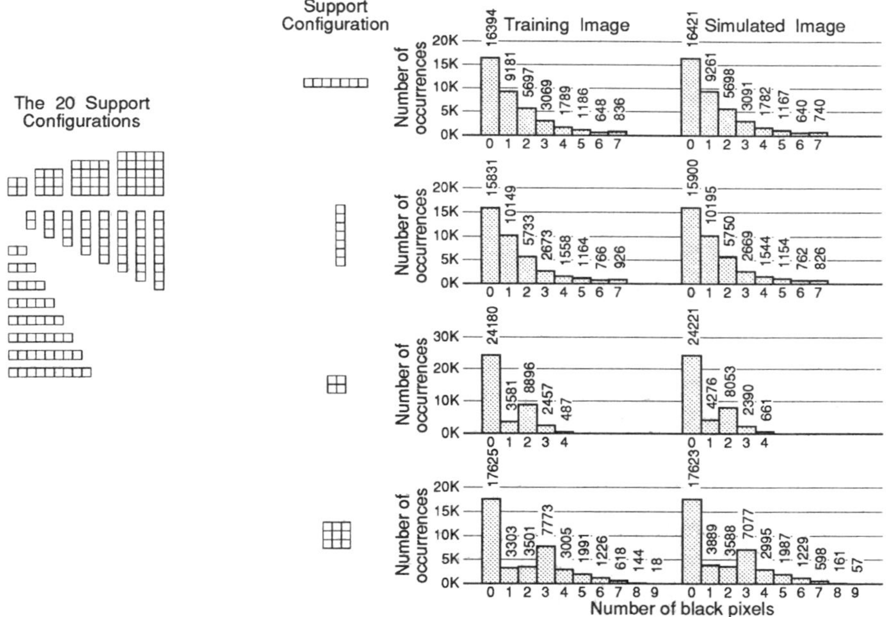

Figure 5. The 20 support configurations used to create the realization shown in Figure 4b and the comparison of the histograms of the simulated image and the training image for a few of these support configurations.

be immediately obvious, and that some isotropic support configurations may be quite useful even when there are evident anisotropies.

In addition to bearing a strong visual similarity to the training image, the simulated fracture network shown in Figure 4b is also quite similar in the way that its connected clusters behave. Figure 6 shows the largest three connected clusters for the training image and for the simulated fracture network. Neither the training image nor the simulated fracture network have any clusters that completely span the region.

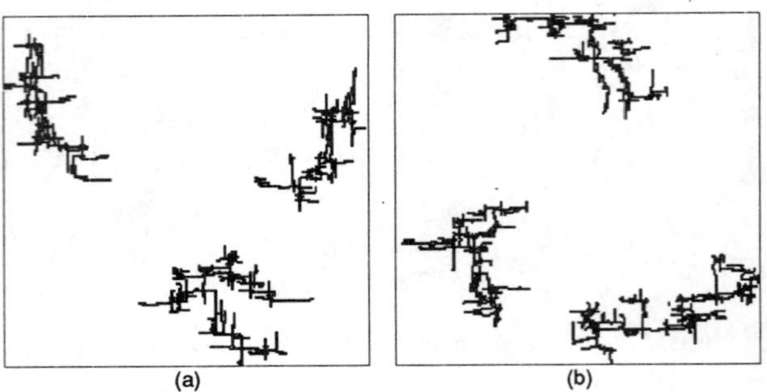

Figure 6. The three largest connected clusters for: (a) the training image and (b) the simulated realization shown in Figure 4.

CONCLUSIONS

Simulated annealing based on honouring change of support statistics from several different configurations holds considerable promise not only for fracture network simulation but for any spatial simulation application.

The main advantages of this new annealing procedure are its speed and its pedagogical simplicity. The simulation shown in Figure 4 was created in slightly less time than it would take to run a conventional sequential indicator simulation. With further improvements in efficiency, this algorithm could rival the fastest of the conditional simulation procedures currently in common use. The simplicity of the two basic concepts, annealing and change of support, make this new technique a very straightforward method for newcomers to grasp.

The single issue that needs further research is the choice of which support configurations the energy function should explicitly incorporate. As more experience is gained on this particular aspect of the method, annealing with experimental change of support statistics should become an important tool in stochastic simulation.

Though geostatistical characterization of fracture networks has traditionally focused on the size and orientation of fractures, it appears that this approach is limited by difficulties with inference and is handicapped by the fact that the key aspect of the model—the planar elliptical shape of the fractures—does not conform to reality.

By shifting our attention from individual fractures to clusters of fractures, we may be able to improve the simulation of fracture networks and therefore to improve the reliability of forecasts from flow and contaminant transport studies that are based on such simulated networks.

The volume density of fractures is a variable that can be defined from field observations, that can be analyzed statistically and that can be modelled in three dimensions. Dealing with volume density also offers the possibility of validating key assumptions or identifying critical parameters through well tests.

ACKNOWLEDGEMENTS

The author would like to thank Carl Renshaw of Stanford University for providing us with data from his research on fracture simulation through deterministic modelling of crack propagation. I am also grateful to Mike Davis for his original insight on how multipoint statistics could be honoured by the annealing procedure implemented in this paper.

REFERENCES

1. AARTS, E., and KORST, J., *Simulated Annealing and Boltzmann Machines,* John Wiley & Sons, New York, (1989).

2. DEUTSCH, C.V., *Annealing Techniques Applied to Reservoir Modelling and the Integration of Geological and Engineering (Well Test) Data,* Ph.D. Thesis, Stanford University, (1992).

3. SHURTZ, R., "The electronic computer and statistics for predicting ore recovery," *Mining Engineering,* vol. 11, no. 10, pp. 1035–1044, (1959).

4. DAVIS, M.W., *Personal Communication,* British Bankers Club, Menlo Park, California, (1992).

5. BARTON, C.C., and LARSEN, E., "Fractal geometry of two-dimensional fracture networks at Yucca Mountain, SW Nevada," *Proceedings of International Symposium on Fundamentals of Rock Joints,* Bjorkliden, pp. 77–85, (1985).

6. SCOTT, R.B., SPENGLER, R.W., DIEHL, S., LAPPIN, A.R., and CHORNACK, M.P., "Geologic character of tuffs in the unsaturated zone at Yucca Mountain, southern Nevada," in *Role of the unsaturated zone in radioactive and hazardous waste disposal,* Ann Arbor Science, Butterworth Group, pp. 289–335, (1983).

7. WILKE, S., GUYON, E., and DE MARSILY, G., "Water penetration through fractured rocks: Test of a tridimensional percolation description," *Mathematical Geology,* vol. 17, pp. 17–27, (1985).

8. KOESTLER, A.G., and MILNES, A.G., "The importance of structural reservoir characterization and reservoir mechanics for enhanced oil recovery," *Proceedings of the First Conference on the Geology of the Arab World,* Cairo, (1992).

9. STOLUM, H.J., KOESTLER, A.G., FEDER, J., JOSSANG, T., and AHARONY, A., "Fractal characterization of a fractured chalk reservoir—the Lägerdorf case," *American Association of Petroleum Geologists Bulletin*, vol. 75, p. 677, (1991).

10. LONG, J.C., *Investigation of equivalent porous medium permeability in networks of discontinuous fracture,* Ph.D. Thesis, LBL-14975, Lawrence Berkeley Laboratory, Earth Science Division, (1983).

11. GRIMMET, G., *Percolation,* Springer-Verlag, (1989).

12. HAMMERSLEY, J.M., and WELSH, J.A., "Percolation theory and its ramifications," *Contemporary Physics,* vol. 21, pp. 593–605, (1980).

13. OLSON, J., and POLLARD, D., "Inferring paleostresses from natural fracture patterns: A new method," *Geology*, vol. 17, pp. 345–348.

DIRECTIONS IN MINING GEOSTATISTICS

THE NEED FOR A CONNECTIVITY INDEX IN MINING GEOSTATISTICS

D. ALLARD,	M. ARMSTRONG	W.J. KLEINGELD
Centre de Geostatistique	Centre de Geostatistique	OED Anlgo–American Corp
35 rue St Honore,	35 rue St Honore,	PO Box 41848
77305 Fontainebleau,	77305 Fontainebleau,	Craighall 2024
FRANCE	FRANCE	SOUTH AFRICA

When orebodies are mined at high cutoff grades (relative to the mean grade), only a fraction of the selective mining units is above cutoff, and worse these blocks are split up into disjoint patches that cannot always be accessed for mining economically. This paper addresses the question of how to characterize the breakup of the reserves above cutoff for the case of 2D orebodies of the Witwatersrand type, firstly by studying a mined out deposit where the true grades are known and then via simulations. Three criteria were found to be important for describing the breakup of the reserves: the proportion of ore above cutoff, the number of connected components (i.e. of clumps of blocks above cutoff) and their size distribution. Looking at images of the reserves above cutoff, we felt that the images needed to be "tidied up" by removing small clumps of ore in the waste or of waste in the ore. Two morphological transformations were found that do this, and more interestingly seem to provide upper and lower bounds for the effective mineable reserves. As this work is just starting, no definitive conclusions can yet be proposed. However, simulation combined with suitably chosen morphological operators seems to be a promising approach for studying the breakup of orebodies with increasing cutoff grades. Having said that, we feel that the work has raised far more questions than it has resolved.

INTRODUCTION

Over recent years, the need for an estimate of the mineable reserve in addition to the geological insitu reserve has become important at the prefeasibility stage and must often be based on limited information. This need results from

- a shift in emphasis from exploration to acquisition of deposits, where the acquisition bid would be influenced by a priori knowledge of mining the reserve estimate, and
- the increased cost of mining and the decreasing metal price resulting in selective mining often at elevated cutoff grades so that the economic mineralised fraction significantly decomposes or breaks up into patches.

In the past it may have been acceptable, during the early prefeasibility phase, to base an estimate only on the spread of values (histogram) and the spatial character (variogram) of

R. Dimitrakopoulos (ed.), Geostatistics for the Next Century, 293–302.

the mineralisation. Available nonlinear geostatistical tools such as the gaussian anamorphosis of a distribution model coupled with a change of support calculation were sufficient to provide grade–tonnage curves of the potential of the reserve, which could be realised during mining. Nowadays, exploiting the reserve at elevated cutoff grades results in the problem of a fragmented mineable component. This often leads to a substantially reduced realisable mining reserve and impacts on the mining selection unit size. In turn this can have economic implications, for example on truck and fleet sizes. Such shifts in equipment can have major cost ramifications especially in third world countries where historically only specific size equipment is manufactured locally, the rest having to be imported with all the foreign currency problems that this poses.

Seen against the above, it has become necessary to find a connectivity index which could be used in conjunction with the classical grade–tonnage curve estimates to arrive at a "mining reserve" estimate on which policy decisions can be based, during the assessment of potential in the prefeasibility phase. Simulation techniques offer a potentially fruitful line of research, but it may well be possible that a more fundamental tool will have to be devised.

The practical problem

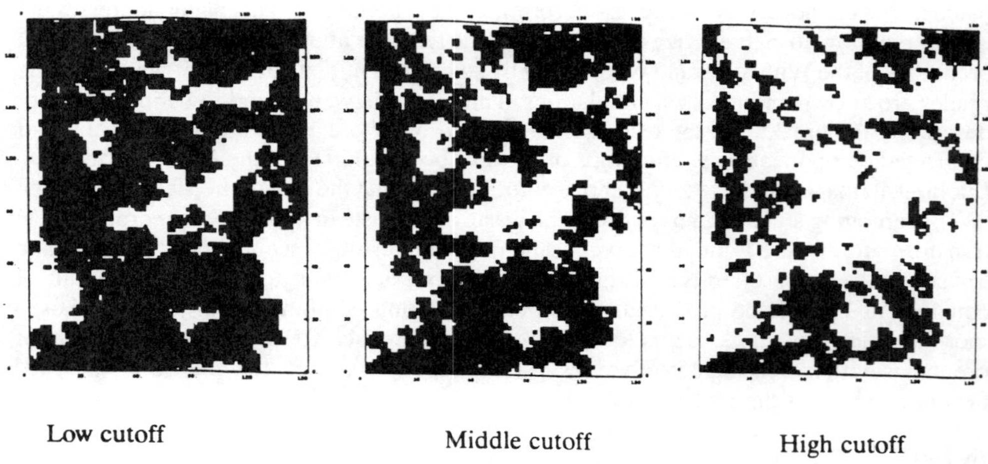

Low cutoff Middle cutoff High cutoff

Figure 1: Blocks above cutoff for 3 cutoffs: low, medium and high.

The impact of the breakup on the recoverable reserves can best be illustrated via an example of South African gold grades. Figure 1 shows the blocks above cutoff in black for three cutoff grades (0.5, 0.75 and 1.0 compared an average that was set to 1.0 to maintain confidentiality). At low cutoffs, most blocks are above cutoff and they tend to be clumped together in "mineable" groups. So the insitu recoverable reserves estimated assuming a free selection are close to the reserves that can actually be recovered. By the middle cutoff, small clumps of recoverable blocks are becoming detached from the main clumps and holes are appearing in the ore. As the cutoff increases further, the insitu reserves fragment into small

groups of blocks which either have to be left behind with the waste or necessitate the mining of waste to access them. The overall impression is one of lacework.
Having visualised the problem, we can now try to describe it mathematically. Clearly the variogram model is a determining factor in the breakup of the reserves. The important factors are in fact, the variogram range, its shape at the origin (linear or quadratic) and the nugget to sill ratio. But is this enough? It soon became apparent that this is not sufficient. Could the indicator variograms contribute the vital missing information? Unpublished work by Lantuejoul showed that this was not the case and so indicator kriging would not solve the problem. In fact, we need to know far more than what can be obtained from any of the different types of variograms because these are essentially 2–point statistics.

No general solution exists

Fairly rapidly those involved in the project realised that although general techniques were available for conditional simulation and for estimating insitu recoverable reserves, this would not be true for orebody breakup. This is very much dependent on the nature of the stochastic process being studied and on the mining method being used. Having said that, there are certain families of deposits that occur commonly and are mined in similar ways (e.g. underground mining of Witwatersrand gold orebodies, opencut exploitation of porphyry copper deposits, etc). It was decided to start out by tackling the problem for 2D orebodies of the Witwatersrand type, by using simulations.

Which critieria to study?

Given the number crunching power of workstations nowadays, nonconditional simulations with Witwatersrand type histograms and variograms can be churned out in great number. The essential question is to decide which of their characteristics to concentrate on. Following earlier work (Allard, 1993) on the connectivity of oil reservoirs, three criteria seem particularly interesting;

- the proportion of ore above cutoff denoted by T by mining engineers, (In petroleum, it corresponds to the porosity.)
- the number of connected components (i.e. of clumps of blocks above cutoff), and
- the distribution of their sizes. We use V_{max} (the ratio of the ore in the largest component to that above cutoff) to characterise this. Alternatively one could consider Q_{max} (the same ratio for the metal).

In his earlier work Allard chose models that are appropriate to petroleum but not necessarily to mining. These were a boolean model, a truncated gaussian with a pure nugget effect variogram and a truncated gaussian with a factorised exponential variogram model. The simulations produced using the factorised exponential look "strange" to miners, as they are too pointed for ore pockets. This is due to the choice of a factorisable model. Consequently, it was critical that the variogram models be better suited to mining data (e.g. ordinary sphericals or exponentials).

CASE STUDY USING WITWATERSRAND GOLD GRADES

Simulations are often criticized as having too much "gaussian" flavour. To avoid this, the initial work was carried out on 1200 blocks in a mined out area where the true gold grades are known. The data have been rescaled to have a mean of 1.0 (thereby protecting their confidentiality). Their histogram was skew with a coefficient of variation of 0.8. The range of the variogram was 15 times the size of the blocks. Ten cutoff grades ranging from 0.25 to 2.50 were selected for study. Table 1 presents the number of components N_{cc} and the percentage of ore in the biggest one, V_{max}.

Cutoff	N_{cc}	V_{max}
0.25	3	98.6
0.50	6	90.2
0.75	6	83.8
1.00	14	45.8
1.25	17	51.7
1.50	12	54.6
1.75	11	48.1
2.00	12	39.5
2.25	13	39.1
2.50	10	43.6

Table 1: the number of components N_{cc} and the ore percentage in the biggest one V_{max}.

As expected at low cutoffs there is one enormous component that takes up virtually all the space, and a few small ones. With increasing cutoff, this big component breaks up into many smaller ones which then "evaporate" as the cutoff rises further. Predictably, a smaller and smaller fraction of the recoverable reserves is contained in the biggest component. No surprises up to this point. Allard had found that the average of N_{cc} (for a large number of simulations) rose monotonically to a maximum then dropped off, and similarly that V_{max}. decreased monotonically, Table 1 shows that this does not necessarily hold for individual cases (or simulations).

How to represent the distribution of component sizes?

One key element in describing orebody breakup is the distribution of the sizes of components. At first one might think of using a histogram. The split up of sizes for the cutoff 0.50 is 752 blocks, 70, 5, 5, 1 & 1. This is why a histogram is inappropriate. A pie diagram is more suitable because we immediately see the relative sizes of components from the areas. The largest one is shown in white. Figure 2 shows this for 6 cutoffs.

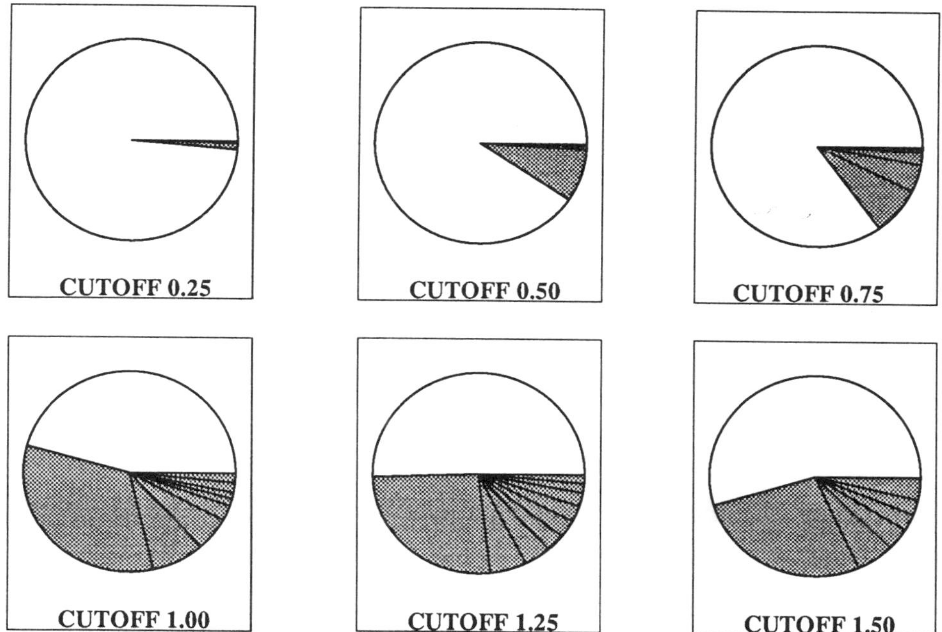

Figure 2 : Pie diagrams representing the distributions of component sizes

Tidying up the shape of the recoverable reserves

Looking back at Figure 1, at higher cutoffs we see small isolated patches of ore in waste and of waste in the ore. Most of these small rich patches would have to be left behind, while the waste enclosed in ore could not be separated from the ore during mining. The first would probably be lost, whereas the second would contribute to dilution. The recoverable reserves need to be "cleaned up" to take account of this and to make them look realistic. This led to the idea of finding morphological operators that would do the job. Clearly these would have to be tailor made to reproduce the characteristics of different mining methods. After some trial and error, two promising morphological operators were found which we call MORFO–1 and MORFO–2. Detailed information on many such operators can be found in Serra (1982); some examples of their use in an underground mining context are given by de Fouquet (1985).

SIMULATION TRIALS

Ten nonconditional simulations of a 200 x 200 pixel grid were made using a model that is typical of Witwatersrand gold. Notably the variogram consisted of a pair of ordinary sphericals (rather than factorised exponentials). The simulated gaussian equivalents were then back transformed to have the same scale as in the case study (i.e. mean 1.0). Only 10 simulations were carried out rather than say 50 or 100, because the resulting curves can then be plotted individually. It would have been difficult to present more than this graphically. Figures 3 and 4 show the tonnages and the metal above cutoff as a function of cutoff grade. As expected the curves fan out at higher cutoffs and this is more marked for metal than for tonnage. This indicates that there is inherently more variability at higher cutoffs, and hence more uncertainty in any predictions.

MORFO–1 and MORFO–2

The number of connected components N_{cc} and the percentage of ore in the biggest one V_{max} were calculated for each simulation. Then each image was transformed using the two operators for each cutoff, and N_{cc} and V_{max} were recalculated. Figure 5 (a) shows N_{cc} as a function of the cutoff grade for the ten simulations while Figures 5 (b) and (c) correspond to the two transformed images of the recoverable reserves. Similarly for V_{max} for Figures 6 (a), (b) and (c). Comparing the three versions of Figure 5, we see that for the simulations N_{cc} reaches a maximum of 800 – 900 components at cutoffs from 1.25 to 1.50. The images obtained using MORFO–1 and MORFO–2 contain far fewer components (their maximums being about 150 for MORFO–1 and 300 for MORFO–2). This proves that both transform–ations do in fact "clean up" small components of ore and waste. But what is particularly interesting, is that MORFO–1 shifts the maximum to the left to a lower cutoff of around 1.00 whereas MORFO–2 moves it right to higher cutoffs from 1.75 to 2.00.

Figures 6 (a), (b) and (c) tell a similar story. The principal feature, the sharp drop in the size of the largest component from nearly 100% to below 50% occurs at cutoffs of 0.5 to 0.75 for the simulated orebodies. For MORFO–1, this occurs almost immediately after 0.0, for most simulations. In contrast to this, for MORFO–2 the percentage is very high until after 0.75. So for both statistics, N_{cc} and V_{max}, the results obtained from the two operators surround those for the deposit. They seem to provide upper and lower bounds for the interesting features. Visual inspection of the images treated using MORFO–1 and MORFO–2 showed good cleaning power. All the small clumps of ore and waste had effectively been removed.

QUESTIONS STILL TO BE RESOLVED

As this work is just beginning, any definitive conclusions would be premature now. The work to date has, however, shown that simulation combined with suitably chosen morphological operators is a promising approach for studying the breakup of orebodies with increasing cutoff grades. Going further, it suggests that appropriate transformations may give upper and lower bounds on the reserves that can effectively be recovered (rather than merely those obtained assuming a free selection). Having said that, the work has raised far more questions than it has resolved. Points requiring further study include the following.

- First and most obviously, the variogram parameters should be modified while keeping the same basic model, i.e. varying its range and its nugget/sill ratio. Thought is required because the sheer number of combinations of parameter values rapidly becomes prohibitive.

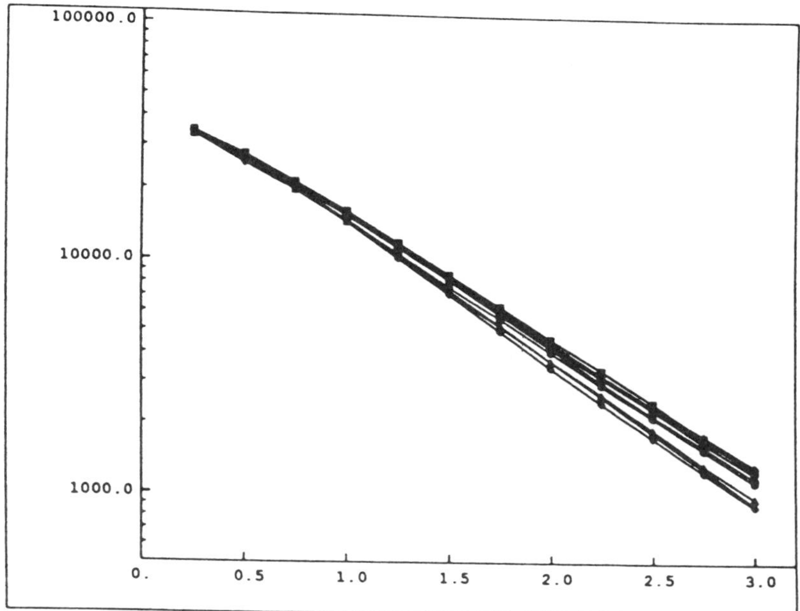

Figure 3: Tonnage above cutoff as a function of cutoff grade

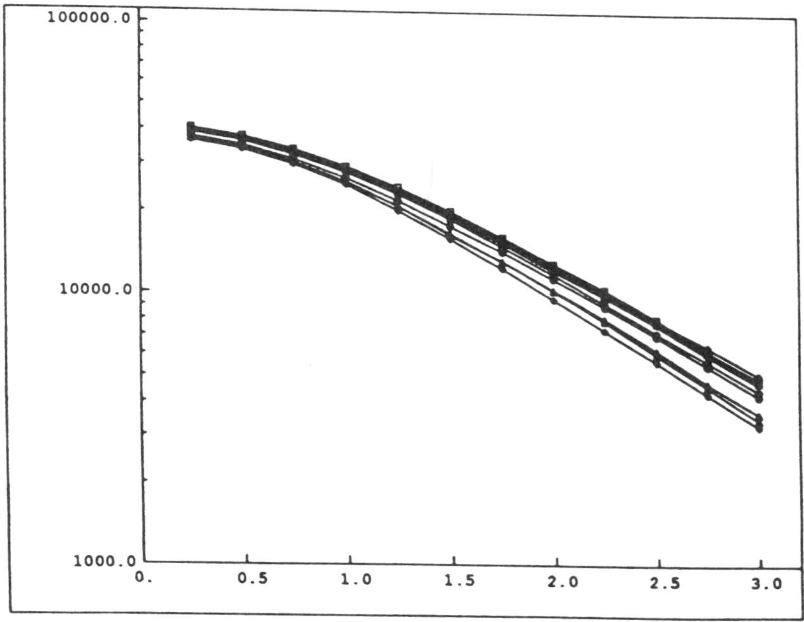

Figure 4: Metal content above cutoff as a function of cutoff grade

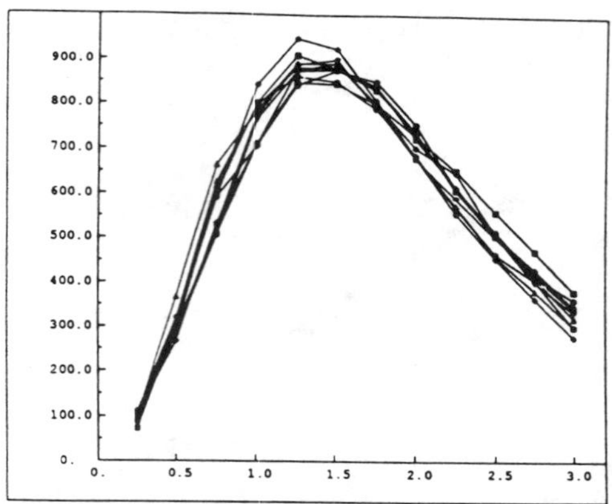

Figure 5a : Number of connected components in simulated orebodies

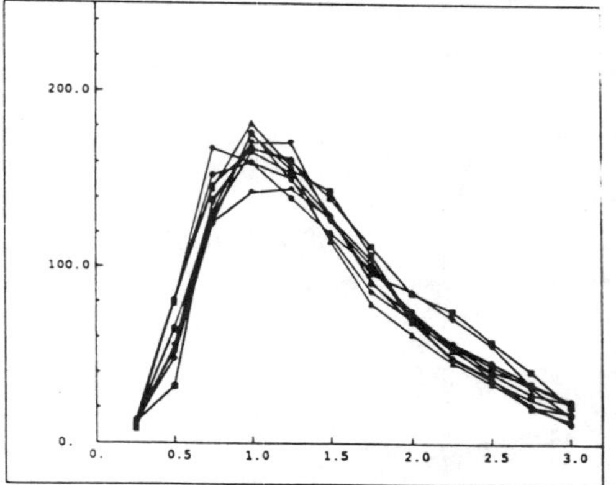

Figure 5b : Number of connected components after using MORFO-1

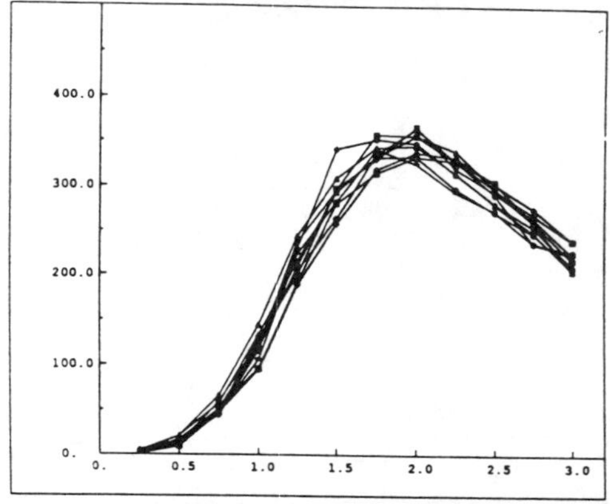

Figure 5c : Number of connected components after using MORFO-2

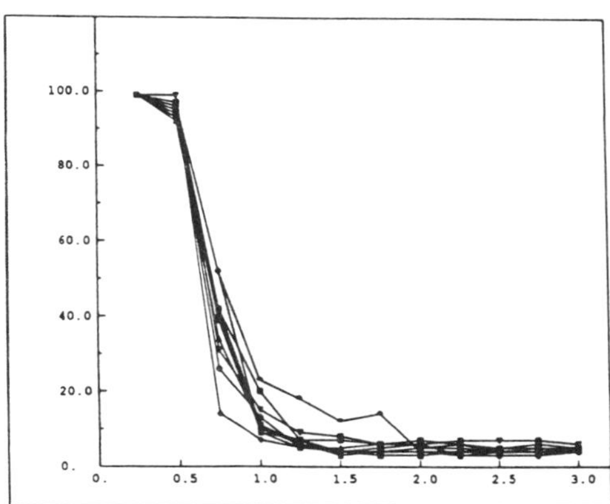

Figure 6a : Percentage in biggest component in simulated orebodies

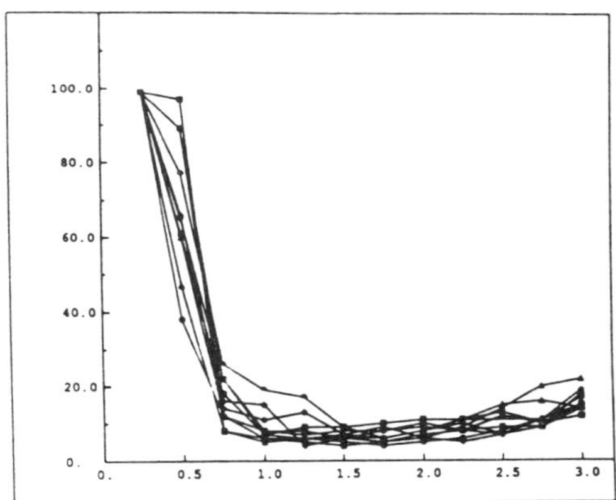

Figure 6b : Percentage in biggest component after using MORFO–1

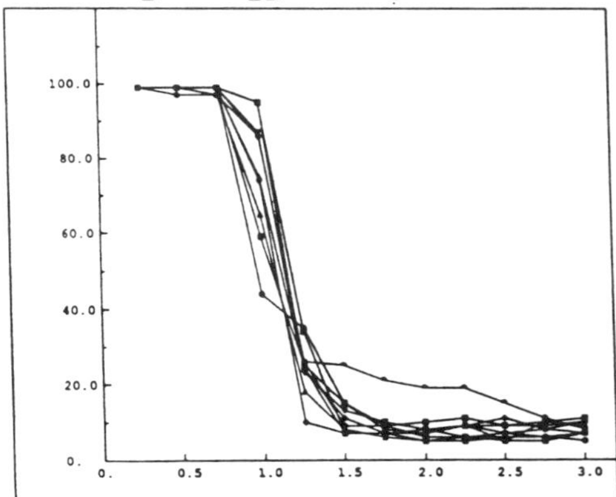

Figure 6c : Percentage in biggest component after using MORFO–2

- Next the variogram type could be changed from sphericals and exponentials (which are linear at the origin) to models that are quadratic at the origin, or alternatively to stable covariances such as those studied by Boulanger (1988) and Lantuejoul (1993). From a scientific point of view this would definitely be of great interest, but it is of less immediate use in mining.
- Another important factor is the size of the selective mining block (i.e. the support size). One could imagine varying both the block size and the variogram parameters, but this is not necessary initially. The important factor is the ratio of the range to the block size, because the size of the structuring element in all the morphological operators considered was taken to be a multiple of the mining support.
- In the long run, it will be vital to tailor morphological operators to suit particular mining methods rather using standard operators. For example, with some more research it may be possible to take account of specific mining constraints such as those due to geotechnical considerations in deep underground mines (e.g. no sharp inward pointing corners).
- One disadvantage of morphological transformations in general, is that they are local rather than regional. For example, erosions eliminate all the clumps of less than a certain width, irrespective of their distance to the nearest connected component. In practice those that are close to a clump already destined to be mined would be likely to be taken whereas distant ones would be left (perhaps even if they were slightly larger). So another potentially interesting line of research would be to find the cost of the cheapest route to each clump of blocks from the larger ones compared to its metal value. This would be like calculating geodesic distances.

REFERENCES

Allard D. (1993) On the Connectivity of Two Random Sets: the Truncated Gaussian and the Boolean, in the proceedings of the 4th International Geostatistics "Geostatistics Troia 92", ed A. Soares, Kluwer Academic Press, Dordrecht, Holland, pp 467 – 478

Boulanger F. (1990) Modelisation et Simulation de Variables Regionalisees par des fonctions aleatoires stables, Docteur en Geostatistique Thesis, Ecole des Mines de Paris, 28 March 1990 386pp

de Fouquet C. (1985) L'estimation des reserves recuperables sur modele geostatistique de gisements nonhomogenes, Doc–Ing thesis, Centre de Geostatistique, ENSMP150pp

Lantuejoul C. (1993) Simulation of multinormal random unctions using the turning band method. Workshop on Geostatistical Simulations held at Fontainebleau, 27–28 May 1993, proceedings to be published by Kluwer Academic Press.

Serra J. (1982) Image Analysis and Mathematical Morphology, Academic Press London 610pp

IS RESEARCH IN MINING GEOSTATS AS DEAD AS A DODO?

Margaret Armstrong
Centre de Geostatistique
35 rue St Honore,
77305 Fontainebleau,
FRANCE

This paper challenges mining geostatisticians to start solving today's problems, in particular to produce more meaningful recoverable reserve estimates for input into financial risk assessment studies. After reviewing several techniques for risk analysis, it focuses on Monte Carlo simulation and encourages geostatisticians to develop a new type of grade–tonnage simulation, a shortcut on traditional conditional simulations. Using high quality interactive graphics in colour, these simulated grade–tonnages curves could be displayed as a 3D cloud with metal price as the variable giving the third dimension. This would allow mining engineers and financial analysts to visualise the inter–relationships between the recoverable reserves and the metal price in a new and appealing way.

AVANT PROPOS

When Michel David started out in geostatistics in the sixties, the name *geostatistics* was virtually synonymous with mining. Most of the major theoretical developments (and there were lots at the time) were made to solve problems in mining. Few applications in other fields appeared in the literature. Over the past decade, geostats congresses have seen an explosion of applications and new techniques in a very wide range of disciplines from fishery, forestry and agriculture, through to the health sciences. But what has been happening in mining geostats? Nothing, well almost nothing.
Fewer and fewer abstracts on mining problems are submitted to congresses, and most of these present drab, uninteresting research. So we can ask whether mining geostats is dead. Clinically dead. Have all the interesting problems been solved so that we, as geostatisticians, can rest on our laurels? This paper argues that this is far from being true. The mining industry, like many others in the world, is passing through a crisis. The companies that will come out as survivors will be those that can better assess the reserves in their deposits, particularly at the feasibilibty stage.
In the author's opinion it is time for mining geostatisticians to get up off their tailends and start solving some of the real problems that surround them. The critical ones include
– the impact of ore reserve evaluation on the subsequent financial analysis,

R. Dimitrakopoulos (ed.), Geostatistics for the Next Century, 303–312.

– the fragmentation of recoverable reserves with increasing cutoff, and
– confidence intervals for grade estimates and for recoverable reserves.

These questions are critical because they link geostatistics (or more generally, ore evaluation) with financial analysis. This paper will focus on the first problem. If solutions to these and other vital problems are found over the next few years, mining geostats will not just be in the doldrums, as it is now; it will be dead, a fossilized vestige of its original self.

INTRODUCTION : DIFFERENT APPROACHES TO RISK ASSESSMENT

When mining companies are considering new projects, one of the important inputs is the predicted ore reserves. If the project looks attractive, the company's financial planners will carry out various risk and sensitivity analyses to ascertain the impact of fluctuations in critical parameters such as the metal price, operating costs, exchange rates, inflation, etc and, of course, of uncertainties in the reserve estimates. There are many ways of quantifying the vulnerability of a project to these fluctuations and uncertainties. A standard textbook on corporate finance (Brealey and Myers, 1991) discusses five approaches:

- Sensitivity Analysis. Experts are asked to give optimistic and pessimistic estimates of all the underlying variables. These are then varied one at a time to assess their impact on the profitability of the project (usually via cashflows or NPV).
- Breakeven Analysis In this variant of sensitivity analysis, instead of studying different scenarios, managers want to know how bad things have to become before the project starts losing money. They are looking for the threshold values that separate a nice healthy project from a Dooms Day one.
- Monte Carlo Simulations. As with geostatistical simulations, the idea is to generate many equally likely "snapshots" of the future project to analyse its vulnerability. Details will be given later. Examples in mining include Billingsley & Yaychuk (1986), Gentry & O'Neil (1984), Munro (1977) and Kajner & Sparks (1992).
- Decision Trees . Financial managers tend to use decision trees for analysing projects involving sequential decisions, such as whether to invest in a new drilling campaign. Brealey and Myers (1991) give examples of this for corporate capital budgeting; Hatchuel & Moisdon (1987) give a case–study on oil reservoir development.
- Option Pricing. This new technique developed initially by financial managers for analysing the value stock options has generated much interest recently because of its possible uses in a wide variety of fields including mining. See for example Brennan & Schwartz (1985), McKnight & Goldie (1990) and Palm et al (1986). The classic references on option pricing, Black and Scholes (1973) and Merton (1973), are quite mathematical, so they are not recommended for the faint of heart.

The common feature of all these techniques is that they attempt to quantify the sensitivity of projects to fluctuations around the base case, by introducing some notion of uncertainty. The first two approaches seek to measure variability in simple ways: cost up 10%, reserves down 20% etc. In Monte Carlo methods the probability distributions are specified explicitly; in decision theory, they are introduced in a less obvious way via prior and posterior probabilities, while the fifth approach is based on sophisticated stochastic process theory. As geostatisticians we should ask ourselves what information we should provide as input into mine planning scenarios, in order for the ultimate financial analyses to be meaningful.

There is no point in hiding behind the "*it's not my responsability*" approach. Enough mining projects have gone broke over the past five years, and many quite spectacularly. It is our responsability to communicate appropriate information to decision makers in a way that they can understand. The aim of this paper is to see how this should be done, and not to make value judgments on these five methods. But in order to show how the ore reserve estimates mesh into the risk analysis, we have to go into some detail. As the Monte Carlo approach is probably the most natural for geostatisticians, we will use this as a vehicle to illustrate the impact of the reserve estimates on the financial analysis. So we start by introducing this technique.

MONTE CARLO FINANCE SIMULATIONS

The recent textbook by Gocht, Zantop and Eggert (1988) gives a clear description of this technique on p109:
Monte Carlo simulation accounts for risk by calculating a probability distribution of possible profits from an investment decision. The figure below illustrates the general approach. The first step is to define profits (NPV) as a function of several underlying determinants. In practice, these would include a variety of factors influencing revenues (reserve estimates and price forecasts, for example) and costs (capital and labor costs, and taxes, for example). The second step is to define probability distributions for each of these (that is, the range and frequency distribution of all possible values). The third and final step is to calculate the probability distribution of profits by, in effect, using every possible combination of the determinants to calculate every possible profit level.

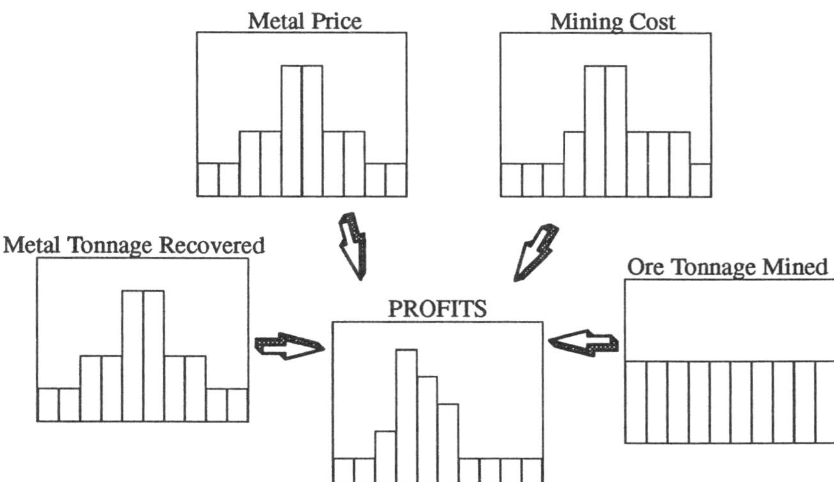

Figure 1 : Schematic Presentation of Monte Carlo Simulations

Once the distribution of profits has been simulated, appropriate financial decisions can be made. As geostatisticians our job is to provide suitable input for these types of simulations

(and for other types of risk analysis). Currently variables are usually assumed to be independent. Kajner and Sparks (1992) made this clear in their paper. Other authors have recognized that this is unrealistic. Tinsley (1985) spelled it out quite in unambiguous terms: *The inter–relationship of risks is difficult to quantify and represent mathematically. Translation: specifying the covariances tends to be the stumbling block in practice.*
Such difficulties have led many financial analysts and project bankers to completely reject mathematical methods of assessing risk (e.g. Tinsley, 1985), but that seems to be throwing out the baby with the bath water. Whether we like it or not, mining companies and lending institutions are going to make risk judgments based on our ore reserve estimates. It is therefore our job to provide these in a meaningful way. Let us review the types of information we can give them, starting out with the most obvious – the kriging variance.

How about the kriging variance as a measure of uncertainty?

A measure of uncertainty, some sort of confidence interval is required for input into financial analyses. But what type? In their initial enthusiasm, many geostatisticians advocated the kriging variance. An example produced by Ravenscroft (1991) highlights the pros and cons of this approach. Suppose we want to krige the two blocks each from 4 samples as shown below. As the data layouts are identical in both cases, the kriging variances are identical and as it happens, so are the kriged estimates (both being 10) but as Ravenscroft commented *As a betting man, I would consider the left–hand scenario a much safer option than the right–hand one, but the ordinary kriging results have apparently provided us with a false level of comfort in the second case.*

11 2
8 ? 9 1 ? 0
12 37

This merely confirms what everyone should know: namely, that the kriged estimate plus / minus twice the kriging standard deviations is not suitable for setting up confidence intervals. But the idea still resurfaces regularly, even in reputable journals (e.g. Woben & Morgan, 1993).

A more interesting approach

By the early 70s, South Africans had realised the necessity for more realistic confidence intervals than those based on the normal distribution. Munro (1977) presented a Monte Carlo risk analysis where great care was taken to choose suitable input values for the ore reserves in the model. Sichel's estimator was used to define the parameters of grade distribution. Although clearly better than the kriging variance approach, this has still two disadvantages. Firstly it is specific to South African gold, and secondly one can only simulate the grade of one ore block. If the production over 12 time periods was to be simulated, then values would be required for the ore tonnage T_i and its metal content Q_i for each time period: (T_1,Q_1), (T_2,Q_2), ... (T_{12},Q_{12}). Going further if selective mining was to be used, these would have to be expressed as functions of z: $(T_1(z),Q_1(z))$, $(T_2(z),Q_2(z))$, ... $(T_{12}(z),Q_{12}(z))$ because of the impact of the metal price on the cutoff grade z.

Conditional simulations

One way of simulating sequences of pairs of ore and metal tonnages like these is by conditionally simulating the deposit (or parts of it). Although this provides suitable data, it would be difficult to carry out 50 or 100 conditional simulations in practice just to obtain 50 or 100 sequences of 12 pairs of data. The computational workload is just too heavy. So we should be looking for a shortcut or an approximation that can give us suitably correlated grade tonnage curves for defined blocks of ore without the ming boggling number crunching currently required. Finding such a shortcut should be one of the priority objectives of research in mining geostatistics.

INTER–RELATIONSHIPS BETWEEN ORE TONNAGE , METAL AND PRICE

In looking for an appropriate shortcut we need to know how the three variables, ore tonnage, metal tonnage and metal price are linked up. This can best be illustrated from a case study. Ten simulations of 12 panels in a Witwatersrand gold deposit were made. Each panel corresponds to a relatively short production period, say 3 months. So for each conditional simulation of the deposit, 12 sets of grade tonnage curves were generated. The question now is how to compare them a) from one simulation to another for a fixed time period (=panel), and b) from one time period/panel to another for a fixed simulation. More interestingly, is there a synthetic graphic way of presenting this information? One quick way of comparing the 10 simulations is by plotting the ore tonnage against the cutoff grade (and similarly the metal content against the cutoff) for all the simulations for each panel. Figures 3 and 4 show these for the 2 panels in the middle row out of the 3 simulated rows. As expected they spread out at higher cutoffs.

When the metal price is high, the minimum grade that can be mined profitably is low and conversely when the price drops, the economic cutoff grade (the breakeven) has to rise. Here we only consider the breakeven grade rather than the optimal cutoff which is a subject in itself. (Lane, 1988; Taylor, 1972, 1985 & 1986). When possible values of Q and T corresponding to one selected cutoff grade are plotted, the points (Q,T) must lie in particular parts of the plane. For a zero cutoff (i.e. a high metal price), T=100%; so all the points lie in the top right corner. Conversely for a very high cutoff, Q=T=0% and so they are down at the origin. What happens for an intermdiate cutoff z? The grade of selective mining units in the recoverable reserves must be, by definition, above cutoff. So the ratio Q/T (i.e. the averaged recovered grade) must also exceed that cutoff. Consequently the points must lie to the left of the straight line representing the cutoff. Moreover they clump roughly along a line with slope m(z). This cloud tends to be more diffuse for large values of Q and T, and more concentrated for small ones.

If the metal price drops and the cutoff z increases, the slope of the line increases as does the averaged recovered grade. But the recovered tonnage T and likewise the metal Q decrease, moving the clump of points down toward the origin. Figures 5 shows the metal content Q plotted against the ore tcnnage T for the same 2 panels for three cutoffs: low, 0.5, medium, 1.0, and high, 1.5, (Note: the mean grade has been set to 1.0, to preserve confidentiality). The 10 points per small figure each correspond to one simulation of that panel.

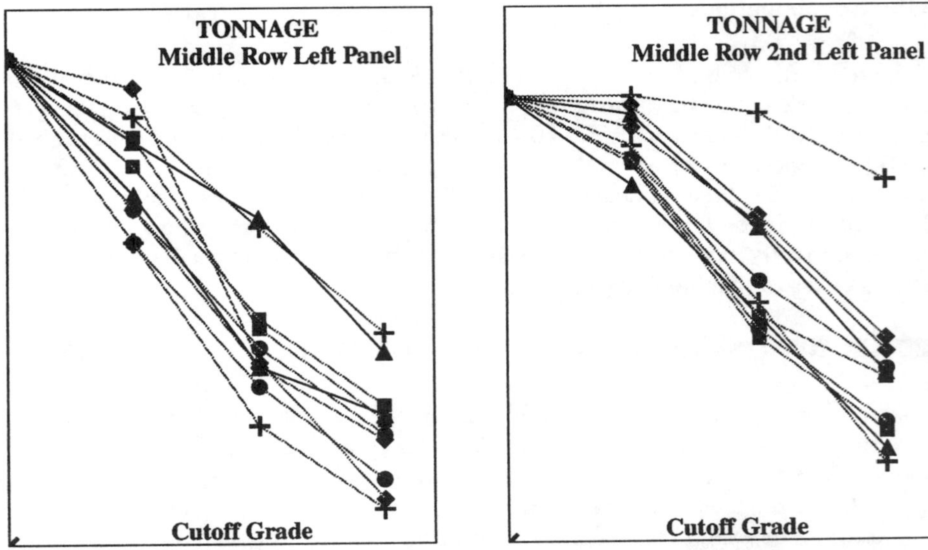

Figure 3 : Recoverable Ore Tonnage as a function of Cutoff Grade for poor panel (left) and a rich one (right).

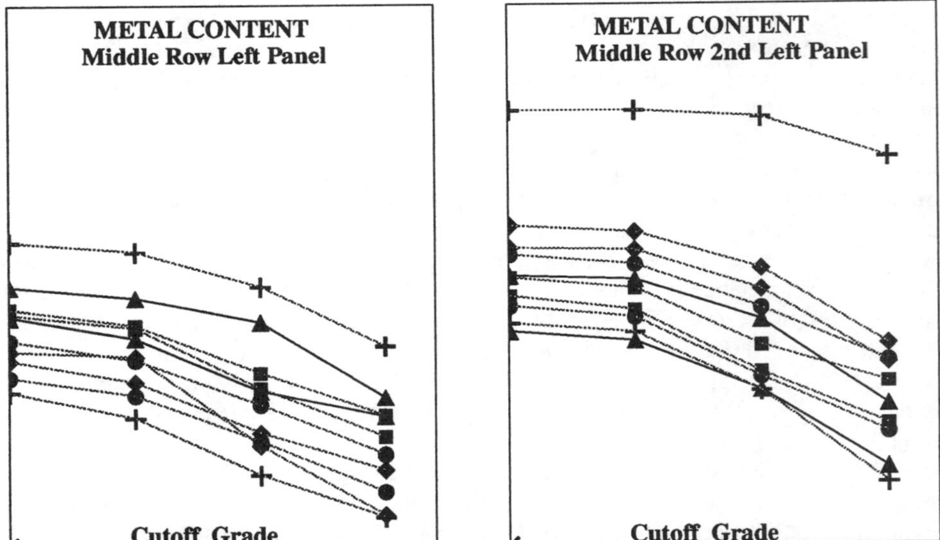

Figure 4 : Metal Content as a function of Cutoff Grade for poor panel (left) and a rich one (right).

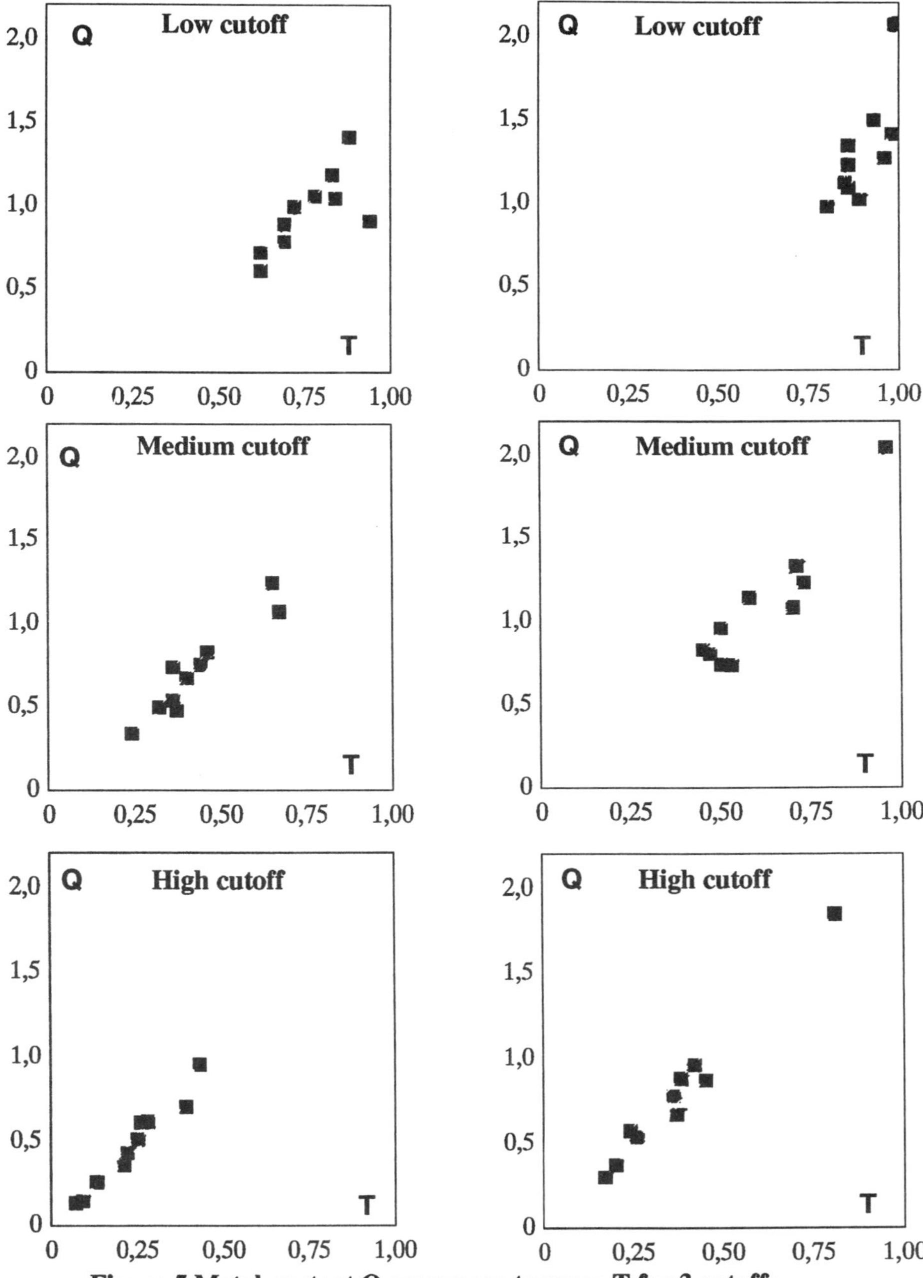

Figure 5 Metal content Q versus ore tonnage T for 3 cutoffs, for 2 panels – 10 simulations.

3D GRADE TONNAGE CURVES WITH HIGH QUALITY COLOUR GRAPHICS

Having seen how the cloud evolves for two different metal prices, we can guess its appearance for other cutoffs and hence other metal prices. Superimposing 3 such clouds as vertical levels gives Figure 6. Metal price which is critical in any financial assessment is shown on the vertical axis. The lowest level corresponds to the lowest metal price and therefore the highest cutoff grade, and so forth going upward. With high quality colour graphics one could represent these clouds with multiple price levels rather than just 3 levels as shown here. The interactive capacity of such programs would allow users to zoom in and examine the ore reserve situation for particular price scenarios.

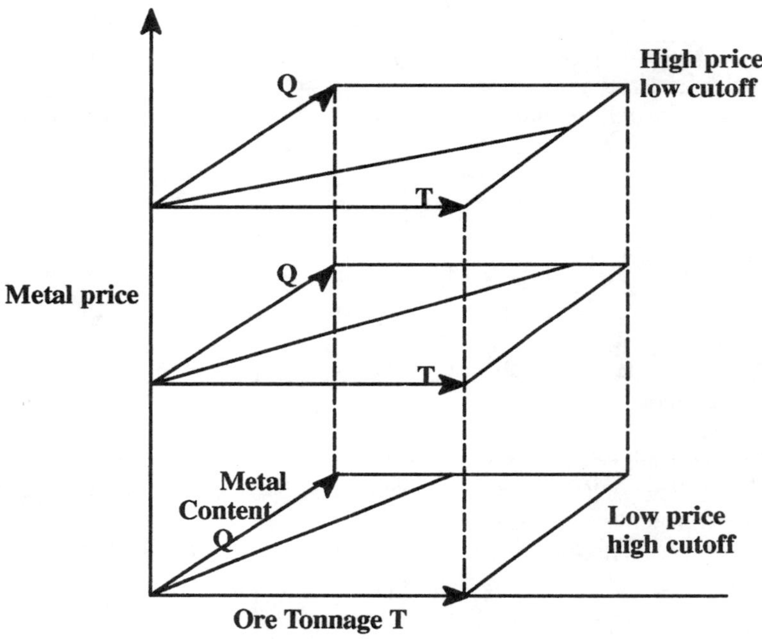

Figure 6: Three vertical levels correspond to different metal prices and hence to different cutoff grades

Individual deposit simulations

Having visualised the whole cloud, it would be interesting to focus in on individual simulations. One point on each level corresponds to each panel simulation. Imagine it. By selecting a particular simulation one could highlight its set of Q–T points in a contrasting colour (say red on a blue–green background). Joining these points from the top level down to the bottom would give the 3D grade tonnage curve for that simulation of the panel. One could easily zap from one simulation to another to see how they vary. If they vary too much then the conditioning on the simulation is poor. It might be adviseable to obtain more drilling and sampling before proceeding further.

Rotating the cloud

Rotating the image around the three axes as can be done with CAD packages, gives three interesting perspectives. Firstly looking at it front on, we see the projection on the price–ore tonnage plane; that is, the evolution of T with increasing metal price. Looking through the other side shows the metal Q as a function of price. Lastly looking down from above we would see the traditional plot of Q versus T .

The advantage of a high quality graphics display of this type is precisely that it would allow mine planners and financial analysts to see the relations between these three (and later other) critical parameters. It would help dispel the reservations that many project analysts currently feel toward statistical risk assessment procedures. Speaking of Monte Carlo techniques, Goode et al (1991) said

The method does not isolate or quantify the most sensitive variables. One result encompasses the combined effect of all variables. If used at all, the Monte Carlo simulation should be developed to supplement the more conventional sensitivity analysis, in which each variable is changed individually through a likely range of values.

CONCLUSIONS

After more than a decade with only minor advances in mining geostatistics, it is time for geostatisticians to start addressing today's real problems rather than chrome–plating techniques developed in the sixties and seventies. Given the large number of mining projects that have failed over the past five years after losing shareholders' funds and defaulting on their bank loans, our priority should be to develop meaningful ways of interfacing geostatistics (or more generaly ore evaluation) with financial analysis. To do this we need to understand the types of methods currently used in both mining and petroleum for project analysis (typically the first four listed in the introduction) and also at new techniques being introduced by financial specialists (e.g. option pricing). Then we need to find new and meaningful ways of presenting this information to potential users. Preferably these should use the graphic potential of workstations and PCs.

As an example of what could be done, this paper focuses on one particular risk analysis method, Monte Carlo simulation. It shows how high quality interactive graphics in colour could be used to present geostatistically simulated grade–tonnages curves as a 3D cloud with metal price as the variable giving the third dimension. This would allow mining engineers and financial analysts to visualise the inter–relationships between the recoverable reserves and the metal price in a new and appealing way. The paper also encourages geostatisticians to find new types of grade–tonnage simulations, that would be a shortcut on traditional conditional simulations. Although the paper has concentrated on Monte Carlo simulations, they are not essential. What is important, is to find meaningful ways of interfacing geostatistics with finance, and in particular with risk analysis, in order to give mining companies appropriate tools for facing the current economic crisis and for the future.

REFERENCES

Billingsley & Yaychuk (1986) Risk analysis related to mineral exploration & mine development, CIM Special Volume No 38 Gold in the Western Shield.

Black F. and Scholes M. (1973) The pricing of options and corporate liabiities J Political Economy 81:637–654 (May–June 1973)

Brealey R.A. and Myers S.C. (1991) Principles of corporate finance, McGraw–Hill Inc New York 950pp

Brennan M. & Schwartz E (1985) Evaluation of natural resource investments, J of Business April 1985 Vol 58 No 2 pp 136–157

Gentry D. & O'Neil T (1984) Mine investment analysis, American Institution of Mining, Metallurgical & Petroleum Engineers New York

Good J.R., Dawe M.J., Smith L.D. & Lattanzi C.R. (1991) Back to bascis – the feasilibilty study, CIM Bulletin Vol 84 No 953 pp53–61

Gotch W.R. Zantop H. & Eggert R.G. (1988) *International Mineral Economics* Springer Verlag 272pp

Hatchuel A. & Moisdon J–C. (1987) Theorie de la decision et pratiques organisationnelles: le cas des investissements petroliers pp251–289

Kajner L. & Sparks G. (1992) Quantifying the value of flexibility when conducting stochastic mine investment analysis. CIM Bulletin Vol 85 No 964 pp 68–71

Lane K.F. (1988) The economic definition of ore – cutoff grades in theory and practice, Mining Journal Books Ltd London 149pp

McKnight R. & Goldie R. (1990) The use of option pricing to value mining projects and mining equities, PAC RIM Congress, Gold Coast Australia, May 1990 pp725–730

Merton R.C. (1973) Theory of rational option pricing, *Journal of Economics & Management Science4:* 141–183 (Spring 1973)

Munro A.H. (1977) Application of risk analysis to a new gold mine, Proceedings of 15th APCOM Brisbane 1977 pp471–480

Palm S.K., Pearson N.D. & Read J.A. (1986) Option pricing: a new approach to mine valuation, CIM Bulletin May 1986 Vol 79 No 889 pp61–66

Ravenscroft (1991) personal communication

Taylor H.K. (1972) General background theory of cutoff grades. Trans Inst Min Metall Sect A July 1972 Vol 81 pp160–179

Taylor H.K. (1985) Rates of working of mines – a simple rule of thumb Trans Inst Min Metall Sect A Vol 95 p203–204

Taylor H.K. (1986) Cutoff grades – some further reflections Trans Inst Min Metall Sect A Vol 96 October pp204–216

Tinsley C.R. (1985) Analysis of risk sharing, in Finance for the minerals industry, Tinsley C.R. et al, eds, SME New York, pp419–426

Wober H.H. & Morgan P.J. (1993) Classification of ore reserves based geostatistical and economical parameters, CIM Bulletin 86: 966 Jan 93 pp73–76

COMMENTS ON "IS RESEARCH IN MINING GEOSTATS AS DEAD AS A DODO?" by M. ARMSTRONG

P.A. Dowd
Department of Mining and Mineral Engineering
University of Leeds, Leeds LS2 9JT, U.K.

The title of Margaret Armstrong's paper offers the promise of an assessment of the current state of research in mining geostatistics. However, the author seems to imply that because the areas of application of geostatistics have now become so diverse and mining is no longer the sole, or main, area of application that research has come to a dead end.

It is perhaps a mistake to categorise geostatistics by prefacing the word with adjectives such as mining, petroleum or environmental. Labelling techniques by the discipline in which they were developed, or first applied, hinders their wider use and mobility into other disciplines. Simply because the recent developments in simulation and modelling of geology were developed by people working in hydrocarbon applications should not hinder their use and adaption to mining or any other field. The solutions to many of the problems posed in other areas, particularly petroleum/hydrocarbons, are equally applicable in mining. The fact that they are solved by a petroleum geologist rather than a mining engineer or geostatistician working on mining applications is irrelevant.

In mining it would appear that we are now going through a period of consolidation and assessment. Much of this work is not published either because of confidentiality of figures used for reconciliation or because it is seen by the research community as rather mundane and of no theoretical interest.

Armstrong suggests three critical areas for research in mining geostatistics :

- the impact of ore reserves on financial analysis
- the fragmentation of recoverable reserves with increasing cut-off
- confidence intervals for grade estimates and recoverable reserves

Most of the paper is devoted to the first of these which is a rather surprising choice. Sensitivity and risk analysis are very well documented techniques in financial analysis and whilst the impact of the unreliability of estimated ore reserves on financial analysis might be loosely classed as "downstream" geostatistics it is difficult to see how it could be classed as geostatistical research.

R. Dimitrakopoulos (ed.), Geostatistics for the Next Century, 313–314.

The fragmentation of ore reserves with increasing cut-off grade is part of the more general problem of estimating recoverable, or minable, ore reserves. Some techniques which might be considered for this can be found in image analysis. However, in a general mining context the problem can be seen as the difference between recoverable and minable reserves and the only important factor is the impact that the fragmentation has on the latter. As an example, a mining engineer might solve the problem simply by designing an open pit on the basis of estimated recoverable reserves at given cut-off grades.

The determination of confidence intervals for grade estimates is still a holy grail of geostatistical research but many workers are reluctantly coming to the conclusion that this is probably a lost cause in any generally accepted sense. The factors contributing to the width and degree of symmetry of confidence intervals are : data errors, estimation errors and modelling errors (eg, representativity of the data when inferring a model variogram from an experimental variogram). By making assumptions which are appropriate to a particular application it is possible to quantify the estimation error component. Whilst the validity of these assumptions can be debated those that are required for the quantification of the remaining two sources of error are such that it is difficult to see how even the ground rules for debate could be established.

In this writer's opinion one major area of research activity in mining geostatistics is the incorporation of geology and geological controls into the estimation procedure. There has already been quite a bit of work in this area, especially in hydrocarbon applications, but there is no unified approach to the problem. In mine production estimates geological control is required on a much more detailed scale than in hydrocarbon applications and is probably the single limiting factor in the use of geostatistics for these types of estimates; approaches to this problem might include (stochastic) geometrical modelling and CAD techniques.

TESTING FOR BIAS BETWEEN DRILLING CAMPAIGNS

B. A. BANCROFT
CONSOL Inc.
1800 Washington Road
Pittsburgh, PA 15317
USA

A geostatistically based test for bias is derived. The relationship of this test to those based on generalized least squares (Aitken) estimators is discussed. Exploration strategy for the purpose of discerning bias is presented as well as diagnostics associated with the bias test.

INTRODUCTION

Frequently coal reserves are drilled in several campaigns where each campaign generally in–fills the previous one. These drilling efforts can be separated by significant amounts of time and conducted by different drilling crews. The cores are logged by different geologists and for the older sets the logging was frequently done by the drilling contractor. Furthermore, since drilling, sample preparation protocols, and laboratory analytical methods have changed over time, the question of bias between drilling campaigns is frequently posed. The existence of biases can critically undermine decisions based on quality forecasts. In this paper a geostatistically based test for bias is derived. The relationship of this test to those based on generalized least squares (Aitken) estimators found in the statistical literature is discussed. Also discussed are diagnostics associated with the bias test and exploration strategy for the purpose of discerning a potential bias.

Geostatistical Test for a Fixed Bias

Let $\{z(\mathbf{x}_{1i}), i=1 \text{ to } n_1\}$ and $\{z(\mathbf{x}_{2j}), j=1 \text{ to } n_2\}$ be the known sets of data values from the two drilling campaigns, and V be an arbitrary block which is spanned by the two sets. Then $D^* = Z^*_{V_1} - Z^*_{V_2}$ is the difference formed from the estimate of block V by data sets one and two respectively, where $Z^*_{V_1} = \Sigma\lambda_i\, Z(\mathbf{x}_{1i})$ and $Z^*_{V_2} = \Sigma\nu_j Z(\mathbf{x}_{2j})$ are the standard linear estimators of block V by the two data sets. Thus the actual but unknown grade of block V is given by $Z_V = Z^*_{V_1} + \epsilon_{V_1} = Z^*_{V_2} + \epsilon_{V_2}$.

R. Dimitrakopoulos (ed.), Geostatistics for the Next Century, 315–322.

Under the null hypothesis of no bias E(D*) = 0 and Var(D*) will depend on spatial variability, n_1, n_2, and the spatial configuration of the data. If D* is large relative to Var(D*), there is evidence of a bias.

The variance of the estimated difference, Var(D*), is given by

$$\mathrm{Var}(D^*) = E(\epsilon_{V_1})^2 + E(\epsilon_{V_2})^2 - 2E(\epsilon_{V_1}\epsilon_{V_2}). \tag{1}$$

$E(\epsilon_{v_1})^2$ and $E(\epsilon_{v_2})^2$ are the estimation variances for estimating block V by data set one and data set two respectively and are given by:

$$E(\epsilon_{v_1})^2 = \overline{C}(V,V) - 2\Sigma\lambda_i\overline{C}(\mathbf{x}_{1i},V) + \sum_i\sum_{i'}\lambda_i\lambda_{i'}C(\mathbf{x}_{1i} - \mathbf{x}_{1i'}) \tag{2}$$

$$E(\epsilon_{v_2})^2 = \overline{C}(V,V) - 2\Sigma\nu_j\overline{C}(\mathbf{x}_{2j},V) + \sum_j\sum_{j'}\nu_i\nu_jC(\mathbf{x}_{2j} - \mathbf{x}_{2j'}) \tag{3}$$

If $E\{Z(\mathbf{y})\} = m$ for $\mathbf{y} \subset V$, then the covariance between the errors of the two estimates is given by:

$$E(\epsilon_{V_1}\epsilon_{V_2}) = \overline{C}(V,V) - \Sigma\lambda_i\overline{C}(\mathbf{x}_{1j},V) - \Sigma\nu_j\overline{C}(\mathbf{x}_{2j},V) + \sum_i\sum_j\lambda_i\nu_j\overline{C}(\mathbf{x}_{1i} - \mathbf{x}_{2j}) \tag{4}$$

Substituting (2), (3), and (5) into (1) gives:

$$\mathrm{Var}(D^*) = \Sigma\Sigma\lambda_i\lambda_{i'}\overline{C}(x_{1i} - x_{1i'}) + \sum_j\sum_{j'}\nu_j\nu_{j'}\overline{C}(\mathbf{x}_{2j} - \mathbf{x}_{2j'}) - 2\sum_i\sum_j\lambda_i\nu_j\overline{C}(\mathbf{x}_{1i} - \mathbf{x}_{2j}). \tag{5}$$

Thus while the dimensions of V impact the magnitude of the estimation variances obtained from estimating V from the individual data sets, it does not affect Var(D*). One can see from (5) that for fixed V, n_1, and n_2 that Var(D*) is minimized, thus facilitating the detection of a bias to the extent:

(1) the points in data set one are close to the points of data set two, and
(2) the greater the interpoint distances between the samples in each of the two data sets.

The weights needed to calculate Var (D*) can be obtained by kriging V with the two data sets and calculating the third term in (1) using (4). In this instance the individual kriging variances are minimized but Var(D*) is not. Alternatively the weights λ_i, $i = 1,2,...n_1$, and ν_j, $j = 1,2,...n_2$ can be obtained by minimizing Var(D*) directly.

To insure E(D*) = 0 the weights are chosen such that:

$$\sum_i^{n_1} \lambda_i = 1 \quad \text{and} \quad \sum_j^{n_2} \nu_j = 1 \tag{6}$$

The Lagrange multiplier method is used to minimize (5) subject to the two constraints given in (6) which results in the following system of equations in matrix form:

$$\underbrace{\begin{bmatrix} [C_{11}(\mathbf{x}_{1i}-\mathbf{x}_{1i}')] & -[C_{12}(\mathbf{x}_{1j}-\mathbf{x}_{2j})] & \begin{matrix}1&0\\1&0\\\vdots&\vdots\\1&0\\0&1\\0&1\end{matrix} \\ -[C_{21}(\mathbf{x}_{2j}-\mathbf{x}_{1i})] & [C_{22}(\mathbf{x}_{2j}-\mathbf{x}_{2j}')] & \begin{matrix}\vdots&\vdots\\0&1\end{matrix} \\ 1\ 1\ \ldots\ 1 & 0\ 0\ \ldots\ 0 & 0\ 0 \\ 0\ 0\ \ldots\ 0 & 1\ 1\ \ldots\ 0 & 0\ 0 \end{bmatrix}}_{[K]} \underbrace{\begin{bmatrix} \lambda_1\\ \lambda_2\\ \vdots\\ \lambda_{n1}\\ \nu_1\\ \nu_2\\ \vdots\\ \nu_{n2}\\ -\mu_1\\ -\mu_2 \end{bmatrix}}_{[\phi]} = \underbrace{\begin{bmatrix} 0\\0\\ \vdots \\ \vdots \\ 0\\1\\1 \end{bmatrix}}_{[c]} \tag{7}$$

The matrix K is composed of:

(a) the $n_1 \times n_1$ submatrix $[C_{11}(\mathbf{x}_{1i} - \mathbf{x}_{1i}')]$, which describes the spatial correlation between the members of data set 1;
(b) the $n_2 \times n_2$ submatrix $[C_{22}(\mathbf{x}_{2j} - \mathbf{x}_{2j}')]$, which describes the spatial correlation between the members of data set 2;
(c) the $n_1 \times n_2$ submatrix $[C_{12}(\mathbf{x}_{1i} - \mathbf{x}_{2j})]$ and its transpose the $n_2 \times n_1$ submatrix $[C_{21}(\mathbf{x}_{2j} - \mathbf{x}_{1j})]$, which describe the cross–variability between the two data sets; and
(d) the remaining 1's and 0's that correspond to the unbiasedness conditions.

The weights recovered from the solution of (7) are used to obtain D^* and $Var(D^*)$. Then if R is the absolute value of $D^*/Var(D^*)^{1/2}$ and R is big, the difference in the estimates is inconsistent with what one would expect from normal sampling variation. If one is willing to assume normality for the distribution of D^* then, for a one–sided test, D^* is statistically significant, i.e., the null hypothesis (H_0) of no difference is rejected at a significance level of α if $R>Z_\alpha$ where Z_α is the $(1-\alpha)$ quantile of the standard normal distribution. Alternatively, and again assuming normality, one could construct a $(1-\alpha)100\%$ confidence interval for D by:

$$D^* - Z_{\alpha/2}\, Var(D^*)^{1/2} < D < D^* + Z_{\alpha/2}\, Var(D^*)^{1/2} \tag{8}$$

and if this interval includes 0 then given the existing data H_0 cannot be rejected.

Exploratory Drilling to Improve Bias Test

In many cases confidence intervals estimated by (8) will cover zero but will be large enough to admit the possibility that a bias of practical consequence exists. Depending on the economic consequences of a bias, one could consider additional exploratory drilling for the purpose of improving ones ability to discern a smaller bias. Since Var(D*) does not depend on knowing grades, but only on locations, the impact of alternative drilling strategies can be investigated prior to the actual drilling.

In such instances one can evaluate alternative drilling strategies by estimating the risks associated with the hypothesis testing decision criterion which will be employed after the drilling is completed. This is accomplished through plots of operating characteristic (O–C) curves. For any α an O–C curve gives the probabilities of accepting the null hypothesis as a function of a true difference (δ). The O–C curve depends on α, δ, and Var(D*) and 1–P(accept $H_0 | D = \delta$) is the power. For the one–sided test with H_0:D = 0 against the alternative hypothesis that D<0, the O–C curve is generated by:

$$P(\text{accept } H_0 : D = 0 | D = \delta < 0) = 1 - \phi \left[\frac{Z_{1-\alpha} \, \text{Var}(D^*)^{1/2} - \delta}{\text{Var}(D^*)^{1/2}} \right]$$

For the one–sided test against the alternative D>0 the O–C curve is generated by:

$$P(\text{accept } H_0 : D = 0 | D = \delta > 0) = \phi \left[\frac{Z_{\alpha} \, \text{Var}(D^*)^{1/2} - \delta}{\text{Var}(D^*)^{1/2}} \right]$$

where ϕ is the cumulative distribution function of the standard normal density.

Shown on Figure 1a are operating characteristic curves based on current drilling, for a one–sided test of H_0:D=0 against H :D<0. In this instance the objective was to determine if sulfur was biased low in an older drilling campaign in a reserve in the Pittsburgh seam. The estimated mean sulfur content of the reserve is 1.7% and the experimental semi–variogram was fit with a spherical model with a sill of .26, a range of 9000 feet, and a nugget of .0125. Sulfur biases on the order of .1 or greater are of practical consequence. As shown on the plot at a significance level of .05 there is a 60% chance that the null hypothesis of no bias will be accepted even if the older drilling campaign contains a –.1 bias.

Equation (5) indicates that the best sampling strategy for increasing the power of the bias test is to locate additional samples as close as possible to the existing samples of the suspect data set. Figure 1b shows how the OC–curves are shifted if an exploratory hole is drilled or a channel sample is taken 500' away from each hole in the suspect data set. Note that the probability of accepting Ho if the bias is –.1 has dropped to .1.

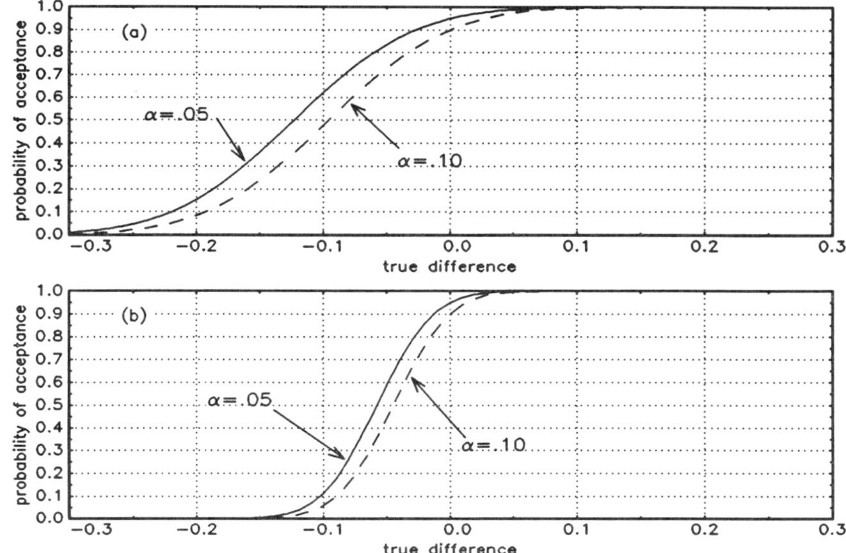

Fig. 1. (a) Current drilling O–C curve (b) Additional sampling O–C curve.

The best strategy obviously depends on sampling costs, the importance of discerning the bias, and the loss in mine product forecasting accuracy that results from additional samples that do not roughly in–fill current drilling. Equation (5) and the subsequent calculation of OC–curves provide for the quantification of the component of this decision–making process related to the detection of a bias.

Generalized Least Squares (Aitken) Estimator of Bias

Assume that the observations in data set k, can be modeled as

$$Z(\mathbf{x}_{ik}) = m_k + \epsilon(\mathbf{x}_{ik}) \quad i = 1, 2, \ldots, n_i \tag{9}$$

where $m_k = E(Z(x_{ik}))$, $\epsilon(\mathbf{x}_{ik})$ is a stochastic error process, and $Var(\epsilon(\mathbf{x}_{ik}))$ exists. Then by (9) the $n = n_1 + n_2$ equations representing all observations can be set out compactly in matrix form as:

$$\mathbf{Z} = \mathbf{X}\,\boldsymbol{\beta} + \epsilon \tag{10}$$

where

$\mathbf{X} = \begin{bmatrix} 1_1 & 0 \\ 0 & 1_2 \end{bmatrix}$ is an n x 2 matrix, 1_1 is an n_1 x 1 vector of ones, and 1_2 is an n_2 x 1 vector of ones;

$\boldsymbol{\beta}$ is an unknown 2 x 1 vector,

ϵ is an n x 1 random vector with mean 0 and covariance matrix **K**, and

Z is a vector with the grades of data set 1 stacked on the grades of data set 2.

It can be shown (Aitken, 1934; Johnston 1972; Quimby, 1986) that the generalized least squares (gls) estimator of β is given by:

$$\hat{\beta} = (\mathbf{X}' \mathbf{K}^{-1} \mathbf{X})^{-1} \mathbf{X}' \mathbf{K}^{-1} \mathrm{Z} \tag{11}$$

is a minimum variance unbiased estimator of $\boldsymbol{\beta}$ with $\mathrm{Var}(\hat{\beta}) = (\mathbf{X}' \mathbf{K}^{-1} \mathbf{X})^{-1}$. The elements of $\hat{\beta}$ are estimates of the means of the two data sets. An estimator of the difference in the means between the two data sets is:

$$D^*_{gls} = \hat{m}_1 - \hat{m}_2 = (1 \;\; -1)\, \hat{\beta} \tag{12}$$

and

$$\mathrm{Var}(D^*)_{gls} = (1 \;\; -1)\, \mathrm{Var}(\hat{\beta}) \, \binom{1}{-1}. \tag{13}$$

If ϵ is distributed as a multivariate normal then D^*_{gls} is a maximum likelihood estimator of the difference and has minimum variance in the class of all unbiased estimators, and a statistical test against the null hypothesis of no difference can be performed using standard normal tables.

It can be shown that while $\hat{m}_1$, and $\hat{m}_2$ differ from $Z^*_{V_1}$ and $Z^*_{V_2}$ that $D^* = D^*_{gls}$. Similarly $\mathrm{Var}(D^*) = \mathrm{Var}(D^*)_{gls}$. Thus, whether one conceptualizes the test as a test of the difference in block estimates or the difference in the mean of two vectors, the result is equivalent.

Variogram Issues

If the two data sets are indeed shifted from each other by a fixed bias, then a variogram cannot be estimated from all the data. In the best case there is enough data in the "good" set to satisfy typically used rules–of–thumb, and if there are not then the experimental variograms from the two data sets can be pooled (Journel and Huijbregts, 1978), although the danger here is that if the bias is related to grade, the resulting variogram will be distorted. Under H_0 this "good" variogram is used to obtain C_{11}, C_{12}, C_{21}, and C_{22} in (7).

As in kriging, the estimate of the Var(D*) assumes the variogram is known when in fact it is estimated. Thus Var(D*) is actually larger then that estimated in (5) by an amount related to the uncertainty in the estimation of the variogram. If the typical rules–of–thumb mentioned above are satisfied, then this low bias should not be significant and could be handled implicitly by the choice of α.
Typically in the generalized least squares setting (Johnston 1972; Quimby, 1986) it is assumed that the covariance matrix is given by σ^2K with K known. Then for the two data set bias case discussed herein, σ^2 is estimated from the data by:

$$\hat{\sigma}^2 = \frac{1}{n-2}(\mathbf{Z} - \mathbf{X}\hat{\beta})' \, \mathrm{K}^{-1}(\mathbf{Z} - \mathbf{X}\hat{\beta}). \tag{14}$$

When σ^2 is estimated, the testing is performed using the t–distribution. Borgman (1988) also discusses the gls two sample t–test and provides a non–parametric two sample test for testing mean equivalence of two correlated vectors.

Diagnostics

One diagnostic is to plot the estimation weights obtained from the solution of (7) against grade for the test (suspect) and the reference data set. Here the objective is to isolate influential points and to insure that this data is good for the purposes of the test. "Bad data" would be data with inconsistent results on the lab sheets or data driven high or low by localized geologic processes.

The test described above is for a fixed bias. Another kind of bias one could postulate is one that is a function of grade. If such a bias exists, then the experimental semi–variograms from the test and reference data sets cannot be pooled. To check for such a bias, each sample in the test data set is estimated by the data in the reference data set. The results of this procedure are plotted as the actual (Z) against the estimated (Z*) or the estimation errors (ϵ) against Z*. If the test data set contains a bias that is related to grade, then one would expect the slope of a line fit to the points Z and Z* to deviate from one or equivalently the slope of a line fit to the points ϵ and Z* to deviate from zero.

It is well known that, in practice, there is no reason why the estimates Z* should satisfy conditional non–biasedness (Journel and Huijbregts, 1978) given by:

$$\mathrm{E}\{\mathrm{Z} \,|\, \mathrm{Z}^* = \mathrm{Z}\} = \mathrm{z}. \tag{15}$$

Therefore for comparative purposes the reference data set is cross–validated and Z, Z* and ϵ, Z* plots generated.

Statistical testing against a null hypothesis of a slope of one for the regression of Z on Z* can be used to explore the existence of a grade related bias. Generalized least squares estimates of the regression coefficients should be used, since ordinary least squares estimates of the fit of Z to Z* are not efficient; they do not account for the correlation between the estimated points. The elements of the covariance matrix for the gls regression are given by:

$$\mathrm{Cov}(\epsilon_i \epsilon_j) = \mathrm{C}(\mathbf{x}_i - \mathbf{x}_j) - \Sigma\lambda_p \mathrm{C}(\mathbf{x}_p - \mathbf{x}_j) - \Sigma\nu_q(\mathbf{x}_q - \mathbf{x}_i) + \sum_p \sum_q \lambda_p \nu_q \, \mathrm{C}(\mathbf{x}_p - \mathbf{x}_q) \tag{16}$$

where the estimate of $\mathrm{Z}(\mathbf{x}_i)$ is given by $\mathrm{Z}(\mathbf{x}_i)^* = \Sigma\lambda_p \mathrm{Z}(\mathbf{x}_p)$ and the estimate of $\mathrm{Z}(\mathbf{x}_j)$ is given by $\mathrm{Z}(\mathbf{x}_j)^* = \Sigma\nu_q \, \mathrm{Z}(\mathbf{x}_q)$. Note that to obtain (16) requires kriging in a global neighborhood in the estimation of the test data set from the reference data set and in the cross–validation of the reference data set, i.e., the estimation set for each point includes all the data.

In several instances it has proven very useful to split the data by grade or by area. Inconclusive results have become conclusive where separate mineralization zones occurred. In one instance a reserve was split in two and bias tests separately conducted. In each area $R>3.5$ and the associated estimated biases for each zone were different by only .02. Subsequent testing of lab procedures which were different for the test and reference data sets verified the magnitude of the estimated bias.

Summary and Discussion

A methodology for testing for bias between drilling campaigns has been developed which is equivalent to conventional generalized least squares tests. The advantage of the geostatistical development is that by the way Var(D*) is expressed one can see directly how to efficiently place exploratory holes for improving the power of the test.

Diagnostics associated with the bias testing have been discussed with a focus on trying to assess whether a bias that depends on grade exists. Such a step is necessary before experimental variograms from the test and reference data sets are pooled. Great care must be taken in this step since under H_0 conditional unbiasedness between Z and Z* is not guaranteed.

There are a number of other issues that must be considered in order to prudently apply this test. In the first place Var(D*) shares a lot of the same features as a global estimation variance. Secondly, no work has been done to assess the performance of this test with highly skewed data. There are variogram validity issues if the reference and test data sets span the entire deposit. If a formal hypothesis testing procedure is employed to determine if R is big enough to reject the null hypothesis of no bias, then the results are sensitive to distributional assumptions on D*. Finally, as always, the geology must be understood. Certainly the data configuration relative to mineralization zones must be examined carefully.

References

Aitken, A. G., 1934, "On Least–squares and Linear Combinations of Observations", Proc. Royal Soc., Edinburgh, vol. 55, pp. 42–48.

Borgman, L. E., 1988, "New Advances in Methodology for Statistical Tests Useful in Geostatistical Studies", Math Geol., Vol. 20, pp. 383–403.

Johnston, J. 1972, Econometric Methods, McGraw–Hill, New York, p. 437.

Journel, A. G., and Huijbregts, C. J., 1978, Mining Geostatistics: Academic Press, London, p. 600.

Quimby, W., 1986, Selected Topics in Spatial Statistical Analysis, Ph.D. Thesis: Statistics Department, University of Wyoming, Laramie, Wyoming.

IMPROVED SAMPLING CONTROL AND DATA GATHERING FOR IMPROVED MINERAL INVENTORIES AND PRODUCTION CONTROL

A. J. Sinclair
Dept. of Geological Sciences,
The University of B. C., Vancouver, B. C.

and

M. Vallée
Géoconseil Marcel Vallée Inc.,
Sainte-Foy, Quebec

Current data acquisition procedures may not be appropriate to provide adequate and efficient estimation of deposits and ore reserves. Several quality control procedures that are both simple and cheap to implement lead to significant improvements in data quality. First, we examine the importance of improved sample acquisition, sample handling and assaying procedures. The second part of this paper examines deposit sampling and geological strategies. Attention to these quality control steps can help increase the validity of geostatistical estimates and the efficiency of mine production quality control.

INTRODUCTION

All mineral inventory studies rely on a data base of location and assay information that normally is vetted and then assumed to be correct and of adequate quality for resource/resource estimation purposes. In many cases this assumption of adequacy either is not met, or is approached to an unknown extent, for lack of adequate quality control procedures (e.g. Pitard, this volume). Thus, deposit data may not be of the best quality attainable, affecting the quality of results regardless of the estimation technique to be used ultimately for mineral inventory evaluation. Data evaluation procedures, particularly at the outset of a mineral inventory study, can contribute to optimizing estimates for a given data base. Such quality control procedures should not be viewed as one-shot measures: they must be planned systematically and maintained during the full course of exploration, deposit development and mine production activities.

Factual mistakes in a data base can be found using a variety of checks and balances, particularly if original data (e.g. assay sheets) are reviewed. For example, it is possible to recognize sample location errors that creep into a data base, as well as errors in assay data such as order of magnitude errors, transposed figures, and so on.

Once exploration is completed, however, it is not possible to improve the quality of individual assays, except, to a limited extent perhaps, by reanalysis of samples. During production, once an ore block or ore zone has been mined and milled, obvious ore selection errors cannot be corrected. However, for projects still in the exploration stage there are controls and procedures that can be implemented to improve the quality of data available for future mineral inventory estimates. Similarly, sampling and assay quality control procedures can contribute to increased selectivity and metal recovery in an operating mine.

R. Dimitrakopoulos (ed.), Geostatistics for the Next Century, 323–329.

There are several ways by which our approach to sampling and data gathering can improve the quality of mineral inventory estimates, whatever the method of estimation (Vallée, 1992). Some of the most important factors are:

i) to reduce and know the magnitudes of errors in our data base (assays, thickness, sample coordinates); in particular to improve and monitor the quality of assays and the representivity of samples,
ii) to develop better correlation between samples of different types that are used in deposit estimation, and
iii) to define assay continuity models for each distinctive geological domain of a mineral deposit, thus providing more opportunity to optimize sampling patterns and grid dimensions.

Minimizing errors in sampling and assaying reduces the nugget effect, the random variability and the sill of an experimental variogram and thus, eases the problem of model fitting to the experimental variogram. This ultimately reduces errors in kriging estimates. Using a single average variogram model for an entire deposit rather than models adapted to the various domains present can lead to an inappropriate application of geostatistics with abnormally large and unrecognized errors in a high proportion of the estimates. Reducing the assay variability related to procedural and instrumental deficiencies also will improve delineation of ore limits and the efficiency of grade quality control in production.

IMPROVING ASSAY QUALITY

Burn (1981) and Vallée (1992) have outlined in a general way, a variety of factors that can influence the quality of assay data. Here we look at some of these factors in greater detail; they are: sampling procedures, sample reduction procedures and assay quality control procedures.

Sampling Methods and Procedures: Main problems

Sampling procedures used in exploration are too commonly accepted routinely without rigorous checking during information gathering for delineating and detailing a specific mineral deposit. This occurs despite the difficulty of recognizing other than glaring errors. Moreover, sampling methods found acceptable for one deposit type might be totally unsuitable for another deposit type. Sample sizes may be different, such as diamond drill cores of various sizes. Commonly, few systematic tests are made to help optimize current procedures: sample size, fragment size, fragment number, etc., despite the low cost of such measures. Too often, we forget to review the implications of the various parameters and procedures associated with a standardized sampling method such as diamond drilling. In some cases, a uniform sample grid size and/or orientation may not be appropriate for all domains or zones of the deposit.

Another source of ambiguity is that commonly several sampling methods are used in the evaluation of a deposit. Evaluation work may involve linear samples: diamond drilling, rotary drill or channel sampling, surface samples: panel samples, large volume samples: trench or mining round samples and bulk samples. In places, some of these samples types may cover different areas of the deposit with little or no overlap between them; hence, systematic procedures to correlate between sampling methods may be lacking. Of course such correlations between sample types present difficulties of interpretation and use, to the point of serious uncertainty if the level of accuracy of the various sample types is not know.

In mine production, quality control of sampling often could be improved. Of course accumulated experience in a producing mine offers a minimal measure of quality control, but too few operations try to improve the quality of routine sampling procedures beyond this starting point. Establishing correlations between different sample types requires a systematic approach involving, in particular:

(i) large enough areas of overlap of the various sampling methods, with numbers of samples sufficient for recognizing correlation; for example, diamond drilling an area of a bulk sample, or, directing drill holes parallel to channel samples.
(ii) systematic statistical and geostatistical controls of each sample set by itself and relative to other data sets involved.

Mine sample test-case

As an example of the potential impact of sampling method on estimates (and thus on productivity) consider the case history of Equity Silver Mine Limited (Giroux et al, 1986) where mine personnel conducted a small but revealing test of various sample methods for production blast hole cuttings.

The standard blasthole sampling procedure in use at the mine, a tube (pipe) sampling method somewhat modified from that described by Ewanchuck (1968), involves shoving a 3-inch diameter tube into 4 spots symmetrically located in the cuttings pile, upending the tube so that material remains in it, and transferring the material in the tube to a sample bag. A second method, channel sampling along 4 symmetrically distributed radii of the cuttings pile to produce 4 separate samples, was to be tested. Finally, the total bulk of remaining cuttings was taken as a sample to provide a means of mass balance for the entire blasthole cuttings pile. Because all the cuttings from a particular drill hole were used as sample material, it is possible to produce a weighted average grade (weighted by sample mass) that is a "best" estimate of the true value of the cuttings pile. Results of individual sampling methods can be compared with this best estimate. Calculations by Giroux et al (1986) indicate that the use of channel sampling at Equity would have produced a dramatic improvement in assay quality and thus, a significant improvement in ore/waste classification and a corresponding significant increase in profit.

Improving Sample Reduction Procedure

Sample reduction schemes used in operating mines commonly can be improved to provide more representative subsamples and, therefore, better quality assays. The normal result of an inadequate sample reduction system is a large random error (sampling plus analytical error) in assays; of course, biases and procedural errors are possible too. These large errors contribute to a high nugget effect and possible masking of the underlying structure (ranges) of the variogram. At the very least they lead to larger than necessary errors in block estimation and delineating ore limits; thus, they contribute to ore/waste misclassification problems and loss of metal.

Gy (1979) addressed this problem at length in developing an empirical sampling equation that has been widely accepted in the mineral industry, with significant modification in the case of some gold deposits (Francois-Bongarcon, 1991). A practical means of applying Gy's principles is through the use of a sampling (or sample reduction) diagram on which a safety line, based on Gy's fundamental sampling error equation, is plotted as a guide to acceptable sample reduction procedures. Where little is known of the detailed characteristics of an ore, Gy's general safety line can be used as a guide; where properties of an ore are reasonably well understood a safety line tailored to a particular ore can be determined.

On a sample reduction diagram the sample path is plotted as a series of connected straight lines representing alternating particle size reduction stages (crushing, grinding, pulverizing) and mass reduction stages (subsampling). It is important that the initial particle size reduction stage be sufficient to move the sample position to the left of the safety line and that subsequent stages not cross to the right of the safety line. Further details of the procedure are provided by Gy (1968; 1979), David (1977) and Francois-Bongarcon (1991).

Assay Quality Control Program

The use of Gy's equation for fundamental error in the form of a safety line on a sampling diagram represents a somewhat idealized expectation that is only approached in practice. A quality control program is necessary to know and to monitor data quality. Laboratories mostly maintain a system of checking subsampling and analytical errors. This normally involves systematically taking duplicate samples, introducing duplicate analytical checks (perhaps one in twenty samples), the use of standards, systematic interlaboratory checks, use of different methods for particular situations (e.g. metallic assays for high grade gold ores), and reassaying of unexpectedly high (or low?) values. It is important that the results of all these quality control measures be provided to those conducting mineral inventory estimates in a concise and informative way. Such information should be demanded prior to payment for commercial assay data. For all this concern, however, errors arising in the laboratory are generally small compared with those that arise from sampling.

A program of duplicate sampling and assaying is undertaken routinely by many mining professionals as part of a sampling program designed to accumulate information upon which to base a mineral inventory. Giroux et al (1986) have outlined a simple procedure for systematically evaluating paired data that arise from such a program. Interactive software for the application of Gy's sampling equation, sampling diagrams and safety lines is available in Radlowski and Sinclair (1993).

IMPROVING SAMPLING STRATEGIES

Basic Sampling Strategies

Random sampling rarely is resorted to in geological appraisals, on account of the influence of geological structures and domains. Instead, grid sampling or stratified sampling are used to optimize sampling patterns within a framework determined by geological structures and domains. The key element of this approach generally consists of orienting the most dense sampling direction perpendicular to the plane of greatest continuity in the geological model developed at the stage in question. Such a strategy can lead to several pitfalls because of the assumptions involved; thus, continual review and refinement of the sampling strategy is required as the knowledge base is increased.

Continuity models

Aspects of continuity, reviewed in a companion paper (Sinclair and Vallée, 1993), distinguish between geometric continuity (form of mineralized volume) and value (grade, thickness, etc.) continuity. Variations in the character of mineralization can lead to dramatic differences in physical and in value continuity as show on the resulting experimental variograms. In many practical cases a deposit can be divided into several domains, each of which is characterized by its own distinctive variogram model. This arises because differences in genesis, lithology or structure have produced differences in the local character of mineralization. The concept of domains is implicit in a discussion by Srivastava (1987) of conditional probability as control for grade estimation; geological data are emphasized as an essential source of conditioning information.

Structural Control

Sinclair and Giroux (1984) demonstrate that two different parts of the South Tail zone of Equity Silver Mine had extremely different variograms for silver grade. In the greater part of the zone a northerly trending, vertically dipping fracture set predominated in controlling mineralization whereas at the northern extremity of the zone an easterly trending fracture set controlled mineralization. The results, summarized as ellipses of relative ranges of anisotropic variogram models, emphasize the very different continuities in the two domains. The negative impact of applying a single, general model incorrectly to the small northern domain is significant (cf. Sinclair and Giroux, ibid).

Lithological Control

Lithologies commonly have a marked correlation with continuity for several reasons:

i) various lithologies respond differently to a structural regime,

ii) ore fluids react variably with wallrocks depending on their textures and compositions, and

iii)some genetic processes combine fundamentally different styles of mineralization in the formation of a deposit (e.g. layered massive sulphide deposit with an underlying stockwork feeder zone).

The characterization of situations such as these requires thorough geological mapping to define the lithological domains that are of significance to continuity. Sinclair et al (1983) demonstrate widely different variogram models for 5 mineralized rock types at the Golden Sunlight deposit in Montana. Use of the corresponding variogram models produced block estimates based on exploration data, that were used for mine planning (Roper, 1986).

Defining Continuity

It is important that some of the early exploration work on a mineral property be directed toward providing sufficient close-spaced data with which to define quantitative, 3-dimensional models of both geological and value continuity. Diamond drill sampling schemes provide continuity information preferentially along drill holes and spacing between holes may not be adequate to measure continuity satisfactorily in the other two dimensions. Sampling patterns and the continuity estimated from them must be reviewed and new sampling designed as required.

Sampling should be closely tied to geology and coded systematically so that data can be easily categorized in domains if the need arises. For example, samples should not cross major lithological boundaries; veins and wallrock should be sampled separately to ascertain the detailed control of metal distribution, and so on. Sampling grid dimensions may have to be adjusted depending on the variability of geological and mineralization parameters.

Following the above strategy will reduce one of the common problems encountered in applying geostatistics at a pre-feasibility stage of exploration, that is, the common scarcity of closely spaced data with which to define both the nugget effect and the short range grade continuity in a confident manner. This detailed knowledge will be essential for the planning required to bring the deposit to the feasibility stage. Of course, if more than one domain has been recognized, control of short range continuity is required for each separate domain.

Additional knowledge of continuity can be attained in various ways. Journel and Huijbregts (1977) recommend that within a larger sampling field, local crosses of closely spaced data be collected to provide some insight into local (short range) continuity. Closely spaced information also can be obtained either in appropriately chosen areas of trenching and stripping or in exploratory underground workings. In some cases, particularly for gold deposits, it may be necessary to obtain a bulk sample for testing of milling parameters and for grade confirmation purposes.

CONCLUSIONS

Current data acquisition procedures may not be adequate in particular cases to quantify properly the geological and value continuities on which mineral inventory estimates depend. For better estimation results the validity of the data base underlying the estimation can be improved by systematic critical evaluation of (i) sample acquisition (including 3-dimensional aspects of deposit sampling patterns), (ii) sample reduction and assaying procedures, and (iii) geological data acqusition, interpretation and modelling.

Sampling patterns and methods, subsampling and sample reduction schemes and assaying procedures must be designed and tested for the local geological substrate. A regular program of duplicate and check sampling and analyses will serve as a continuing measure of quality control. Attention to these matters will reduce the nugget effect and improve the quality, and therefore the ease of interpretation, of experimental variograms.

Detailed geological mapping and 3-dimensional deposit reconstruction is essential and must be integrated fully into models of continuity of value measures (grade, thickeness, etc.) that are based on well-designed sampling patterns. Many deposits can be divided into two or more geological domains (lithological, stuctural, genetic), each of which is characterized by its own grade continuity model. In such cases, the indifferent application of a single "average" continuity model (variogram) to determine mineral inventory can lead to substantial local errors in block estimation.

These improved sampling and geological procedures also improve the potential for conditional or stochastic simulation procedures. In mining operations, the continual critical application of quality control measures will directly contribute to improved metal recovery and income. Many geologists and geostatisticians still are shy of demanding the quality improvement measures necessary, even where costs and efforts involved are modest. Managers must be informed of the required procedures and convinced of their necessity.

Data quality problems must be set in a broader perspective such as Total Quality or Continuous Improvement Management methods that are being adopted by increasing numbers of mining companies. The essential element of these philosophies, whatever the name, is a continuous process of review and change that is oriented to an overall improvement in profitibility. For deposit/reserve estimation the systematic scrutiny fostered by these methods will apply to all stages from data acquisition and editing to mine planning and production control. To be effective, such a review process must be carried out in a systematic and holistic manner, avoiding haphazard procedures.

ACKNOWLEDGEMENTS

AJS thanks the Science Council of B. C. for funding of projects that have contributed to the ideas expressed here. Much of the work was done through the Mineral Deposit Research Unit; the paper is Mineral Deposit Research Unit contribution no. 026.

REFERENCES

Burn, R.G. (1981) "Data reliability in ore reserve assessments", Mining Magazine, Oct, 289-299.

David, M. (1977) "Geostatistical ore reserve estimation", Elsevier Publishing Co, Amsterdam, 386 p.

Ewanchuck, H. G. (1968) "Grade control at Bethlehem Copper", in Spec. Vol. 9, Can. Inst. Min. Metall., p. 302-307.

Francois-Bongarcon, D. (1991) "Geostatistical determination of sample variances in the sampling of broken gold ores", Can. Inst. Min. Metall. Bull. 84, 46-57.

Giroux, G. H., Sinclair, A. J. and Miller, J. H. (1986) "Production quality control experiments at Equity Silver Mines Ltd.", in David, M., et al (eds.), Ore reserve estimation: methods, models and reality; Proc. Symp. of the Can Inst. Min. Metall., Montreal, P. Q., 238-260.

Gy, P. (1979) "Sampling of Particulate Materials, Theory and Practice", Elsevier Publishing Co., Amsterdam, Netherlands, 431 p.

Gy, P. (1968) "Theory and practice of sampling broken ore", in Spec. Vol. 9, Can. Inst. Min. Metall., 5-10.

Journel, A., and Huijbregts, C. (1978) "Mining geo-statistics", Academic Press, Dordrech, Netherlands, 600 p.

Pitard, F. (1993) "Exploration of the nugget effect", in Proc. Symp. on Geostatistics for the Next Century, June 3-5, Montreal, Quebec.

Radlowski, Z.A., and Sinclair, A. J. (1993) "Mineral inventory of precious metal deposits in British Columbia:GYSAMPLE 0.11, a computer program for the evaluation of sample reduction schemes", Tech. Report MT-6, Mineral Deposit Research Unit, Dept. of Geological Sciences, The University of British Columbia, Vancouver, Canada.

Roper, M. W. (1986) "Geostatistics in planning and production, Golden Sunlight mine, Whitehall, Montana", in David, M., et al (eds.), Ore reserve estimation: methods, models and reality; Proc. Symp. of the Can Inst. Min. Metall., Montreal, P. Q., 238-260.

Sinclair, A. J., and Giroux, G. H. (1984) "Geological controls of semi-variograms in precious metal deposits; in Verly, G., et al, (eds) Geostatistics for Natural Resource Characterization, NATO ASI series. D. Reidel Pub. Co., Holland, 965-978.

Sinclair, A. J., and Vallee, M. (1993) "Reviewing continuity: an essential element of quality control in deposit and re-serve estimation" (abs), Can. Inst. Min. Metall. Bull. 86, 60.

Sinclair, A. J., Lonergan, E., and McConechy, W. (1983) "Geostatistics at the feasibility stage, Golden Sunlight deposit, Montana", Nev. Bur. Mines and Geol., Rept. 36, 165-172.

Sinclair, A. J., Radlowski, Z. A., Srivastava, R. M. and Samis, A. (1993) "Geostatistical estimation of grade and dilution, Snip Gold Mine, Northern British Columbia", To be presented at Mining Symposium, Queen's University, August 1993.

Srivastava, R. M., (1987) "Minimum variance or maximum profitability?". Inst. Min. Metall. Bull. v. 80, p. 63-68.

Vallée, M., (1992) "Guide to the Evaluation of Gold Deposits", Can. Inst. Min. Metall. Spec. Vol. 45, 299 p.

Vallée, M., and Cote, D. (1992) "The guide to the evaluation of gold deposits: integrating deposit evaluation and reserve inventory practice", Can. Inst. Min. Metall. Bull. 85, 50-61.

IMPROVING PREDICTIONS BY STUDYING REALITY

VIVIENNE SNOWDEN
Joint Managing Director
Snowden Associates, P O Box 77
West Perth, WA 6872
Australia

Estimation is a case-specific exercise in that no two orebodies are ever identical. How then is it possible to improve predictions in the general case when we can only ever study reconciliations in specific cases? The real problem is to establish correct geostatistical procedures and provide adequate estimates before the event, i.e. when data are scarce and the exact nature of geological controls is not well understood. Thereafter ongoing reconciliation allows corrective action to be taken if necessary.

Knowledge of the results of mining from 30 out of 400 cases studied over a period of six years suggests a degree of predictability with respect to estimation problems and their causes. It is believed that an awareness of the relative freqency of each type of problem is of benefit when evaluating a new deposit. Details on corrective action and its degree of success are described with a view to prioritising remedial investigations and focussing geostatistical training.

INTRODUCTION

Table 1 ranks the most common reconciliation problems encountered amongst 30 of the 400 cases studied over a period of six years. These 30 cases are the mines for which significant production has been undertaken or for which mining is complete.

Most estimation and reconciliation problems boil down to a few significant theoretical explanations revolving around distribution statistics and the volume-variance relationship and the failure of the mining industry professional to take these into account when calculating ore reserves and during delineation of grade control blocks. Table 2 ranks the most common causes of the problems documented in Table 1.

R. Dimitrakopoulos (ed.), Geostatistics for the Next Century, 330–337.

Table 1 Main Reconciliation Problems Ranked By Frequency Of Occurrence	
Problem	**Frequency %**
Head grade lower and tonnage higher than predicted	50%
Low grade stockpiles higher grade than predicted	24%
Tonnes and grade are both lower or higher than predicted	21%
Head grade is higher and tonnage lower than predicted	6%
Operation reconciles pit to production at one cutoff but fails to reconcile when cutoff is raised	3%

Table 2 Main Causes Of Reconciliation Problems Ranked By Frequency Of Occurrence	
Cause	**Frequency %**
Incorrect interpolation technique	54%
Sample bias	30%
Incorrect geological/boundary recognition	24%
Insufficient drilling	18%

INTERPOLATION TECHNIQUE

By far the most common issue is that of incorrect interpolation. Polygonal and sectional reserve estimations and polygonal grade control tend to underestimate tonnes and overestimate grade in orebodies with gradational or zonal boundaries because they do not cater for the volume variance relationship. They also result in significant low grade stockpiles at higher grades than expected. Overestimation of grade results if there is no compensation for data skewness or mixed populations using either polygonal estimation or methods such as inverse distance weighting or kriging.

Case histories are drawn from three mines displaying gradational geological boundaries. Macraes Gold Mine in Otago, New Zealand, is a complex intermixture of structured and zonal environments illustrating skewed and mixed assay distributions. At Copperhead Mine in Western Australia, plunging stratabound mineralisation in folded mafics and banded iron is modified by cross fractures and shear zones which enhance mineralisation. Pit A at Boddington Gold Mine (BGM) in Western Australia has four laterite profile layers - gravel, hardcap, B zone and clay. Hardcap and B Zone assays approximate lognormality. Gravel and clay both represent mixed populations.

Volume-Variance Relationship

During the initial stage of mining at Macraes, truckloads of low grade material, delineated by polygonal grade control on blastholes of 0.7 g/t - 0.9 g/t in the pit, were routinely re-sampled. It was found that these bulk samples consistently averaged 1.1 g/t, that is 0.25 g/t higher in grade than the in-pit blast hole average. Grade control kriging of selective mining units (5m x 5 m) confirmed that the cutoff on assays should be dropped about 0.25 g/t in order to achieve the production cutoff required (Figure 1). This action corrects for the volume-variance effect responsible for the observed regression relationship whereby low grade blocks are underestimated by sample grades.

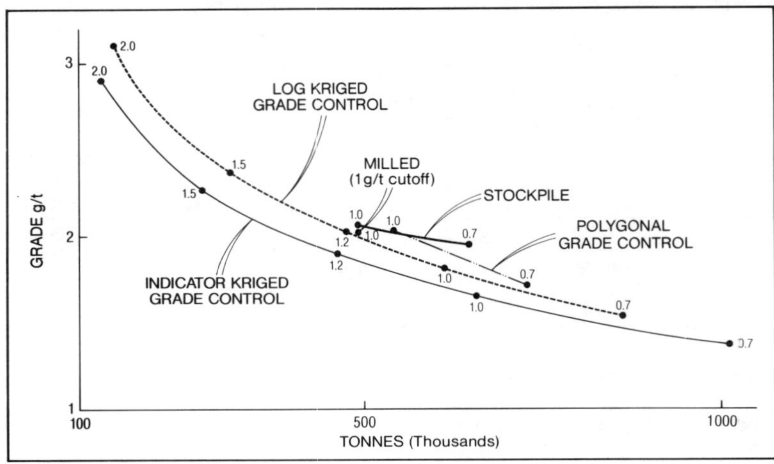

Figure 1 Grade/Tonnage Reconciliation Diagram at Macraes

A similar relationship was observed at Copperhead where exploration and grade control kriged results were compared with production estimates based on polygonal grade control block outlines. The effective production cutoff grade on selective mining units of 5 m x 5 m on a 2.5 m bench is 1.2 g/t compared with an assay cutoff of 1 g/t on blastholes. This is likely to be the major reason for material below the assay cutoff of 1 g/t (block cutoff 1.2 g/t) returning higher than expected stockpile grades at Copperhead.

Skewed and Mixed Distributions

Lognormal kriging (Journel, 1980; Journel and Huijbregts, 1978; David, 1977) was chosen at Copperhead because the data is perfectly lognormally distributed. All available assays were used in kriging each domain but using the specific parameters for a given domain. This was designed to avoid edge effects at the domain boundaries since these are gradational and were not interpreted with any degree of exactness. The bulk exploration model reconciled well with the grade control kriging in the mined-out area and thus gave confidence in the remaining reserves.

At Boddington anisotropy was established by logarithmic semivariograms and nested indicator semivariograms were then calculated for each of the indicator cutoffs along the main structural directions. Logarithmic semivariograms display ranges and nugget/variance ratios similar to the first (median) indicator.

The block model was interpolated using nested indicator kriging (Dagbert, 1990) employing 10 ordinary kriging runs on the zeros and ones corresponding to the 10 indicators for each of the three zones separately and using the median instead of the mean of assays to allocate the grade in the top indicator category. Testwork showed that using the mean instead of the median inflated the average block grade by 0.3 g/t due to the presence of significantly high grade outliers. The recovery above cutoff was corrected to the minimum selective mining unit.

Lognormal kriging was also undertaken. The kriging parameters were defined from logarithmic semivariograms after normalising the total sill to the logarithmic variance of the input grade composites for each zone. The lognormal shortcut (David, 1977) was used to estimate the selective mining estimate.

The comparative estimates are all displayed on the grade/tonnage diagram in Figure 2.

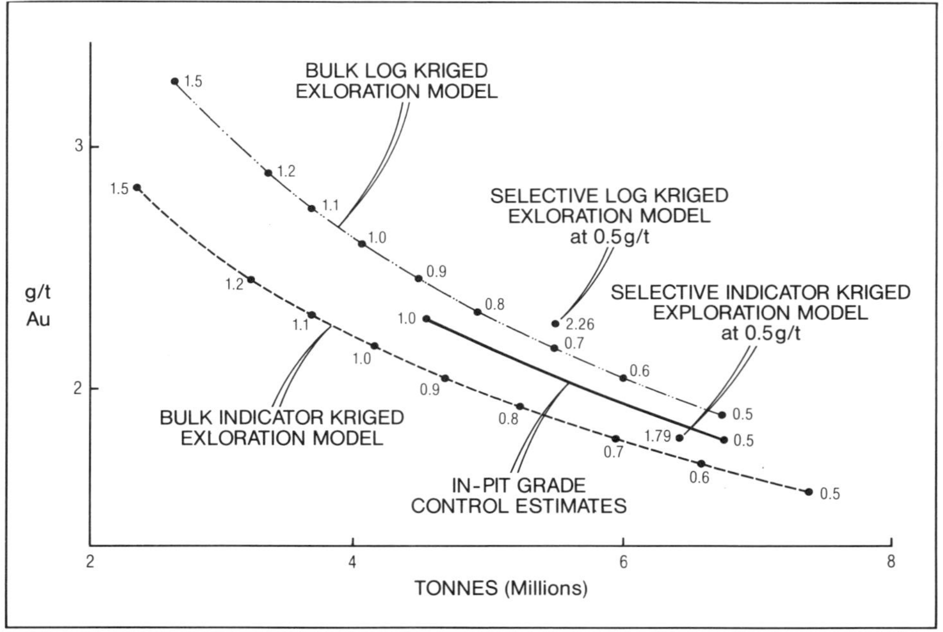

Figure 2 Comparative Grade/Tonnage Relationships At Boddington

The selective indicator estimate for a cutoff of 0.5 g/t reconciles closely with the in-pit estimate at the same cutoff. The log kriged results give higher grades at lower tonnages for a substantial increase in estimated metal content at all cutoffs compared with the indicator model. The reason for this is believed to be the pronounced bimodality evident in the clay horizon and the spatial behaviour of the separate populations. This inflates the logarithmic variance which in turn inflates the kriged estimate. Certainly, in the case of the clay zone, it is expected that the indicator model will provide a more appropriate response than the log kriged model because of the distinct bimodality of the distribution.

BOUNDARY CONDITIONS

Girilambone mine is a new copper mine in New South Wales, Australia. Assays are lognormally distributed. This deposit was originally modelled using strict geological boundaries as well as assay boundaries to constrain ordinary kriging without cutting high grades. As soon as mining commenced it became evident that there was a good deal of material around 0.5% Cu which had not been interpolated in the model. A review of the block model showed that an artificial bimodality had been imposed on the model by virtue of the 1% Cu assay boundary used to constrain high grade blocks (Figure 3).

Relaxing the assay constraint and re-interpolation using nested indicator kriging with the logarithmic semivariogram (which was similar to the median indicator semivariogram in nugget ratio and range of influence) changed the grade/tonnage curve quite significantly.

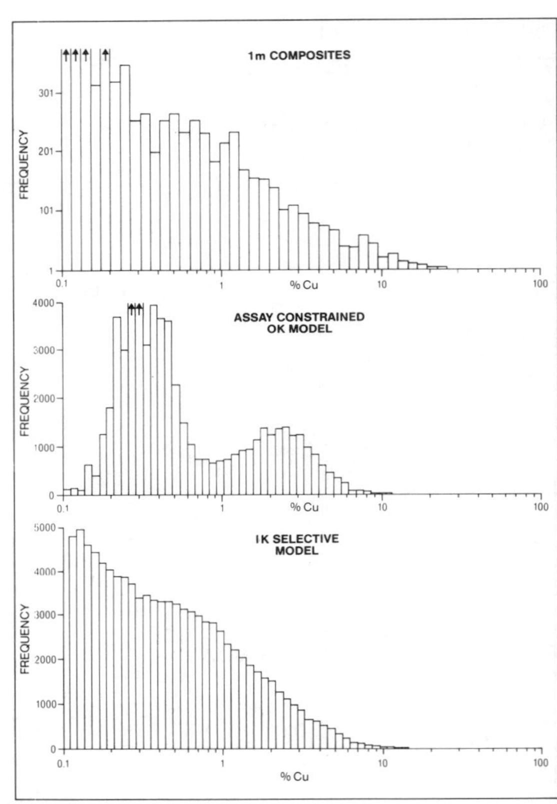

Figure 3 Artificial Bimodality Caused by Use of Hard Assay Boundary

The effect on the reserve above a 0.5% cutoff is to double the reserve tonnage. The grade hardly suffers because of the flat gradient of the grade/tonnage curve (Figure 4). Production schedules for this mine had to be adjusted to reflect the increased bench ore tonnage and the life of mine has been extended.

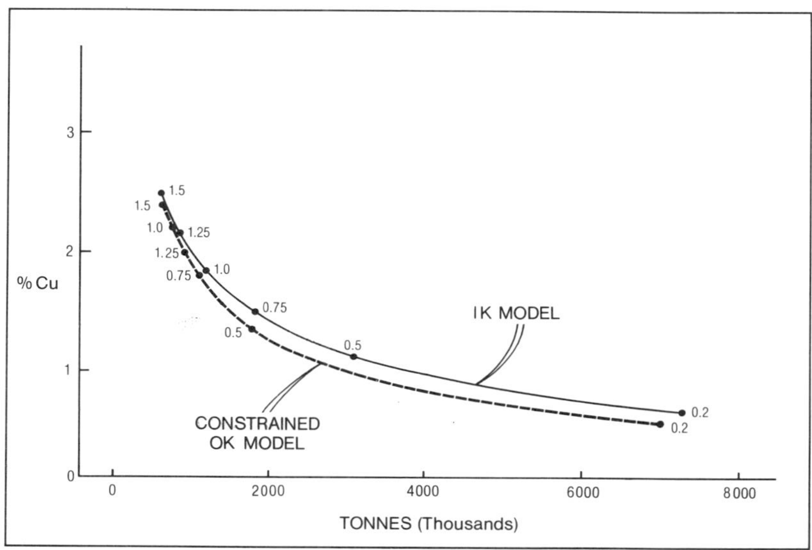

Figure 4 Grade/Tonnage Relationship For Reconciliation Area At Girilambone

SAMPLE BIAS AND/OR INSUFFICIENT DRILLING

The most fundamental of all errors is that of sample bias. Inappropriate or inadequate sample collection, crushing, splitting and analysis techniques give rise to non-representative assays. It is very necessary to correct sampling problems before statistical and interpolation errors combine with sampling errors to increase uncertainty.

After more than a year's mining, grade bias became a problem during grade control sampling at Macraes. Figure 5 shows the tonnage mined at each elevation (mRL), annotated with the production grade. Also plotted is the comparative data from the resource model at both the 0.9 g/t and 1.0 g/t cutoffs. Pit to mill reconciliation, which had previously been good, gradually began to deviate at about the time the production estimates began to exceed the 0.9 g/t model tonnage. This inadvertent change in tonnage reconciliation from model to pit resulting in overestimation of grade from pit to mill was caused by a high grade sample bias. The bias was tracked to the loss of low grade fines from the cyclone of the grade control drilling rig. Once the rig was replaced the bias was removed and good reconciliation was restored.

Detailed reconciliation at Copperhead showed that the log kriged model underestimated tonnes and grade where drilling was limited and at depth due to an apparent low grade bias in the drilling. Management was alerted to drill target holes to validate the results of the earlier deep drilling program. The block model overestimated grade in other domains known to suffer a sampling bias. Thus, although on average the model performs well, in detail there are various imperfections, none of which could be overcome by modelling technique alone.

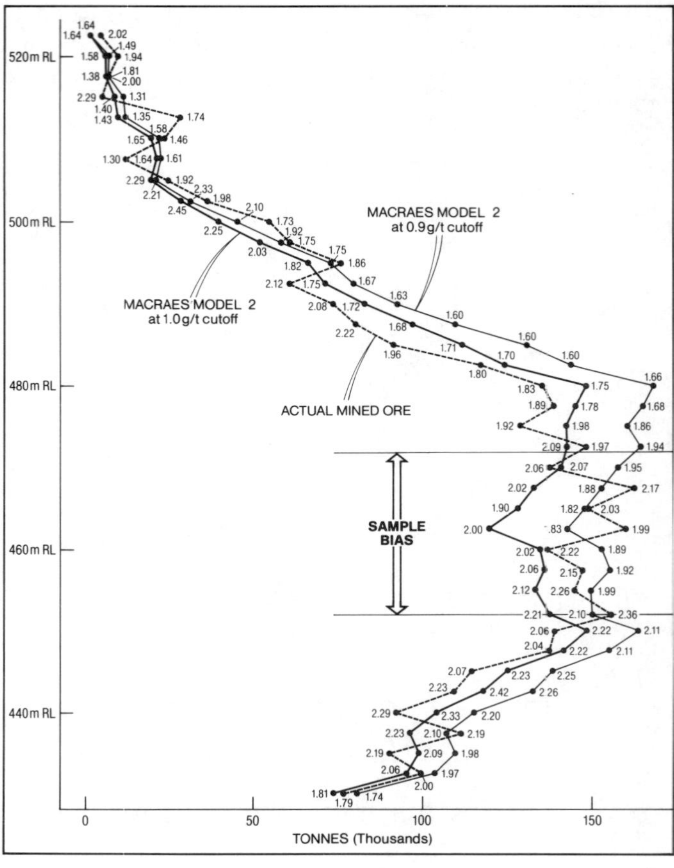

Figure 5 Grade Trend Showing Sampling Bias at Macraes

CONCLUSIONS

There is no single approach to orebody modelling which can be guaranteed of getting the right answer. Sample bias and low drilling density, for example, lead to significant estimation errors. The style of mineralisation and statistical distribution dictates which modelling technique is most appropriate.

Experimentation with various interpolation techniques shows that where geological interpretation is simple and there are single statistical distributions, different methods can give similar results. Even lognormal kriging, which is taboo to so many in the industry, and was once described to me by Michel David as "riding a wild horse", can give remarkably accurate results if applied correctly to a well defined lognormally distributed assay population. Indeed it is more effective than other techniques in mapping cutoffs correctly as it is sensitive to the presence of low grade boundaries. However, lognormal kriging displays a reluctance to be bullish where input data is too wide spaced (Copperhead, for example) and does, of course, demand that the sill of the semivariogram accurately reflects the population variance.

Indicator kriging is a more universally applicable practical technique as it can deal with mixed distributions (as at Boddington) as well as single skew distributions (as found at Girilambone). Where indicator kriging shows most sensitivity is in the top grade category. Should assays in the top category be cut and if so to what level? We have demonstrated remarkable sensitivity to this top category and experience at Boddington suggests that better reconciliation is likely with a top cut.

The use of ordinary kriging for exploration data within assay constrained boundaries simply does not give an accurate grade/tonnage curve for skew data. The average grade within the boundary is correct, but if the boundary does not happen to coincide with the mining cutoff, the interpolated grade distribution cannot be used to map cutoffs accurately.

ACKNOWLEDGEMENTS

The management and production staff of Macraes Mining Company, Copperhead Mine, Boddington Gold Mine and Girilambone Copper Company are thanked for their assistance in project work and for permission to publish the results presented in this paper. Colleagues at Snowden Associates are thanked for their valuable contributions to this work - Mark Thomas for analyses on Girilambone and Annick Manfrino-Thomas for analyses on Macraes and Boddington.

REFERENCES

David, M. (1977) Geostatistical Ore Reserve Estimation, Elsevier, Amsterdam, 364p.

Journel, A G and Huijbregts, C J (1978). Mining Geostatistics, Academic Press, London, 600p.

Journel, A G (1983). "Nonparametric estimation of spatial distributions", Math. Geology, 15, 445-468p.

Journel, A J (1980). "The lognormal approach to predicting local distributions of selective mining unit grades", Math. Geology, 12, 283-301p.

Dagbert, M (1990). "Nested indicator approach for ore reserve estimation in highly variable mineralization", 92nd Gen. Ann. Meeting, Can. Inst. Mining, Ottawa, Canada.

DEALING WITH HETEROGENEITY, UNCERTAINTY AND FLUID FLOW

CLEOPATRA'S NOSE AND THE DIAGRAMMATIC APPROACH TO FLOW MODELLING IN RANDOM POROUS MEDIA

G. CHRISTAKOS, C. T. MILLER and D. OLIVER
Department of Environmental Sciences and Engineering, CB#7400
University of North Carolina
Chapel Hill, N.C. 27599
USA

"Cleopatra's nose: if it had been shorter, the face of the whole world would have been changed." Mentioning B. Pascal's famous passage here aims at emphasizing the power of an image and the important consequences that it may have in real world situations. *Mutatis mutandis*, the idea behind the use of diagrammatic representations in groundwater flow modelling is based on the concept that the qualitative grasp of form, shape and geometric order may go deeper than the quantitative grasp of abstract mathematical symbol and number. In this spirit, the proposed approach uses topological diagrams which reduce the original stochastic groundwater flow problem to a closed set of equations for the statistical moments. Random integral forms of flow are considered in the light of porous media description operators and graphic Green's functions. Graphic visualizations of the underlying flow processes allow previously undetected features to be seen and can yield more general and accurate results than traditional methods. Depending upon the choice of a porous medium description operator, the diagrammatic approach can handle both cases of small and large fluctuations and can work at long as well as short range correlation scales.

INTRODUCTION

Among a variety of random phenomena encountered in stochastic subsurface hydrology, those which are related with groundwater flow and solute transport in random porous media occupy a central position (e.g., Gutjahr and Gelhar, 1981; Gelhar and Axness, 1983; Cushman, 1987, 1990; Sposito *et al.*, 1990; Rubin, 1991; Neuman and Orr, 1993; Dimitrakopoulos and Desbarats, 1993).

A porous medium is characterized by hydrologic processes that have complex spatial and/or temporal variabilities, which can only be represented in terms of random fields. Then, the porous medium Π under consideration is assumed to be one possibility from an ensemble of porous media $\Pi(u)$, which is a function of the elementary events u defined on the sample space Ω. A σ-field F of subsets of Ω and a probability measure P on F is also assigned to $\Pi(u)$, leading to a probability space denoted by the triplet (Ω, F, P). Then, the physical properties of the porous medium will be functions of the event u, the spatial position $s \in D$ ($D \subset R^3$ is the spatial domain of the medium) and/or the temporal instant $t \in T$ (T is the temporal domain). In mathematical terms each property is

R. Dimitrakopoulos (ed.), Geostatistics for the Next Century, 341–358.

represented by a *random field (RF)* $X(s,t,u)$ (Christakos, 1992a). For simplicity the argument u is usually omitted, and one writes $X(\mathbf{s})$ to denote a purely *spatial* RF and $X(\mathbf{s},t)$ to denote a *spatiotemporal* RF.

The stochastic flow problems are formulated using the same equations and conditions as the dynamic problems of traditional groundwater flow, only now the equations and conditions represent individual realizations of the RFs modelling the flow processes, and the operators, parameters and coefficients are random entities. The formulation of a broad class of stochastic flow problems can be reduced to the following principal scenarios: (a) random porous media characteristics; (b) random flow and transport sources; or (c) random boundary and/or initial conditions. Problems dealing with combinations of (a), (b) and (c) are also possible. Nevertheless, this work is devoted to problems coming under the heading of scenarion (a) above.

Assume that the distribution in space-time of the hydrologic process $X(\mathbf{s},t)$ is described by the *linear operator* (e.g., differential operator) L, so that $X(\mathbf{s},t)$ obeys the inhomogeneous dynamic equation

$$L[X(\mathbf{s},t)] = S(\mathbf{s},t), \tag{1.1}$$

where $S(\mathbf{s},t)$ denotes a source function. In some situations, we may deal with *vector* RFs $\mathbf{X}(\mathbf{s},t)$ and $\mathbf{S}(\mathbf{s},t)$; then, L will be an operator matrix (tensor). Due to its dependence on both spatial **s** and temporal coordinates t, Eq. (1) is often called an *unsteady-state* model. If there is dependence on the spatial coordinates **s** only, one is dealing with purely spatial processes and Eq. (1) turns into a *steady-state* model. If the porous medium domain D contains no sources, $S(\mathbf{s},t)=0$, and Eq. (1) becomes a *homogeneous* one. Some media are restricted by finite boundaries which impose *boundary conditions* on Eq. (1). Similarly, in the unsteady-state case Eq. (1) may need to satisfy certain *initial conditions*, as well.

An *exact* stochastic solution of the flow situation described by Eq. (1), if it existed, should be expressed in terms of various realizations of the random field $X(\mathbf{s},t)$ in the form

$$X(\mathbf{s},t) = L^{-1}[S(\mathbf{s},t)], \tag{1.2}$$

where the integral operator L^{-1} should involve Green's functions meeting all auxiliary conditions (e.g., boundary and initial conditions).

However, in the vast majority of hydrologic applications the exact representation (1.2) is not feasible, even in the case of a purely deterministic single realization (e.g., Unny, 1992). It is then necessary to turn to *approximate* methods. An interesting characteristic of these methods is that they are problem oriented and that their applicability depends on a number of factors closely associated with the specific flow situation. A class of methods which is particularly suitable for a wide variety of groundwater flow situations is that of *diagrammatic* or *graphic perturbation* techniques considered earlier by Christakos *et al.* (1993a and b). Diagrammatic perturbation has certain attractive mathematical and physical properties, which will be discussed below in the context of a three-dimensional groundwater flow problem.

STEADY-STATE GROUNDWATER FLOW IN RANDOM POROUS MEDIA

Consider a specific case of Eq. (1.1), namely the *three-dimensional, steady-state saturated* groundwater flow in a *statistically isotropic* porous medium. Random characteristics of the medium affect the propagation of flow through the medium. One of these characteristics is the *hydraulic conductivity field* denoted by the spatial RF K(s), $\mathbf{s} = (s_1, s_2, s_3) \in R^3$ is the position vector. It is assumed that the field K(s) is isotropic and log-normally distributed and the random gradient vector $\mathbf{W}(\mathbf{s}) = \nabla \log K(\mathbf{s})$ can be decomposed as $\mathbf{W}(\mathbf{s}) = \overline{\mathbf{W}(\mathbf{s})} + \mathbf{w}(\mathbf{s})$, where $\overline{\mathbf{W}(\mathbf{s})}$ is the mean value of $\mathbf{W}(\mathbf{s})$ over all possible realizations of the porous medium $\Pi(\mathrm{u})$ and $\mathbf{w}(\mathbf{s})$ is the zero mean hydraulic gradient vector fluctuation. The $\mathbf{J}(\mathbf{s})$ denotes the gradient vector of the hydraulic head H(s), viz. $\mathbf{J}(\mathbf{s}) = -\nabla H(\mathbf{s})$.

The groundwater flow equations of the porous medium $\Pi(\mathrm{u})$ are the continuity equation and Darcy's law (e.g., Bear, 1972). In the isotropic case, these equations lead to the *stochastic partial differential equation (SPDE)* of flow within the flow domain *D*

$$\nabla \cdot \mathbf{J}(\mathbf{s}) + [\overline{\mathbf{W}(\mathbf{s})} + \mathbf{w}(\mathbf{s})] \cdot \mathbf{J}(\mathbf{s}) = 0, \tag{2.1}$$

where the hydraulic gradient vector $\mathbf{J}(\mathbf{s})$ together with the hydraulic conductivity and head potential processes meet the necessary conditions across the boundary *B*.

As was mentioned in the preceding section, there is not a general stochastic method to study groundwater flow in media with randomly fluctuating properties. An exact stochastic solution would require the derivation of the associated Green's function for any realization of the RFs involved, which is practically not feasible. Hence, such problems are generally handled using *approximate* methods. The choice of the appropriate approximate method depends upon the specific flow situation; this includes:

(i) The size and the amount of random fluctuations of the porous medium properties (large or small fluctuations, slow or abrupt fluctuations, etc.). A common fluctuation measure is the standard deviation σ_w of the RF $\mathbf{w}(\mathbf{s})$.

(ii) The spatial variability characteristics of the porous media processes (long-range or short-range correlations, strong or weak interactions, etc.). A spatial correlation measure is provided by the correlation range ε_w of $\mathbf{w}(\mathbf{s})$.

When the assumption can be made that the fluctuations $\mathbf{w}(\mathbf{s})$ are sufficiently small (e.g., $\sigma_w << 1$ or $\sigma_w << \overline{\mathbf{W}}$), a widely used approximate method is the *low-order ordinary perturbation* technique (see following section; references therein). This technique seeks to express the hydraulic gradient vector $\mathbf{J}(\mathbf{s})$ as a series in $\mathbf{w}(\mathbf{s})$, or (which is the same) as a series in σ_w. If the small fluctuation assumption is not valid, one can apply the *diagrammatic perturbation* method (see below), which uses geometrical porous medium description operators and topological diagrams of stochastic flow equations. In fact, the diagrammatic method has the unique advantage that, depending upon the proper choice of a so-called porous medium description operator, it is valid for both small and large fluctuation cases; and can be applied in both cases of long and short range correlation scales.

THE ORDINARY PERTURBATION METHOD AND THE SMALL FLUCTUATION ASSUMPTION

Eq. (2.1) can be written as the inhomogeneous SPDE

$$\nabla \cdot \mathbf{J}(\mathbf{s}) + \overline{\mathbf{W}(\mathbf{s})} \cdot \mathbf{J}(\mathbf{s}) = -\mathbf{w}(\mathbf{s}) \cdot \mathbf{J}(\mathbf{s}). \tag{3.1}$$

The solution of Eq. (3.1) can be expressed as

$$\mathbf{J}(\mathbf{s}) = \mathbf{J}_0(\mathbf{s}) - \int_V \mathbf{G}_0(\mathbf{s},\mathbf{s}')[\mathbf{w}(\mathbf{s}') \cdot \mathbf{J}(\mathbf{s}')]d\mathbf{s}', \tag{3.2}$$

where $\mathbf{J}_0(\mathbf{s})$ is the *unperturbed* hydraulic gradient vector which satisfies

$$\nabla \cdot \mathbf{J}_0(\mathbf{s}) + \overline{\mathbf{W}(\mathbf{s})} \cdot \mathbf{J}_0(\mathbf{s}) = 0; \tag{3.3}$$

$\mathbf{G}_0(\mathbf{s},\mathbf{s}')$ is the associated vector Green's function obeying equation

$$\nabla \cdot \mathbf{G}_0(\mathbf{s},\mathbf{s}') + \overline{\mathbf{W}(\mathbf{s})} \cdot \mathbf{G}_0(\mathbf{s},\mathbf{s}') = \delta(\mathbf{s}-\mathbf{s}'); \tag{3.4}$$

and V is the domain of the fluctuations $\mathbf{w}(\mathbf{s})$. The non-random vectors $\mathbf{J}_0(\mathbf{s})$ and $\mathbf{G}_0(\mathbf{s},\mathbf{s}')$ satisfy the corresponding boundary conditions, as well. The integral Eq. (3.2) is equivalent to the differential Eq. (2.1); it also satisfies (through $\mathbf{G}_0$) all the boundary conditions *B*. In order to keep the theoretical analysis as general as possible, no mention of the (*B*-dependent) specific form of $\mathbf{G}_0$ has been made.

From Eq. (3.2) the perturbation expansion is obtained by iteration; in terms of the *Neumann* series

$$\mathbf{J}(\mathbf{s}) = \mathbf{J}_0(\mathbf{s}) - \int_V \mathbf{G}_0(\mathbf{s},\mathbf{u}_1) \sum_{j=1}^{\infty} \Theta_j(\mathbf{u}_1)\, d\mathbf{u}_1, \tag{3.5}$$

where $\Theta_1(\mathbf{u}_1) = \mathbf{w}(\mathbf{s}) \cdot \mathbf{J}_0(\mathbf{u}_1)$, and $\Theta_{j+1}(\mathbf{u}_1) = -\int_V [\mathbf{w}(\mathbf{u}_1) \cdot \mathbf{G}_0(\mathbf{u}_1,\mathbf{u}_{j+1})]\Theta_j(\mathbf{u}_{j+1})d\mathbf{u}_{j+1}$. Eq. (3.5) can be also written as the *perturbed* hydraulic gradient vector field

$$\mathbf{J}_p(\mathbf{s}) = \mathbf{J}(\mathbf{s}) - \mathbf{J}_0(\mathbf{s}) = \sum_{k=1}^{\infty} \mathbf{J}_k(\mathbf{s}). \tag{3.6}$$

Each perturbed term of the series (3.6) is given by

$$\mathbf{J}_k(\mathbf{s}) = (-1)^k \int_V \mathbf{G}_0(\mathbf{s},\mathbf{s}')[\mathbf{w}(\mathbf{s}') \cdot \mathbf{J}_{k-1}(\mathbf{s}')]d\mathbf{s}'; \tag{3.7}$$

i.e., each term $\mathbf{J}_k(\mathbf{s})$ is the result of a direct, linear interaction of the previous term $\mathbf{J}_{k-1}(\mathbf{s})$ with $\mathbf{w}(\mathbf{s})$. In fact, the $\mathbf{J}_1(\mathbf{s})$ is the only term directly excited by $\mathbf{J}_0(\mathbf{s})$; the $\mathbf{J}_2(\mathbf{s})$ is in turn excited by $\mathbf{J}_1(\mathbf{s})$; etc.

Low-order perturbation assumes that the fluctuations $\mathbf{w(s)}$ are sufficiently small (e.g., $\sigma_w << 1$). Then, only the first perturbed term in Eq. (3.6) can be retained and the perturbed hydraulic gradient vector $\mathbf{J}_p(\mathbf{s}) \approx \mathbf{J}_1(\mathbf{s})$ is a linear function of $\mathbf{w(s)}$. Its statistical moments are expressed linearly in terms of the corresponding moments of $\mathbf{w(s)}$. Since $\mathbf{J}_0(\mathbf{s})$ is a constant vector and $\overline{\mathbf{w(s)}} = 0$,

$$\overline{\mathbf{J(s)}} = \mathbf{J}_0(\mathbf{s}) + \overline{\mathbf{J}_1(\mathbf{s})} = \mathbf{J}_0(\mathbf{s}), \tag{3.8}$$

and

$$\mathbf{c_J}(\mathbf{s},\mathbf{s}') = \overline{\mathbf{J}_1(\mathbf{s})\mathbf{J}_1(\mathbf{s}')^T} = \int\int \mathbf{G}_0(\mathbf{s},\mathbf{u}_1)\mathbf{J}_0(\mathbf{u}_1)^T \mathbf{c_w}(\mathbf{u}_1,\mathbf{u}_2)\mathbf{J}_0(\mathbf{u}_2)\mathbf{G}_0(\mathbf{s}',\mathbf{u}_2)^T d\mathbf{u}_1 d\mathbf{u}_2 \,. \tag{3.9}$$

However, many times in applications one is dealing with highly heterogeneous media where the small fluctuations assumption is violated (e.g., Sudicky, 1986; Zeitoun and Breister, 1991; Neuman and Orr, 1993). To overcome these difficulties, a deeper understanding of the various flow terms and their spatial correlations must be obtained. The means to accomplish this are provided by methods such as (Christakos *et al.*, 1993a and b):

(i) *Stochastic interactive perturbation (SIP)* ; this is a mathematically powerful but rather abstract method, which uses Laplace transformation and formal power series inversion operators.

(ii) *Stochastic diagrammatic perturbation (SDP)* ; this is a more physical method which takes advantage of *geometrical* porous media representations and *graph theory* techniques.

Despite their mathematical similarity these two methods of flow analysis are not necessarily physically equivalent. Since the second approach is more useful in the intuitive understanding of the flow phenomena, it is the one to be discussed next.

THE STOCHASTIC DIAGRAMMATIC PERTURBATION METHOD

The Diagrammatic Formalism

As we saw above, ordinary perturbation techniques seek to express the hydraulic gradient vector $\mathbf{J(s)}$ as a series in $\mathbf{w(s)}$ (or σ_w) and the unperturbed gradient vector $\mathbf{J}_0(\mathbf{s})$. If the series converges, its summation is a calculable physical quantity. However, it is not generally known whether the series converges; or even under what conditions it may converge. If the series diverges, there is no mathematically sound and physically meaningful method of determining the flow process that is supposed to be represented by the series. Certainly, solutions can be obtained by keeping only a limited number of low-order terms, but it is uncertain how large an error was incurred by truncating the series. Christakos *et al.* (1993a and b) have discussed groundwater flow situations where the finite sums of the series representing hydraulic head lead to solutions that diverge towards infinity, which is physically unacceptable. In these situations, however, valuable information can be obtained from the formalism of stochastic flow modelling by an appropriately arranged expansion series, which differs markedly from the ordinary perturbation approach. These methods are not purely abstract mathematical techniques. Instead, they provide new physical insight regarding the expansion terms by means of appropriate geometrical interpretations of the flow parameters. One-dimensional groundwater flow situations were considered, which supported the above theoretical claims. In this work the analysis is extended in a multidimensional groundwater flow setting.

The motivation behind the use of diagrammatic representations in groundwater flow modelling is based on the concept that the qualitative grasp of form, shape and geometric order can go deeper than the quantitative grasp of abstract mathematical symbol, number and magnitude. The proposed approach to the stochastic flow problem uses *topological diagrams* and *graph theory* (mathematical descriptions of these techniques may be found, e.g., in March *et al.*, 1967; and Thulasiraman and Swamy, 1992).

Generally, a *diagram* (or a *graph*) is a collection of *points* (*vertices*) and *links* (*edges*). Note that the word graph will be here interchangeably used instead of diagram. Several types of diagrams can be constructed, viz. single or double, reducible or irreducible, etc. (definitions are given below), which can be used to represent abstract mathematical quantities according to some well-defined set of ***rules***. There may be a variety of such rules chosen so that: they adequately reflect the specific physical characteristics, spatial correlations, etc. of the flow problem under consideration; and satisfy certain topological constraints. Properly selected diagrams provide valuable geometrical representations of groundwater flow which can then lead to novel interpretations of porous media properties, extend one's physical knowledge and create intuitive justification of the observed flow processes.

The Random Integral Operator Form of Flow

Before we proceed with the graphic treatment of the flow problem, the solution to the flow Eq. (2.1) above is expressed in the ***random integral operator*** form (Christakos, 1992b)

$$\mathbf{J}(\mathbf{s}) = \mathbf{G}_0(\mathbf{s},0) - \int_V \mathbf{G}_0(\mathbf{s},\mathbf{s}')[\mathbf{w}(\mathbf{s}')\cdot\mathbf{J}(\mathbf{s}')]d\mathbf{s}' = \sum_{\lambda=0}^{\infty}[-\mathbf{G}_0\mathbf{L}]^{\lambda}\mathbf{G}_0, \tag{4.1}$$

where L is the operator defined as $\mathbf{L}: \mathbf{f} \rightarrow \int \mathbf{w}\cdot\mathbf{f}$ and $\mathbf{G}_0$ is the non-random vector Green's function governed by Eq. (3.4) (i.e., $\mathbf{G}_0$ may be viewed as the solution to Eq. (2.1) when there is no randomness). An equivalent expression may be obtained in terms of the random Green's function **G** given by

$$\mathbf{G}(\mathbf{s},\mathbf{s}_0) = \mathbf{G}_0(\mathbf{s},\mathbf{s}_0) - \int_V \mathbf{G}_0(\mathbf{s},\mathbf{s}')[\mathbf{w}(\mathbf{s}')\cdot\mathbf{J}(\mathbf{s}')]d\mathbf{s}', \tag{4.2}$$

where **G** obeys

$$\nabla\cdot\mathbf{G}(\mathbf{s},\mathbf{s}_0) + [\overline{\mathbf{W}(\mathbf{s})} + \mathbf{w}(\mathbf{s})]\cdot\mathbf{G}(\mathbf{s},\mathbf{s}_0) = \delta(\mathbf{s},\mathbf{s}_0). \tag{4.3}$$

Green's function **G** is a symmetric function, viz. $\mathbf{G}(\mathbf{s},\mathbf{s}_0) = \mathbf{G}(\mathbf{s}_0,\mathbf{s})$, which implies that if the observation and the source locations within a porous medium are interchanged the flow field will remain unchanged.

Since the random log-conductivity fluctuation **w(s)** is assumed to be zero mean Gaussian, all odd-order statistical moments are zero and the even-order moments can be written as sums of terms consisting of products of second-order moments. It is worth-noticing that the diagrammatic approach is general and does not require any Gaussian assumption. Nevertheless, the latter may be a reasonable assumption in practice, for only certain second-order statistical information is usually available about the random vector gradient $\mathbf{w}(\mathbf{s})$. With the covariances $c_{\mathbf{w}}(\mathbf{u}_i, \mathbf{u}_j)$ correlation ranges $\varepsilon_{\mathbf{w}}$ are associated. The $\varepsilon_{\mathbf{w}}$ gives a measure of the separation distance between any two points $\mathbf{u}_i$ and $\mathbf{u}_j$ over which the $\mathbf{w}(\mathbf{s})$ has nonnegligible spatial correlations.

The Mean Flow Equation and its Diagrammatic Representation

By taking the mean value of Eq. (4.1) above one finds

$$\overline{J(s)} = \sum_{\lambda=0}^{\infty} [-\overline{G_0 L}]^{\lambda} G_0. \tag{4.4}$$

One possible graphic representation of Eq. (4.4) is derived by means of the first three rules of Table 1.

TABLE 1: Diagrammatic rules for steady-state groundwater flow

Rule 1: The term G_0 will be represented by a straight line.

Rule 2: The operator L will be denoted by a dot.

Rule 3: A wavy line connecting two dots denotes expectation between any two points in space.

Rule 4: The PMDO-1 $\aleph$ will be represented by a triangle △

Rule 5: The $\overline{J(s)}$ will be represented by a thick line.

Rule 6: The PMDO-2 is denoted by the symbol ⊗.

Rule 7: The PMDO-3 will be represented by the symbol ☒.

Rule 8: The $c_J(s,s')$ is denoted by the rhombus ◇.

The above rules lead to a representation of the mean head gradient Eq. (4.4) in terms of a set of simple and illuminating *single* diagrams as follows

$\overline{J(s)} =$ (1) + (2) + (3) + (4) + (5) + (6) + (7) +

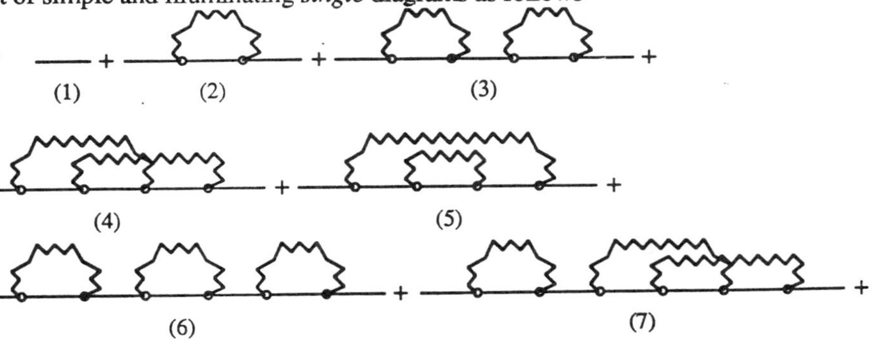

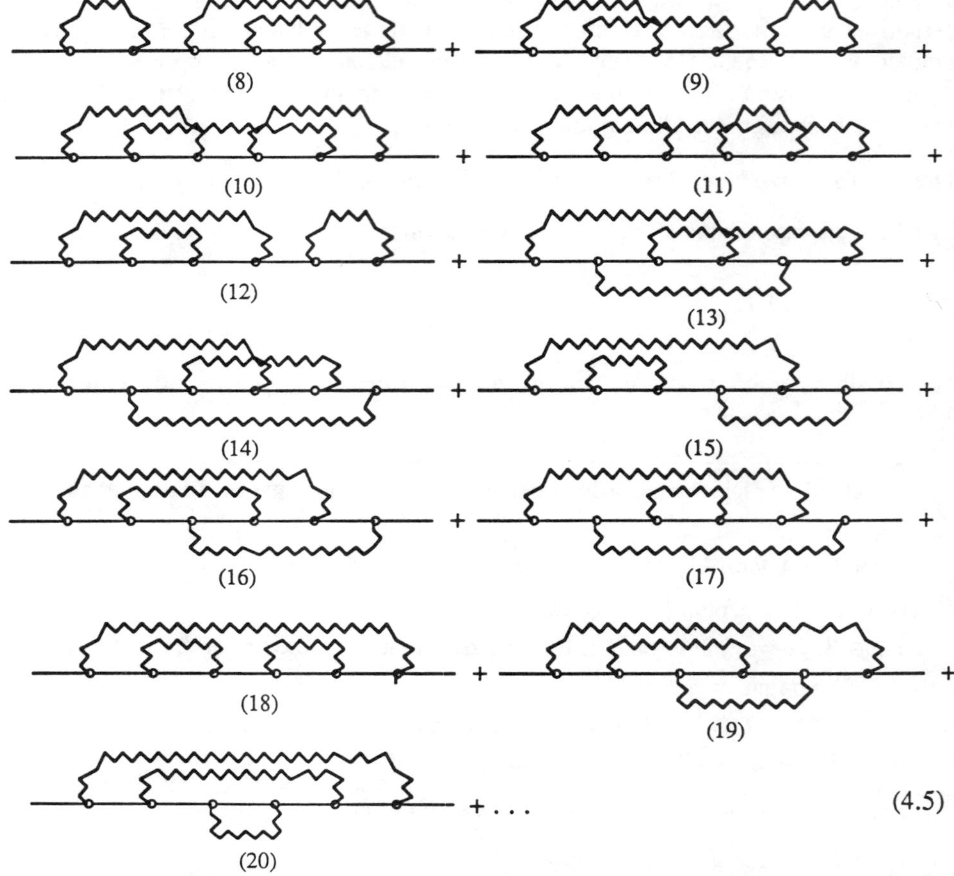

Some interesting observations can be made regarding the geometric representation of Eq. (4.5):

(a) Since the Green's function $\mathbf{G}_0$ is not random, the expectation of its diagramatic representation -i.e. the straight line- is again the same straight line.

(b) The terms consist of all possible combinations of dots by two (i.e., the inner vertices are connected pairwise by wavy lines in all possible ways).

(c) An *nth-order diagram* consists of 2n dots (inner vertices $\mathbf{u}_1, \mathbf{u}_2, \ldots, \mathbf{u}_{2n}$), 2n+1 lines of Green's functions $\mathbf{G}_0$ and (2n-1)!! covariances $\mathbf{c}_w$. There exist $(2n-1)!! = 1 \times 3 \times \ldots \times (2n-1)$ diagrams of n-order (2n dot-diagram); e.g., 1 diagram of order 1 (2 dot-diagram; see diagram no. 2); 3 diagrams of order 2 (4 dot-diagram; no. 3, 4 and 5); 15 diagrams of order 3 (6 dot-diagram; no. 6 through 20).

(d) High-order diagrams may contain low-order diagrams. E.g., diagram no. 19 contains no. 4.

(e) The single diagrams above provide a simple means for representing complex integrals as follows:

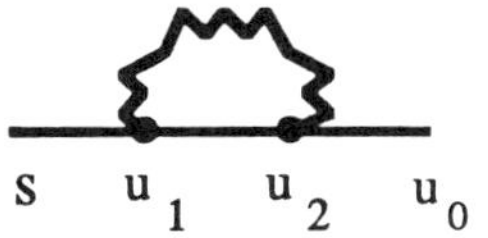

$$= \int_V\int_V G_0(\mathbf{s},\mathbf{u}_1)G_0(\mathbf{u}_1,\mathbf{u}_2)^T\mathbf{c}_w(\mathbf{u}_1,\mathbf{u}_2)G_0(\mathbf{u}_2,\mathbf{u}_0)d\mathbf{u}_1 d\mathbf{u}_2; \quad (4.6)$$

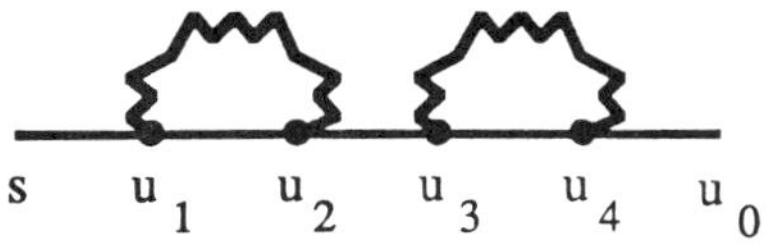

s u1 u2 u0

$$= \int_V\int_V\int_V\int_V G_0(\mathbf{s},\mathbf{u}_1)G_0(\mathbf{u}_1,\mathbf{u}_2)^T\mathbf{c}_w(\mathbf{u}_1,\mathbf{u}_2)G_0(\mathbf{u}_2,\mathbf{u}_3) G_0(\mathbf{u}_3,\mathbf{u}_4)^T\mathbf{c}_w(\mathbf{u}_3,\mathbf{u}_4)G_0(\mathbf{u}_4,\mathbf{u}_0)d\mathbf{u}_1 d\mathbf{u}_2 d\mathbf{u}_3 d\mathbf{u}_4; \quad (4.7)$$

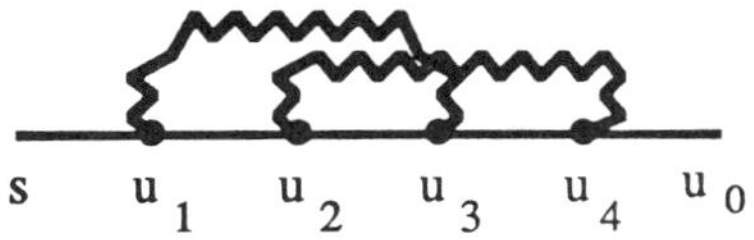

s u1 u2 u3 u4 u0

$$= \int_V\int_V\int_V\int_V G_0(\mathbf{s},\mathbf{u}_1)G_0(\mathbf{u}_1,\mathbf{u}_2)^T\mathbf{c}_w(\mathbf{u}_1,\mathbf{u}_3)G_0(\mathbf{u}_2,\mathbf{u}_3) G_0(\mathbf{u}_3,\mathbf{u}_4)^T\mathbf{c}_w(\mathbf{u}_2,\mathbf{u}_4)G_0(\mathbf{u}_4,\mathbf{u}_0)d\mathbf{u}_1 d\mathbf{u}_2 d\mathbf{u}_3 d\mathbf{u}_4; \quad (4.8)$$

s u1 u2 u3 u4 u0

etc. There is actually no need to indicate the coordinates $\mathbf{u}_1$, $\mathbf{u}_2,\ldots,\mathbf{u}_{2n}$ of the inner vertices on the diagrams, for the corresponding analytical representations are independent of these coordinates.

(f) There is an one-to-one correspondence between the diagrams and the analytical expressions. Moreover, the diagrams have a compact form as compared to analytical representations; they simplify the treatment significantly; and they enable one to classify terms more easily by means of their topological properties and then evaluate the series approximately by selecting the most important types of terms in it.

Each diagrammatic term describes spatial correlations between flow terms at points in space whose coordinates are the numeric indices $\mathbf{u}_1,\mathbf{u}_2$, etc. of the dots in Eqs. (4.6), (4.7), etc. The length of the lines connecting the dots of points $\mathbf{u}_1,\mathbf{u}_2$, etc. where covariances are considered should not be larger than the correlation ranges ε_w of the random gradients $\mathbf{w}(\mathbf{u})$, for if any two points are further apart than ε_w, the corresponding diagrammatic terms will be negligible.

The interior of the closed diagram in Eq. (4.6), defined by the straight line and the wavy line which connect the two circles, may be associated with the porous medium region within which important spatial correlations occur. More complex correlation regions of the porous medium are represented by sets of overlapping diagrams, such as in Eq. (4.8). Related to these observations are the following definitions. After the external solid lines have been eliminated, the remaining diagrams may be: (i) either *reducible,* i.e. they can be reduced to simpler diagrams by cutting solid lines (e.g., the diagrams 3, 6 through 9 and 12 in Eq. (4.5)); (ii) or *irreducible,* i.e. they cannot be reduced to simpler diagrams by cutting solid lines (e.g., the diagrams 2, 4, 5, 10, 11 and 13 through 20). The closed regions defined by these diagrams may be extended to a temporal, spatiotemporal, or some other abstract space. In such a case, of course, the overlapping of the corresponding diagrams are not spatial.

POROUS MEDIA DESCRIPTION OPERATORS FOR THE MEAN VALUE

The Porous Medium Description Operator of the First Kind

In order to determine the mean $\overline{J(s)}$, the iteration series (4.5) might be used. But then, only a limited number of terms could be calculated and it would be uncertain how large an error was incurred by truncating the series. However, if $\overline{J(s)}$ could be shown to satisfy a known mean flow equation of a closed form, the latter might be solved. Indeed, as we shall see below such an integrodifferential mean flow equation exists and can be established in terms of so co-called *porous medium description operators (PMDOs).*

The irreducible diagrams are especially important in this respect. Let us consider the subsequence **F** of irreducible diagrams. On the basis of the analysis above and since each of the diagrams belonging to **F** begins and ends with the line $\mathbf{G}_0$, **F** can be written as

$$\mathbf{F}(\mathbf{s},\mathbf{u}_0) = \int_V\int_V \mathbf{G}_0(\mathbf{s},\mathbf{u}_1)\aleph(\mathbf{u}_1,\mathbf{u}_2)\overline{\mathbf{G}(\mathbf{u}_2,\mathbf{u}_0)}d\mathbf{u}_1 d\mathbf{u}_2, \tag{5.1}$$

where $\aleph$ is the so-called *PMDO of the first kind (PMDO-1)*, which is defined as the sum of all irreducible diagrams. This can be expressed in a compact form with the help of the fourth rule of Table 1, namely,

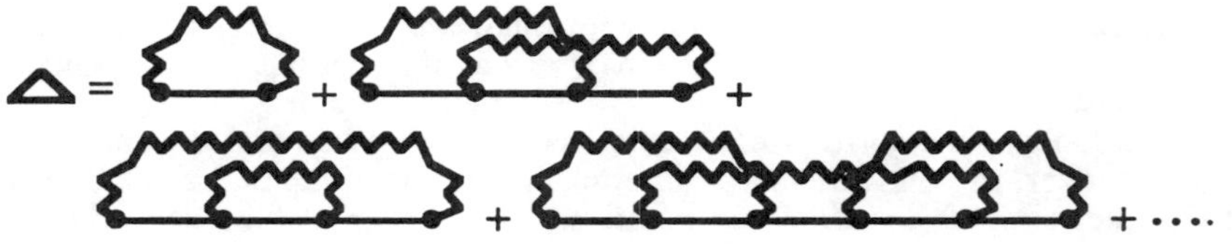

(5.2)

The reducible diagrams which contain only m irreducible parts occur in the sum of graphs shown below

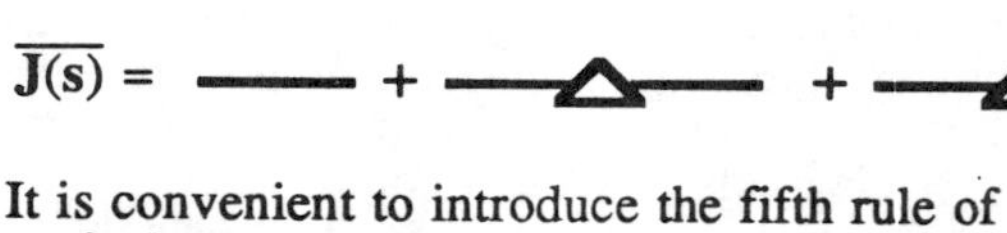

(5.3)

m-times

where m=2, 3,... For example, the diagram no. 3 in Eq. (4.5) is obtained by taking the first term of $\aleph(\mathbf{u}_1,\mathbf{u}_2)$ in Eq. (5.2). Consequently, the $\overline{J(s)}$ can be expressed as

$\overline{J(s)}$ = + + + ... (5.4)

It is convenient to introduce the fifth rule of Table 1 so that Eq. (5.4) is topologically equivalent to

= + (5.5)

(Note that Eq. (5.4) is immediatelly obtained from Eq. (5.5) by successive iterations.) Using Eq. (5.1), Eq. (5.5) can be expressed in analytic form by

$$\overline{\mathbf{J(s)}} = \mathbf{G}_0(\mathbf{s},0) + \mathbf{F}(\mathbf{s},0) = \mathbf{G}_0(\mathbf{s},0) + \int_V\int_V \mathbf{G}_0(\mathbf{s},\mathbf{u}_1)\aleph(\mathbf{u}_1,\mathbf{u}_2)\overline{\mathbf{G}(\mathbf{u}_2,0)}d\mathbf{u}_1 d\mathbf{u}_2. \tag{5.6}$$

The process that leads to Eq. (5.6) replaces one of the $\mathbf{G}_0$ terms in the original series expansion (4.5) by the mean value $\overline{\mathbf{G}}$ of the random vector Green's function. This process is sometimes called *renormalization*. If the gradient $\mathbf{w(s)}$ is *spatially homogeneous,* viz. $\mathbf{c_w}(\mathbf{u}_1,\mathbf{u}_2) = \mathbf{c_w}(\mathbf{u}_1 - \mathbf{u}_2)$, then $\aleph(\mathbf{u}_1,\mathbf{u}_2) = \aleph(\mathbf{u}_1 - \mathbf{u}_2)$ and Eq. (5.6) is of the convolution-type. Hence, useful solutions can be derived in terms of Fourier transforms.

Various approximations to Eq. (5.5) are established by keeping a finite number of terms in the operator (5.2). In the case of small fluctuation only the first term could be retained. Then, the resulting series expansion accounts for second-order interactions only, leading to the *first-order approximation*

$$\overline{\mathbf{J(s)}} = \mathbf{G}_0(\mathbf{s},0) - \int_V\int_V \mathbf{G}_0(\mathbf{s},\mathbf{u}_1)\mathbf{G}_0(\mathbf{u}_1,\mathbf{u}_2)^T\mathbf{c_w}(\mathbf{u}_1,\mathbf{u}_2)\overline{\mathbf{G}(\mathbf{u}_2,0)}d\mathbf{u}_1 d\mathbf{u}_2; \tag{5.7}$$

or in a diagrammatic form

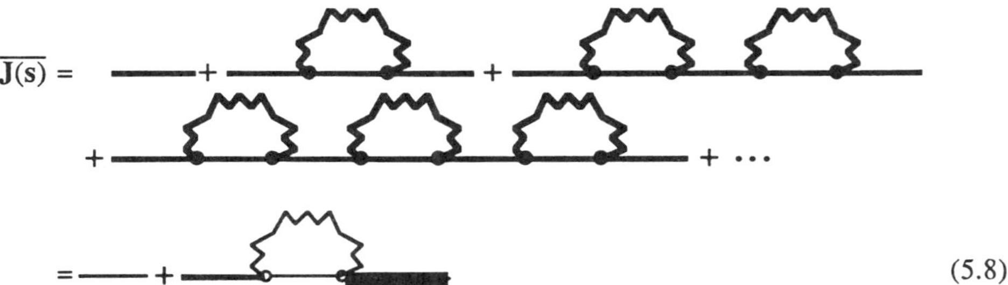

(5.8)

Eq. (5.8) is an infinite subset of the Neumann series expressed in closed form, taking advantage of the selective summing process introduced by the diagrammatic method.

Using the geometrical properties of the graphic representations and depending upon the spatial variability structure of the porous medium, other interesting approximations can be obtained, as well. For example, assume that the correlation range $\varepsilon_{\mathbf{w}}$ is significantly larger than the $|\mathbf{s}|$ values of importance in the particular flow problem under consideration. Then the three fourth-order integrals in Eq. (4.5), which correspond to diagrams no. 3, 4 and 5, can be calculated approximatelly by replacing $c_{\mathbf{w}}(\cdot)$ with $\sigma^2_{\mathbf{w}}$. As a consequence, all the fourth-order integrals will be approximatelly equal to each other. By following a similar procedure for the higher-order integrals, the diagrammatic expansion of $\overline{\mathbf{J(s)}}$ can be obtained by multiplying each diagrammatic term in Eq. (5.8) by the corresponding number of diagrams of order n (i.e., by (2n-1)!!); namely,

$\overline{J(s)}$ = ——— + ——— + 3 × ——— +

15 × ——— + · · ·

(2n-1)!! × ——— · · · ——— + ··· · (5.9)

← 2n dots →

Example 1

Consider the *one-dimensional* steady-state groundwater flow in a semi-infinite domain with constant log-conductivity spatial mean and $\overline{J(0)} = -1$. Assume that the fluctuation gradient w(s) is spatially homogeneous with an exponential covariance $\rho_w(r) = \exp[-\frac{|r|}{\beta}]$, $r = s - u_1$, and a correlation range $\varepsilon_w = 3\beta$. Under the appropriate stochastic auxiliary assumptions discussed in Christakos *et al.* (1993b) the exact mean flow solution is

$$\overline{J(s)} = -\exp\{-\tau^2\beta^2[s\beta^{-1} + \exp(-s\beta^{-1}) - 1]\}, \tag{5.10}$$

where τ is a normalization constant. In the one-dimensional case, Eq. (5.8) simplifies to

$$\frac{d}{ds}\overline{J(s)}^* = \tau^2 \int_0^s \rho_w(s - u_1)\overline{J(u_1)}^* du_1 . \tag{5.11}$$

If $\tau\beta << 1$, the approximate solution obtained from Eq. (5.11) is $\overline{J(s)}^* \approx -\exp[-\tau^2\beta\, s]$, which is in very good agreement with the exact solution (5.10). However, when larger $\tau\beta$ values are considered the approximate solution obtained from Eq. (5.11) is not in good agreement with (5.10) (the extent of this disagreement will depend upon the magnitude of $\tau\beta$ in a continuum sense; see, also, Fig. 1 later).

The diagrammatic approach can handle large fluctuations as well. This can be achieved in a number of ways. For example, the formulation suggested by Eq. (5.9) above may work in some cases. However, it seems that in the case of groundwater flow the most powerful and general way that applies for any type of fluctuations (large and small) can be developed in terms of the second kind PMDO to be discussed next.

The Porous Medium Description Operator of the Second Kind

The choice of the appropriate PMDO is a very important part of the diagrammatic approach. There may be various mathematically acceptable PMDOs for the specific flow problem. It is then a matter of physical knowledge and intuition to choose the one that describes most appropriately the flow characteristics.

In this section the *PMDO of the second kind (PMDO-2)* will be presented. This operator, which is particularly suitable for the analysis of steady-state flow problems in heterogeneous geologic media, replaces the set of "unperturbed" flow ($\mathbf{G}_0$) diagrams by a set of irreducible "actual" mean-flow ($\bar{\mathbf{J}}$) diagrams where, whenever more than one wavy lines exist, each wavy line must cross another wavy line. In the light of the sixth rule of Table 1 the PMDO-2 admits the graphic representation

$$\otimes = \ldots + \ldots + \ldots + \ldots, \tag{5.12}$$

where

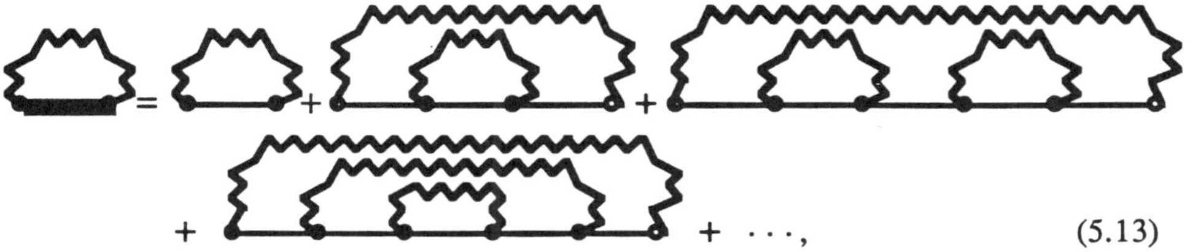

$$\ldots = \ldots + \ldots + \ldots + \ldots + \cdots, \tag{5.13}$$

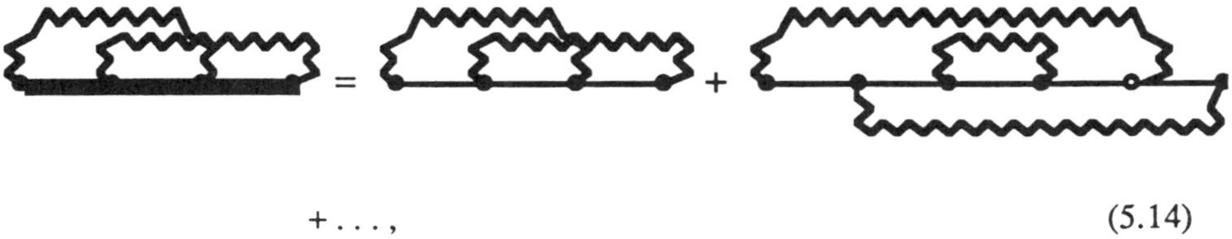

$$\ldots = \ldots + \ldots + \ldots, \tag{5.14}$$

etc. The PMDO-2 enables us to represent the mean $\overline{\mathbf{J(s)}}$ graphically as

$$\overline{\mathbf{J(s)}} = \ldots = \ldots + \ldots \otimes \ldots + \ldots \otimes \ldots \otimes \ldots + \ldots \tag{5.15}$$

Topologically, Eq. (5.14) is equivalent to the more concise form

$$\overline{\mathbf{J(s)}} = \ldots = \ldots + \ldots \otimes \ldots. \tag{5.16}$$

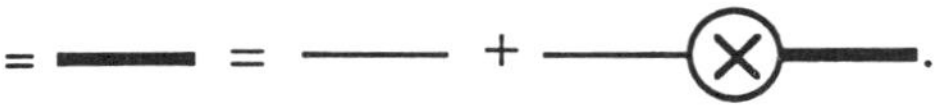

Expansion (5.16) accounts for the correlation structure of the log-conductivity gradient, as well as the spatial interactions between head gradients. Practically useful approximations to Eq. (5.16) can be established by retaining a finite number of terms in the PMDO-2. For example, if only the first term is kept in the PMDO-2, Eq. (5.16) yields

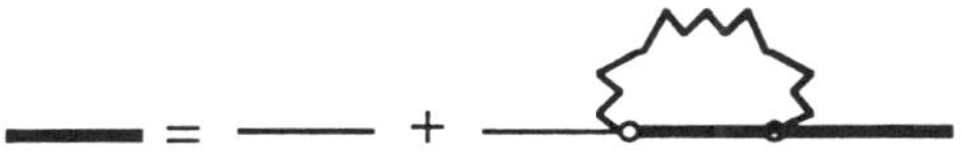

$$\ldots = \ldots + \ldots \tag{5.17}$$

Example 2

Consider the one-dimensional steady-state flow situation of Example 1 above. Eq. (5.17) reduces to

$$\frac{d}{ds}\overline{J(s)}^{\#} = \tau^2 \int_0^s \rho_w(s-u_1)\overline{J(s-u_1)}^{\#}\ \overline{J(u_1)}^{\#}\, du_1. \tag{5.18}$$

Eq. (5.18) can be solved numerically. The PMDO-2 approximate solution, $\overline{J(s)}^{\#}$ for $\beta=1$ and $\tau=2$ is plotted in Fig. 1. The exact solution, $\overline{J(s)}$, the PMDO-1 approximation $\overline{J(s)}^{*}$,

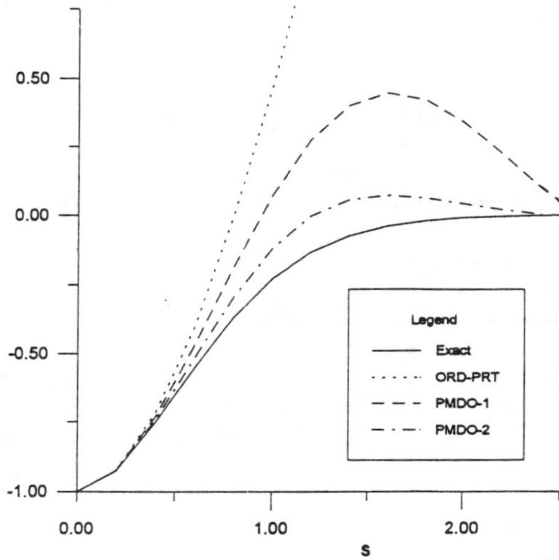

Figure 1: The exact solution $\overline{J(s)}$ (—) vs. the approximate solutions, $\overline{J(s)}^{\#}$ (PMDO-2; – · –), $\overline{J(s)}^{*}$ (PMDO-1; – –), and $\overline{J(s)}^{0}$ (ord. pert. series; · · ·); $\beta=1$ and $\tau=2$.

and the ordinary perturbation solution $\overline{J(s)}^{0}$ (which was obtained by keeping the first three terms of the corresponding power series) are also plotted for comparison. The $\overline{J(s)}^{\#}$ not only provides the best approximation to the exact solution $\overline{J(s)}$, but it also fulfills essential physical requirements (it is bounded for all s, accounts for important interactions between hydraulic gradients, etc.), which the ordinary perturbation approximation $\overline{J(s)}^{0}$ failed to satisfy. A physical interpretation of the difference between approximations (5.8) and (5.17) may be attributed to the fact that while the former approximation considers unperturbed flow among the two fixed points (dots) of the single porous medium diagram in Eq. (5.8), the latter approximation considers a perturbed flow process (i.e., a series of spatial correlations among inbetween points as represented by the series of diagrams in Eq. (5.13)). This may explain the fact that Eq. (5.8) does not yield good approximations for large fluctuations where the spatial correlations represented by the series of diagrams in Eq. (5.13) are important. More detailed results and plots for a wide range of $\tau\beta$ values were obtained in Christakos *et al.* (1993b), where it was found that $\overline{J(s)}^{\#}$ offers very good approximations even for large $\tau\beta$ values. Eq. (5.17) has led to good approximations, despite the fact that it is only a first-order approximation. Higher-order approximations can be considered by keeping more than one terms in Eq. (5.17).

POROUS MEDIA DESCRIPTION OPERATORS FOR THE COVARIANCE FUNCTION

The Porous Medium Description Operator of the Third Kind

Consider the (non-centered) covariance function of the hydraulic head gradient vector $\mathbf{J}(\mathbf{s})$, viz.

$$\mathbf{c}_{\mathbf{J}}(\mathbf{s},\mathbf{s}') = \overline{\mathbf{J}(\mathbf{s})\mathbf{J}(\mathbf{s})^{T}} \tag{6.1}$$

To obtain a graphic representation of Eq. (6.1), Eq. (4.1) must be multiplied by its transpose and then take the average of the resulting expression. This expression contains certain quantities that can be represented diagrammatically. This can be done easily in terms of rules 1, 2 and 3 above. Then, since a set of diagrams is multiplied by another, the result is a set of *double* diagrams, viz.

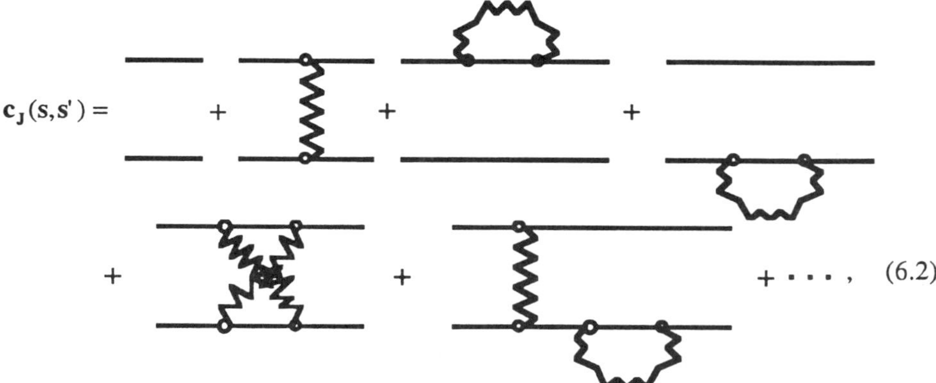

(6.2)

where, e.g.,

s u$_1$ 0

s' u'$_1$ 0

$$= \int_V \int_V \mathbf{G}_0(\mathbf{s},\mathbf{u}_1)\mathbf{J}_0(\mathbf{u}_1)^T \mathbf{c}_w(\mathbf{u}_1,\mathbf{u}_1')\mathbf{J}_0(\mathbf{u}_1')\mathbf{G}_0(\mathbf{s}',\mathbf{u}_1')^T d\mathbf{u}_1 d\mathbf{u}_1' ; \tag{6.3}$$

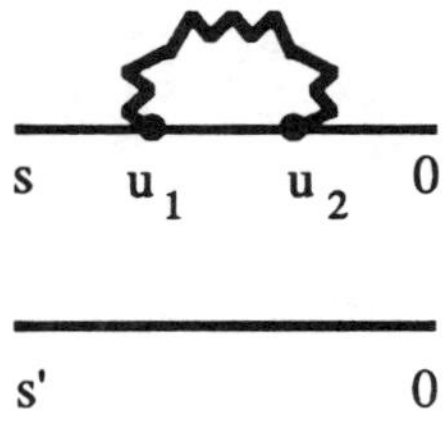

s u1 u2 0

s' 0

$$= \int_V \int_V G_0(s,u_1) G_0(u_1,u_2)^T c_w(u_1,u_2) J_0(u_2) G_0(s',0)^T du_1 du_2 \quad (6.4)$$

etc. The discussion of the previous section regarding reducible and irreducible single diagrams applies to double diagrams, as well. Then, analogous to the PMDOs defined above is the so-called *porous medium description operator of the third kind (PMDO-3)* defined as the sum of all irreducible double diagrams. To proceed in a graphic manner the rules no. 7 and 8 of Table 1 are introduced. Then PMDO-3 can be read as

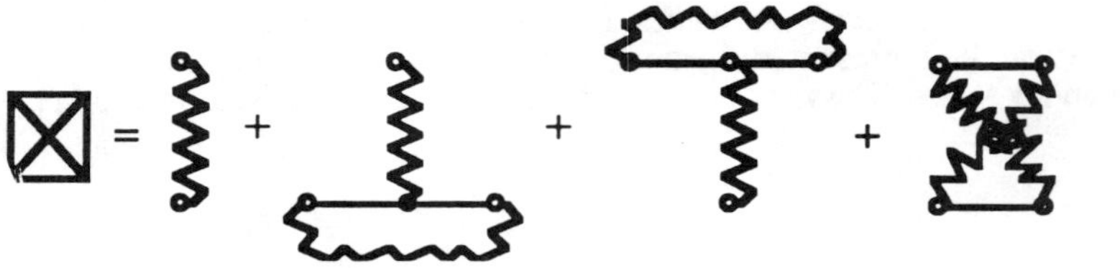

= + + + + ··· ; (6.5)

and Eq. (6.2) becomes

$c_J(s,s') =$ = + + + · · ·, (6.6)

which is topologically equivalent to the more concise graphic expression

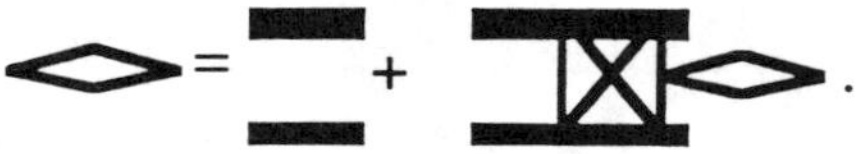

= + . (6.7)

The foregoing analysis, which led to the fundamental Eqs. (5.17) and (6.7), was based on the expansion (4.1). (However, we could arrive at the same results starting from Eq. (4.2).) Approximations to Eq. (6.7) are obtained by retaining a finite number of terms in the PMDO-3. For example, keeping only the first term in Eq. (6.6) leads to the *first-order approximation*

= - . (6.8)

In the case of constant "unperturbed" Green's functions G_0 and spatially homogeneous gradients $w(s)$, certain equivalences between the single and the double diagram representations can be shown, thus leading to simple covariance expressions.

Example 3

Consider the one-dimensional flow situation of Example 1 above, where the fluctuation

RF w(s) is now assumed to have infinite correlation range so that it degenerates to a random variable. The first-order approximation $\mathbf{c_J}(\tau s, \tau s')^{\#}$ was calculated and compared to the exact covariance $\mathbf{c_J}(\tau s, \tau s')$. The maximum difference $\Delta = \mathbf{c_J}(\tau s, \tau s')^{\#} - \mathbf{c_J}(\tau s, \tau s')$ which is plotted in Fig. 2 shows that the approximation $\mathbf{c_J}(\tau s, \tau s')^{\#}$ provides a very good fit to the exact covariance $\mathbf{c_J}(\tau s, \tau s')$.

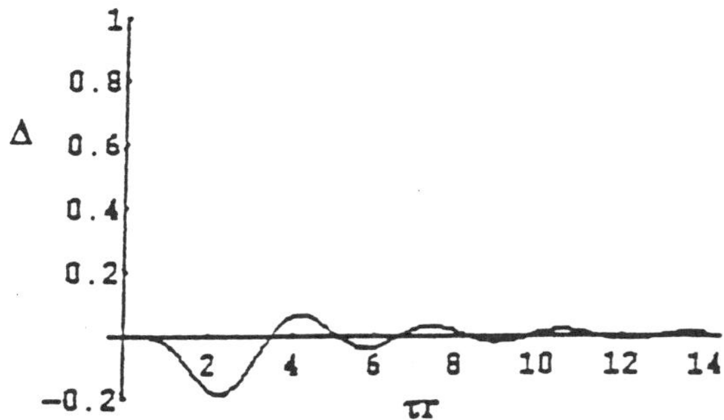

Figure 2: The maximum difference $\Delta = \mathbf{c_J}(\tau s, \tau s')^{\#} - \mathbf{c_J}(\tau s, \tau s')$ $(r = |s - s'|)$.

CONCLUSIONS

The diagrammatic approach replaces the original stochastic flow model, which may suffer from problems like closure and inadequate approximations, by a closed set of integro-differential equations (5.17) and (6.7) for $\overline{J(s)}$ and $\mathbf{c_J}(s,s')$, respectively. The solution of these equations depends on the stochastic and physical properties of the specific flow problem under consideration as they are reflected by the appropriate PMDOs, the choice of which plays an important role in the analysis.

ACKNOWLEDGEMENTS

This work has been supported by grants from the National Institute of Environmental Health Sciences (contract no. P42 ES05948-02), and the Army Research Office (contract no. DAAL03-92-G-0111).

REFERENCES

Bear, J. (1972): *Dynamics of Fluids in Porous Media*. Elsevier, Amsterdam, The Netherlands.

Christakos, G. (1992a): *Random Field Models in Earth Sciences:* Academic Press. San Diego, CA.

Christakos, G. (1992b): "A Treatise on the Stochastic Interactive Perturbationn Analysis of Groundwater Flow". Res. Notes SM/3.92, *Dept. of Envrir. Sci. and Engin.*, Univ. of North Carolina, Chapel Hill, 63 p.

Christakos, G., C.T. Miller and D. Oliver (1993a): "The development of stochastic space transformation and diagrammatic perturbation techniques in subsurface hydrology." *Stoch. Hydrol. and Hydraul.* 7(1), 14-32.

Christakos, G., C.T. Miller and D. Oliver (1993b): "Stochastic perturbation analysis of groundwater flow: Spatially variable soils, semi-infinite domain and large fluctuations."*Stoch. Hydrol. and Hydraul.* 7; in press.
Cushman, J.H. (1987): "Development of stochastic partial differential equations for subsurface hydrology". *Stochastic Hydrol. Hydraulics* 1, 241-262.
Cushman, J.H. (1990): *Dynamics of Fluids in Hierarhical Porous Media*.. Acad. Press, N.Y.
Dimitrakopoulos, R. and A.J. Desbarats (1993): "Geostatistical modeling of gridblock permeabilities for 3D redervoir simulators." *SPE Reservoir Engineering*, 13-18.
Gelhar, L.W. and C.L. Axness (1983): "Three-dimensional stochastic analysis of macrodispersion in a stratified aquifer".*Water Resources Research*, 19(1), 161-180.
Gutjahr, A.L. and L.W. Gelhar (1981): "Stochastic models of subsurface flow: Infinite vs. finite domains and stationarity."*Water Resources Research* 17(2), 337-350.
March, N.H., W.H. Young and Sampanthar, S. (1967): *The Many-Body Problem in Quantum Mechanics*. Cambridge Univ. Press, Cambridge, UK.
Neuman, S.P. and S. Orr (1993): "Prediction of steady-state flow in nonuniform geologic media by conditional moments: exact nonlocal formalism, effective conductivities, and weak approximation".*Water Resources Research* 29(2), 341-364.
Rubin, Y. (1991): "Transport in heterogeneous porous media: Prediction and uncertainty". *Water Resources Research* 27(7), 1723-1738.
Sposito, G., D.A. Barry and Z.L. Kabala (1990): "Stochastic differential equations in the theory of solute transport through inhomogeneous porous media". In *Advances in Porous Media* 2, M.Y. Corapcioglu (ed.), Elsevier, Amsterdam.
Sudicky, E.A. (1986): "A natural gradient experiment on solute transport in a sand aquifer: Spatial variability of hydraulic conductivity and its role in the dispersion process".*Water Resources Research* 22(13), 2069-2082.
Thulasiraman, K. and M.N.S. Swamy (1992): *Graphs: Theory and Applications*. J. Wiley, N.Y.
Unny, T.E. (1989): "Stochastic partial differential equations in groundwater hydrology". *Stochastic Hydrol. Hydraul.* 3(2), 135-153.
Zeitoun, D.G. and C. Braester (1991): “A Neumann expansion approach of flow through heterogeneous formations”. *Stoch. Hydrol. and Hydraul.* 5, 207-226.

GEOSTATISTICAL ANALYSIS OF WATER SATURATION IN HETEROGENEOUS UNSATURATED MEDIA UNDER CAPILLARY-GRAVITY EQUILIBRIUM

A.J. DESBARATS
Geological Survey of Canada
601 Booth st., Ottawa, Ontario, K1A 0E8
CANADA

ABSTRACT

This study investigates the spatial distribution of water content and the upscaling of capillary pressure-saturation curves in heterogeneous porous media under static conditions. It extends the work of Kocberber and Collins (1990) to randomly heterogeneous formations characterized within a geostatistical framework. Lab-scale capillary pressure-saturation curves are normalized using the Leverett "J" function and are represented by a Corey-Brooks power model. These curves, giving saturation as a function of capillary pressure, are parametrized in terms of permeability, porosity and irreducible water saturation. Simulated cross-sectional fields of these three coregionalized variables are generated using geostatistical methods. The corresponding saturation fields are then calculated from the parameter fields and the capillary pressure field, using the capillary pressure-saturation model. Under static conditions, capillary pressures at any elevation above the phreatic surface are uniform and are given by the equilibrating gravity forces. A series of numerical experiments are used to investigate the effect of parameter heterogeneity on the spatial distribution of water saturations. Results show that, in heterogeneous media, uniform capillary pressures give rise to sharp discontinuities in saturation. For each simulated saturation field, megascopic-scale capillary pressure-saturation curves are obtained by averaging saturation over horizontal layers within the field. These upscaled curves are found to reduce to the same non-dimensional form when normalized using a Leverett "J" function approach. This result indicates a promising direction for further research on upscaling of capillary pressure-saturation relationships.

INTRODUCTION

Societal concerns regarding groundwater contamination and the subsurface disposal of hazardous substances have stimulated considerable new research on fluid flow in unsaturated heterogeneous porous media (Gee et al., 1991). A particularly important aspect of this complex problem is the relationship between capillary pressure and water saturation. This relationship is required in order to close the mathematical model describing unsaturated groundwater flow (Bear, 1979) : Water saturation determines

R. Dimitrakopoulos (ed.), Geostatistics for the Next Century, 359–370.

relative permeability which, in turn, determines flow rates and capillary head. The capillary pressure-saturation relationship is a characteristic of the porous medium and is usually obtained experimentally at the macroscopic or laboratory scale. However, in numerical flow models, this relationship must be specified at the scale of grid blocks discretizing the flow field although, as yet, no simple or practical upscaling approach has been proposed. Therefore, this paper is concerned with the determination of capillary pressure curves in heterogeneous media at the larger, megascopic or formation scale. The problem is addressed within a geostatistical framework, considering the simple but enlightening case of the vertical distribution of water saturations in heterogeneous media under capillary-gravity equilibrium.

The upscaling of capillary pressure-saturation curves from the pore scale to the laboratory scale has been studied in numerous works reviewed in Ferrand and Celia (1992). In these studies, percolation models were used to describe fluid drainage or imbibition processes in the pore space and to synthesize macroscopic scale capillary pressure-saturation relationships. Kueper and McWhorter (1992) extended the percolation approach to model capillary drainage and imbibition at a larger scale, using it to calculate capillary pressure-saturation curves for heterogeneous formations discretized by nodal grids. For incremental levels of capillary pressure, fluid was allowed to percolate through the lattice until equilibrium was reached, at which point average fluid saturations in the field were recorded. In a series of numerical experiments, the authors investigated the sensitivity of megascopic capillary curves to the statistical parameters of the permeability field. In contrast to the dynamic modelling approach of Kueper and McWhorter, Kocberber and Collins (1990) used an approach based on capillary-gravity equilibrium to study the initial distribution of fluids in heterogeneous petroleum reservoirs. The physical principals behind their approach are well known (Collins, 1961; Bear, 1979) yet have not been widely applied in studies of heterogeneous porous media. These authors also used a parametric Leverett-type model to describe laboratory-scale capillary pressure-saturation relationships however they considered only simple, deterministic models of formation heterogeneity. This paper extends the approach of Kocberber and Collins (1990) to randomly heterogeneous media while incorporating a more sophisticated Leverett-type macroscopic capillary pressure-saturation model than that used by Kueper and McWhorter (1992). The goals are to investigate the effects of heterogeneity on megascopic capillary pressure-saturation curves and to assess the possibility of normalizing these curves using an extension of the Leverett "J" function approach.

The following section provides a review of basic theory. Section 3 describes a model for laboratory-scale capillary pressure-saturation curves, parametrized in terms of permeability, porosity and irreducible water saturation. The method used to simulate heterogeneous fields of these variables is described in section 4. Through the capillary pressure-saturation model, these fields are used to calculate corresponding saturation profiles of the vadose zone, as shown in the numerical experiments of section 5. Results of these experiments are discussed in section 6.

THEORY

This section reviews the equations describing the distribution of water saturation in heterogeneous porous media under capillary-gravity equilibrium. A more complete treatment of the subject can be found in Collins (1961) or Bear (1979).

When two immiscible fluids coexist within the void space of a porous medium there exists a pressure difference across the interface between the two phases. At the macroscopic scale of a core sample, this pressure difference, known as the capillary pressure, is given by the average pressure in the non-wetting fluid p_a minus $p_w$the average pressure of the fluid wetting the matrix. Here, the non-wetting fluid is air and the wetting fluid is water. Capillary pressure is a function of the interfacial tension τ between the two fluids, the pore-size distribution and the fraction of pore volume occupied by each fluid. This macroscopic relationship is expressed by a capillary pressure p_c versus saturation S_W curve obtained experimentally, for a given sample (Figure 1). Capillary pressure-saturation relationships depend strongly on pore space characteristics and therefore may be expected to exhibit significant spatial variability in natural geological media.

In a partially saturated heterogeneous formation under static conditions, the piezometric head $\phi = z + p_w/\rho g$ must be uniform throughout the unsaturated zone. Assuming that air pressure is atmospheric ($p_a \simeq 0$), it follows that the capillary pressures at two points $\mathbf{x_0}$ and $\mathbf{x_1}$ are related by :

$$p_{c1} - p_{c0} = (p_a - p_{w1}) - (p_a - p_{w0}) = \rho g\,(z_1 - z_0) \tag{1}$$

If the point $\mathbf{x_0}$ is located on the phreatic surface (where $S_W = 1$ and $p_c = 0$, by definition) and this surface is taken as the datum ($z_0 = 0$), then :

$$p_{c1} = \rho g\, z_1 \tag{2}$$

Considering an additional point $\mathbf{x_2}$ such that $z_2 = z_1 = z$, it follows from equilibrium conditions that water and capillary pressures are uniform at any given elevation z above datum :

$$-\,p_{w1} = -p_{w2} = p_{c1} = p_{c2} = \rho g z \tag{3}$$

Since p_c is a function of S_W, the water saturations at points $\mathbf{x_1}$ and $\mathbf{x_2}$ are $S_{W1} = p_{c1}^{-1}\,(\rho g z)$ and $S_{W2} = p_{c2}^{-1}\,(\rho g z)$, respectively. In a homogeneous medium, $p_{c1}(S_W) = p_{c2}(S_W)$, and water saturations are also uniform at a given elevation above the phreatic surface. However, if the capillary pressure-saturation curves vary from point to point as in the examples of Figure 1 then, in general, $S_{W1} \neq S_{W2}$. Therefore, in a heterogeneous medium, water saturation at any point $\mathbf{x}$ is a function not only of elevation z but also of the local pore properties :

$$S_W(\mathbf{x}) = p_{c(\mathbf{x})}^{-1}\,(\rho g\, z(\mathbf{x})) \tag{4}$$

In order to calculate S_W at every point, the capillary pressure-saturation relationship must be known throughout the medium. Clearly this is impractical unless a common functional form can be assumed for the local capillary curves. One possible model is described next.

CAPILLARY PRESSURE-SATURATION MODEL

Leverett (1941) showed that when suitably normalized, experimental capillary pressure data from different samples plotted on the same curve. He defined the dimensionless capillary pressure "J" function, $J(S_W^*(\mathbf{x}))$, given by :

$$J(S_W^*(\mathbf{x})) = \frac{p_c(S_W^*(\mathbf{x}))}{\tau}\,\frac{\sqrt{k(\mathbf{x})}}{\sqrt{n(\mathbf{x})}} \tag{5}$$

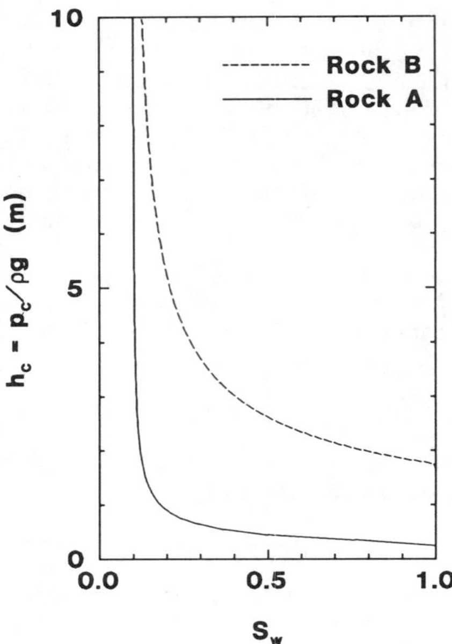

Figure 1: Typical capillary pressure-saturation curves : coarse-grained rock A; fine-grained rock B.

where $k(\mathbf{x})$ and $n(\mathbf{x})$ are permeability and porosity at $\mathbf{x}$, respectively. The quantity $\sqrt{k(\mathbf{x})}/\sqrt{n(\mathbf{x})}$ can be interpreted as a characteristic length scale of the pore space. Dimensionless capillary pressure is expressed as a function of normalized water saturation $S^*_W(\mathbf{x})$ defined by :

$$S^*_W(\mathbf{x}) = \frac{S_W(\mathbf{x}) - S_{Wi}(\mathbf{x})}{1 - S_{Wi}(\mathbf{x})} \tag{6}$$

where $S_{Wi}(\mathbf{x})$ is the irreducible water saturation at point $\mathbf{x}$.

Brooks and Corey (1964) suggest the following model for the drainage capillary pressure-saturation relationship :

$$\begin{aligned} S^*_W(\mathbf{x}) &= \left(\frac{p_c}{p_e}\right)^{-b} = \left(\frac{J(S^*_W)}{p_{eD}}\right)^{-b} && \text{if } p_c > p_e \\ &= 1.0 && \text{otherwise} \end{aligned} \tag{7}$$

where the pore entry pressure p_e is defined as the limit of p_c as S_W tends towards 1.0 and p_{eD} is its dimensionless form according to (5). Rose and Bruce (1949) have shown that p_{eD} is related to the tortuosity t of the pore space by $p_{eD} = 1/\sqrt{t}$. In granular media, p_{eD} is of the order of 0.2 whereas in fractured media p_{eD} is in the range of 0.6. The empirical parameter b in (7) is an index reflecting the distribution of pore sizes and is assumed equal to 2.0 unless otherwise stated.

Finally, by combining (5), (6) and (7), an explicit model is obtained for (4) :

$$\begin{aligned} S_W(\mathbf{x}) &= S_{Wi}(\mathbf{x}) + (1 - S_{Wi}(\mathbf{x})) \left(\frac{p_{eD}\,\tau}{\rho\, g z(\mathbf{x})}\right)^b \left(\frac{k(\mathbf{x})}{n(\mathbf{x})}\right)^{-b/2} && \text{if } p_c > p_e \\ &= 1.0 && \text{otherwise} \end{aligned} \tag{8}$$

This relationship gives water saturation at any point $\mathbf{x}$ in terms of $z(\mathbf{x})$, the elevation above the phreatic surface, and three petrophysical variables $k(\mathbf{x})$, $n(\mathbf{x})$ and $S_{Wi}(\mathbf{x})$. In a geological medium, these quantities vary in space and are, in addition, correlated amongst themselves. The following section describes the method used to simulate heterogeneous fields of permeability, porosity and irreducible water saturation from which corresponding fields of water saturation can be derived using (8).

SIMULATION OF COREGIONALIZED VARIABLES

This section describes the geostatistical method used to simulate auto- and cross-correlated fields of permeability, porosity and irreducible water saturation discretized on a grid. The simulation of coregionalized variables is discussed in Journel and Huijbregts (1978) and Luster (1985).

Typically, permeability $k(\mathbf{x})$ and porosity $n(\mathbf{x})$ are positively correlated (Collins, 1961) whereas permeability and irreducible water saturation $S_{Wi}(\mathbf{x})$ are usually negatively correlated (Wyllie and Rose, 1950; Collins, 1961). Therefore, in addition to their respective spatial autocorrelation structures, these variables also exhibit cross-correlation. For simplicity, a linear, intrinsic model of coregionalization (Journel and Huijbregts, 1978) is assumed to represent auto- and cross-correlation relationships amongst the three variables. Because permeability is often lognormal and porosity and irreducible water saturation are bounded, the following transformations are applied to the original variables so that they may be simulated using a multivariate Gaussian approach :

$$\begin{aligned} Z_1 &= \ln k(\mathbf{x}) \\ Z_2 &= \ln\left(\frac{1 - n(\mathbf{x})}{n(\mathbf{x})}\right) \\ Z_3 &= \ln\left(\frac{1 - S_{Wi}(\mathbf{x})}{S_{Wi}(\mathbf{x})}\right) \end{aligned} \tag{9}$$

These variables are further standardized to mean 0 and variance 1.0 according to $Z_i' = (Z_i - m_i)/\sigma_i$ where m_i and σ_i are the specified mean and standard deviation for variable Z_i, respectively . The variables Z_i' are then standard Normal, with positive-definite correlation matrix :

$$[C(h)] = C_0(h)\ [C] = C_0(h) \begin{bmatrix} 1.0 & c_{12} & c_{13} \\ c_{12} & 1.0 & c_{23} \\ c_{13} & c_{23} & 1.0 \end{bmatrix} \tag{10}$$

All auto- and cross-covariances are proportional to the same basic spatial covariance $C_0(h)$ described by a single structure, exponential model with principal integral ranges λ_X, λ_Y and λ_Z in the three coordinate directions, respectively. The three variables Z_i' are obtained from three independent standard Gaussian variables Y_i, each with autocovariance $C_0(h)$, according to the linear transformations :

$$\begin{aligned} Z_1' &= a_{11}Y_1 \\ Z_2' &= a_{12}Y_1 + a_{22}Y_2 \\ Z_3' &= a_{13}Y_1 + a_{23}Y_2 + a_{33}Y_3 \end{aligned} \tag{11}$$

where the coefficients a_{ij} ensure that the covariance matrix (10) is reproduced :

$$[Z'].[Z']^T = [A].[Y].[Y]^T.[A]^T = [A].[I].[A]^T = [C] \tag{12}$$

Realizations of the independent Gaussian variables Y_i are generated using the turning bands method (Journel and Huijbregts, 1978; Luster, 1985) although other methods could also be used.

RESULTS

This section presents the results of a series of numerical experiments investigating the effect of heterogeneity on megascopic capillary pressure-saturation relationships.

Using the techniques described above, water saturations are calculated over discretized two-dimensional fields representing vertical cross-sections of the unsaturated zone. For each layer S parallel to the phreatic surface, upscaled porosities $n(S)$, and water saturations $S_W(S)$ and $S_{Wi}(S)$ are calculated by spatial averaging over all points $\mathbf{x}$ within S such that $z(\mathbf{x}) = z$:

$$\begin{aligned} n(S) &= \frac{1}{S}\int_S n(\mathbf{x})\,d\mathbf{x} \\ S_W(S) &= \frac{1}{S}\int_S n(\mathbf{x})S_W(\mathbf{x})\,d\mathbf{x}\ /\ n(S) \\ S_{Wi}(S) &= \frac{1}{S}\int_S n(\mathbf{x})S_{Wi}(\mathbf{x})\,d\mathbf{x}\ /\ n(S) \end{aligned} \tag{13}$$

Under static conditions, the capillary pressure head h_c for each layer is given by $h_c = p_c/\rho g = z$ where z is the elevation above the phreatic surface. Therefore, vertical profiles of layer-averaged water saturations can be used as a convenient representation of megascopic capillary pressure-saturation relationships. This is illustrated in Figure 2 which shows a digital image of water saturations across the unsaturated zone, and the corresponding profile of layer-averaged values.

Basic parameters used in the following numerical experiments are summarized in Table 1. In individual experiments a single parameter is varied while all others are held constant.

Figure 3 shows the effect of varying α, the mean log permeability. Lower permeability rocks exhibit a much gentler decrease in saturation with increasing h_c than high permeability rocks.

Figure 4 shows the effect of σ^2, the log variance of permeability. Somewhat surprisingly, this effect seems quite small. These results are clarified by the results of

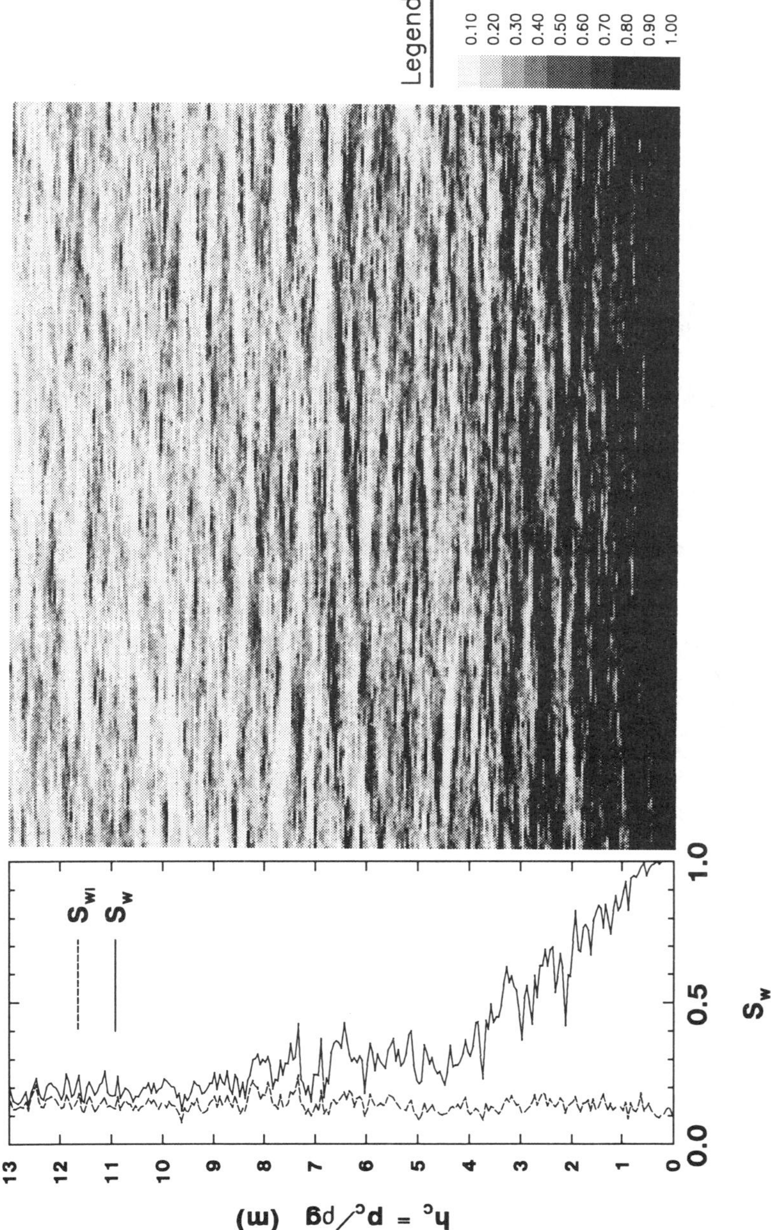

Figure 2: Digital image of simulated water saturations in a heterogeneous medium with corresponding average saturation profile.

Parameter	Value
$N_X \times N_Y \times N_Z$	$1000 \times 1 \times 200$
$\Delta_X \times \Delta_Y \times \Delta_Z$	$1 \times 1 \times 0.1$ (m)
$\lambda_X \times \lambda_Y \times \lambda_Z$	$20 \times 20 \times 0.2$ (m)
α	-30.0
σ^2	3.0
$\overline{n}$	0.15
$\overline{S_{Wi}}$	0.10
c_{12}	-0.60
c_{13}	0.50
c_{23}	0.30
p_{eD}	0.20
τ	0.045 N/m

Table 1: Parameters for base case numerical experiments.

an additional experiment shown in Figure 5. Here, the quantity $(\exp(\alpha+\sigma^2/2))$ is held constant while α and σ^2 are varied. These results show that with mean permeability held constant, permeability variance does indeed affect megascopic curves. In other experiments (results not shown), it was found that the anisotropy ratio λ_X/λ_Z had no significant effect on megascopic curves.

Figure 6 investigates the effect of the relationship between $k(\mathbf{x})$, $n(\mathbf{x})$ and $S_{Wi}(\mathbf{x})$. Curve C (fixed values of $n(\mathbf{x})$ and $S_{Wi}(\mathbf{x})$) is closer to Curve B (Base case) than Curve A ($k(\mathbf{x})$, $n(\mathbf{x})$ and $S_{Wi}(\mathbf{x})$ uncorrelated). Differences between these curves are largest in the critical region where saturations approach irreducible levels. Other experiments (results not shown) in which the c_{ij} of (10) were varied, showed similar differences in the same transition region of the curves.

The previous results show that megascopic capillary pressure-saturation relationships depend strongly on the statistical paramaters of permeability, and those of porosity and irreducible water saturation. If these megascopic relationships are to be allowed to vary from block to block in numerical models of unsaturated flow, practical considerations dictate that they share some common functional form, parametrized in terms of a small number of variables. Following the approach of Leverett (1941) used at the macroscopic scale in section 3 above, a dimensionless capillary pressure is defined at the megascopic scale of a horizon S :

$$J_S(S_W^*(S)) = p_c \, \frac{d(S)}{\tau} \tag{14}$$

where $p_c = \rho g z$ and $d(S) = (k(S)/n(S))^{0.5}$ represents a characteristic length scale for the pore space in layer S, yet to be defined. The upscaled normalized water

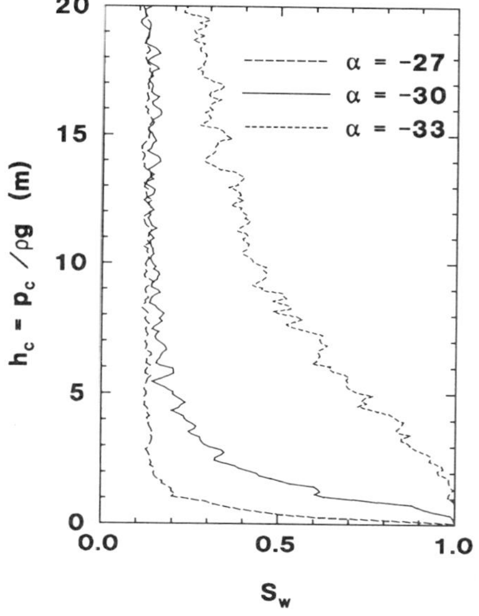

Figure 3: Effect of mean log permeability.

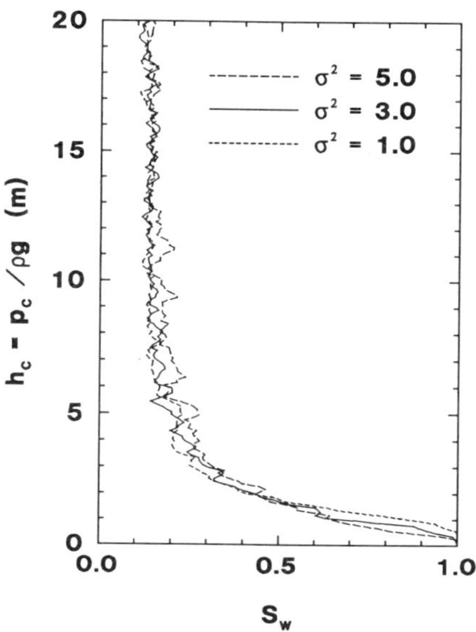

Figure 4: Effect of log permeability variance.

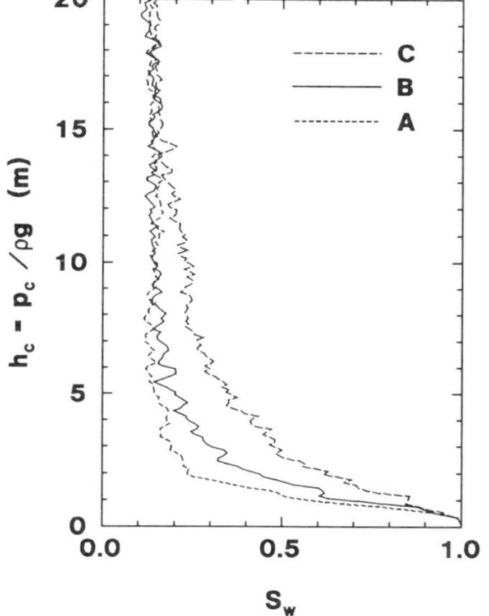

Figure 5: Curves for combinations (α, σ^2) : curve A (-29, 1.0) ; curve B (-30, 3.0) ; curve C (-31, 5.0).

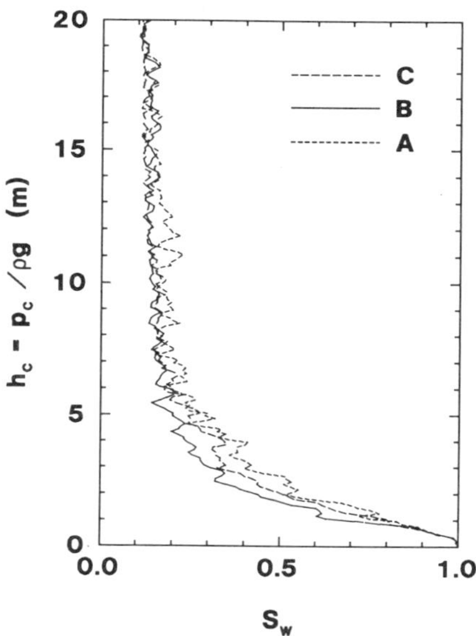

Figure 6: Effect of $n(\mathbf{x})$ and $S_{Wi}(\mathbf{x})$: curve A (uncorrelated); curve B (base case); curve C ($n(\mathbf{x})$ and $S_{Wi}(\mathbf{x})$ fixed at mean values).

saturation $S^*_W(S)$ is defined by analogy with (6) as :

$$S^*_W(S) = \frac{S_W(S) - S_{Wi}(S)}{1 - S_{Wi}(S)} \tag{15}$$

Inspection of (8) suggests that the quantity $k(S)/n(S)$ may be defined by the following spatial power average over horizon S :

$$\left(\frac{k(S)}{n(S)}\right)^{-b/2} = \frac{1}{S_f} \int_{S_f} \left(\frac{k(\mathbf{x})}{n(\mathbf{x})}\right)^{-b/2} d\mathbf{x} \tag{16}$$

In this equation, S_f is the set of points $\mathbf{x}$ within layer S invaded by the non-wetting fluid, i.e. such that $p_c > p_e$ or, alternatively, such that :

$$\frac{k(\mathbf{x})}{n(\mathbf{x})} > \left(\frac{\tau\, p_{eD}}{\rho g}\right)^2 \frac{1}{z^2(\mathbf{x})} \tag{17}$$

Thus, the subregion S_f depends on a threshold, function of elevation z, and therefore also depends on the probability density function of $k(\mathbf{x})/n(\mathbf{x})$ in layer S.

Figure 7 a) shows capillary pressure-saturation curves for three combinations of the pair $(\alpha,\ \sigma^2)$. When capillary pressures and saturations for each layer are normalized according to (14) and (15) using (16), the results shown in Figure 7 b) are obtained. This figure also shows the non-dimensional curve for point-scale values given by (7). Figure 7 c) shows the fraction $f = S_f/S$ of layer S invaded by the non-wetting fluid at different elevations above the phreatic surface.

DISCUSSION AND CONCLUSIONS

The results of Figures 3-6 support the earlier work of Kueper and McWhorter (1992) demonstrating the dominant effect of permeability statistics on megascopic capillary pressure-saturation curves. To a large extent this is a consequence of the model (5) for point-scale curves and the fact that the permeability coefficient of variation is typically 2 to 3 orders of magnitude greater than that of $n(\mathbf{x})$ or $S_{Wi}(\mathbf{x})$.

While $k(\mathbf{x})$ and $n(\mathbf{x})$ determine the non-dimensional scaling of capillary pressure curves, $S_{Wi}(\mathbf{x})$ and the dimensionless pore entry pressure p_{eD} are important end points on these curves. Here, p_{eD} was set to 0.2 mainly because there is very little published information on its spatial variability or its relationship to the other petrophysical variables. It does, however, have an important effect on the overall shape of capillary curves and the fraction f of layer S invaded by the non-wetting fluid, as shown by (17). On the other hand, this paper does incorporate the spatial variability of $S_{Wi}(\mathbf{x})$ and its cross-correlation with $k(\mathbf{x})$ and $n(\mathbf{x})$. The effect of $n(\mathbf{x})$ and $S_{Wi}(\mathbf{x})$ variability on megascopic capillary pressure curves is not as dramatic as that of $k(\mathbf{x})$. However, it may be significant for saturation values approaching irreducible levels.

In light of the preceding results, the relatively sophisticated model for $k(\mathbf{x})$, $n(\mathbf{x})$ and $S_{Wi}(\mathbf{x})$ variability adopted here may be simplified somewhat by simulating $k(\mathbf{x})$ alone, and either fixing $n(\mathbf{x})$ and $S_{Wi}(\mathbf{x})$, or making them deterministic functions of $k(\mathbf{x})$ as in Kocberber and Collins (1990). A poor choice would be to assume that

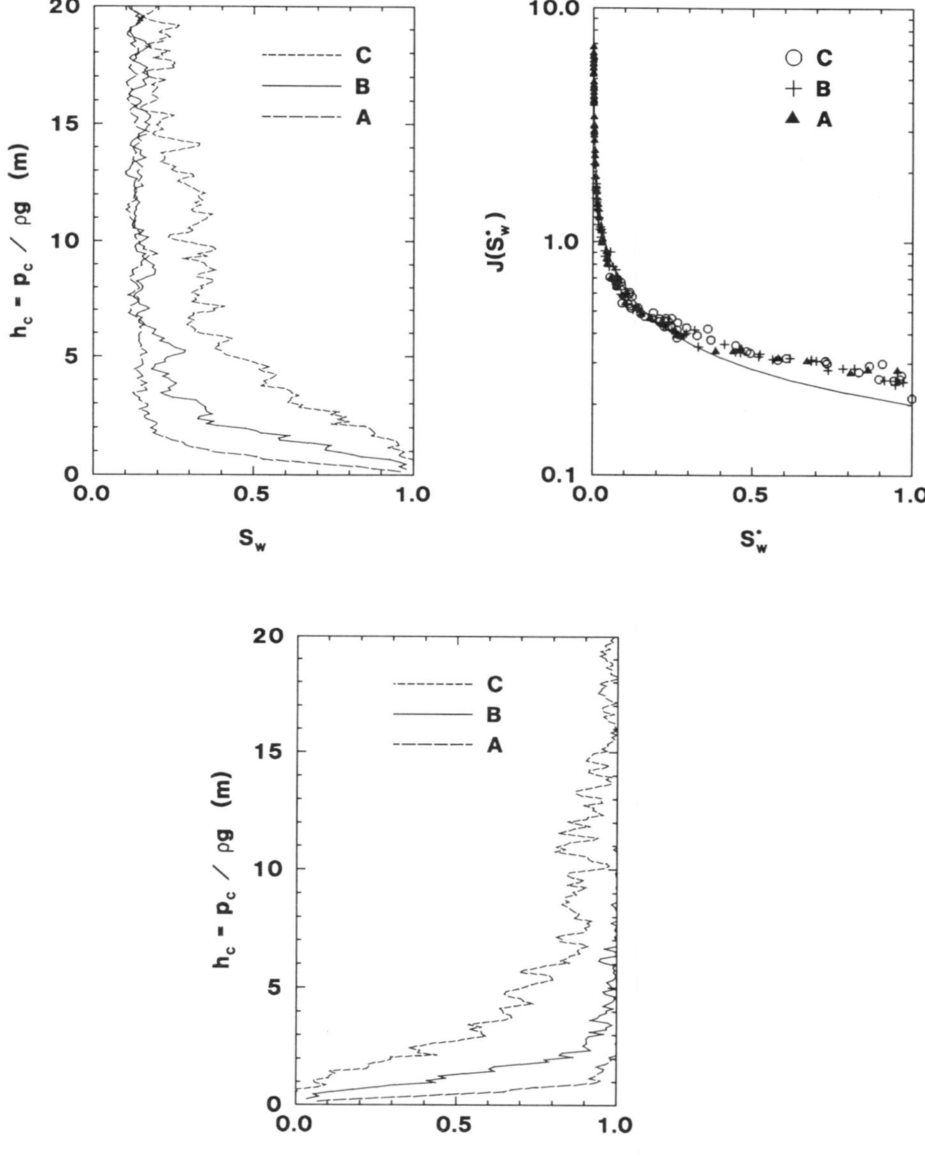

Figure 7: Standardization experiment : a) original curves; b) normalized results; c) invaded fraction.

the three petrophysical variables are uncorrelated as this could give rise to unrealistic combinations of values.

The results shown in Figure 7 b) clearly demonstrate that widely different capillary pressure curves can be reduced to the same non-dimensional form using an appropriate scaling process based on the Leverett "J" function approach. Unfortunately, the non-dimensional form for megascopic curves does not coincide with the non-dimensional form for point curves and it has yet to be determined. Although the scaling process is successful, it is not simple since it requires knowledge of the permeability *pdf* within the averaging region rather than a single average value.

Unavoidably, the results of this paper depend on the model assumed for point-scale capillary pressure-saturation relationships. Other models exist and could also be investigated. Most importantly however, the simulation results presented here should be compared against field data. This paper has explored some avenues that may be fruitful in such future research.

ACKNOWLEDGMENTS

The author wishes to thank F.P. Agterberg for reviewing the original manuscript. This paper is dedicated to Michel David, who introduced me to Geostatistics back in 1978. Geological Survey of Canada contribution no. 13793.

REFERENCES

Bear, J. (1979) Hydraulics of Groundwater, McGraw-Hill, New York.

Brooks, R.H. and A.T. Corey (1964) Hydraulic Properties of Porous Media, Hydrol. Pap. 3, Colo. State Univ., Ft. Collins.

Collins, R.E. (1961) Flow of Fluids Through Porous Materials, Van Nostrand- Reinhold, New York.

Ferrand, L.A. and M.A. Celia (1992) "The effect of heterogeneity on the drainage capillary pressure-saturation relation", Water Resour. Res., 28 (3), 859-870.

Gee, G.W., C.T. Kincaid, R.J. Lenhard and C.S. Simmons (1991) "Recent studies of flow and transport in the vadose zone", in US Nat. Rep. to Int. Union of Geodesy and Geophysics, Contributions in Hydrology, Reviews of Geophysics, Suppl., 227-239.

Journel, A.G. and C. Huijbregts (1978) Mining Geostatistics, Academic, London.

Kocberber, S. and R.E. Collins (1990) "Impact of reservoir heterogeneity on initial distributions of hydrocarbons", SPE paper 20547, presented at the 65th Ann. Tech. Conf., New Orleans LA, Sept. 23-26.

Kueper, B.H. and D.B. McWhorter (1992) "The use of macroscopic percolation theory to construct large-scale capillary pressure curves", Water Resour. Res., 28(9), 2425-2436.

Leverett, M.C. (1941) "Capillary behavior in porous solids", Trans. Am. Inst. Min. Metall. Pet. Eng., 142, 152-169.

Luster, G.R. (1985) Raw Materials for Portland Cement : Applications of Conditional Simulation of Coregionalization, Ph.D. dissert., Stanford Univ., Stanford CA.

Rose, W.D. and W.A. Bruce (1949) "Evaluation of capillary character in petroleum reservoir rock", AIME Pet. Trans, 186, 127-142.

Wyllie, M.R.J. and W.D. Rose (1950) "Some theoretical considerations related to the quantitative evaluation of the physical characteristics of reservoir rocks from electrical log data", AIME Pet. Trans., 189, 105-118.

ADDITIVE LOGRATIO ESTIMATION OF REGIONALIZED COMPOSITIONAL DATA: AN APPLICATION TO CALCULATION OF OIL RESERVES

V. PAWLOWSKY
Dept. de Matemàtica Aplicada III
E.T.S. d'Enginyers de Camins, Canals i Ports, Universitat Politècnica de Catalunya
Gran Capitán s/n, E-08034 Barcelona
Spain

R.A. OLEA and J.C. DAVIS
Kansas Geological Survey
The University of Kansas
1930 Constant Avenue, Lawrence, Kansas 66047
U.S.A.

To jointly estimate values of a coregionalization with spatially correlated components either cokriging or separate kriging of each component can be used. Cokriging is an attractive alternative, because it allows integration of the available information. However, when data are subject to a constant sum constraint at every point of a sampling region $\mathcal{F}$, they constitute a sample of a regionalized composition. In this circumstance, there are problems with the application of cokriging: because of the constant sum constraint the spatial covariance structure is affected by spurious spatial correlation and may be misinterpreted. In addition, the cross-covariance matrices of the regionalized composition are singular for each distance h, as is the coefficient matrix of the cokriging system of equations. This prevents use of cokriging as a method of estimating regionalized compositions.

The additive logratio transformation of Aitchison (1986) can be used to obtain a suitable spatial covariance structure for regionalized compositions. It also will allow the use of cokriging, and will permit statistical analyses of compositional data.

Here, this transformation is used for the first time to solve a common problem in petroleum geology and engineering—estimating the initial volume of oil in place at every location within a reservoir. The example uses measurements of porosity, water saturation, and thickness of the oil column in the Lyons West oil field in Kansas, U.S.A.

INTRODUCTION

Often in petroleum geology and engineering it is necessary to estimate the initial volume of oil in place at every location x within a reservoir $\mathcal{F}$. For this purpose the spatial distributions of properties such as porosity, water saturation, and thickness of the oil column have to be estimated. From these, the initial oil in place in the reservoir can be calculated directly by

$$R = 10^{-4} \int_{\mathcal{F}} \phi(x) S_o(x) T(x) dx \,,$$

R. Dimitrakopoulos (ed.), Geostatistics for the Next Century, 371–382.

where $T(x)$ is the thickness in feet of the oil-saturated interval within the reservoir at any location x inside the horizontal projection of the field, $\mathcal{F}$; $\phi(x)$ is the average reservoir porosity over the saturated thickness $T(x)$, in percent of the rock volume; and $S_o(x)$ is the average oil saturation, or initial average proportion of hydrocarbon over the saturated thickness $T(x)$, in percent of the pore volume. Core analyses typically report the water saturation, $S_w(x)$, which is the complement of $S_o(x)$. Traditionally, each of these variables is estimated separately by conventional methods, such as ordinary kriging, and then are combined to obtain the desired estimate R^*.

These variables must satisfy several restrictions: saturated thickness $T(x)$ must be positive, and both average porosity $\phi(x)$ and average oil saturation $S_o(x)$ must lie between 0% and 100%. Using kriging there is no guarantee that these restrictions will be satisfied. In addition, the distributional assumptions of normality that underly kriging as an estimation method are difficult to justify for strictly positive variables, and even more difficult to justify for variables that are restricted to an interval. Recall that the sample space associated with a random variable that follows a normal distribution is the real line $\mathbb{R}$. This means that, at least hypothetically, the random variable can assume any value between $-\infty$ and $+\infty$.

To improve the estimation of initial oil in place in a reservoir, the direct use of cokriging is not advisible, because it is based on an analogous hypothesis. However, if percent volume of oil

$$V_o(x) = \frac{(100 - S_w(x))\phi(x)}{100},$$

percent volume of water

$$V_w(x) = \frac{S_w(x)\phi(x)}{100}$$

and percent volume of rock

$$V_r(x) = 100 - \phi(x)$$

are considered, a regionalized composition,

$$\mathbf{V}(x) = (V_o(x), V_w(x), V_r(x)),$$

is obtained. If the reservoir contains a gas component, an additional term can be added easily to account for its presence.

Although a regionalized composition also cannot be estimated directly using cokriging—because the matrix of weights in the cokriging system of equations is necessarily singular as a consequence of the constant sum constraint—it is possible to use the additive logratio approach to obtain suitable estimates. This method allows for the integration of all available information and, at the same time, takes into account the interval restrictions.

BASIC CONCEPTS AND PROPERTIES

Regionalized composition

A *regionalized composition* (Pawlowsky, 1984) is by definition both a *composition* (Aitchison, 1986) or *closed system* (Chayes, 1960, 1971)) and a *coregionalization* (Matheron, 1965, 1971; Journel and Huijbregts, 1978; Olea, 1991). At each location in the reservoir $\mathcal{F}$, the regionalized composition is a random vector whose components add to a constant and also are a set of two or more regionalized variables defined over the same spatial domain. Here, the term *regionalized composition* is used both for the vector random function which constitutes a composition and for the realization which is modeled.

Formally, a *regionalized composition* (r-composition for short) is defined (Pawlowsky, 1986, 1989, 1991; Pawlowsky and Burger, 1992) as a vector random function that satisfies at every point the following conditions:

i) all components are strictly positive; and

ii) the sum of all components is a constant.

In the particular case discussed here, $\mathbf{V}(x)$, $x \in \mathcal{F}$, clearly satisfies both conditions:

i) the three components $V_o(x)$, $V_w(x)$ and $V_r(x)$ are strictly positive inside the reservoir $\mathcal{F}$; and

ii) the sum of all three components at every point inside the reservoir is 100%, and therefore a constant:

$$V_o(x) + V_w(x) + V_r(x) = 100\%\,, \qquad x \in \mathcal{F}\,.$$

Consequently, $\mathbf{V}(x)$ is an r-composition.

Remarks:

1. There are no zero values for any component inside the reservoir; if zero values occur, they must be excluded in order to avoid unnecessarily complicating the presentation. For the manipulation of data sets in which null values occur, see Aitchison (1986).
2. Indicator kriging has been used to assess the reservoir boundaries, as described in Pawlowsky, Olea, and Davis (1991, 1992, 1993).

Since early work by Pearson (1897) it has been known that—as a consequence of the constant sum constraint—statistical analysis of compositions is difficult because the covariances are subject to essential nonstochastic controls. Numerically induced covariances and correlations also arise within r-compositions, and are indicated as spurious spatial correlations (Pawlowsky, 1984, 1986, 1989). They falsify the picture of the spatial covariance structure and may lead to misinterpretations. Furthermore, it is not possible to use standard techniques such as cokriging for estimation purposes. This is because the variance-covariance matrices are necessarily singular and, consequently, the matrices of weights in the resulting systems of equations also are singular.

The additive logratio transform

One way to avoid these difficulties is to transform the r-composition. Aitchison (1986) suggests use of the additive logratio transformation, alr(.).

For an r-composition with D components

$$\mathbf{y}(x) = (\,y_1(x) \quad y_2(x) \quad \cdots \quad y_d(x) \quad y_D(x)\,)'\,,$$

where the prime indicates transposition, the additive logratio transform is defined as (Pawlowsky, 1986, 1989):

$$\begin{aligned} \text{alr} : \mathcal{S}^d &\longrightarrow \mathbb{R}^d \\ \mathbf{y}(x) &\longmapsto \mathbf{z}(x) = \text{alr}(\mathbf{y}(x)) = \mathbf{F}\ln(\mathbf{y}(x))'\,, \end{aligned}$$

where $d = D - 1$, and

$$\mathbf{F} = \begin{pmatrix} 1 & 0 & \cdots & 0 & -1 \\ 0 & 1 & \cdots & 0 & -1 \\ \vdots & \vdots & \ddots & \vdots & \vdots \\ 0 & 0 & \cdots & 1 & -1 \end{pmatrix},$$

where $\mathbf{F}$ is a $d \times D$ matrix. Note for each component of $\mathbf{z}(x)$ that $z_i(x) = \ln(y_i(x)/y_D(x))$.

The inverse transformation of alr(.) is called the *additive generalized logistic transformation* and is denoted by agl(.). In the regionalized case this is written:

$$\begin{aligned}
\text{agl} : \mathbb{R}^d &\longrightarrow \mathcal{S}^d \\
\mathbf{z}(x) \longmapsto \mathbf{y}(x) &= \text{agl}(\mathbf{z}(x)) \\
&= \frac{\exp(\mathbf{z}(x))}{1 + \sum_{i=1}^{d} \exp(z_i(x))} .
\end{aligned}$$

Consequently, for each component of $\mathbf{y}(x)$:

$$\begin{aligned}
y_i(x) &= \frac{\exp(z_i(x))}{1 + \sum_{j=1}^{d} \exp(z_j(x))}, \qquad i = 1, 2, \ldots, d; \\
y_D(x) &= \frac{1}{1 + \sum_{i=1}^{d} \exp(z_i(x))} .
\end{aligned}$$

In the particular case of the r-composition $\mathbf{V}(x)$,

$$\begin{aligned}
\mathbf{W}(x) = \text{alr}(\mathbf{V}(x)) &= \begin{pmatrix} 1 & 0 & -1 \\ 0 & 1 & -1 \end{pmatrix} \begin{pmatrix} \ln(V_o(x)) \\ \ln(V_w(x)) \\ \ln(V_r(x)) \end{pmatrix} \\
&= \begin{pmatrix} \ln(V_o(x)) - \ln(V_r(x)) \\ \ln(V_w(x)) - \ln(V_r(x)) \end{pmatrix} \\
&= \begin{pmatrix} \ln\left(\frac{V_o(x)}{V_r(x)}\right) \\ \ln\left(\frac{V_w(x)}{V_r(x)}\right) \end{pmatrix} = \begin{pmatrix} W_1(x) \\ W_2(x) \end{pmatrix} ,
\end{aligned}$$

and $\mathbf{V}(x) = \text{agl}(\mathbf{W}(x))$ with

$$\begin{aligned}
V_o(x) &= \frac{\exp(W_1(x))}{1 + \exp(W_1(x)) + \exp(W_2(x))} ; \\
V_w(x) &= \frac{\exp(W_2(x))}{1 + \exp(W_1(x)) + \exp(W_2(x))} ; \\
V_r(x) &= \frac{1}{1 + \exp(W_1(x)) + \exp(W_2(x))} .
\end{aligned}$$

The components of the resulting coregionalization are unrestricted, and not constrained to be positive or to fall within a specific interval. Thus, standard techniques such as cokriging can be applied for analysis and estimation of the spatial structure. The description of cokriging methods can be found in Journel and Huijbregts (1978), Myers (1982), Pawlowsky (1986, 1991) and Samper and Carrera (1990).

As in other instances, the direct use of cokriging to estimate $\mathbf{W}(x)$ is advisible if there are no reasons to suppose that the deviation of the variables from multivariate normality is statistically significant. In the multivariate case, the simple cokriging estimator is "best" in the sense that it coincides with the conditional expectation (Pawlowsky, 1986; see Journel and Huijbregts, 1978, and Samper and Carrera, 1990, for the univariate case. The multivariate case is an extension.)

It is clear that multivariate normality as required for cokriging is impossible to verify, especially since only one partial realization of the coregionalization is available. Only clear deviations from multivariate normality may be detected. Consequently, if there are indications of deviations from multivariate normality it may be advisible to use some other transformation than the one suggested here.

If the resulting coregionalization—after applying the alr(.) transformation—has independent components, kriging may be used instead of cokriging.

The additive log-ratio approach used here to estimate the initial volume of oil in place includes the following steps: (i) compute the r-composition $\mathbf{V}(x)$ from measured variables $S_w(x)$ and $\phi(x)$ at each location x of the field $\mathcal{F}$ using the equations:

$$V_o(x) = \frac{(100 - S_w(x))\phi(x)}{100}, \quad V_w(x) = \frac{S_w(x)\phi(x)}{100}, \quad \text{and} \quad V_r(x) = 100 - \phi(x),$$

to obtain $\mathbf{V}(x) = (V_o(x), V_w(x), V_r(x))$;

(ii) compute the coregionalization $\mathbf{W}(x) = (W_1(x)\, W_2(x))'$ as:

$$\ln\left(\frac{V_o(x)}{V_r(x)}\right) = W_1(x), \quad \ln\left(\frac{V_w(x)}{V_r(x)}\right) = W_2(x);$$

(iii) determine the spatial structure of $\mathbf{W}(x)$; (iv)) perform cokriging of $\mathbf{W}(x)$ or kriging if the components are independent; (v) apply the back-transformation agl(.) to obtain estimates $V_o^*(x)$, $V_w^*(x)$ and $V_r^*(x)$:

$$V_o^*(x) = \frac{100\exp(W_1^*(x))}{1+\exp(W_1^*(x))+\exp(W_2^*(x))};$$
$$V_w^*(x) = \frac{100\exp(W_2^*(x))}{1+\exp(W_1^*(x))+\exp(W_2^*(x))};$$
$$V_r^*(x) = \frac{100}{1+\exp(W_1^*(x))+\exp(W_2^*(x))};$$

(vi) estimate $T(x)$ using kriging; (vii) use $R^* = \int_{\mathcal{F}} V_o^*(x)T^*(x)dx$ to obtain a global estimate of initial oil in place.

Advantages of using $\mathbf{W}(x)$ instead of the original variables are:

i) It is not necessary to consider the numerical constraints. Consequently, standard methods such as cokriging can be applied directly, especially if there is no evidence of strong departure from multivariate normality.
ii) All available information is integrated in the estimation process, leading—at least from a theoretical point of view—to better estimates.

Disadvantages that may be important include:

i) Spatial structure and estimation errors of the original variables are not computed and cannot be derived from the results.
ii) Back-transforming estimated components of the coregionalization to estimate other variables such as the components of the r-composition $\mathbf{V}(x)$ does not yield estimators whose properties, including unbiasedness and minimum variance, are known.
iii) Interpretation of results must be based on variables which are not always familiar, and consequently the meaning of the results may not be clear to the user.

THE LYONS WEST OIL FIELD: A CASE STUDY

The Lyons West oil field is located at 98°15′ west longitude and 38°20′ north latitude in west-central Kansas, near the center of the United States. The reservoir occurs in Mississippian (Lower Carboniferous) rocks that were deposited in the shallow epeiric seas that covered much of North America in the late Paleozoic. The field was discovered somewhat accidentally in 1963, during the drilling of a deeper Ordovician

Table 1: Basic statistics of original variables, components of the r-composition, and components of the coregionalization (number of samples = 76).

Variable	Mean	Std. dev.
T, ft	15.6	8.5
ϕ, %	11.9	1.7
S_w, %	32.0	5.7
S_o, %	68.0	5.7
V_w, %	3.8	0.7
V_o, %	8.1	1.6
V_r, %	88.1	1.7
W_1	–2.40	0.20
W_2	–3.17	0.19

prospect. The main area of the field, measuring approximately 5 miles long by 2 miles wide, was estimated by decline curve analysis to initially contain 22 million stock tank barrels of oil (MMSTB).

The reservoir is composed of carbonate-cemented sands; its genesis is interpreted as an offshore bar enclosed in marine shales. Regional uplift tilted the sand body, which was truncated along the western margin by the unconformity marking the base of the Pennsylvanian (Upper Carboniferous). The sands interfinger with marine shales to the east, but the eastern margin of the reservoir is established by the intersection of the oil-water contact with the shale seal at the top of the reservoir interval (Ehm, 1965).

Lyons West field data

As part of an engineering study of the field by Kewanee Oil Company, data were collected on core samples from 76 production wells in the Lyons West field, and from 34 additional dry holes drilled outside the margins of the field. These data were used by Sampson (1988) to demonstrate grid manipulation in contour mapping.

Core analyses typically report $S_w(x)$, the water saturation; $T(x)$, the saturated thickness; and $\phi(x)$, average reservoir porosity over $T(x)$ at different locations x. These variables are known only where wells have been drilled inside the field $\mathcal{F}$.

The location of the margin of the reservoir is known only approximately because of the wide spacing between holes drilled outside the productive area. As a consequence, there will be uncertainties in the calculated volume. Indicator kriging has been used, as decribed in Pawlowsky, Olea, and Davis (1991, 1992, 1993), to estimate the location of the field boundary.

Basic statistics

Basic statistics, computed on variables measured inside the field $\mathcal{F}$ as well as on the components of the r-composition $\mathbf{V}(x)$ and on the coregionalization $\mathbf{W}(x)$ are given in Table 1. Covariance and correlation matrices are given in Table 2.

Kolmogorov-Smirnov tests indicate that the oil-saturated thickness T and water saturation S_w can be considered univariate normally distributed. The univariate distribution of porosity ϕ is not so clearly normal, as it has a 2-tailed probability $\alpha = 0.013$ for the Kolmogorov-Smirnov test. However, neither S_w nor ϕ can be really normally distributed, as both variables are restricted to an interval. T also is subject

Table 2: Covariance (upper triangles) and correlation (lower triangles) matrices of the original variables (top), r-composition (lower left), and coregionalization (lower right).

	T	ϕ	S_o	S_w
T		4.97	3.56	−3.56
ϕ	0.34		3.77	−3.77
S_o	0.07	−0.38		−32.39
S_w	−0.07	−0.38	−1	

	V_o	V_w	V_r
V_o		-0.04×10^{-4}	-2.58×10^{-4}
V_w	−0.04		-0.43×10^{-4}
V_r	−0.92	−0.36	

	W_1	W_2
W_1		0.003
W_2	0.081	

Table 3: Semivariograms for the coregionalization $\mathbf{W}(x)$ and the sum of the components.

Variable	Model	Sill	Range (miles)
T	exp.	71 ft^2	1
W_1	exp.	0.042	2.3
W_2	exp.	0.036	0.95
$W_1 + W_2$	exp.	0.078	1.4

to the restriction that it must be positive. Similar arguments can be made that components of the r-composition cannot be strictly normal. This is not true for the coregionalization $\mathbf{W}$, the α-values of which are 0.144 for the first component, W_1, and 0.919 for the second component, W_2.

Spatial structure of $\mathbf{W}(x)$

Experimental semivariograms and cross-semivariograms for coregionalization $\mathbf{W}(x)$ were calculated and modeled, and cross-validation was performed. No anisotropy nor asymmetry were detected in the directional experimental semivariograms and cross-semivariograms. Results are summarized in Tables 3 and 4. Cross-semivariograms were modeled using upper and lower bounds derived from the Cauchy-Schwarz inequality, as described in Kirsch and Pawlowsky (1985). Figure 1 shows the semivariograms of $W_1(x)$, $W_2(x)$ and $W_1(x)+W_2(x)$, and the cross-semivariogram $\gamma_{1,2}(h)$ between regionalized variables $W_1(x)$ and $W_2(x)$. Note that the cross-semivariogram $\gamma_{1,2}(h)$ is nearly identical to 0, demonstrating that $W_1(x)$ and $W_2(x)$ can be considered spatially independent. A similar procedure using the semivariogram of $W_1(x) - W_2(x)$ produces the same results.

As can be seen in Table 4, cross-validation results both for cokriging and ordinary kriging are nearly identical and can be considered equivalent, again confirming that $W_1(x)$ and $W_2(x)$ can be regarded as spatially independent. Cross-validation was performed by suppressing one sampled point at a time.

Estimation of oil reserves

Because the components $W_1(x)$ and $W_2(x)$ can be considered to be spatially independent, both components can be estimated simultaneously by cokriging or each can

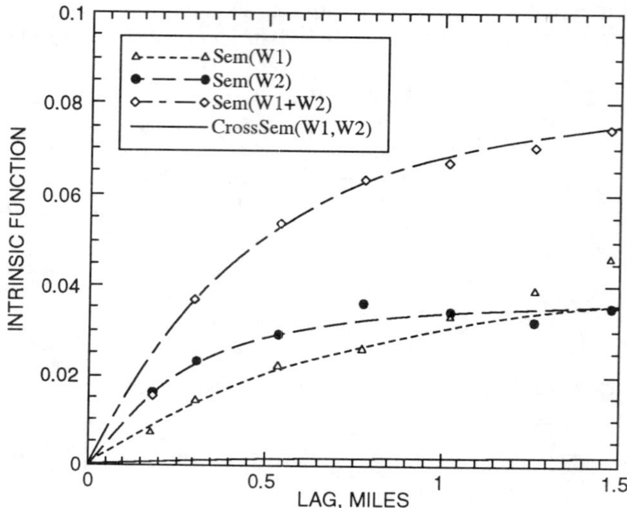

Figure 1: Semivariogram models for $W_1(x)$, $W_2(x)$, $W_1(x) + W_2(x)$ and resulting cross-semivariogram model $\gamma_{o,w}(h)$.

Table 4: Cross-validation results using ordinary kriging and cokriging.

Variable	Cokriging W_1	Cokriging W_2	Ordinary Kriging W_1	Ordinary Kriging W_2
$E[Z - Z^*]$	0.0044	0.0020	0.0044	0.0021
$Var[Z - Z^*]$	0.0153	0.0241	0.0153	0.0241
$E[(Z - Z^*)/CSTD]$	0.0198	0.0051	0.0199	0.0053
$Var[(Z - Z^*)/CSTD]$	1.137	1.134	1.135	1.132

be estimated separately by kriging. Cokriging was performed using the program COKRIG, an extension of a program developed originally by Carr, Myers, and Glass (1985). Results are shown in Figure 2a for $W_1(x)$ and Figure 2b for $W_2(x)$. Estimates obtained with ordinary kriging are nearly identical and are not shown.

The agl(.) transformation was applied to estimate $V_o(x)$ (Fig. 3a):

$$V_o^*(x) = \frac{\exp(W_1^*)(x)}{1 + \exp(W_1^*(x)) + \exp(W_2^*(x))}.$$

Finally, using estimated values of the net thickness, $T(x)$, obtained from kriging (Fig. 3b), the equivalent thickness of oil was estimated by

$$T_o^*(x) = V_o^*(x)T^*(x)$$

at each point x of the field $\mathcal{F}$ (Fig. 4).

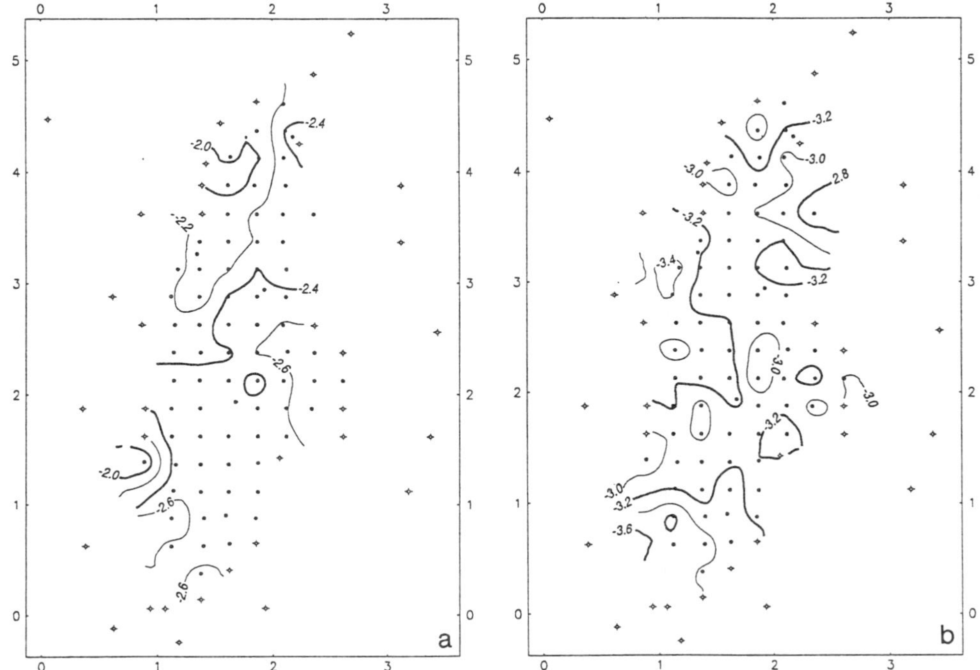

Figure 2: Contour maps estimated using cokriging of (a) $W_1(x)$ and (b) $W_2(x)$.

Integration of $T_o^*(x)$ over $\mathcal{F}$ leads to an estimate of 30.482 MMBBL for the initial volume of oil in place at reservoir conditions, or 25.402 MMSTB at surface conditions. This compares to engineering estimates of 22 MMSTB predicted from decline curve analysis.

CONCLUSIONS

Cokriging of alr variables allows the forcing of interval contraints, simultaneously taking advantage of potential cross-correlations between variables. Additional investigations must address important properties of the backtransformed estimates, such as unbiasedness and the spatial structure of the estimation errors.

REFERENCES

Aitchison, J., 1986, *The Statistical Analysis of Compositional Data:* Chapman & Hall Ltd., London, 416 p.

Carr, J.R., Myers, D.E., and Glass, Ch., 1985, Co-kriging—A computer program: *Computers and Geosciences,* v. 11, p. 111–120.

Chayes, F., 1960, On correlation between variables of constant sum: *Jour. Geophys. Res.,* v. 65, p. 4185–4193.

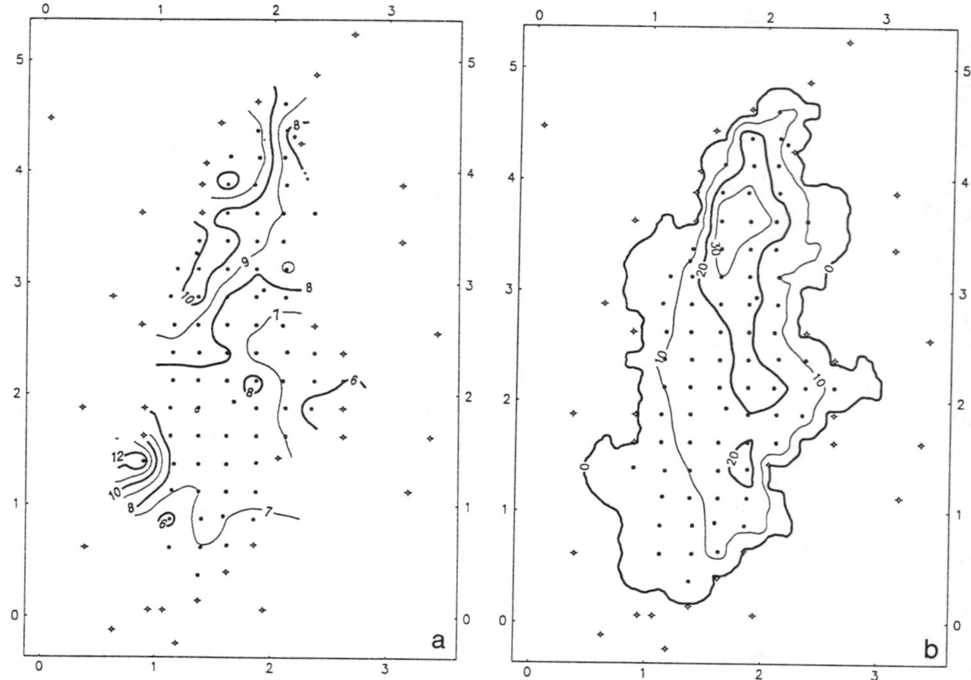

Figure 3: Contour maps of (a) volume of oil $V_o(x)$ estimated using the agl(.) back-transformation, and (b) net thickness $T(x)$ of the saturated interval estimated using ordinary kriging.

Chayes, F., 1971, *Ratio Correlation:* University of Chicago Press, Chicago, 99 p.

Ehm, A.E., 1965, *Lyons West field:* Kansas Oil and Gas Fields, Vol. 4, Kansas Geological Society, Wichita, Kansas, p. 146–156.

Journel, A.G., and Huijbregts, Ch. J., 1978, *Mining Geostatistics:* Academic Press, London, 600 p.

Kirsch, Ch., and Pawlowsky, V., 1985, Modelisation de variogrammes croisées: comment satisfaire a l'inegalité de Cauchy-Schwarz: *Sci. de la Terre,* Sér. Inf., v. 24, p. 54–62.

Matheron, G., 1965, *Les Variables Régionalisées et Leur Estimation—Une Application de la Théorie des fonctions aléatoires aux Sciences de la Nature:* Masson et Cie., Paris, 305 p.

Matheron, G., 1971, *The Theory of Regionalized Variables and Its Applications:* C–5, Centre de Morphologie Mathématique, École Nationale Supérieure des Mines de Paris, Fontainebleau, France, 211 p.

Myers, D.E., 1982, Matrix formulation of cokriging: *Mathematical Geology,* v. 14, p. 249–257.

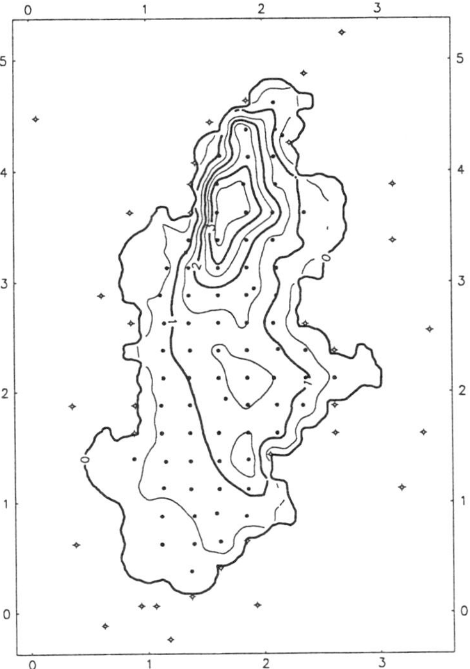

Figure 4: Contour map of equivalent thickness of oil obtained as $T_o^*(x) = V_o^*(x)T^*(x)$.

Olea, R.A., Ed., 1991, *Geostatistical Glossary and Multilingual Dictionary:* Oxford University Press, New York, 192 p.

Pawlowsky, V., 1984, On spurious spatial covariance between variables of constant sum: *Sci. de la Terre,* Sér. Inf., v. 21, p. 107–113.

Pawlowsky, V., 1986, *Räumliche Strukturanalyse und Schätzung ortsabhängiger Kompositionen mit Anwendungsbeispielen aus der Geologie:* Dissertation, FB Geowissenschaften, Freie Universität Berlin, Berlin, 170 p.

Pawlowsky, V., 1989, Cokriging of regionalized compositions: *Mathematical Geology,* v. 21, p. 513–521.

Pawlowsky, V., 1991, Concepts of null correlation for regionalized compositions: *Proceedings of the XXX Symposium "The Mining Příbram in Science and Technology - International Section on Mathematical Methods in Geology",* Příbram, Czechoslovakia, p. 327–342.

Pawlowsky, V., and Burger, H., 1992, Spatial structure analysis of regionalized compositions: *Mathematical Geology,* v. 24, p. 675–691.

Pawlowsky, V., Olea, R.A., and Davis, J.C., 1991, Determinación automática de fronteras de cuerpos geológicos: cuantificación de la incertidumbre: *Actas del XII CEDYA (Congreso de Ecuaciones Diferenciales y Aplicaciones), II Congreso de*

Matemática Aplicada, Servicio de Publicaciones de la Universidad de Oviedo, Oviedo/Gijón, Spain, p. 555–560.

Pawlowsky, V., Davis, J.C., and Olea, R.A., 1992, Indikator-Kriging zur Festlegung von Grenzen geologischer Körper (Eine Fallstudie): *in* Peschel, G. J. (ed.) *Beiträge zur Mathematischen Geologie und Geoinformatik, Band 3, Anwendung geostatistischer Verfahren,* Verlag Sven von Loga, Cologne, Germany, p. 55–60.

Pawlowsky, V., Olea, R.A., and Davis, J.C., 1993, Boundary assessment under uncertainty: a case study: *Mathematical Geology,* v. 25, p. 125–144.

Pearson, K., 1897, Mathematical contributions to the theory of evolution. On a form of spurious correlation which may arise when indices are used in the measurement of organs: *Proc. Royal Society,* v. 60, p. 489–498.

Samper, F.J., and Carrera, J., 1990, *Geoestadística—Aplicaciones a la hidrogeología subterránea:* CIMNE, Barcelona, Spain, 484 p.

Sampson, R.J., 1988, *SURFACE III User's Manual:* Interactive Concepts, Inc., Lawrence, Kansas, 277 p.

REGIONALIZATION OF SOME HYDROGEOLOGICAL PROCESSES AND PARAMETERS BY MEANS OF GEOSTATISTICAL METHODS - CURRENT STATUS AND FUTURE REQUIREMENTS

MARIA-TH. SCHAFMEISTER[1] and GHISLAIN DE MARSILY[2]

[1]Freie Universität Berlin,
Institut für Geologie, Geophysik und Geoinformatik
Malteserstr. 74-100, D-12249 Berlin, Germany

[2]Université Pierre et Marie Curie, Paris
Laboratoire de Géologie Appliquée
4, place Jussieu, F-75252 Paris CEDEX 05, France

This paper discusses specific problems that arise when geostatistics are used to deal with hydrogeological regionalized variables. Compared with other geostatistical problems, the most significant differences are: (i) that many variables vary both in space and time and (ii) that independent and dependent variables are linked through a set of partial differential equations, the groundwater flow and transport equations. Whereas time-constant parameters describing the characteristics of an aquifer material vary enormously in space and thus, strongly affect groundwater flow and contaminant transport, variables such as hydraulic heads or contaminant concentrations are also subject to temporal variations. Different regionalizations are needed, aiming either at providing spatial distributions capable of reflecting the whole spectrum of variations (by simulation techniques) or providing the most likely interpolation (by kriging). This paper presents a spectrum of studies on these topics, mostly on regionalized hydrogeological variables from very heterogeneous, glaciogene porous aquifers, which form the most important groundwater reservoirs in northern Germany.

1. DEALING WITH AQUIFER PROPERTIES, 2-DIMENSIONAL EXAMPLES

For both groundwater flow and mass transport, the most sensitive parameter is the hydraulic conductivity K. According to DARCY´s law, K depends on the physical properties of both the medium and the water, the former being dominant. Other parameters determining the flow velocity and mass exchange within an aquifer are saturated aquifer thickness b and effective pore volume. For 2-dimensional flow problems K and b appear as a product, transmissivity T, both in the flow equation and in the field measurements by pump tests. Both K and T are second-rank tensors, and there are still many unanswered questions about how to measure them in the field, and how to derive effective values for a given support area. Even when measurements are available (e.g. from grain-size distribution analysis, pump or slug tests, tracer experiments), their number is, as a rule, rather limited, and the selection of the best structural model is, in general, difficult. Spherical and exponential variograms are often found with a significant nugget-effect. Although simple kriging has been very much in use for spatial interpolation of T´s, in recent years this lack of data has led to the use of other intercorrelated variables to achieve better estimates of T. These can be geophysical measurements (e.g. Ahmed et al., 1988), or simply, the

R. Dimitrakopoulos (ed.), Geostatistics for the Next Century, 383–392.

specific capacity of wells, which have been found to be well-correlated with T (e.g. Aboufirassi & Marino, 1984). These estimated maps of T´s are used as input for groundwater flow models, but these estimates must subsequently be modified to calibrate the flow model with the available head measurements. By calculating the kriging estimation error, a measure of reliability of the estimate is also provided; this information can then be used by modellers when calibrating, by trial and error, their groundwater flow model, as a guide to where the largest changes in T´s are still compatible with the measured values.

Whereas smooth estimates of K and T, produced by kriging, are adequate for groundwater flow modelling, this is no longer true for contaminant transport problems, which are strongly affected by small-scale variability of the velocity vector, resulting in what is described as "macro-dispersion". More realistic images of spatial variability of hydraulic conductivity are required. To produce such images, several stochastic simulation techniques, based on the knowledge of the statistical moments of the hydraulic parameters, have been found appropriate. These simulations will reflect the whole spectrum of variations of the hydraulic parameters, but are not meant to represent reality. The outcome of such studies will therefore generate alternative distributions in space of the contaminant, and not the "most likely" one; however, this reflects the actual uncertainty in contaminant transport by groundwater, if the simulations can be conditioned on all available measurements.

1.1 Monte Carlo Simulation

The simplest method of Monte Carlo simulation is to just assume that the p.d.f.´s of the parameters to be simulated are known, ignoring any spatial correlation or intercorrelation. The flow equation is then solved for each realization, resulting in a p.d.f. of the hydraulic head at every nodal point in the domain. By this technique, confidence intevalls for hydraulic heads can be calculated and compared with pre-defined threshold values of acceptable uncertainty. Freeze (1975) was among the first to use this technique with T.

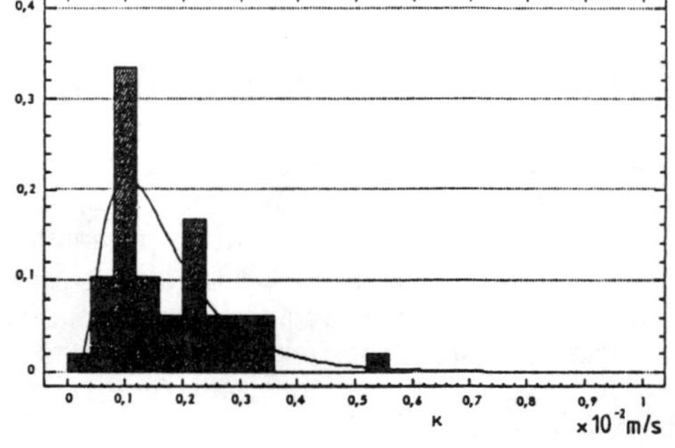

Fig. 1: Histogram of hydraulic conductivities derived from 48 short term pump tests in a glaciofluvial aquifer in Berlin/Germany.

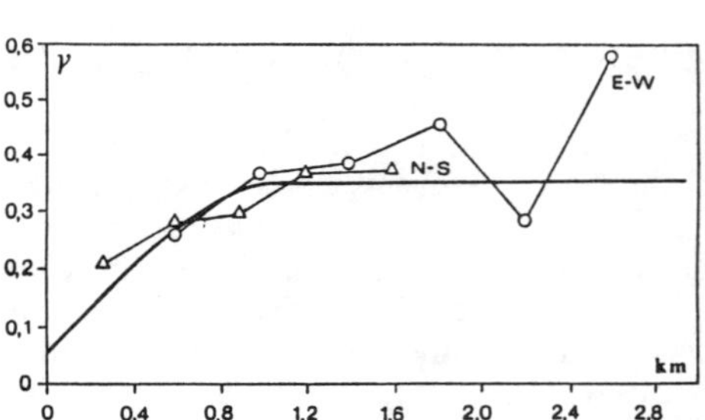

Fig. 2: Directional experimental variograms of ln-(K) and fitted spherical model.

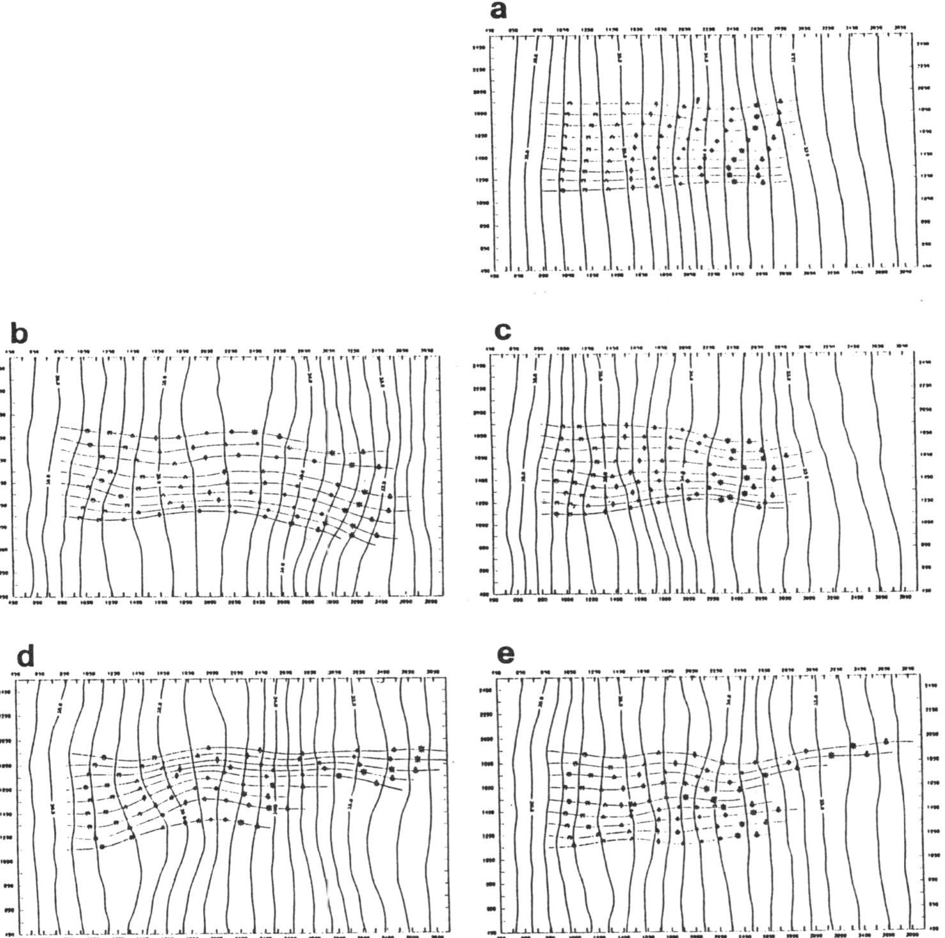

Fig. 3: Results of flow model, piezometric heads and flow paths (symbols indicate flow time of 1 year) based on hydr. conductivity maps provided by a) kriging, b,d) unconditional simulation, c,e) conditional simulation.

1.2 Unconditional and conditional simulation of spatially correlated fields

A number of techniques were developed to incorporate the spatial structure of random variables, e.g. the Turning Band method (Matheron, 1973), the nearest neighbour (Smith & Freeze, 1979), the covariance matrix decomposition (Binsariti, 1980, Davis, 1987), the Sequential Simulation (Journel & Alabert, 1989, Gomez-Hernandez & Srivastava, 1990). These techniques use different algorithms and therefore demand varying amounts of computer storage and time, but they all have in common that they produce realizations that quite precisely respect not only the p.d.f. of a random variable but also its spatial structure. Conditioning these realizations on the measured values at the observation points can easily be achieved, see e.g. Delhomme (1979).

We present here an example of hydraulic conductivity simulation (unconditioned and conditioned) from a glacio-fluvial aquifer in Berlin, Germany (Schafmeister, 1990). Fig. 1 shows the histogram of 48 hydraulic conductivities. Although the K-values are derived from short term pump tests and thus have a volume support, the values could be treated as point values with respect to the extension of the study area. The p.d.f. was taken as lognormal, and a spherical variogram with an isotropic horizontal range of 1,1 km was fitted (Fig. 2). This information was used to generate 2 unconditioned realizations of K from which groundwater flow and travel paths originating from a line source were simulated in 2 dimensions (horizontal) by means of a conventional FD-model (MOC, Konikow & Bredehoeft, 1978).

Fig. 3 shows how groundwater flow is affected by the heterogeneity of the aquifer. A comparison with the results obtained when hydraulic conductivity is kriged (fig. 3a) gives the the following results: the kriged K field gives a travel time through the domain (1 km) ranging from 4 y to 5.5 y, with an average flow velocity of 0.8 m/d. For the unconditioned K fields (Fig. 3b,d) a minimum of 2.7 y and a maximum of 6 y were obtained. This means that kriging severely underestimates the range of the velocity variations and also narrows the spread of the travel path directions. This is particularly true here since the range of the variogram is on the order of the length of the flowpath; had the flow path been on the order of e.g. 10 or more times the range, then the realizations would have been closer to the single average given by kriging, because ergodic behaviour would have been reached by spatial averaging. However, in practical problems, the length of the travel path is usually shorter than, or of the same order as, the range of the variogram.

In order to more accurately describe the flow field, the same simulated realizations were also conditioned on the measured values of K at the 48 wells. Note (Fig. 3c,e) that the variation spectrum of travel times is conserved (minimum 3.3 y, maximum 7.2 y), but that a high conductivity zone in the NE part of the domain is now always present, resulting in a slight deviation of travel paths into that direction.

1.3 Indicator approach in stochastic simulation

Although the standard simulation techniques seem appropriate for dealing with many flow and transport problems in aquifers, they often suffer from the strong restriction that the p.d.f. of the variable (or its logarithm) must be multigaussian in space. Recent research on aquifer material properties has shown that often the observed type of heterogeneity cannot be represented by a simple Gaussian or lognormal model. In fact, aquifers in glacio-fluvial or fluvial sediments show variations in K ranging from 2 to 6 orders of magnitude, with extremely polymodal p.d.f.´s (Jussel, 1989, Schafmeister, 1990).

To deal with these extremely heterogeneous media, one must use simulation algorithms which can represent more general p.d.f´s. The indicator transformation, which has been incorporated into the Sequential Indicator Simulation (SIS) technique (Journel & Alabert, 1989, Gomez-Hernandez & Srivastava, 1990) significantly improves the modelling of such aquifers. The idea is to choose a set of threshold values, from the observed shape of the p.d.f. and from hydraulic considerations, and to calculate seperate variograms for each indicator class. It is thus possible to incorporate different correlation structures for the high-K zones and the low-K zones, that better reflect the differing sedimentation conditions under which such coarse or fine sediments were deposited.

This method was applied to a glacio-fluvial porous medium in the Bornhöved area, in northern Germany. Hydraulic conductivity measurements were derived from grain-size distribution analysis on an outcrop. The SIS technique was used to provide an ensemble of 20 realizations which were by construction conditioned on the 132 measurements, to run a numerical tracer experiment. A nested structure of variograms is observed, with some of them even showing a nugget-effect, implying the existence of another small-scale structure, not identified with the present sampling pattern. A first average variogram range of around 10 m can be derived from a coarse sampling pattern (average horizontal distance between sampling points: 2.5 m), but a second range of 0.5 m has also been detected with a fine sampling pattern in the center of the region (average distance between sampling points: 0.15 m). It is interesting to note that the range of each indicator class is systematically decreasing when the hydraulic conductivity of the class increases (from 2.4 to 0.7 m for the first structure, for a three-fold increase in K, and from 15 to 6 m for the second structure, for a two-fold increase in K). The vertical anisotropy of the variogram range was found to be $a_h/a_v = 4$, which was respected in the simulation by an anamorphosis of the coordinate system.
Fig. 4 shows the results of the numerical tracer experiment which was simulated on one realization of hydraulic conductivity on a 2-d vertical cross-section. Note that the tracer plume is split up into small fractions, which clearly shows the impact on the transport of this type of heterogeneity. Nineteen other simulations were performed and further illustrated the uncertainty in the flow path and travel time in this site.

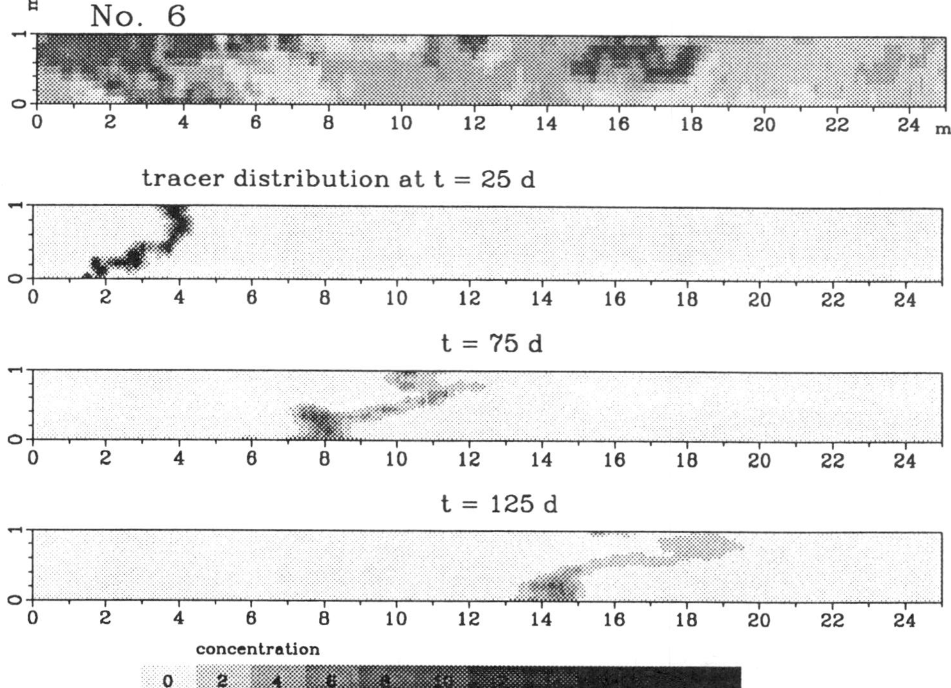

Fig. 4: One realization of hydraulic conductivity within a vertical model domain generated by SIS and evolution of modelled tracer plume with time.

1.4 Conditioning simulations on additional hydraulic information

The simulation techniques, described so far, for hydraulic conductivity and transmissivity are only conditioned on the p.d.f. of the variable, the variogram and the measurements at the wells. But it is also possible to condition them on additional, easily obtainable information such as head measurements.

Most studies along this line have been limited to estimation by co-kriging, using heads and transmissivities together (e.g. Hoeksema & Kitanidis, 1984, Rubin & Dagan, 1987, Dong et al., 1990, Ahmed & Marsily, 1993), where the cross-covariance between heads and transmissivities or conductivities has been derived by solving analytically the stochastic groundwater flow equation. Other authors dealing with the so-called inverse problem (i.e. the automatic calibration of the hydraulic conductivity of an aquifer based on head measurements) have mentioned the possibility of simulating conditional conductivity fields as a post-optimal sensitivity study (e.g. Neuman, 1982, Carrera & Neuman, 1986). An example of generating simulations conditioned also on head measurements is given here. This technique is an extension of the "Pilot Point" inverse method (Marsily et al., 1984, Certes & Marsily, 1991).

The principle of the inverse methods is, in general, to provide a groundwater flow model with an initially defined T field, use it to calculate the heads with the help of the appropriate boundary conditions and source terms, in steady state or transient conditions, and then compare these heads with the observed ones. If the match is not considered adequate, a minimization algorithm is set up to alter the initial T field to reduce the difference in heads. In general, the initial T field is kriged and the alteration is done by "zoning". The resulting calibrated T fields do not have much similarity with the variability of the real T field, and they are much smoother and sometimes zoned. In the technique presented here the initial T field is a simulated field conditioned on all T measurements. The alteration is then done by adding to the real field fictious "pilot points", which are used as new conditioning points in the simulation, but the value of T assigned to them is progressively altered by the minimization algorithm. Since many initial T fields can be simulated, each one of them, is progressively altered to match the head data, thus generating as many fields as desired. These fields have the advantage of preserving the correct variogram of T, of being exact at the wells where T has been measured and to be consistent with both steady-state and transient head measurements when they are used in a digital flow model of the site. The example concernes a dolomite aquifer at the WIPP site, New Mexico, potentially a nuclear waste disposal site. The simulated fields were used to calculate the uncertainty in travel path and travel time in the aquifer. Fig. 5 gives the ensemble of potential pathways generated in this way at the WIPP site for a hypothetical release point, and fig. 6 gives the distribution of the travel time (Lavenue & Ramarao, 1992).

2. DEALING WITH HYDROGEOLOGICAL MAGNITUDES THAT VARY BOTH IN TIME AND SPACE

Whereas aquifer properties such as hydraulic conductivity, porosity and storativity are subject only to variation in space and are constant in time, variables controlled by dynamic processes, such as hydraulic heads or concentrations of contaminants in groundwater, are subject to variations both in space and time. Spatially, hydraulic heads are in general non-stationary, due to a general trend in the direction of flow. In the time domain, heads are in general pseudo-periodic random functions, due to seasonal effects and climatic variability. It is clear, however, that there is a very strong continuity of heads in time, leading to an excellent correlation in time. To estimate the head field at a given time t, it would therefore make sense to use measurements made at different times.

From a geostatistical point of view the most striking problem lies in the way data are collected. Whereas there is only a limited number data points, more or less randomly distributed over space, a large number of regularly-spaced "time locations" is often available. If one does not take into account the fact that, for piezometric heads, there is often a strong periodicity, the amount of information contained in the time domain may be greatly overestimated. Much of the present research work has been done in order to overcome these problems and to take advantage of the additional information provided by temporal measurements.

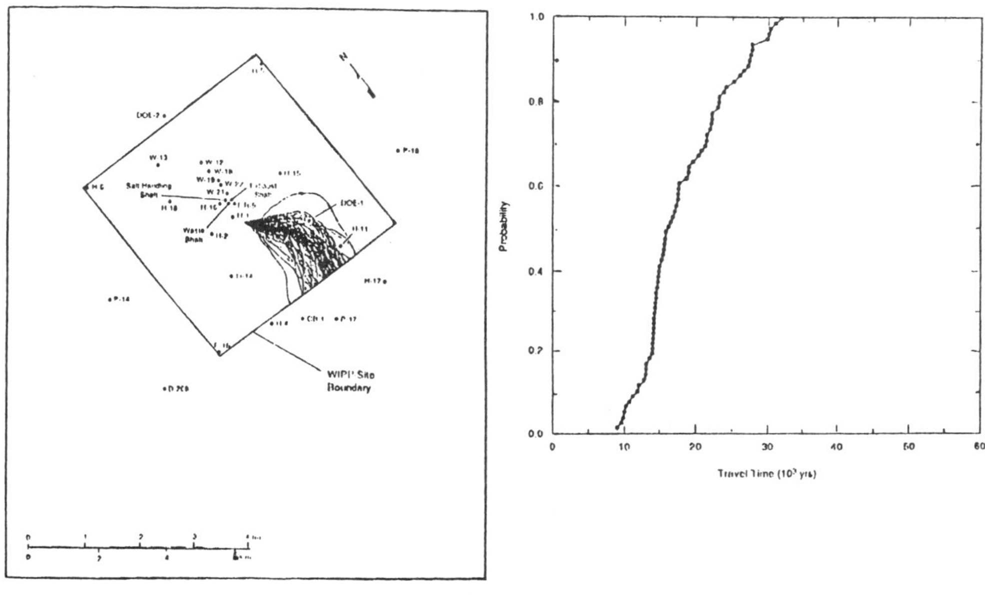

Fig. 5: Travel paths corresponding to the travel times contained in the c.d.f.. From Lavenue & Ramarao (1992).

Fig. 6: Travel time c.d.f. determined from the 70 calibrated fields. From Lavenue & Ramarao (1992).

2.1 Kriging in a spatio-temporal domain

If heads are considered as RF $Z(x,t)$ with t being an additional dimension, the temporal kriging could be used. It is clear, however, that the time axis cannot be treated in the same way as the space domain: while Z may exhibit some directional dependencies, e.g. anisotropy, the time axis follows a strict ordering from past to future. The kriging estimator takes the form

$$Z(x_0,t_0) = \Sigma\lambda_i \, Z(x_i,t_i)$$

$Z(x,t)$ satisfies the intrinsic hypothesis in both time and space for increment (h,τ) when (Bilonick, 1985, Rouhani & Myers, 1990):

$$E[Z(x+h,t+\tau) - Z(x,t)] = 0 \text{ for any } x,t,h \text{ and } \tau$$

and

$$0.5\, Var[Z(x+h,t+\tau) - Z(x,t)] = \gamma(h,\tau)$$

does not depend on x or t and is only a function of h and τ; this function is a space-time variogram $\gamma(h,\tau)$. The problem of deriving the kriging equations to obtain the λ´s can be treated in several ways, but the main problem is to define a permissible structural model, see e.g. Buxton & Pate (1993).

2.2 Co-Kriging

Let $Z_i(x)$ and $Z_j(x)$ be piezometric head measurements at time t_i and t_j, they can be treated as intercorrelated RF´s. The time dependence is then ignored. The estimation is given by Co-kriging. The variograms of each variable must be estimated from the data, as well as the cross-variogram. The latter can be estimated by calculating the variogram of the sum or the difference of Z_i and Z_j.

$$\gamma(Z_i+/-Z_j) = \gamma(Z_i) + \gamma(Z_j) +/- \gamma(Z_i,Z_j)$$

This approach can be generalized to more than two dates, but does not make use of the time correlation of the data, except the cross-correlation at any two dates. Co-kriging applied to time dependent variables appears therefore to be much more intricate than classical coestimation of intercorrelated ore concentrations whithin a deposit, due to the fact that there are usually many more time situations which are now treated as variables, and that a large number of variograms and cross-variograms must be determined.

2.3 Kriging from a "basic outline" with an external-drift

The use of an external drift in kriging has been suggested first by Delfiner et al. (1983). If it can be assumed that the spatial shape of the piezometric remains approximately the same with time, one can take either a well-known situation, or the average of several well-known situations, as the "basic outline" of the head distribution Z(x) at time of interest t_0 (Chiles, 1992). The RF Z can then be considered as a known function of this basic outline, plus a residual:

$$Z(x) = a + bE(x) + R(x)$$

where E(x) is the basic outline, and R(x) is the residual, assumed to have a zero expected value. Z and R are time-dependent, but E is unique. The coefficients a and b of the basic outline are different for each time situation. Universal kriging is then made at any time t_0 using only measurements at time t_0, and the basic outline E as an external drift. The measurements at time t_0 are then respected but away from the measurements, the interpolation is shaped by the basic outline E. This method would be well-suited for situations where dense sampling is done from time to time (to determine E), and only sparce sampling in between. Again this method does not make use of the temporal correlation structure of the head data.

3. CONCLUSIONS AND PROSPECTS

In this paper, some recent developments in the application of geostatistics to hydrogeological processes have been summerized: we don´t claim that this overview is comprehensive in any way; many other authors should have been mentioned. We have tried, however, to stress the differences and specificities which must be considered when one is dealing with hydrogeological regionalized variables, compared to more common variables considered in geostatistics, e.g. in mining or reservoir engineering.

One additional difficulty which has not yet been mentioned, and is also specific to this domain, is the general paucity of data in real site-specific problems, while a very large number of such cases are treated annually world-wide. Collection of data is often restricted by economic considerations, by impossibility of drilling e.g. in urban areas or by fears of creating new routes for contamination by perforation impervious layers between aquifers, or by disturbing waste disposal sites. Finally, it is often required that decisions on aquifer protection or remediation be taken quickly, prior to any in-depth reconnaissance of the site. In this case, the uncertainty associated with the planned decision must be clearly stated.

Stochastic techniques seem to be very well suited for making such predictions with their corresponding uncertainty; various simulation codes have been developed, are easy to use and can provide initial answers; when additional data become available, the simulations can be repeated with conditioning on the new information, thus reducing the uncertainty in the predictions.

It is, however, of utmost importance to have reasonable prior information on the spatial variability of the major parameters (i.e. p.d.f.´s, variograms). It may be possible here to consider the transfer of information from one well-known area to one, where few data are available, if they have similar geological features. This is particularly relevant since a very large number of sites are investigated annually, world-wide. It would therefore be of great interest to establish a "catalogue"of spatial variability of typical geologic formations, characterized by a few general features easy to collect on a new site. Geographical Information Systems could greatly contribute to the task of gathering such data, combining them with other information and making the results available to the public. Some work along this line has already started (e.g. Schafmeister & Pekdeger, 1993).

4. REFERENCES

Aboufirassi, M. & Marino, M.A.: Cokriging of aquifer transmissivities from field measurements of transmissivity and specific capacity.- Math. Geol. 16(1), 19-35, 1984.

Ahmed, S. & Marsily, Gh. de, Talbot, A.: Combined use of hydraulic and electrical properties of an aquifer in a geostatistical estimation of transmissivity.- Groundw. 26(1), 1988.

Ahmed, S. & Marsily, Gh. de: Cokriged estimation of aquifer transmissivity as an indirect solution of the inverse problem: A practical approach.- WRR 29(2),. 521-530, 1993.

Bilonick, R.A.: The space-time distribution of sulphate deposition in the Northeastern United States.- Atmospheric Environment 19(11), 1829-1845, 1985.

Binsariti, A.A.: Statistical analysis stochastic modeling of the Cortaro aquifer in Southern Arizona.- Ph.D. diss., Dept. of Hydr. and Water Res., Univ. of Ariz., Tucson, 1980.

Buxton, B.E. & Pate, A.D.: Joint temporal-spatial modelling of concentrations of hazardous pollutant concentrations.- Forum Geost. for the Next Cent., Montreal, June 3-5, 1993, (this issue).

Carrera, J. & Neuman, S.P.: Estimation of aquifer parameters under transient and steady-state conditions: 2. Uniqueness, stability and solution algorithms.- WRR 22(2), 221-227, 1986.

Certes, C. & Marsily, Gh. de: Application of the pilot point method to the identification of aquifer transmissivities.- Adv. Water Resources 14(5), 284-300, 1991.

Chiles, J.P.: The use of external-drift kriging for designing a piezometric observation network.- in :A.. Bárdossy (ed.) Geostatistical Methods: Recent Developments and Applications in Surface and Subsurface Hydrology, pp. 11-20, UNESCO, Paris, 1992.

Davis, M.: Production of conditional simulations via the LU decomposition of the covariance matrix.- Math. Geol. 19, 91-98, 1987.

Delfiner, P., Delhomme, J.P., Pelissier-Combescure, J.: Application of geostatistical analysis to the evaluation of petroleum reservoirs with well logs.- SPWLA 24th Annual Symposium, Calgary, June 27-30, 1983, Paper WW, 1993.

Delhomme, J.P.: Spatial variability and uncertainty in groundwater flow parameters: a geostatistical approach.- WRR 15(2), 269-280, 1979.

Dong, A., Ahmed, S., Marsily, Gh. de: Developments of geostatistical methods dealing with the boundary conditions problem.- 2nd ECMOR,Arles, Sept. 11-14´90, 21-30, 1990.

Freeze, R.A.: A stochastic-conceptual analysis of one-dimensional groundwater flow in non-uniform, homogeneous media.- WRR 11(5), 725-741, 1975.

Gomez-Hernandez, J.J. & Srivastava R.M.: ISIM3D: An ANSI-C three-dimensional multiple indicator conditional simulation program.-Comp. & Geosc. 16(4),395-440, 1990.

Hoeksema, R.J. & Kitanidis P.K.: An application of geostatistical approach to the inverse problem in two-dimensional groundwater modelling.- WRR 20(7),1009-1020, 1984.

Journel, A.G. & Alabert, F.: Non-Gaussian data expansion in the earth sciences.- Terra Nova 1(2), 123-134, 1989.

Jussel, P.: Stochastic description of typical inhomogeneities of hydraulic conductivity in fluvial gravel deposits.- in: Kobus & Kinzelbach (eds.) Contaminant Transport in Groundwater, pp. 221-228, Balkema, Rotterdam ISBN 90 6191 879 0, 1989.

Konikow, L.F. & Bredehoeft J.D.: Computer Model of Two- Dimensional Solute Transport and Dispersion in Groundwater.- Techniques of Water Resources Investigations of the US Geological Survey, Book 7, C2, Scient. Publ. Comp., Washington, 1978.

Lavenue, M.A. & Ramarao, B.S.: A modelling approach to address spatial variability within the Culebra Dolomite transmissivity field.- Sandia National laboratories, Report SAND 92 - 7306, 1992.

Marsily, Gh. de, Lavedan, G., Boucher, M., Fasanino, G.: Interpretation of interference tests in a well field using geostatistical techniques to fit the permeability distribution in a reservoir model.- *Geostatistics for Natural Resources Characterization*, pp. 831-849, D. Reidel, Hingham, Mass, 1984.

Matheron, G.: The intrinsic random functions and their applications.- Adv. Appl. Prob. 5, pp. 438-468, 1973.

Neuman, S.P.: Statistical characterization of aquifer heterogeneities: an overview. (P.A. Witherspoon´s 60th birthday, Berkely, 1979).- in: Narasimham, T.N. (ed.) Recent Trends in Hydrogeology, Spec. Pap. Geol.. Soc. Are., 189, pp. 81-102, Boulder, Colorado, 1982.

Rouhani, S. & Myers D.E.: Problems in space-time kriging of geohydrological data.- Math. Geol. 22(5), 1990.

Rubin, Y. & Dagan, G.: Stochastic identification of transmissivity and effective recharge in steady groundwater flow, 1. Theory.- WRR 23(7), 1185-1192, 1987.

Schafmeister M.-Th.: Geostatistische Simulationstechniken als Grundlage der Modellierung von Grundwasserströmung und Stofftransport in heterogenen Aquifersystemen.- Dissertation, Verlag Schelzky & Jeep, p. 143, Berlin, 1990.

Schafmeister, M.-Th. & Pekdeger A.: Spatial structure of hydraulic conductivity in various porous media - problems and experiences. in: A. Soares (ed.) Geostatistics Tróia ´92, Quant. Geology and Geost., Vol. 5, p. 733-744, Kluwer Acad. Press, Dordrecht, 1993.

Smith, L. & Freeze R.A.: Stochastic analysis of steady-state groundwater flow in a bounded domain. 1. One-dimensional simulations.- WRR 15(3), 521-528, 2. Two-dimensional simulations.- WRR 15(6), 1543-1559, 1979.

COMBINING GEOPHYSICAL DATA WITH GEOLOGICAL PRIOR KNOWLEDGE IN A GEOSTATISTICAL FRAMEWORK: THE EXAMPLE OF CROSS-WELL TOMOGRAPHIC INVERSION.

A. TRACK, J.H. MEYER, S. ZURQUIYAH and J.P. DELHOMME
Interpretation Engineering Department
Etudes et Productions Schlumberger
26, Rue de la cavee, Clamart, France.

In most geophysical surveys (e.g. seismic, electro-magnetic, or piezometric), inverse modelling is needed: values of model parameters that describe the subsurface have to be inferred from values of some observable parameters. Because of coverage limitations of the surveys but also due to the nature of the physical laws involved, it is often necessary to use a priori information about the model parameters in order to get geologically meaningful results while stabilizing the inversion.

The example of a cross-well seismic survey is taken to illustrate the discussion: model parameters are slownesses (i.e., the reciprocals of seismic velocities in rocks) whereas the observable parameters are cross-well traveltimes. The prior knowledge mainly comes from geology and standard well logging. Dipmeter surveys and gamma ray logs are used for well-to-well correlation; combined with structural geology laws, this information lead to the a priori layer geometry between wells. Sonic logs give the a priori layer slownesses at both well locations; sedimentological considerations give a clue to slowness variability along and across layering.

Several ways of translating the prior information into an a priori model and a model error covariance matrix have been investigated. This is illustrated by showing the effect of different approaches on the resulting cross-section. Some theoretical results are also presented: e.g., it is shown that, under some assumptions, the inverse error model covariance matrix which is involved in the inversion is sparse and can be calculated analytically, thus saving storage space and computation time.

R. Dimitrakopoulos (ed.), Geostatistics for the Next Century, 393–404.

BASIC PRINCIPLES

The objective of a cross-well tomography experiment is to infer the spatial distribution of the seismic velocity in the subsurface rocks. To this end, a source of seismic waves is shot at various locations in a first borehole. For each source location, several receivers placed in a second borehole record the arrival of the waves (Fig.1). Given a set of source and receiver locations, we assume that the traveltimes of the first wave from each source to each receiver across the unknown formation can be picked. Because of the large amount of data, an automatic time picking algorithm is used. In Fig.2, the solid line connects the times obtained by such an automatic picking for one source location and various receiver locations; the two dashed lines indicate the attached error bars.

The objective is to find the velocity distribution that explains –or rather best explains– the observed times. The exact meaning of "best" will become clearer in the following sections.

FORWARD MODELING

Rather than going through many equations, we will proceed by reviewing the underlying ideas. The domain to be imaged (i.e., a cross-section of the subsurface) is first decomposed into a regular mesh of identical cells. A constant (unknown) velocity is attached to each cell. The problem is to find the unknown velocities using the traveltimes.

Each source-receiver pair defines a traveltime and hence an equation of the form

$$T(S_i, R_j) = \sum_{n=1}^{N(i,j)} D_n \mathbf{s}_n$$

where:

$T(S_i, R_j)$ is the known observed traveltime from source S_i to receiver R_j,

$N(i,j)$ is the number of cells crossed by the ray connecting the source S_i to the receiver R_j,

D_n is the length of the ray in cell number n,

and $\mathbf{s}_n$ is the unknown slowness (reciprocal of velocity) in cell number n.

The above equation simply states that the total traveltime from a source to a receiver is the sum of partial times spent in each of the cells crossed by the ray connecting the given source and receiver.

The set of sources and receivers provides as many equations of the type shown above as there are source-receiver pairs. As a first approximation, one can assume the raypath to be a straight-line connecting the source-receiver pair. The advantage of the straight-ray assumption is the linearity of the resulting equation system. It is only valid for small slowness variations.

Contrary to the case of X-Ray tomography, the raypaths in cross-well tomography are a priori unknown because they depend upon the actual structure of the slownesses. Generally, the raypath will be complicated, involving refractions at boundaries of cells with different slownesses. Given a slowness distribution, the actual raypath can be obtained using Fermat's theorem. As the raypath depends on the slownesses, time is a non-linear function of slowness. The equation system that describes the forward problem is non-linear in this case.

DETERMINISTIC INVERSE PROBLEM SOLVING

With the straight ray approximation, the error functional to be minimized can be expressed as

$$F(M) = (T - AM)^t(T - AM)$$

where:

A is a Jacobian matrix

T is the data vector (in this case, times)

M is the model vector (in this case, slownesses).

The least squares solution is obtained through the pseudo inverse matrix of A, which leads to invert the square matrix A^tA. More precisely, the linear system to be solved for during the inversion process has the following form:

$$A^tAM = A^tT$$

In the non-linear case, the use of a Newton or quasi-Newton method implies the solution of such a linear system at each iteration.

Such systems can theoretically be solved as such. However, for a given cell, the result reliability dramatically increases with the number of rays passing through this cell. This number depends directly on the acquisition geometry and leads to the notion of tomographic coverage. A "hit-count map" of rays gives an idea of the ray coverage of the imaged area. In the case of cross-well surveys, this map typically has a butterfly-shape.

In addition, the ability to obtain a well resolved image of closely spaced cells increases as rays pass through a given cell in different directions. Thus, useful indicators of good resolution are the angular aperture of rays through a given cell and the average ray direction for the cell. Roughly speaking, the expected resolution in the reconstructed image is perpendicular to the average ray directions. As a consequence, in the cross-well geometry, the direction of best resolution is vertical rather than horizontal (i.e., horizontal boundaries are better resolved than near-vertical faults).

Tomographic inversion problems are generally ill-posed due to the limited coverage inherently present in most experimental setups. The system is somehow both overdetermined for part of the cells (i.e., more equations than unknowns) and underdetermined for cells which are not correctly "illuminated". This leads to a nonunique solution to the inverse problem and it is desirable and sometimes compulsory to stabilize the inversion (Schweppe, 1973).

In such cases, the matrix A^tA is numerically singular, even when the number of equations (i.e. number of rays) is greater than the number of unknowns (i.e. number of cells). This "unequal resolution" can be evaluated quantitatively through the eigenvalues obtainable with a singular value decomposition of A^tA.

One classical solution to this situation is through numerical damping whereby a diagonal matrix is added to A^tA prior to inversion. This is sometimes referred to as the Levenberg-Marquardt approach (Marquardt, 1970). More sophisticated damping schemes using derivation operators are also considered. Thus the dimension of the null space is reduced and the matrix inversion stabilized.

The system is then rewritten as:

$$(A^tA + \lambda I + \mu H^tH)M = A^tT$$

where:

H is a linear operator, usually a (first, second...) derivation operator

λ and μ are dimensional damping coefficients.

GEOSTATISTICAL INVERSE PROBLEM SOLVING

The approach we propose leads to a physical interpretation of the numerical damping introduced, thus enabling to relate its value to the physics of the problem at hand. We also extend it to include all the geological knowledge within a geostatistical framework. The particular case of a diagonal damping for instance then appears as a way of accounting for traveltime measurement errors.

This approach, which can be called a geostatistical or Bayesian one, uses covariance matrices together with an a priori model in order to stabilize the inversion, both by relaxing the need to perfectly honor the observed traveltimes and by introducing a priori knowledge about slownesses (Tarantola, 1987).

Uncertainties on traveltime picks, tool positioning, and well geometries are accounted for via a time error covariance matrix C_T. A priori geological knowledge is incorporated via a priori velocity model M_0 and a model covariance matrix C_M to describe allowed deviations of the solution from M_0. Should M_0 be obtained by kriging, C_M would be the covariance matrix of the kriging errors at the various grid nodes.

The diagonal elements correspond to allowed deviations from the measured times (for C_T) or from the a priori slowness values (for C_M). The off-diagonal terms link the errors in picked time for two different source-receiver pairs in the case of C_T, and relate to an interdependancy between the allowed deviation from the a priori model of slownesses at two different cells in the case of C_M.

Using the covariance matrices described briefly above, the problem amounts, in algebraic terms, to replace the least square norm (L2-norm) by a norm defined by the quadratic forms C_T and C_M. Equivalently, in statistical terms, it amounts to assume that the measurement uncertainties and the deviation of the solution from an a priori model are Gaussian random variables.

With covariance matrices, the error functional to be minimized now becomes:

$$F(M) = (T - AM)^t C_T^{-1}(T - AM) + (M - M_0)^t C_M^{-1}(M - M_0)$$

This leads to

$$(A^t C_T^{-1} A + C_M^{-1})M = A^t C_T^{-1} T + C_M^{-1} M_0$$

where:

C_T and C_M are covariance matrices respectively for the data and for the model;

M_0 is the a priori model.

It appears in the expressions above that both the inverse of C_T and C_M have to be computed. C_T is almost always taken as a diagonal matrix since the picking errors are in general uncorrelated from shot to shot. As to C_M, it can be a full matrix. If C_T is a diagonal matrix with constant entries σ_T^2, then $(\sigma_T^2 C_M^{-1})$ in this formulation compares to $(\lambda I + \mu H^t H)$ in the numerical damping one. The appendix shows a specific full covariance matrix C_M, such that C_M^{-1} is sparse and can be computed analytically and compared to the sum of a diagonal and a derivation operator damping.

APPLICATION AND RESULTS

The various approaches have been tested on a synthetic case study. The cross-section to reconstruct was chosen with both horizontal and highly slanted sharp velocity contrasts: it consists of a multiply faulted ramp anticline (Fig.1), also known as a "fault-bend fold" case in structural geology textbooks.

41 source and 41 receiver locations in two vertical wells were used to simulate a cross-well experiment (1681 equations). The cross-section has been discretized using a 30 $\times$ 30 grid (900 cells i.e., 900 unknowns). Synthetic waveforms were generated from the "true" slowness distribution, using a finite-difference forward modelling scheme. Source-receiver traveltimes were automatically picked. Error bars were assessed, based on local picking accuracy (Fig.2); they directly lead to the time error covariance matrix.

The synthetic data set also comprises dipmeter surveys and squared sonic logs in both wells. These data were manually correlated from well to well so as to create "geologically meaningul" a priori distributions of the slownesses. Three a priori models were successively considered: the true model, a model derived from the sonic and dipmeter data and a model derived from the sonic data only (see cross-section maps in Figures 1, 3 and 4).

An example of model covariance matrix is shown in Fig.6 for a simplified case (Fig.5). Diagonal terms which reflect the confidence in the a priori values, mainly as a function of distance from the wells, can be displayed in form of a cross-section map or in 3D (as in Fig.7). Those terms control the tolerance of variations away from the a priori model. Off-diagonal terms can be displayed for a fixed node also in form of a map or in 3D (Fig.8); each line or column of the matrix corresponds to such a map. Those terms control how variations away from the a priori model may vary from a given cell to the neighboring ones.

Fig.9 shows the resulting tomogram when the true velocity model (Fig. 1) is used to constrain the inversion, starting from a homogeneous velocity model. This ideal situation is used to check the quality of the subsequent inversions. In Fig. 10, the inversion is constrained using the dipmeter structural information together with the sonic logs at well locations (Fig. 3). Although of lower quality as compared to Fig. 9, the result is very good and the true model has been very well recovered. Fig. 11 shows the result when less a priori information is used in the constraints, namely when only the sonic logs are available (Fig. 4). As expected, the image quality is lower but the main features of the true model are still reasonably reproduced. Finally, in Fig. 12, no a priori information is used: the inversion is driven solely by the data, starting from a homogeneous background velocity model. Artifacts are obvious and dominate the image as no extra information compensates for the under-determination of the inverse problem.

The use of a geostatistical framework provided the spatial consistency criterion which has enabled the inversion process to converge to quite acceptable images by making the best use of the traveltime information available, although the last two a priori slowness models were in some places quite far from the truth.

Besides stabilizing the inversion, the technique provided a simple way of building a solution that agrees with predefined geological constraints. In this respect it can be viewed as a useful synergetic tool between seismic and geological information.

References

Marquardt, D.W. (1970) Generalized inverses, ridge regression, biased linear estimation and non-linear estimation, *Technometrics*, vol.12, pp. 591-612.

Schweppe, F.C. (1973) *Uncertain Dynamic Systems*, Prentice-Hall, Englewood Cliffs.

Tarantola, A. (1987) *Inverse Problem Theory*, Elsevier, Amsterdam.

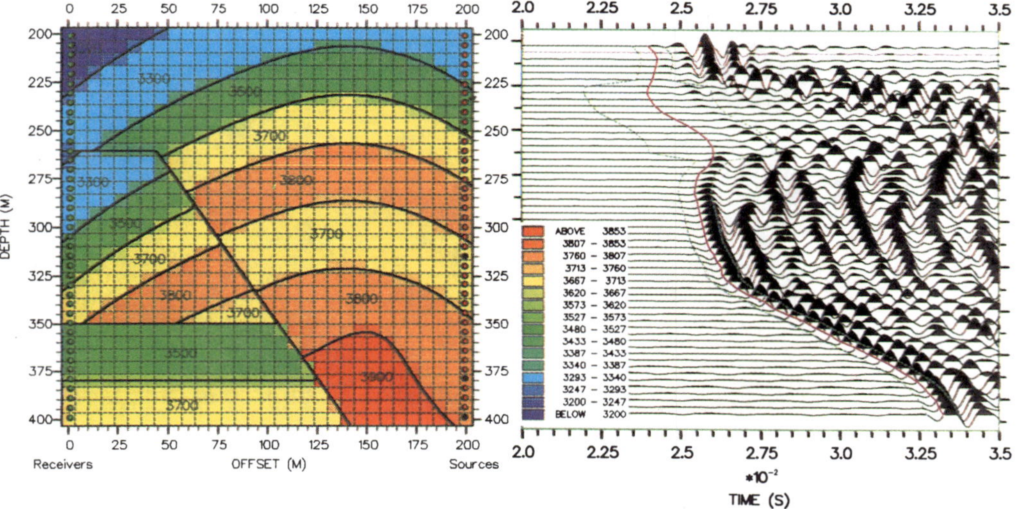

Fig. 1 Geological model consisting of a multiply faulted ramp anticline also known as a fault-bend fold. Red dots on the right side denote sources deployed in one borehole, green dots on the left refer to receivers in a second borehole.

Fig. 2 Common-source gather of a real-data example with traveltimes of the direct waves connected by the red solid line. The two dashed lines indicate the error bars attached to the time picking.

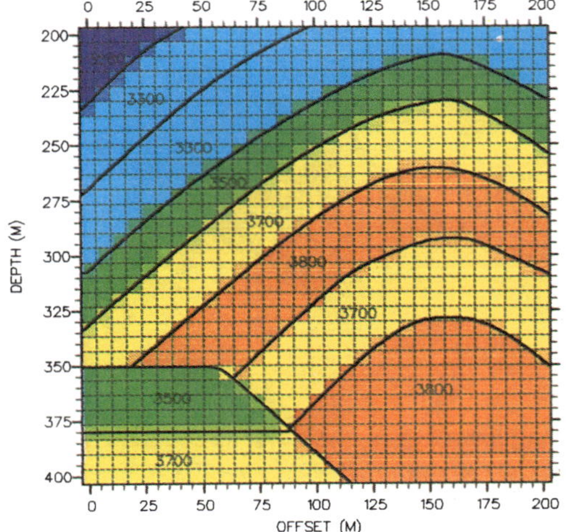

Fig. 3 A priori model derived from the geological cross-section (Fig. 1) based on sonic and dipmeter data.

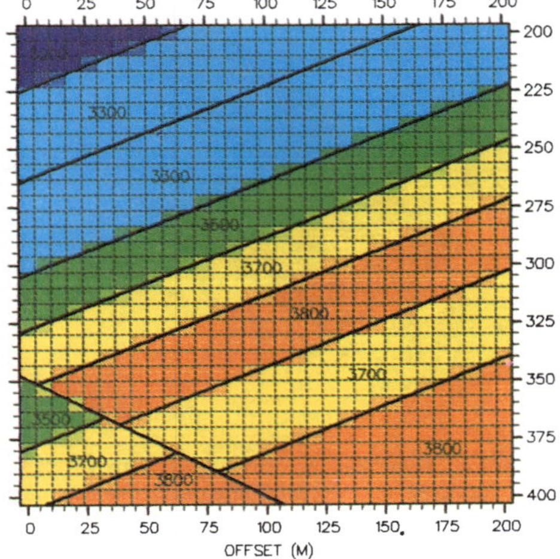

Fig. 4 A priori model derived from the geological model (Fig. 1) assuming to have only sonic log data.

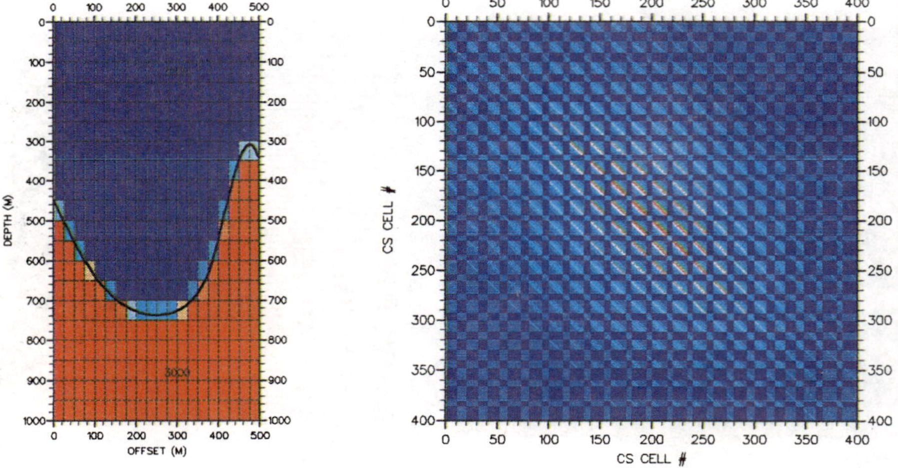

Fig. 5 A simple 2-layer syncline model containing 400 cells to illustrate the Bayesian approach of constraints in form of model covariances.

Fig. 6 Entire model covariance matrix for the syncline model (Fig. 5).

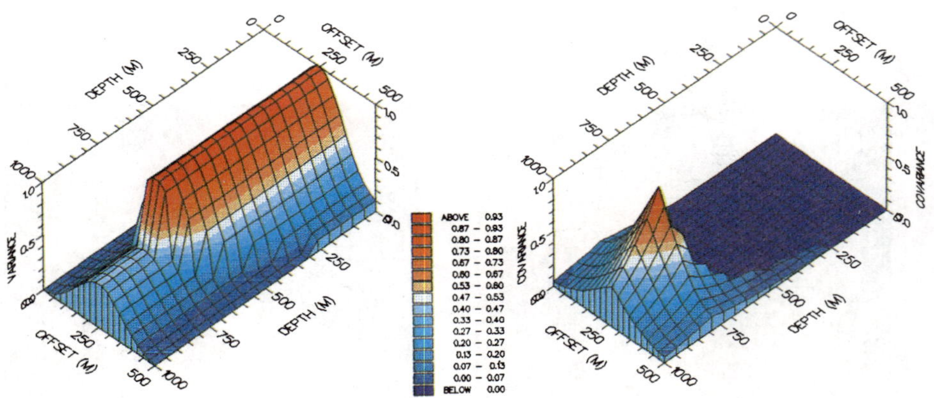

Fig. 7 3-D cross-section map of the covariance matrix *diagonal* terms reflecting the confidence in the a priori values as a function of distance from the wells. Higher confidence at the wells refer to smaller values in this Gaussian-distributed map.

Fig. 8 3-D cross-section map of the covariance matrix *off-diagonal* terms which corresponds to one line of the covariance matrix (Fig. 5). Those terms control how variations away from the a priori model may vary from a given cell to the neighboring ones. Note that only cells of a same layer are correlated in this case.

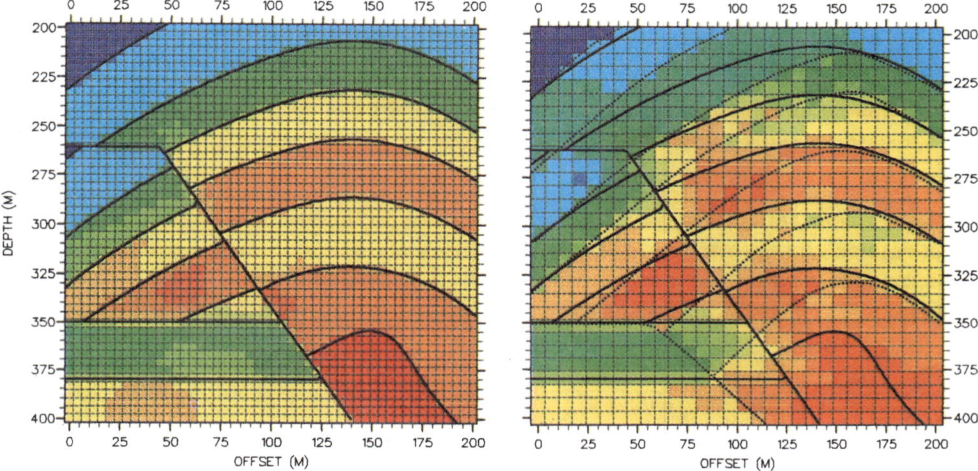

Fig. 9 Tomogram as a result of a constrained but under-determined inverse problem (40*40 equations inverted for 50*50 cells). This image also serves as a check for the method used.

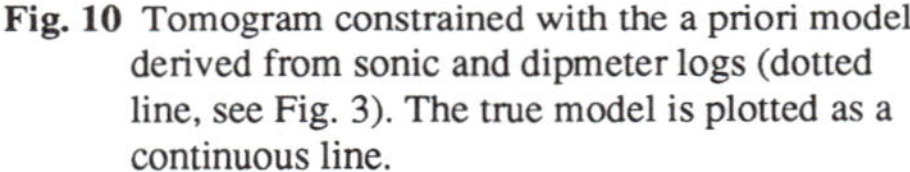

Fig. 10 Tomogram constrained with the a priori model derived from sonic and dipmeter logs (dotted line, see Fig. 3). The true model is plotted as a continuous line.

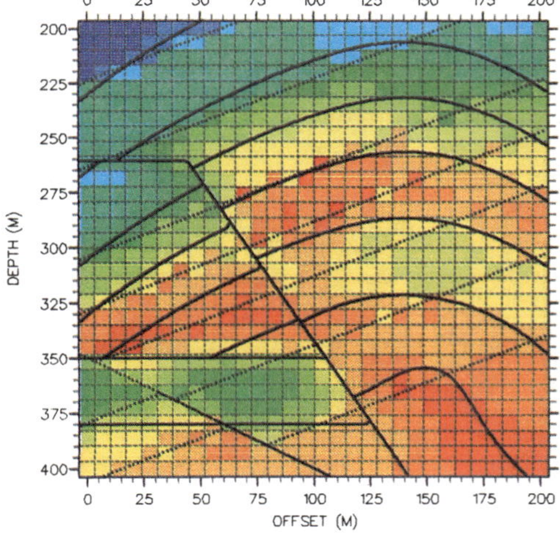

Fig. 11 Tomogram constrained with the a priori model derived from sonic logs only (dotted line, see Fig. 4). The true model is plotted as a continuous line.

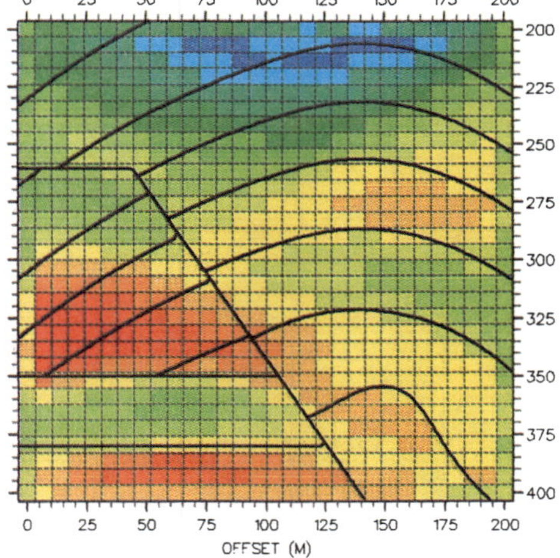

Fig. 12 Unconstrained tomogram starting from a constant averaged background velocity model. This image does not retrieve the true solution (continuous line), is affected by significant artifacts, and represents the worst scenario of the seismic cross-well inversion.

APPENDIX

This appendix presents a method for constructing a model covariance matrix such that its inverse is analytical and sparse. This choice leads to better understand the link between geostatistical constraints and damping in tomographic inversions.

Let us consider a rectangular tomographic cross-section domain divided into cells by a regular grid with n_x nodes in the x direction and n_y nodes in the y direction. The spacing in between the nodes is respectively Δx in x and Δy in y.

Let $Z(N_i)$ be the deviation between the true model value and the a priori model value at node N_i (x_i, y_i) and let $Z(N_j)$ be the deviation at node N_j (x_j, y_j).

Let us assume the separability of the two directions of space for the covariance and exponentially decreasing correlations in both directions. Let σ^2 be the variance, constant in the domain, and let a and b be the correlation distances respectively in the x and y direction:

$$C_{ij} = cov(Z(N_i), Z(N_j)) = \sigma^2 e^{-\frac{|x_j - x_i|}{a} - \frac{|y_j - y_i|}{b}}$$

The model covariance matrix C that describes the relationship between the deviations from the a priori model for the different cells is of dimension $n_x n_y \times n_x n_y$. If the nodes are numbered by column and from top to bottom, this matrix can be written as follows, with σ^2 taken equal to unity in the following for the sake of simplicity:

$$C = \begin{bmatrix} 1 & e^{-\frac{\Delta y}{b}} & \dots & e^{-\frac{\Delta x}{a}} & e^{-\frac{\Delta x}{a} - \frac{\Delta y}{b}} & \dots \\ e^{-\frac{\Delta y}{b}} & 1 & \dots & e^{-\frac{\Delta x}{a} - \frac{\Delta y}{b}} & e^{-\frac{\Delta x}{a}} & \dots \\ \vdots & & & & & \\ \vdots & & & & & \\ e^{-\frac{\Delta x}{a}} & e^{-\frac{\Delta x}{a} - \frac{\Delta y}{b}} & \dots & 1 & e^{-\frac{\Delta y}{b}} & \dots \\ e^{-\frac{\Delta x}{a} - \frac{\Delta y}{b}} & e^{-\frac{\Delta x}{a}} & \dots & e^{-\frac{\Delta y}{b}} & 1 & \dots \\ \vdots & & & & & \\ \vdots & & & & & \end{bmatrix}$$

It can be observed that C is the tensorial product of two Toeppliz matrices Φ_x and Φ_y, respectively of dimensions $n_x \times n_x$ and $n_y \times n_y$.

$$\Phi_x = \begin{bmatrix} 1 & e^{-\frac{\Delta x}{a}} & \dots & e^{-\frac{(n_x-1)\Delta x}{a}} \\ e^{-\frac{\Delta x}{a}} & 1 & \dots & e^{-\frac{(n_x-2)\Delta x}{a}} \\ \vdots & & & \\ e^{-\frac{(n_x-1)\Delta x}{a}} & e^{-\frac{(n_x-2)\Delta x}{a}} & \dots & 1 \end{bmatrix}$$

and

$$\Phi_y = \begin{bmatrix} 1 & e^{-\frac{\Delta y}{b}} & \dots & e^{-\frac{(n_y-1)\Delta y}{b}} \\ e^{-\frac{\Delta y}{b}} & 1 & \dots & e^{-\frac{(n_y-2)\Delta y}{b}} \\ \vdots & & & \\ e^{-\frac{(n_y-1)\Delta y}{b}} & e^{-\frac{(n_y-2)\Delta y}{b}} & \dots & 1 \end{bmatrix}$$

Since Φ_x and Φ_y are Toeppliz matrices, their inverses are tridiagonal matrices and they can be derived analytically. Using generic notations:

$$\Phi^{-1} = \begin{bmatrix} \alpha_1 & \beta & 0 & 0 & 0 & \dots & \dots & 0 \\ \beta & \alpha_2 & \beta & 0 & 0 & \dots & \dots & 0 \\ 0 & \beta & \alpha_2 & \beta & 0 & \dots & \dots & 0 \\ \vdots & & & & & & & \\ \vdots & & & & & & & \\ 0 & \dots & \dots & 0 & \beta & \alpha_2 & \beta & 0 \\ 0 & \dots & \dots & 0 & 0 & \beta & \alpha_2 & \beta \\ 0 & \dots & \dots & 0 & 0 & 0 & \beta & \alpha_1 \end{bmatrix}$$

where:

$$\alpha_1 = \frac{1}{1-r^2} \qquad \alpha_2 = \frac{1+r^2}{1-r^2} \qquad \beta = \frac{-r}{1-r^2}$$

with either $r = r_x = e^{\frac{-\Delta x}{a}}$ or $r = r_y = e^{\frac{-\Delta y}{b}}$.

The inverse of a tensorial product of matrices is the tensorial product of their inverses, thus:

$$\boxed{C^{-1} = (\Phi_x \otimes \Phi_y)^{-1} = (\Phi_x^{-1} \otimes \Phi_y^{-1})}$$

Since Φ_x^{-1} and Φ_y^{-1} are tridiagonal matrices, their tensorial product is a 9-diagonal matrix. So, in this case, the inverse of the model covariance matrix C can be calculated analytically and the result only contains nine diagonals, which is an interesting result as far as storage is concerned.

Let us now look at some properties of the inverse covariance matrix C^{-1} in terms of its equivalent operator: $C^{-1} = \Phi_x^{-1} \bigotimes \Phi_y^{-1}$ corresponds to the composition of two linear operators, one in the x direction, and one in the y direction.

Let us consider the product of C^{-1} and a vector $z \in \Re^{n_x \times n_y}$, for an interior node. It can be shown that:

$$C^{-1}z = L_x \circ L_y(z)$$

where L_x and L_y are two 3 point- operators defined as follows:

$$L_x(z_k) = \alpha_{2x}\ z_k + \beta_x\ z_{k-n_y} + \beta_x\ z_{k+n_y}$$

and

$$L_y(z_k) = \alpha_{2y}\ z_k + \beta_y\ z_{k-1} + \beta_y\ z_{k+1}$$

The subscripts x and y correspond to the operator in the x direction and in the y direction. These two operators can be written as the sum of two terms: the first term corresponds to the diagonal part of the operator, and the second term corresponds to its second derivative part.

Using generic notations:

$$L(z_k) \quad = \quad (\alpha_2 + 2\beta)\ z_k - \beta\ (-z_{k-1} + 2z_k - z_{k+1})$$

When $r \rightarrow 0$, L tends to a unit weight diagonal damping (the second term goes to zero). When $r \rightarrow 1^-$, L behaves like a second-order derivative operator (of increasing weight).

$r \rightarrow 0$ means that the correlation length a is small with respects to Δx and that the deviations from the a priori model are uncorrelated. In such a case, we only have a diagonal damping of weight $\frac{1}{\sigma^2}$ if σ^2 is no longer taken as equal to unity. Thus the damping coefficient λ defined in the main text corresponds to $\frac{\sigma_T^2}{\sigma^2}$ with σ_T being the standard deviation of the error on the data.

$r \rightarrow 1^-$ means that the correlation length a is large with respects to Δx and that the deviations from the a priori model are strongly correlated. We have a second-order derivative damping of weight $\frac{1}{\sigma^2} \times \frac{a}{2\Delta x}$. Thus the damping coefficient μ defined in the main text corresponds to $\frac{\sigma_T^2}{\sigma^2} \times \frac{a}{2\Delta x}$

NEW METHODS, ALTERNATIVE FRAMEWORKS AND DIRECTIONS IN MODELLING

CONDITIONAL fBm SIMULATION WITH DUAL KRIGING

JINCHI CHU and ANDRE G. JOURNEL
Geology and Environmental Sciences Department
Stanford University
Stanford, CA 94305
U.S.A.

Stochastic fractal simulation, such as derived from a fractional Brownian motion (fBm) model, are typically extremely fast as long as they are not made conditional to real data. If conditioning is done by a traditional (primal) kriging, that speed advantage is lost. If dual kriging is used instead, the conditional simulation can still be represented as an analytical function of the coordinates vector, which can be read at any location where a simulated value is required.

Comparison with conditional simulations generated by sequential Gaussian simulation confirms the speed advantage if the number of conditioning data is small.

Various novel implementations of the fBm algorithm are suggested.

INTRODUCTION

A fractal model is one in which the variation of the phenomenon is assumed to be self-affine over a wide range of scales. If one examines a fractal structure at different scales, one repeatedly encounters the same fundamental elements. The repetitive pattern defines the fractional, or fractal, dimension of the structure. Although the assumption of scale invariance may not be accurate in geological sciences (whenever data are available, patterns of variability are seen to change with scales), it is a much better model than lumping all short scale variabilities into an undifferentiated nugget effect (white noise) and assuming no large scale structure beyond the largest revealed by the present data. Also fractal models used to describe reservoir heterogeneneities have been shown to allow a good reproduction of flow behavior (Hewett, 1986).

Unlike Euclidean geometry in traditional mathematics, fractals have dimensions which are greater that their topological dimension. Fractal models are expressed with algorithms, or sets of mathematical procedures. These algorithms are then translated into geometric forms with the aid of computers. The class of algorithms to generate fractals is very large and can be subdivided into two categories: deterministic fractals and random fractals. One particular model of random fractals which has been applied to geostatistical simulation is fractional Brownian motion (fBm) (Mandelbrot, 1968). Typically, in the fractal literature stochastic fractal images are not conditioned to local data. Most geostatistical applications command that the stochastic images honor local data at their locations. This paper proposes an algorithm for producing fBm conditional simulations without losing the main attraction of a fractal model which is the ability to read the simulated image as a function of the coordinates vector ($\mathbf{u}$). Various novel implementations relate to simulation of 3-D anisotropy and nested fractals with different fractal dimensions.

R. Dimitrakopoulos (ed.), Geostatistics for the Next Century, 407–421.

FRACTIONAL BROWNIAN MOTION

The most frequently used model for random fractals is that of fractional Brownian motion (fBm), which is an extension of the traditional Brownian motion stochastic process. Denoted as $Z(\mathbf{u})$, the fBm-derived random function (RF) is a single-valued function of the coordinates vector, $\mathbf{u} = \{x, y, z\}$. Its increments $Z(\mathbf{u}_2) - Z(\mathbf{u}_1)$ have a stationary Gaussian distribution with variance proportional to $|\mathbf{u}_2 - \mathbf{u}_1|^{2H}$. The single parameter H, called fractal co-dimension, is a measure of spatial similarity of the phenomenon represented by $Z(\mathbf{u})$ since similarity increases with H:

- $H = 0$, pure nugget effect.
- $H < 1/2$, the increments are negatively correlated.
- $H = 1/2$, the increments are uncorrelated Gaussian white noise.
- $H > 1/2$, the increments are positively correlated.
- $H = 1$, the phenomenon is differentiable and smooth, i.e., deterministic.

Most random function-related stochastic simulation algorithms, such as sequential Gaussian simulation(sGs), and most of the spectral techniques, rely on the assumption that spatial correlation exists only over a finite range and that no global trend exists. The fractal model, conversely, assumes that the structure of dependence persists over all scales. In geostatistical terms, the process has a power law semivariogram model:

$$\gamma(\mathbf{h}) = V_H |\mathbf{h}|^{2H}, \quad \text{with } \mathbf{h} = \mathbf{u}_2 - \mathbf{u}_1 \tag{1}$$

where V_H is a scaling factor. This variogram has no sill and range unless $H = 0$.

An alternative to the semivariogram measure is the autocovariance:

$$C(\mathbf{h}) = \sigma_Z^2 - \gamma_Z(\mathbf{h}), \text{ for any } \mathbf{h} = \mathbf{u}_2 - \mathbf{u}_1$$

where σ_Z^2 is the variance of $Z(\mathbf{u})$. The Fourier transform of this covariance is the spectrum $S(\mathbf{f})$, or spectral density function, defined as:

$$C(\mathbf{h}) = \int_0^\infty S(\mathbf{f}) cos(\mathbf{fh})\, d\mathbf{f}.$$

If the spectrum $S(\mathbf{f})$ is constant, then $C(\mathbf{h}) = \sigma_Z^2 \delta(\mathbf{h})$, corresponding to a pure nugget effect.

A random function $Z(\mathbf{u})$ with isotropic power variogram $V_H|\mathbf{h}|^{2H}$ has spectrum $S(\mathbf{f}) = 2\pi/|\mathbf{f}|^\beta, 1 < \beta = 1 + 2H < 3$, and autocovariance $C(\mathbf{h}) = \sigma_Z^2 - V_H|\mathbf{h}|^{\beta-1}$, with σ_Z^2 representing the spatial variance of $Z(\mathbf{u})$ over a finite size field. Note that this covariance is field size-dependent.

GENERATING UNCONDITIONAL FRACTALS

Fractional Brownian motion (fBm) can be generated with various spectral analysis-related algorithms. The particular algorithm considered here relates to the "Weierstrass-Mandelbrot"(WM) random function model(Falconer, 1990), which is a simplified version of the more CPU-time demanding Fast Fourier Transformation (FFT) algorithm (Gutjahr, 1989). Instead of an equally-spaced sampling of the frequency axis as done with FFT's, a geometric sequence of frequences with $r (r < 1)$ as the scale ratio is used, allowing more lower frequency components to be sampled. The generalized Weierstrass-Mandelbrot function with fractal co-dimension H (Falconer, 1990) is expressed as

$$Z(\mathbf{u}) = \sum_{n=1}^{\infty} C_n r^{nH} sin(\mathbf{u} \cdot \mathbf{f}_n + \Phi_n) \tag{2}$$

where $\mathbf{u} = (x, y, z)$ is the coordinate vector of the location being simulated and

$$\mathbf{f}_n = (f_x, f_y, f_z)_n, n = 1, ..., \infty$$

are discretization frequency vectors. The amplitudes $C_n, n = 1, ..., \infty$, are independent random variables normally distributed with zero mean and unit variance. The 'phases' Φ_n, $n = 1, ..., \infty$ are independent random variables uniformly distributed in $[0, 2\pi]$. In practice, the discretization is finite: n = 1, ..., N.

Clearly: $E[Z(\mathbf{u}+\mathbf{h}) - Z(\mathbf{u})] = 0$.

Furthermore,

$$\begin{aligned} E\{[Z(\mathbf{u}+\mathbf{h}) - Z(\mathbf{u})]^2\} &= E\{[\sum_{n=1}^{\infty} C_n r^{-nH} 2sin(\frac{1}{2}r^n\mathbf{h})cos(r^n(\mathbf{u}+\frac{1}{2}\mathbf{h}) + \Phi_n)]^2\} \\ &= 2\sum_{n=1}^{\infty} r^{-2nH} sin^2(\frac{1}{2}r^n\mathbf{h}) \end{aligned}$$

Since $C_n, C_m (m \neq n)$ are independent with unit variance, and the mean of $cos^2(\alpha + \Phi_n)$ is $\frac{1}{2}$.

The discretization level N is taken such that $r^{-(N+1)} \leq |\mathbf{h}| < r^{-N}$, then

$$\begin{aligned} E\{[Z(\mathbf{u}+\mathbf{h}) - Z(\mathbf{u})]^2\} &\approx \frac{1}{2}\sum_{n=1}^{N} r^{-2nH} r^{2n}|\mathbf{h}|^2 + 2\sum_{n=N+1}^{\infty} r^{-2nH} \\ &\approx V r^{-2NH} \\ &\approx V|\mathbf{h}|^{2H}, \text{ with } 0 < V < \infty. \end{aligned}$$

Thus the random field (2) has approximately the properties of a fractional Brownian motion (fBm) of dimension H, i.e., a field with Gaussian increments of mean zero and power variogram (1).

Since a single co-dimension value H is available to characterize the spatial variability, the model (2) can be used only to generate isotropic random fields.

Sampling the Frequencies

Theoretically the frequencies $\mathbf{f}$ should run from 0 to infinity, in practice they are bounded by the accuracy and scale of the data available. In practice, one must set a lower and upper limit (L_{min} and L_{max}) on the size of the correlated features to be generated in any particular direction. This amounts to defining a maximum and a minimum frequency (f_{max} and f_0 respectively). The distribution of frequency terms within the range $[f_0, f_{max}]$ could be a uniform sequence. However, whenever large scale features (with lower frequencies) are deemed more important than smaller ones (with higher frequencies), a geometric sequence of frequencies should be preferred. Let $f_0 = \frac{2\pi}{L_{max}}$, then a geometric frequency sequence can be defined as:

$$f_n = f_0/r^n, n = 1, ..., N-1$$

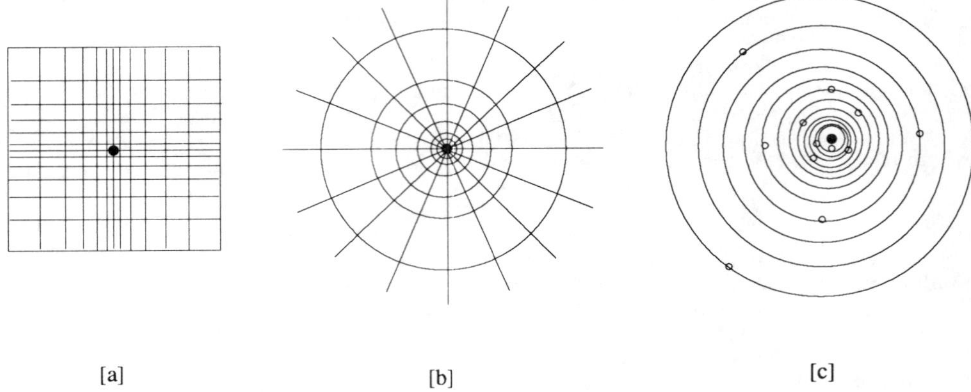

Figure 1: Three alternatives for 2D frequency sampling. The density of sampling decrease in geometric sequence.

where $r < 1.0$ is the scaling ratio for the geometric sequence and N is the number of frequency terms:

$$N = \text{integer part of } [\frac{log(L_{min}/L_{max})}{log(r)}] \tag{3}$$

Values of L_{max} much larger than the size of the study area would add a trend, whereas values of L_{min} much lower than the grid spacing add random variations which average to zero at the grid scale.

In general, a value of r closer to 1.0 results in a more complete sampling (N larger) of the frequency range. The choice of r depends also on the shape of the frequency distribution. A more exhaustive sampling is required to characterize a white noise or pure nugget effect (corresponding to a uniform spectrum) than a more peaked spectrum.

Sampling a geometric sequence in higher dimensions is not as straightforward as in 1D. There are at least three ways for doing it:

1. Frequencies are sampled on a rectangular grid, see Figure 1a.

2. Frequencies are sampled on a circular (isotropic) grid, see Figure 1b. The grid is defined by radial vectors at equal azimuth interval, with the grid size increasing in a geometric sequence.

3. A sampling scheme similar to the second one, but sampling is done along random azimuths, see Figure 1c. Each frequency term is associated with a *vector* $\mathbf{f}$ along a random direction θ. The discretization of the inner product $\mathbf{u} \cdot \mathbf{f}$ in expression (2) is calculated prior to the summation:

$$Z(x, y) \approx \sum_{n=1}^{N} r^{nH} sin(x\ |\mathbf{f_n}|\ sin\theta_n + y\ |\mathbf{f_n}|\ cos\theta_n + \phi_n) \tag{4}$$

In a 3D case, another random angle (the dip) is involved.

The first sampling strategy tends to generate artifact anisotropies in the diagonal directions. The third alternative is the one implemented hereafter.

Standardized Weierstrass-Mandelbrot Fractal Function

The distribution of the realizations generated using random function (2) is approximately normal per the Central limit Theorem. For a given discretization level N, one can make it exactly standard normal N(0, 1) using a normal score transform. When the dimension of the field is much greater than L_{max}, i.e., in an ergodic situation, the mean and standard deviation of the random function (2) can be theoretically calculated:

$$\begin{aligned} E[Z(\mathbf{u})] &= \sum_{n=1}^{N} r^{nH} E\{sin(\mathbf{u}\cdot\mathbf{f}_n+\phi_n)\} = 0 \\ Var\{Z(\mathbf{u})\} &= \sum_{n=1}^{N} (r^{nH})^2 Var\{sin(\mathbf{u}\cdot\mathbf{f}_n+\phi_n)\} \\ &= \sum_{n=1}^{N} r^{2nH}\cdot\frac{1}{2} \\ &= \begin{cases} \frac{N}{2} & \text{, if H=0;} \\ \frac{r^{2H}(1-r^{2NH})}{2(1-r^{2H})} & \text{, otherwise} \end{cases} \end{aligned}$$

since, for any random variable X,

$$Var\{sin(X)\} = E\{sin^2(X)\} - 0 = \frac{1}{2}E\{1-cos(2X)\} = \frac{1}{2}.$$

Therefore, a standardized Weierstrass-Mandelbrot model can be defined as:

$$Z(\mathbf{u}) = \begin{cases} \sqrt{\frac{2}{N}}\ \sum_{n=1}^{N} r^{nH} sin(\mathbf{u}\cdot\mathbf{f}_n+\Phi_n) & \text{, if H=0} \\ \sqrt{\frac{2(1-r^{2H})}{r^{2H}(1-r^{2NH})}}\ \sum_{n=1}^{N} r^{nH} sin(\mathbf{u}\cdot\mathbf{f}_n+\Phi_n) & \text{, otherwise} \end{cases} \quad (5)$$

Most realizations generated using this latter expression, with L_{max} small enough, do present distributions close to N(0, 1).

SIMULATION WITH NESTED FRACTALS

Most fractal (fBm) models presented in the literature are fractal structures characterized by only *two* statistics: the overall variance at some reference resolution (V_H) and the fractal co-dimension (H). The corresponding spatial variability is then modeled by the power variogram:

$$\gamma(\mathbf{h}) = V_H \mathbf{h}^{2H}$$

Although this model is sufficient to generate some visually appealing images (Voss, 1985), experimental variograms calculated from real data tend to be more complex: the log-log plot of variance versus resolution need not be linear. Examples of such more complex features include:

1. Cases where the pattern of spatial variation is different from one scale to another. For example, isotropic short scale structures associated with a larger scale anisotropic variability.

2. Anisotropic spatial variability of geologic phenomena would call for parameters V_H and H varying with direction. Hewett and Behrens, 1986, detected and modeled such anisotropy of well log properties in stratified reservoirs with two different types of fractals: a random fractal, similar to that described above in the horizontal direction, and a fractional Gaussian noise (fGn) in the vertical direction. It is not clear yet if Hewett's combination would be enough to model complex zonal anisotropies; moreover, the generation of fGn depends on how the derivative of actually non-differentiable random fractal is calculated.

To alleviate this limitation, we could consider a nested sum of power variogram models, i.e., the RF $Z(\mathbf{u})$ is modeled as the sum of several independent random functions $Y_i(\mathbf{u}), k = 1, ..., K$, each with a different power variogram:

$$\gamma_k(\mathbf{h}) = V_k \mathbf{h}^{H_k}, \quad k = 1, ..., K$$

The various fractal fields $Y_k(\mathbf{u}), k = 1, ..., K$, with appropriate contributions to the total variance are added together to define the field $Z(\mathbf{u})$. If the same frequency discretization procedure is used for all K structures with the same discretization parameters r and N, the simulation at each location can be done in a single run. In the 2D case, the model is rewritten as:

$$Z(x, y) = \sum_{n=1}^{N} \sum_{k=1}^{K} C_n r^{nH_k} sin(x \ f_n \ sin\theta_n + y \ f_n \ cos\theta_n + \phi_n) \tag{6}$$

This "nested fractals" model overcomes the major limitation of present fractal techniques, i.e., scale invariance. Indeed:

- any experimental variogram which does not appear affine over the study area can be approximated by a nested sum of power variograms with different variance contributions (V_k) and different fractal co-dimensions H_k and
- any anisotropy, no matter whether geometric or zonal, could be modeled much the same way as currently done in the practice of anisotropic variogram modeling (Isaaks and Srivastava, 1989, p.369).

Figure 3 shows the well-published reference Berea data set (Giordano *et al*, 1985, Journel, 1989), a set of 1600 permeability values taken from a slab of Berea sandstone. The corresponding exhaustive variograms in the two major directions of continuity (N35E or across banding and N55W) are shown in Figure 3. The variogram across banding reveals a nugget effect not seen on the variogram along the banding. Such feature could not be modeled with a single fractal model. Instead consider the following nested power variogram model, whose fit is shown on Figure 3:

$$\begin{aligned} \gamma(\mathbf{h}) &= 70.0 \ [\sqrt{h_x^2 + (\frac{h_y}{\infty})^2}]^\epsilon + 30.0 \ |\mathbf{h}|^{0.58} \\ &= 70.0|h_x|^\epsilon + 30.0|\mathbf{h}|^{0.58} \end{aligned}$$

with x and y denoting the coordinates along the least and most continuous directions. The first structure models the anisotropic nugget effect with H_1=0.0 and V_1 = 70.0. The second structure models the isotropic structure with H_2 = 0.29 and V_2 = 30.0.

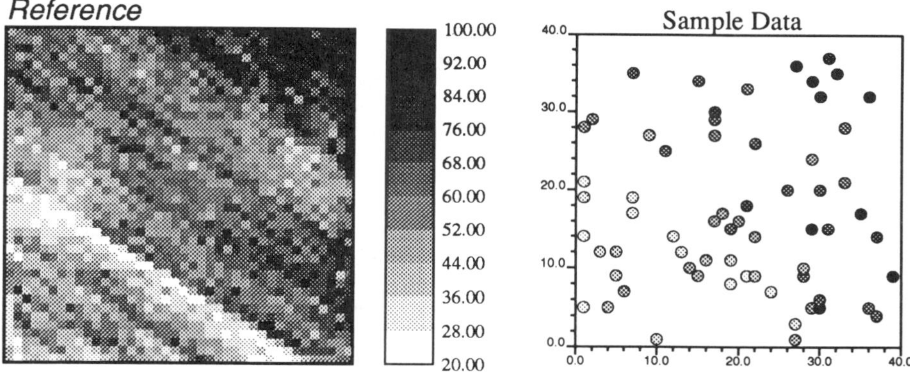

Figure 2: Berea Data: the reference (on the left) and 64 sampled data (on the right).

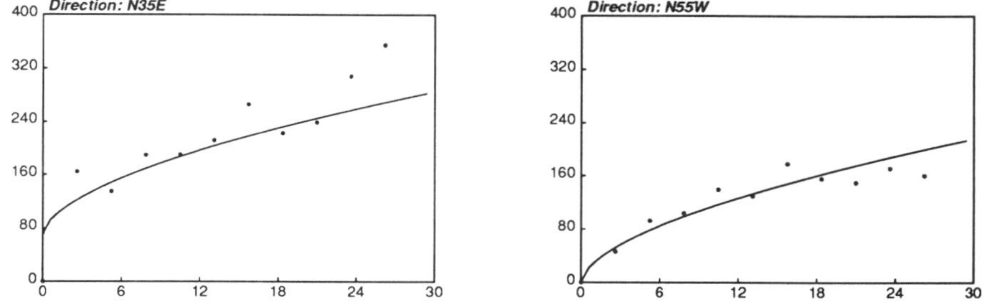

Figure 3: Experimental variograms with zonal nugget effect and their models.

ANISOTROPIC FRACTAL SIMULATION

A concern common to all spectral analysis approaches is that of anisotropy. These techniques do not deal well with anisotropy in two and three dimensions. Since many geological phenomena such as porosity in a stratified cross section do present a high anisotropy ratio, we must find a practical yet consistent way to cope with this problem.

As presented in the previous section, an anisotropic variogram model with fractal co-dimension varying with direction could be considered. However, the common definition of anisotropy ratio, as the ratio of correlation ranges in the directions of maximum and minimum continuity, does not apply because fractal variograms have no correlation range.

Suppose that the fractal variograms in the two principal directions have the same power, H, but differ in the coefficient V, say one is V_1 and the other $V_2 = \nu V_1$. The factor ν cannot be used as the anisotropy ratio since the anisotropy relation $\gamma_2(h) = \gamma_1(\nu h)$ would not hold, indeed:

$$V_2 h^{2H} = \nu V_1 h^{2H} \neq V_1(\nu h)^{2H}$$

However, if the anisotropy ratio is defined as

$$\rho = \nu^{1/2H}, \qquad 0 < H \leq 1, \tag{7}$$

then

$$V_2 h^{2H} = \nu V_1 h^{2H} = \rho^{2H} V_1 h^{2H} = V_1(\rho h)^{2H},$$

i.e, the variability over distance h in one direction is the same as that over distance ρh in the other direction, regardless of h. Similar to the conventional anisotropy ratio,

- $\rho = 1$ if the variogram is isotropic,
- $\rho > 1$ if the random variable is more continuous along the direction being corrected and $\rho < 1$ otherwise.

Simulation with the anisotropy ratio ρ consists of a rescaling of the simulation grid by the factor ρ. Assuming that the major directions of anisotropy are aligned with the simulation grid, the simulated value at a location, e.g., (x, y) in 2D, is that of an isotropic field assigned to location $(x', y') = (x, \rho y)$:

$$Z(x,y) = \sum_{n=1}^{\infty} r^{nH} sin(x' \; |\mathbf{f}| \; sin\theta_n + y' \; |\mathbf{f}| \; cos\theta_n + \phi_n) \tag{8}$$

In words, the anisotropic field is obtained as a geometric rescaling of an isotropic field.

CONDITIONAL SIMULATION

According to the general practice of conditional simulation (Journel and Huijbregts, 1978, p.495), a conditionally simulated value at any location $\mathbf{u}$ is obtained as:

$$z_{cs}(\mathbf{u}) = z_K^*(\mathbf{u}) + [z_S(\mathbf{u}) - z_{KS}^*(\mathbf{u})]$$

where $z_K^*(\mathbf{u})$ is the kriging estimate at location $\mathbf{u}$ using the conditioning data, $z_S(\mathbf{u})$ is the unconditionally simulated value and $z_{KS}^*(\mathbf{u})$ is the kriging estimate using the unconditionally simulated values at data locations. This algorithm allows for honoring the data values at data locations and reproducing the second-order statistics.

Furthermore, since $z_K^*(\mathbf{u})$ and $z_{KS}^*(\mathbf{u})$ have the same data configuration, one single kriging system is enough for both estimates. We actually need only an estimate of their difference:

$$\delta(\mathbf{u}) = z_K^*(\mathbf{u}) - z_{KS}^*(\mathbf{u}) = \sum_{\alpha=1}^{n} \lambda_\alpha [z(\mathbf{u}_\alpha) - z_s(\mathbf{u}_\alpha)]$$

Then the conditional simulation at location $\mathbf{u}$ is the sum of the unconditional simulation and this difference term:

$$z_{CS}(\mathbf{u}) = z_S(\mathbf{u}) + \delta(\mathbf{u}) \tag{9}$$

While the algorithm (9) can be applied to any non-conditional simulation, it may be slow even if a fast unconditional simulation algorithm, such as the fBm algorithm, is used. Indeed a kriging system must be solved at each location $\mathbf{u}$ on a usually very dense simulation grid. The algorithm would be fast if $\delta(\mathbf{u})$ is made an analytical function of $\mathbf{u}$ as expression (2) for the non-conditional term $z_S(\mathbf{u})$; then expression (9) for the conditional simulation would be *literally* read as a function of the coordinates vector $\mathbf{u}$. Dual kriging allows such a representation.

Dual Kriging

Dual kriging is just another presentation of the traditional or "primal" kriging (Galli *et al*, 1984, Journel, 1989). In the traditional simple kriging (SK), the estimated residual of an attribute $Z(\mathbf{u})$ at an unsampled location $\mathbf{u}$ is a linear combination of the n surrounding residual data:

$$z^*(\mathbf{u}) - m = \sum_{\alpha=1}^{n} \lambda_\alpha [z(\mathbf{u}_\alpha) - m]$$

the primal SK weights λ_α are solutions of a linear system involving the covariance function $C(\mathbf{h})$,

$$\sum_{\beta=1}^{n} \lambda_\beta C(\mathbf{u}_\alpha - \mathbf{u}_\beta) = C(\mathbf{u} - \mathbf{u}_\alpha), \qquad \alpha = 1, 2, ..., n$$

The weights λ_α can be seen as linear functions of the right hand side covariance values $C(\mathbf{u} - \mathbf{u}_\alpha), \alpha = 1, ..., n$. Consequently, the SK estimate $z^*(\mathbf{u})$ is also a linear combination of these right hand side covariance values, leading to the "dual" kriging form:

$$z^*(\mathbf{u}) - m = \sum_{\alpha=1}^{n} b_\alpha C(\mathbf{u}_\alpha - \mathbf{u}) \tag{10}$$

The dual weights $b_\alpha, \alpha = 1, ..., n$ are obtained by solving the n by n dual kriging system:

$$\sum_{\beta=1}^{n} b_\beta C(\mathbf{u}_\alpha - \mathbf{u}_\beta) = z(\mathbf{u}_\alpha) - m, \qquad \alpha = 1, 2, ..., n \tag{11}$$

This dual kriging system can be interpreted as a process of identification of the data values at data locations: $z^*(\mathbf{u}_\alpha) = z(\mathbf{u}_\alpha)$. The dual representation (10) allows some interesting remarks:

- The dual kriging estimate appears as a linear combination of n interpolation functions $C(\mathbf{u} - \mathbf{u}_\alpha)$ anchored at the n data locations. These interpolation functions need not be covariance functions and need not be stationary (e.g., the anisotropy directions can be made varying in space). It suffices that the left hand side matrix of coefficients in system (11) be invertible, which is a relaxation of the more demanding condition of positive definiteness for the full nxn covariance matrix.

- The location $\mathbf{u}$ being estimated does not appear in the dual kriging system (11). Therefore if the same data set is used to estimate Z at different locations $\mathbf{u}$, the dual kriging weight b_α associated with each sample datum does not depend on $\mathbf{u}$.

It is that second property that is of particular importance here. If the entire data set (not too large) is used in the estimation process at any location $\mathbf{u}$, the dual kriging system needs to be solved only once to determine the n dual weights. The estimate at any location $\mathbf{u}$ is just a weighted linear combination of the covariance interpolation functions, $C(\mathbf{u}_\alpha - \mathbf{u}), \alpha = 1, ..., n$, as in (10).

Similarly, ordinary kriging can be expressed in its dual form as:

$$z^*(\mathbf{u}) = \sum_{\alpha=1}^{n} b_\alpha C(\mathbf{u}_\alpha - \mathbf{u}) + s$$

where the (n+1) unknown variables ($b_\alpha, \alpha = 1, ..., n$ and s) are provided by the dual OK system:

$$\begin{cases} \sum_{\beta=1}^{n} b_\beta C(\mathbf{u}_\alpha - \mathbf{u}_\beta) + s = z(\mathbf{u}_\alpha), & \alpha = 1, 2, ..., n \\ \sum_{\beta=1}^{n} b_\beta = 0 \end{cases} \tag{12}$$

Once again, provided that all n data are used for each location $\mathbf{u}$ being estimated, the dual weights b_α, s, are independent of $\mathbf{u}$.

Procedures for Conditional Simulation

Building on the dual kriging formalism, one can read relation (9) as a function of location $\mathbf{u}$. The whole process of simulating a field with honoring of the data can then be summarized as follows.

1. Read in the parameters and conditioning data from file

2. The specific fractal-generating algorithm defines the non-conditional simulation as a specific function of $\mathbf{u}$: $z_S(\mathbf{u}) = \tau(\mathbf{u})$, see, e.g., expression (2) for fBm

3. Retrieve the unconditionally simulated values at the original data locations: $z_S(\mathbf{u}_\alpha)$, $\alpha = 1, ..., n$

4. Build and solve the (possibly large) dual kriging system (11) or (12) to obtain the dual weights b_α, and the kriging difference function

$$\delta(\mathbf{u}) = z_K^*(\mathbf{u}) - z_{KS}^*(\mathbf{u})$$

5. Deduce the conditional simulation expression:

$$z_{CS}(\mathbf{u}) = z_S(\mathbf{u}) + [z_K^*(\mathbf{u}) - z_{KS}^*(\mathbf{u})] = \tau(\mathbf{u}) + \delta(\mathbf{u}),$$

and read the value $z_{CS}(\mathbf{u})$ at any location $\mathbf{u}$ to be simulated.

Note that although a grid may be used, it is not required. The algorithm can be used to simulate z_{CS} at any location $\mathbf{u}$; for example, one can zoom over any section of the study area and retrieve the simulated values at any scale, provided that the fractal variogram model is deemed correct at all such scales.

A 2D CASE STUDY

A 2D fractal conditional simulation of the Berea data set is presented. The most important feature of that data set, as seen on Figure 2, is the strong anisotropy along direction $N57^oW$. Sixty four conditioning data values were randomly sampled from the reference set and their normal score variograms in various directions were calculated in order to evaluate the fractal dimension.

Figure 4 shows the normal score variograms in the two principal directions: N57W (h'_x) and N33E (h'_y). Also given is the fit by the anisotropic model:

$$\gamma(h'_x) = 0.6 + 0.04|h'_x|^{0.7} \text{ for the N33E direction and,}$$

$$\gamma(h'_y) = 0.1 + 0.08|h'_y|^{0.7} \text{ for the direction orthogonal.}$$

In words, two nested fractal models were considered with co-dimensions 0.0 and 0.35. The anisotropy ratios are 1.66 and 0.2 respectively, as defined in relation (7).

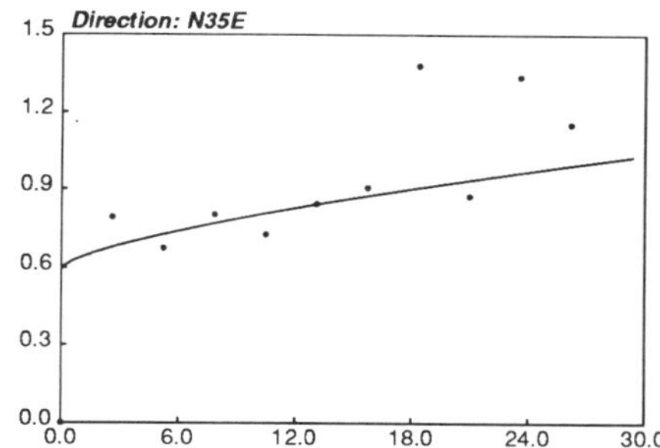

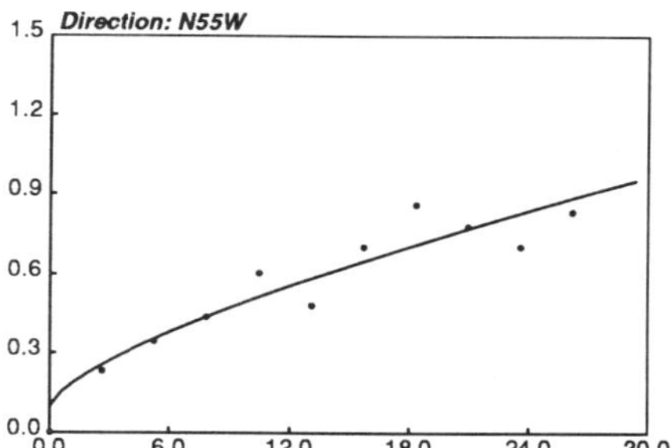

Figure 4: Sample normal score variograms and corresponding nested fractals model

Three conditional simulation realizations are shown in Figure 5, to be compared to the reference image of Figure 2. The conditioning data are honored and the major structures are seen to be reasonably well reproduced.

From Figure 4, one could argue that the sample points could have been fit within a finite sill variogram model, not calling for any fractal co-dimension. Indeed, Figure 6 gives the same sample variogram values fit by the following nested spherical structures with zonal anisotropy and total sill 1:

$$\gamma(\mathbf{h}) = .1 + .5Sph(h'_y/40) + 0.4Sph(|\mathbf{h}|/25), \quad \forall\mathbf{h} > 0,$$

where

$$Sph(h) = \begin{cases} \frac{3}{2}h - \frac{1}{2}h^3 & , \ \forall \text{ h} < 1 \\ 1 & , \ \text{if h} \geq 1 \end{cases}$$

This normal score variogram model was input into a sequential Gaussian simulation algorithm (Deutsch and Journel, 1992, p.141) to generate the two conditional simulation realizations of Figure 7. The large range structures, in particular the diagonal banding of low permeability values, are seen to be well reproduced.

The fractal simulation realizations of Figure 5 are seen to be no better (and no worse) than the realizations obtained from sequential Gaussian simulations. Either algorithm has only available one variogram model common to all classes of permeability values where

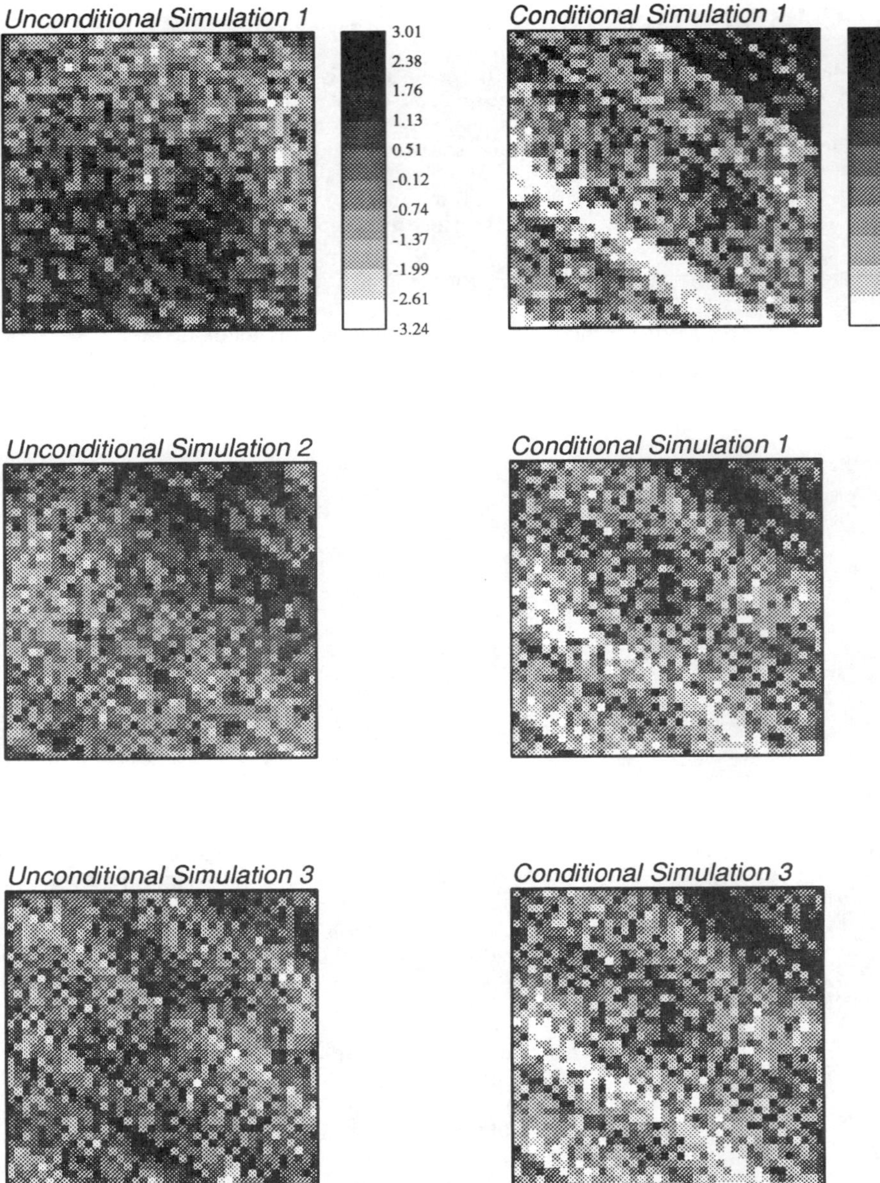

Figure 5: Three realizations of conditional fractal simulation

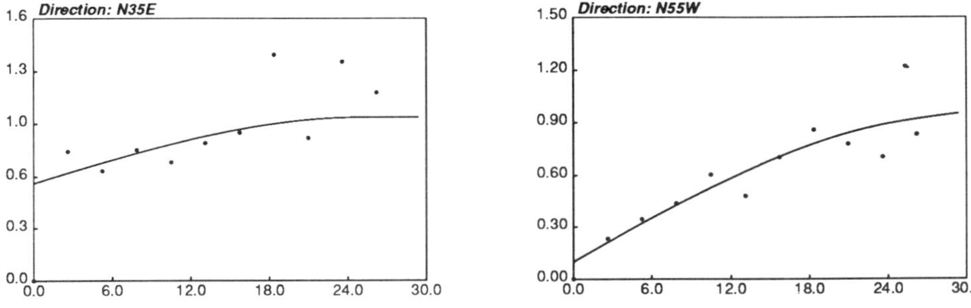

Figure 6: The nested spherical model used for Gaussian simulation.

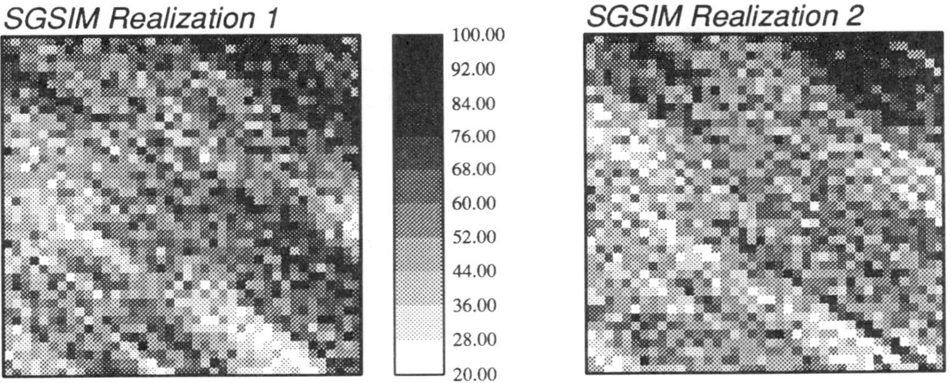

Figure 7: Two conditional realizations using a sequential Gaussian simulation algorithm.

it is apparent on the reference image of Figure 2 that low values are more banded than high permeability values. The indicator simulation formalism (Deutsch and Journel, 1992, p.149) calls for as many indicator variogram as there are classes of permeabilities minus one, hence allows reproduction of structures specific to each class of the continuous attribute being simulated, see Journel and Alabert, 1989.

Analysis of cpu time

Considering the two terms of the conditional fractal expression (9), the step of unconditional simulation is the fastest even when nested power variograms and rotations are involved and a high discretization of frequencies is used. Since calculation of sine functions is expensive, one can speed up the simulation by utilizing a pre-calculated sine function lookup table

The second step of conditioning involving dual kriging with a global search neighborhood is very sensitive to the number(n) of conditioning data. Not only does a large system need to be solved, the dual estimate calls for summation of two vectors of, possibly, very long length at each node $\mathbf{u}$:

$$z^*(\mathbf{u}) = \sum_{\alpha=1}^{n} b_\alpha C(\mathbf{u}_\alpha - \mathbf{u})$$

Redundant, repetitive calculation of the covariance values is avoided by relocating the conditioning data to the simulation grid nodes and building a covariance lookup table prior to the conditional simulation. This approximation is acceptable because in the simulation grid is usually much denser than the original data layout.

In Table 1 the CPU times of several fractal simulations on a 3D grid of 40x40x40 nodes using different numbers of conditioning data are shown and are compared to those of sequential Gaussian simulations using program **sgsim** of GSLIB (Deutsch and Journel, 1992, p.164). The fractal algorithm with dual kriging is faster than SGSIM only when the number (n) of data is less than 200. More data, e.g., 500, would make dual kriging impractical, because of the corresponding single but very large system. Also a linear system solver, using iterative methods, and designed to reduce the effect of rounding errors when handling large matrices, would be needed.

Number of	CPU time (seconds)	
cond. data	fBm Fractal	Sequential Gaussian
0	15	142
100	49	136
200	75	134
300	106	131

Table 1: CPU time comparison between fractal simulation and sequential Gaussian simulation on a 40x40x40 grid. Runs were done on a DEC5000/240 workstation.

CONCLUSIONS

An algorithm for the simulation of realizations of the fractional Brownian motion (fBm) model has been proposed. The algorithm allows for nested fractals, as modeled by the sum of power variogram models with different power values (or fractal co-dimensions) and different anisotropy directions and ratios. The fBm model is represented as a sum of sine functions with frequencies oversampling the short wave lengths, i.e., the large range structures; in this regard the fBm model falls into the class of spectral representation of (pseudo-)stationary random function models.

The main contribution of this paper is the implementation of dual kriging to condition the fBm realization to local data. Provided there are not too many such data, dual kriging with a global data search neighborhood, allows representing the conditional simulation as an analytical function of the coordinates vector $\mathbf{u}$ which, then, can be read at any location $\mathbf{u}$ whether gridded or not and at any spacing.

fBm simulation shares the limitations of sequential Gaussian and spectral simulation algorithms. They assume multiGaussianity (after normal score transform) of the phenomenon being studied. Soft data connot be easily incorporated. The fBm simulation is less flexible than sequential Gaussian simulation in that it is limited to power variogram models, while the sequential Gaussian algorithm can accommodate power-type looking sample variograms with models with large but finite ranges and zonal anisotropy. The case study presented did not show any clear advantage of the fBm model over the more traditional multiGaussian model (with zonal anisotropy).

References

Deutsch, C. V. and Journel, A. G. (1992) GSLIB: Geostatistical Software Library and User's Guide, Oxford University Press, 340 p.

Falconer, K.J. (1990) Fractal Geometry: Mathematical Foundation and Application, Wiley & Sons Publ., 288 p.

Galli, A., Murillo, E. and Thomann, J., (1984) "Dual kriging – its properties and its uses in direct contouring", In Geostatistics for Natural Resources Charactererization, ed. Verly et al, publ. Reidel, Dordrecht, Holland, part2, pp. 621-634.

Giordano, R., Salter, S. and Mohanty, K. (1985) "The effects of permeability variations on flow in porous media", SPE paper no. 14365.

Gutjahr, A. (1989) "Fast Fourier Transforms for Random Fields", Technical Report No. 4-R58-2690R, Los Alamos, NM.

Hewett, T. A., (1986) "Fractal Distributions of Reservoir Heterogeneity and Their Influence on Fluid Transport", SPE paper no. 15386.

Hewett, T. A. and Behrens, R. A. (1988) "Conditional Simulation of Reservoir Heterogeneity with Fractals", SPE paper no. 18326, 1988.

Isaaks, E. and Srivastava, R.M. (1989) An introduction to Applied geostatistics, Oxford University Press, New York, NY.

Journel, A. G. (1985) "Universal and Dual Kriging", unpublished course note.

Journel, A. G. (1989) Geostatistics in Five Lessons, Short course in Geology: Volume 8, American Geophysics Union, 40 p.

Journel, A. G. and Alabert, F. (1989) "Non-Gaussian data expansion in the earth sciences", Terra Nova, 1:123-134.

Mandelbrot, B. B. and Ness, J. V. (1968) "Fractional Brownian Motions, Fractional Noises and Applications", SIAM Review, V. 10, No. 4.

Voss, R. F. (1985) "Random Fractal Forgeries", Proceeding of the NATO ASI. Fund. Algorithms in Comp. Graphics, Ilkley, U.K.

ALGORITHMICALLY-DEFINED RANDOM FUNCTION MODELS

CLAYTON V. DEUTSCH
Exxon Production Research Co., P.O. Box 2189, Houston, TX 77252

Three criteria are proposed to select an appropriate stochastic simulation technique: 1) the implementation of the technique must allow realizations to be generated with a reasonable amount of human-involvement and CPU time, 2) all relevant prior information must be honored, and 3) the realizations should generate the largest space of uncertainty. These criteria are independent of the mechanism or random function (RF) underlying the stochastic simulation technique. This paper compares algorithmically-defined random function models, such as those based on simulated annealing, to analytically defined random function models, such as the multiGaussian model.

Stochastic simulation techniques based on these different random function models are presented. An example is presented which illustrates the relative advantages of algorithmically-defined and analytically defined techniques. The mathematical cleanliness and internal consistency of analytical models must be balanced against the flexibility of algorithmically-defined models.

INTRODUCTION

The generation of alternative numerical models or images of spatially-varying attributes that account for the known aspects of the spatial distribution is generally referred to as *stochastic simulation* or *stochastic imaging*. For practical problems, the stochastic realizations must be further processed by a *transfer function*, e.g., a program that simulates the mining operation or the fluid flow in an aquifer or petroleum reservoir. Generating the stochastic realizations is the first step; it is the performance of the models that matters. This paper will focus on the petroleum reservoir context; many of the conclusions and examples, however, apply to other areas of application.

In practice, the spatial distribution of porosity and permeability must be modeled using information from many sources: a limited number of good quality well data, a greater number of indirect seismic data, knowledge of the geological setting, and interpretations from a limited number of well tests. The reservoir management problem is to assess the performance of a number of alternative production scenarios.

Given sparse sampling and uncertainty in the available data there should not be a unique model of the spatial distributions of porosity and permeability. The idea behind stochastic reservoir modeling is to generate a number of alternative numerical

R. Dimitrakopoulos (ed.), Geostatistics for the Next Century, 422–435.

models (called *realizations*) that are all consistent with the known data. Uncertainty in the prediction due to uncertainty in the rock/fluid properties can be quantified by running a flow simulation program on a number of alternate numerical models.

This concept is contrasted with the reality of a single true distribution of rock properties in Figure 1. The first step in a reservoir modeling exercise is to establish the spatial distribution of rock and fluid properties (upper portion of Figure 1). Although, there is only one true distribution of these properties in reality, yet, there can be many stochastic realizations of the spatial distribution, each of which is consistent with the available data.

The next step is to consider the proposed recovery scheme (central portion of Figure 1). The actual recovery scheme, symbolized by the drilling rig, can be implemented only once in the actual reservoir. A flow simulation program, symbolized by the computer, provides a numerical model of the recovery scheme for each realization.

Finally, as illustrated at the bottom of Figure 1, there is only one true value for each response variable (e.g., hydrocarbon recovery, breakthrough time, flow rate, bottom hole pressure, etc.). Each realization potentially yields a different response providing a probability distribution for each response variable. In practice, the true response remains unknown until it is too late to alter the recovery scheme. The simulated distributions of response variables can be used to assess the risk involved with any particular recovery scheme.

There are many techniques for stochastic simulation. Given a choice between two techniques which one should be preferred? The first practical criterion is that all potentially *good* methods must be feasible, i.e., they must generate plausible realizations in a reasonable amount of time (both human and CPU time). If two candidate techniques pass this first criterion (see the bottom of Figure 1) the spread of the output distribution should be maximized to increase the chance that this output distribution includes the true value (of bottom Figure 1). This relates to the *maximum entropy* concept used in information theory [11,18].

An important aspect of the maximum entropy approach is to consider an output distribution *conditional to all the information available*; the spread or entropy of the output distribution should not be artificially expanded by geologically implausible realizations or those inconsistent with observed data. An appreciation for the plausibility of a model, based on experience and an understanding of the geological processes that created the reservoir, is yet another piece of information that constrains the output distribution.

Another important point is that the entropy to be maximized is that of the response (output) distributions and **not** that of the input realizations. The output response variables are related to the input spatial distributions through a specific transfer function (flow simulator); however, that transfer function is usually very complex and non-linear. Even though spatial entropy of the input realizations can be defined and predicted (see [15]), in general, its relation to the entropy of the response or output distribution is not known a priori.

The spread of a response distribution is sometimes referred to as a *space of uncertainty*. The extent of that space, i.e., the uncertainty about the output is necessarily related to the model for input uncertainty. To summarize, a good technique:

1. Must generate plausible realizations in a reasonable amount of time. The time refers to the human and the CPU time required for the initial set up and the repeated application of the technique.

2. Allows the maximum prior information to be accommodated. This is the only

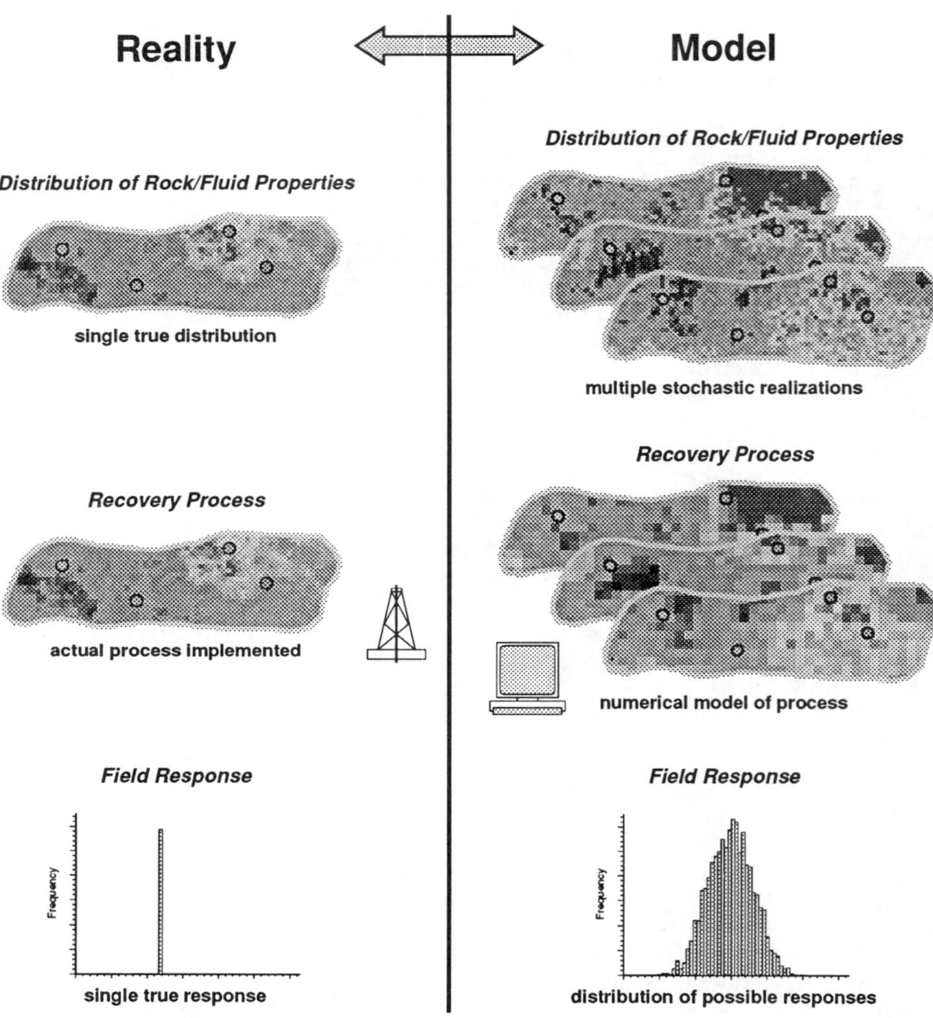

Figure 1: A schematic illustration of stochastic reservoir modeling. The first step consists of establishing the spatial distribution of rock and fluid properties. In reality, there is only one true distribution of these properties, yet, there can be many stochastic realizations of that distribution. The next step is the implementation of the recovery scheme. In reality, the recovery scheme may be implemented only once in the field. A flow simulator provides a numerical model of the recovery scheme using each alternate input model. Finally, there is only one true reservoir response, but there is a distribution of possible responses given the alternate stochastic realizations which can be generated.

direct way to ensure that the output distribution is as accurate as possible.

3. Explores the largest space of uncertainty for the output, i.e., one that generates a maximum entropy distribution of response variables.

These criteria provide the basis for comparison of different simulation techniques and RF models.

RANDOM FUNCTION MODELS

A random function (RF), defined over some field of interest, e.g., $\{Z(\mathbf{u}), \mathbf{u} \in \text{study area } A\}$, is characterized by the set of all its K-variate cdf's for any number K and any choice of the K locations $\mathbf{u}_k, k = 1, \ldots, K$:

$$F(\mathbf{u}_1, \ldots, \mathbf{u}_k; z_1, \ldots, z_K) = Prob\{Z(\mathbf{u}_1) \leq z_1, \ldots, Z(\mathbf{u}_K) \leq z_K\} \tag{1}$$

Just as a univariate probability distribution (cdf) may be used to characterize uncertainty about a single value $z(\mathbf{u})$, the multivariate cdf (1) may be used to characterize joint uncertainty about the K values $z(\mathbf{u}_1), \ldots, z(\mathbf{u}_K)$.

Stochastic simulation is the process of drawing alternative, equally probable, joint realizations from a RF model. The (usually gridded) realizations $\{z^{(l)}(\mathbf{u}), \mathbf{u} \in A\}$ $l = 1, \ldots, L$ represent L possible images of the spatial distribution of the attribute values $z(\mathbf{u})$ over the field A. Each realization reflects the properties imposed on the RF model $Z(\mathbf{u})$. As mentioned above, the more properties that are inferred from the sample data and incorporated into the RF model $Z(\mathbf{u})$, the better that RF model.

An analytical RF is one where the multivariate distribution is known analytically and may be written in a mathematically concise expression. An algorithmically-defined RF model is one where the multivariate probability distribution is observed by generating alternative realizations. In the case of algorithmically-defined RF models, there is no need for a mathematical definition of the distribution; the only requirement is a repeatable algorithm to generate stochastic realizations that honor the conditioning data and spatial statistics. The obvious advantage of analytically-defined RF models is that they may be studied theoretically; the mathematical consistency allows proofs, theorems, and limit (stationary and ergodic) properties to be evaluated a priori.

The best example of an analytically-defined RF model is the Gaussian RF model. Most analytically-defined RF models are related to the Gaussian model in some way.

The Gaussian RF Model

The Gaussian RF model is unique in statistics for its analytical simplicity and for being the limit distribution of many analytical theorems known as "central limit theorems" [2,12]. Some *characteristic* properties of the multivariate Gaussian (normal) RF model $Y(\mathbf{u})$ are:

- all subsets $\{Y(\mathbf{u}), \mathbf{u} \in B \subset A\}$ are also multivariate normal.
- all linear combinations of the components of $Y(\mathbf{u})$ are (univariate) normally distributed, e.g.,

$$X = \sum_{\alpha=1}^{n} \omega_\alpha Y(\mathbf{u}_\alpha) \text{ is normally distributed,}$$

$$\forall \text{ the weights } \omega_\alpha, \text{ as long as } \mathbf{u}_\alpha \in \mathbb{A}.$$

- zero covariance (or correlation) entails full independence: If $Cov\{Y(\mathbf{u}), Y(\mathbf{u}')\} = 0$, the two RV's $Y(\mathbf{u})$ and $Y(\mathbf{u}')$ are not only uncorrelated, they are independent.
- all conditional distributions of any subset of the RF $Y(\mathbf{u})$ are (multivariate) normal.

The Gaussian RF is the only analytically-defined RF model considered in this paper. The following discusses two different algorithmically-defined RF models. The indicator RF model, discussed first, has a better theoretical pedigree [13,16] than the annealing RF model, discussed last.

The Indicator RF Model

Indicator kriging of a continuous variable is *not* aimed at estimating the indicator transform

$$i(\mathbf{u}; z_k) = \begin{cases} 1, & \text{if } z(\mathbf{u}) \leq z_k \\ 0, & \text{otherwise} \end{cases} \tag{2}$$

Indicator kriging provides a least-squares estimate of the conditional cumulative distribution function (ccdf) at cutoff z_k:

$$\begin{aligned} [i(\mathbf{u}; z_k)]^* &= E\{I(\mathbf{u}; z_k|(n)\}^* \\ &= Prob^*\{Z(\mathbf{u}) \leq z_k|(n)\} \end{aligned} \tag{3}$$

where (n) represents the conditioning information available in the neighborhood of location $\mathbf{u}$.

The IK process is repeated for a series of K cutoff values $z_k, k = 1, \ldots, K$, which discretize the interval of variability of the *continuous* attribute z. The conditional cdf, built from assembling the K indicator kriging estimates represents a probabilistic (RF) model for the uncertainty about the unsampled value $z(\mathbf{u})$.

If $z(\mathbf{u})$ is a continuous variable, then the optimum selection of the cutoff values z_k at which indicator kriging takes place is essential: too many cutoff values and the inference and computation becomes needlessly tedious and expensive; too few, and the details of the distribution are lost.

In indicator kriging the K cutoff values z_k are usually chosen so that the corresponding indicator covariances $C_I(\mathbf{h}; z_k)$ are different one from another. There are cases, however, when the sample indicator covariances/variograms appear proportional to each other, i.e., the sample indicator correlograms are all similar in shape. The corresponding continuous RF model $Z(\mathbf{u})$ is the so-called "mosaic" model [14] such that:

$$\rho_Z(\mathbf{h}) = \rho_I(\mathbf{h}; z_k) = \rho_I(\mathbf{h}; z_k, z_{k'}), \ \forall z_k, z_{k'} \tag{4}$$

where $\rho_Z(\mathbf{h})$ and $\rho_I(\mathbf{h}; z_k, z_{k'})$ are the correlograms and cross correlograms of the continuous RF $Z(\mathbf{u})$ and its indicator transforms.

Indicator kriging under the mosaic model (4) is called "median indicator kriging" [13]. It is a particularly simple and fast procedure since it calls for a single easily inferred variogram (often the median indicator variogram) that is used for all K cutoffs.

The Annealing RF Model

In the "annealing" approach to stochastic simulation there is no explicit random function model. Rather, the creation of a simulated realization is formulated as an optimization problem to be solved with a numerical optimization technique (seminal references for the application of these techniques to spatial problems include [8,9,17, 19]). The first requirement of this class of methods is the construction of an objective (or energy) function which is some measure of difference between the desired spatial characteristics and those of a candidate realization.

The global optimization technique most often used to obtain such realizations is based on an analogy with the metallurgical process of annealing. Annealing is the process by which a material undergoes extended heating and is slowly cooled. Thermal vibrations permit a reordering of the atoms/molecules to a structured lattice, i.e., a low energy state. In the context of 3-D numerical modeling, the *annealing* process may be simulated through the following steps:

1. An initial 3-D numerical model (analogous to the initial metal in true annealing) is created, for example, by assigning a random value at each grid node by drawing from the population distribution.

2. An energy function (analogous to the Gibbs free energy in true annealing) is defined as a measure of difference between desired spatial features and those of the realization. For example, the energy or objective function could be the sum of the squared difference between the variogram of the realization and a model variogram over a predefined set of lag distances.

3. The image is perturbed, for example, by swapping pairs or sets of values taken at random locations in the 3-D numerical model (this mimics the thermal vibrations in true annealing).

4. The perturbation (thermal vibration) is always accepted if the energy is decreased; it is accepted with a certain probability if the energy is increased (the Boltzmann probability distribution of true annealing). Technically, the name "simulated annealing" applies only when the acceptance probability is based on the Boltzmann distribution [1,17] Through common usage, however, the name "annealing" is used to describe the entire family of methods that are based on this optimization principle.

5. Continue the perturbation procedure while reducing the probability with which unfavorable swaps are accepted (lower the temperature parameter of the Boltzmann distribution) until a low energy state is achieved.

6. Low energy states correspond to plausible numerical models (realizations).

At first glance this approach appears terribly inefficient; millions of perturbations may be required to obtain an image having the desired spatial structure. These methods, however, are more efficient than they might seem as long as only a few arithmetic operations are required to update the objective function after each perturbation. Virtually all conventional spatial statistics (e.g., covariances/correlations) may be updated locally (considering only a few locations) rather than recalculated globally (considering all locations).

The objective function is defined as some measure of difference between a set of reference properties and the corresponding properties of a candidate realization. The reference properties could consist of any *quantified* geological, statistical, or engineering property. Some examples include two-point transition probabilities [8, 5], seismic data [7], mutiple-point statistics [3,10], and well test-derived effective properties [4]. Traditional two-point variogram/covariance functions and correlation coefficients with a secondary attribute will be considered in this paper. The real advantage of annealing is this ability to integrate many disparate sources of data.

A modified version of the public domain `sasim` source code documented in [6] was used for the examples presented below.

AN EXAMPLE

A cross section bounded by two wells is shown at the top of Figure 2. The rock is a binary mixture of sandstone (1000md) and shale (0.01md). The gray-shaded profile next to each well represents the shale proportion for the corresponding vertical interval; white corresponds to 0% shale and black to 100% shale. The absolute permeability of each vertical interval was taken as the geometric average of the component sandstone and shale fractions. The histogram of shale proportion, 50 values from each well, is given on the lower-left of Figure 2. The variogram model of shale proportion (standardized to a unit sill) is shown on the lower-right of Figure 2. Note that the vertical extent of the cross-section is 10 meters (5 times the vertical variogram range) and the horizontal extent is 25 meters (2.5 times the horizontal variogram range).

The stochastic simulation problem is to construct representative 2-D models of the shale proportion and convert them to elementary grid-block permeabilities (here using a geometric average of the component shale and sandstone).

Three candidate stochastic simulation techniques based on three different RF models were considered: multiGaussian (`sgsim` program in GSLIB [6]), indicator (`sisim` [6]), and simulated annealing (`sasim` [6]). The objective function in `sasim` was the variogram for 50 lags ($n_h = 50$) covering all directions:

$$O = \sum_{i=1}^{n_h} [\gamma^*(h_i) - \gamma(h_i)]^2 \tag{5}$$

where n_h is the number of variogram lags to be honored $\gamma^*(h_i)$ is the variogram of the realization for lag h_i, and $\gamma(h_i)$ is the model variogram.

One hundred realizations were generated by each technique. The CPU time requirements are shown on Table 1. The time to generate 100 realizations varies from 15 minutes for `sgsim` to 4 hours for `sisim` (10 cutoffs). Two realizations from the Gaussian, indicator, and annealing RF models are shown on Figure 3; the realizations appear quite different due to characteristics of the RF model beyond the bivariate (variogram) level. All of the methods, however, appear plausible and meet the first criterion, i.e., they generate plausible realizations in a reasonable amount of time.

A transfer function is required to judge the space of uncertainty that is sampled by each RF model. The response variable selected here is the effective absolute permeability in the horizontal direction. Many other flow-related response variables

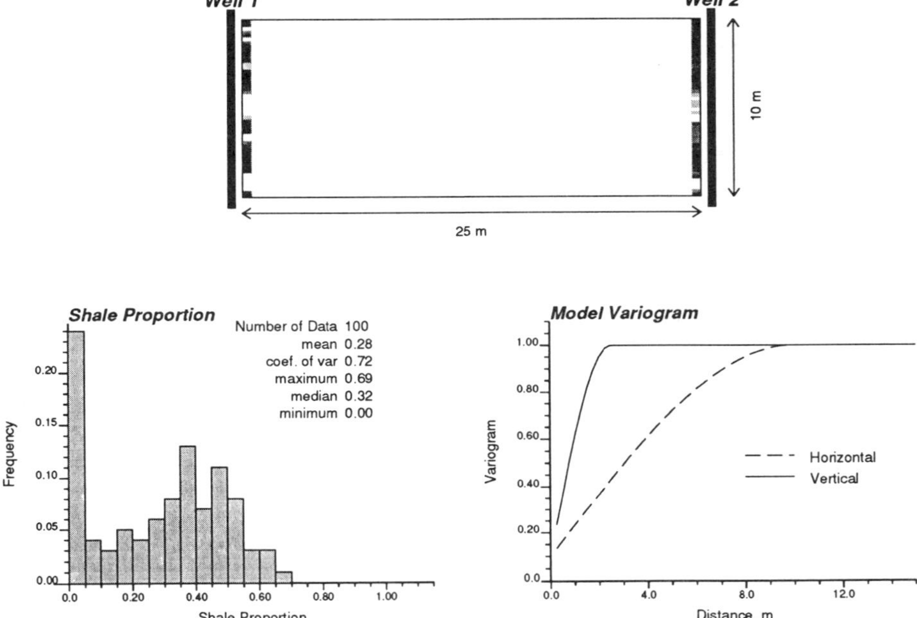

Figure 2: A gray scale plot of the two wells (at the far right and far left sides), a histogram of the 100 data and a model variogram.

Algorithm	Program	Total CPU Time(min)	Time (sec/node)	Relative Time
Gaussian	`sgsim`	15.4	0.0037	1.0
Indicator (median approx.)	`sisim`	30.0	0.0072	2.0
Indicator (10 cutoffs)	`sisim`	238.7	0.0573	15.5
Annealing (50 lags)	`sasim`	109.2	0.0262	7.1

Table 1: The CPU time, measured on a Silicon Graphics Crimson workstation, required to generate 100 simulations each with 2500 nodes. The time to simulate one node is also shown. The relative times are with respect to the `sgsim` program.

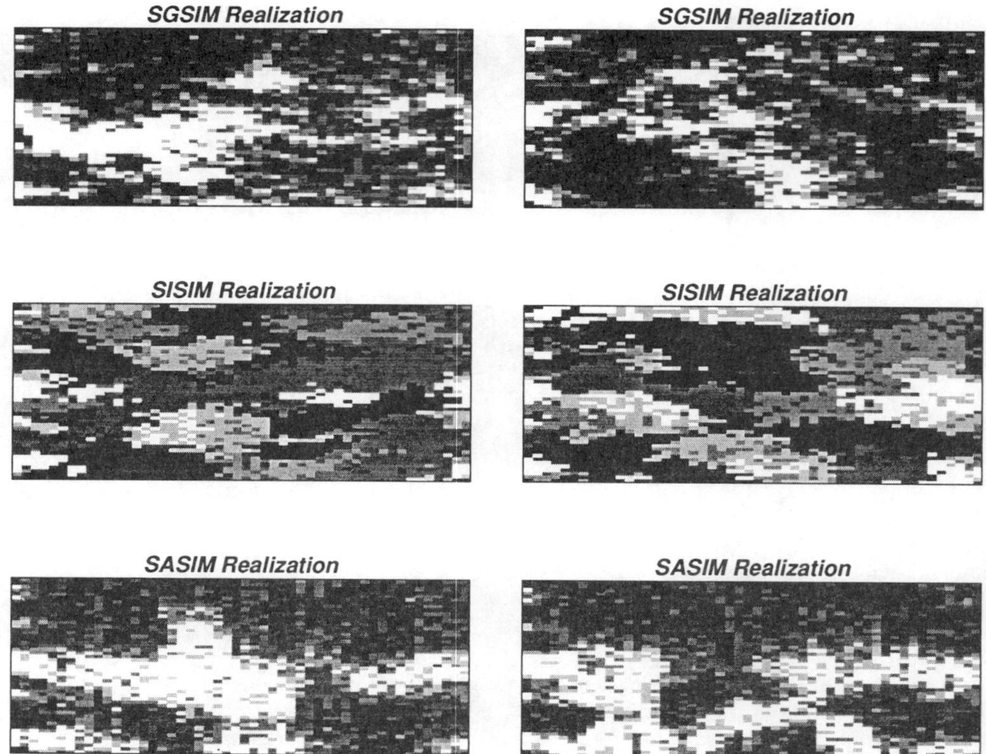

Figure 3: Two realizations of the multiGaussian RF (top), the indicator RF (center), and the annealing RF (bottom).

depend on the permeability. Histograms of the horizontal effective absolute permeability are shown on Figure 4.

The center of these distributions (the most likely response) and the spread (a measure of uncertainty) are significantly different for each simulation technique. The measures of the spread of the distribution are the standard deviation and the width of the 95% probability interval (shown below each histogram).

To extend this example further, suppose that we knew that the sandstone units (low shale proportion) had a greater continuity in the horizontal direction and less continuity vertically. The overall continuity is represented by the variogram (Figure 2), however, we now have the added indicator variogram model, see Figure 5, for the shale proportions less than 5%.

The Gaussian RF cannot directly account for this information. One could consider a mixture of two Gaussian RF models. The indicator RF model can integrate this information. However, the typical order relations corrections make it difficult to honor an indicator variogram for an extreme threshold and the overall variogram simultaneously. The annealing RF model can integrate this information by simply adding a component to the objective function:

$$O = \sum_{i=1}^{n_h} [\gamma^*(h_i) - \gamma(h_i)]^2 + \sum_{j=1}^{n_c} \sum_{i=1}^{n_h} \left[\gamma_j^*(h_i) - \gamma_j(h_i)\right]^2 \tag{6}$$

where there are n_c indicator variograms (one in this case). The number of lags $n_h = 50$ will be constant for both the variogram and the single indicator variogram. Two realizations using this objective function are shown on Figure 6. Notice the increased continuity of the sand (white).

One hundred realizations were generated and the horizontal effective permeability was computed, see Figure 7. Notice that the center of the distribution has changed significantly; it has more than doubled from the previous runs (Figure 4). Further, the spread of this distribution has increased. The uncertainty should decrease as more information becomes available. The implication is that the uncertainty shown on Figure 4 is optimistic. It is a fairly general observation, that conventional measures of uncertainty used in geostatistics tend to increase rather than decrease as more information becomes available.

REMARKS AND CONCLUSIONS

The multivariate probability law of RF models implicit to most simulation algorithms, aside from the multiGaussian RF, is usually too complex to be defined and understood analytically. The main advantages of an analytically-defined RF, mathematical cleanliness and internal consistency, are of no benefit when the response variable is obtained from a non-linear transfer function, such as a flow simulator. The output distributions of uncertainty, in general, are obtainable only by generating multiple realizations.

RF models which are inferred by generating a number of realizations and observing the multivariate probability law are refered to as algorithmically-defined RF models. The main advantage of these RF models is the flexibility to integrate additional data from various sources.

The RF models discussed in this paper are a fair sample of commonly used techniques. There are many other RF models that could have been discussed, e.g.,

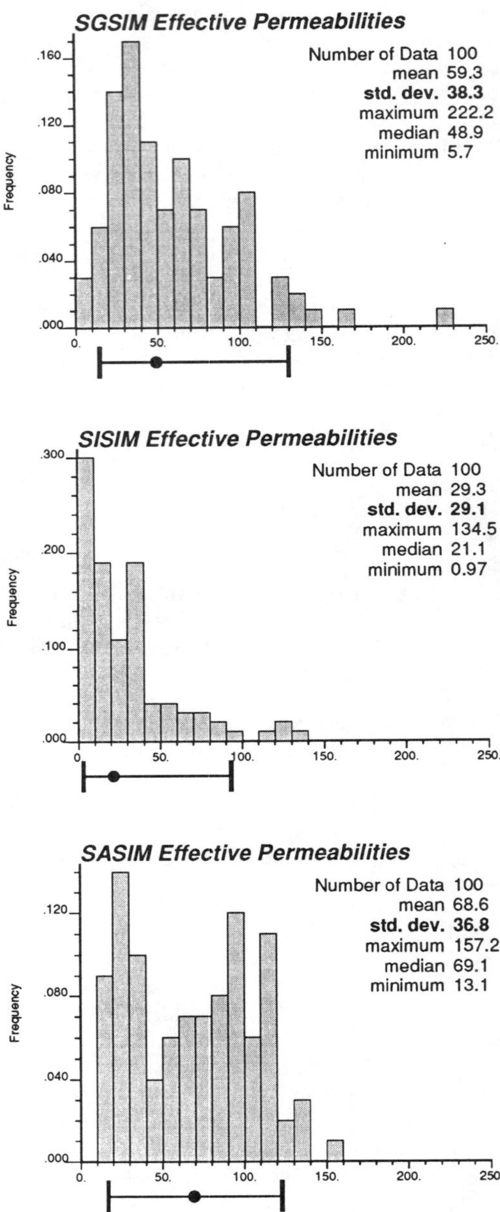

Figure 4: Histograms of the horizontal effective permeability for 100 realizations of the tiGaussian RF (top), the indicator RF (center), and the annealing RF (bottom). The 95% probability interval and the median is shown below each histogram.

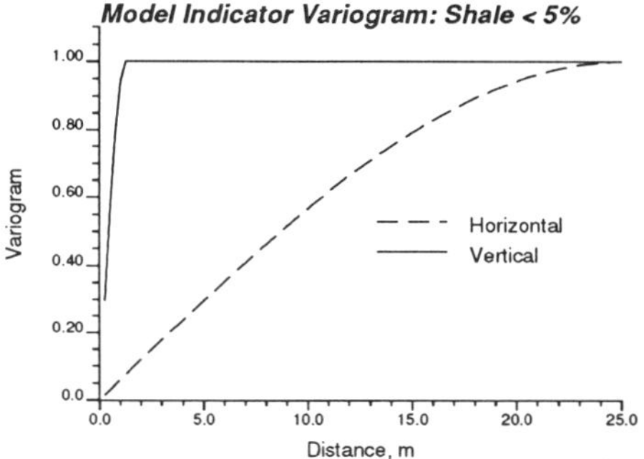

Figure 5: The model indicator variogram for the sandstone (low shale proportion).

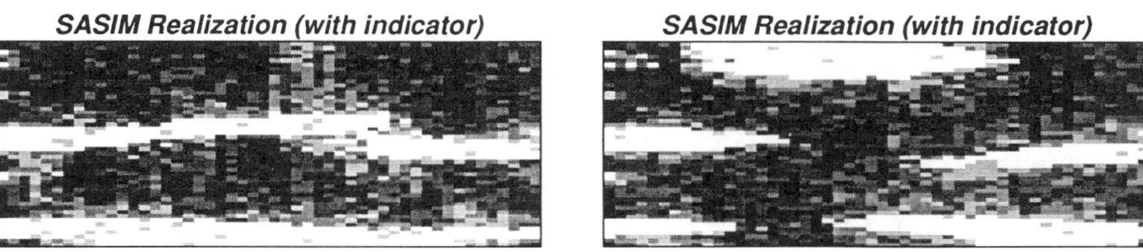

Figure 6: Two realizations of the annealing RF honoring the variogram and an indicator variogram.

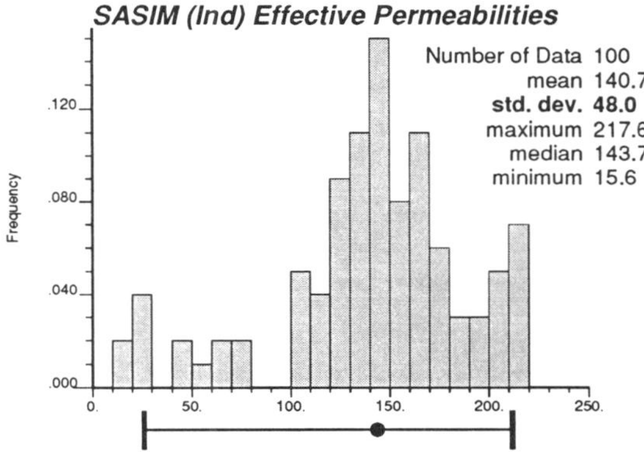

Figure 7: Histogram of the horizontal effective permeability for 100 realizations of the annealing RF honoring the variogram and an indicator variogram. The 95% probability interval and the median are shown below the histogram.

- The RF model implicit in the use of fractals is multiGaussian. The implementation is significantly different from that for the sequential algorithm adopted in GSLIB [6].
- The RF model underlying the new, and fairly popular, probability field simulation is also algorithmically-defined; its multivariate probability characteristics are understood with repeated applications.
- Boolean algorithms and their extension, marked point processes are generated by the distribution of geometric objects in space according to some probability laws. The multivariate distribution of most marked point processes is usually too complex to be analytically defined and understood.

Generating the stochastic realizations is the first step; it is the performance of the models that matter. Processing each stochastic realization with the a flow simulator allows the uncertainty to be quantified. Since reality is certain, it is difficult to validate our quantification of uncertainty; we cannot actually measure *real* uncertainty. Nevertheless, the concept of quantifying uncertainty is useful if only to evaluate its impact on the project at hand.

Perhaps the goal should be to obtain a limited number of realizations, constrained by **all** the data, to predict the center of the output response variable distribution. All models of uncertainty are but models; there is no full or largest measure of uncertainty in absolute.

ACKNOWLEDGEMENTS

The author would like to thank the management of Exxon Production Research Company for permission to publish this paper.

REFERENCES

[1] E. Aarts and J. Korst. *Simulated Annealing and Boltzmann Machines.* John Wiley & Sons, New York, NY, 1989.

[2] T. Anderson. *An Introduction to Multivariate Statistical Analysis.* John Wiley & Sons, New York, NY, 1958.

[3] C. Deutsch. *Annealing Techniques Applied to Reservoir Modeling and the Integration of Geological and Engineering (Well Test) Data.* PhD thesis, Stanford University, Stanford, CA, 1992.

[4] C. Deutsch. Conditioning reservoir models to well test information. In *Fourth International Geostatistics Congress*, Troia, September 1992.

[5] C. Deutsch and A. Journel. The application of simulated annealing to stochastic reservoir modeling. *Submitted to J. of. Pet. Tech.*, January 1991.

[6] C. Deutsch and A. Journel. *GSLIB: Geostatistical Software Library and User's Guide.* Oxford University Press, New York, NY, 1992.

[7] P. Doyen and T. Guidish. Seismic discrimination of lithology: a Bayesian approach. In *Geostatistics Symposium*, Calgary, AB, May 1990.

[8] C. Farmer. Numerical rocks. In P. King, editor, *The Mathematical Generation of Reservoir Geology*, Clarendon Press, Oxford, 1992. (Proceedings of a conference held at Robinson College, Cambridge, 1989).

[9] S. Geman and D. Geman. Stochastic relaxation, Gibbs distributions, and the Bayesian restoration of images. *IEEE Transactions on Pattern Analysis and Machine Intelligence*, PAMI-6(6):721–741, November 1984.

[10] F. Guardiano and R. M. Srivastava. Multivariate geostatistics: beyond bivariate moments. In *Fourth International Geostatistics Congress*, Troia, September 1992.

[11] E. Jaynes. Where do we go from here? In C. Smith and W. Gandy Jr., editors, *Maximum Entropy and Bayesian Methods in Inverse Problems*, pages 21–58, Reidel, Dordrecht, Holland, 1985.

[12] R. Johnson and D. Wichern. *Applied Multivariate Statistical Analysis.* Prentice Hall, Englewood Cliffs, NJ, 1982.

[13] A. Journel. Non-parametric estimation of spatial distributions. *Math Geology*, 15(3):445–468, 1983.

[14] A. Journel. The place of non-parametric geostatistics. In G. Verly et al., editors, *Geostatistics for natural resources characterization*, pages 307–355, Reidel, Dordrecht, Holland, 1984.

[15] A. Journel and C. Deutsch. Entropy and spatial disorder. *Math Geology*, 1993.

[16] B. Kedem. *Binary Time Series.* Marcel Dekker, New York, NY, 1980.

[17] S. Kirkpatrick, C. Gelatt Jr., and M. Vecchi. Optimization by simulated annealing. *Science*, 220(4598):671–680, May 1983.

[18] S. Kullback. *Information Theory and Statistics.* Dover, New York, NY, 1968.

[19] N. Metropolis, A. Rosenbluth, M. Rosenbluth, A. Teller, and E. Teller. Equation of state calculations by fast computing machines. *J. Chem. Phys.*, 21(6):1087–1092, June 1953.

MEASURING THE UNCERTAINTY IN KRIGING

MARK S. HANDCOCK *
Department of Statistics & Operations Research
Leonard N. Stern School of Business
New York University
New York, NY 10012-1126 U. S. A.

Spatial phenomena is commonly modelled as a realization from a stochastic process. Even when the reality is unique such models can usefully represent the uncertainty the modeler has about the phenomena. This paper is concerned with predicting for Gaussian random fields in a way that appropriately deals with uncertainty in the covariance function. To this end, we analyze the best linear unbiased prediction procedure (kriging) within a Bayesian framework. Particular attention is paid to the treatment of parameters in the covariance structure and their effect on the quality, both real and perceived, of the prediction. We show how this model can be improved by accounting for the uncertainty in the model parameters.
Key words: Bayesian statistics; Interpolation; Robustness; Spatial Statistics.

1. INTRODUCTION

In this paper we consider a modeling approach for spatially distributed data. As an illustration of the types of problems addressed consider the spatially distributed data in Figure 1 . The data are 52 topological elevations over a small area on the northern side of a hill. The data were measured by a surveying class, using a plane table and alidade. Davis (1973) was interested in the analysis of maps and used the survey to produce contours of the region. An important feature is the small streams running northward down the hill and joining together at the base of the region.

How should we analyze the data if our objective is to predict the elevations within the region surveyed? The perspective taken is that the actual elevations at each possible survey location (i.e. northing and easting from a reference point) taken together are a 'realization' from a particular stochastic process. The general process is described in the next section. Based on the observed data at the 52 locations and this statistical model a prediction of the elevation at unobserved locations can be made. Just as importantly from our perspective, an estimate of the uncertainty of that prediction can be derived from the model.

The statistical model is usually estimated from the very same data from which the predictions are made. The objective of this paper is to assess the effect of the fact that the model is estimated, rather than known, on the prediction and the associated prediction uncertainty. We describe a method for achieving this objective.

*The author thanks Michael Stein and a referee for numerous useful comments that have greatly improved the paper.

R. Dimitrakopoulos (ed.), Geostatistics for the Next Century, 436–447.

For example, suppose we wish to predict the elevation where the streams join at the base of the hill. We actually have an observation there, indicating that the elevation is 705 feet, but we will ignore it at this point except to use it to check our prediction. The commonly used method of maximum likelihood for estimating the best model suggests that the true elevation has a 95% chance of being in the interval (699, 707) feet. This interval only represents the uncertainty in the prediction given this particular estimated model and does not represent the uncertainty in estimating the model itself. When this is accounted for, using the method described in the next sections, the Bayesian 95% prediction interval is (694, 713) feet. The posterior probability content, incorporating the model uncertainty, of the maximum likelihood interval is 73%. Similar inaccuracies are to be excepted when the estimated uncertainties are based on alternative point estimates of the model.

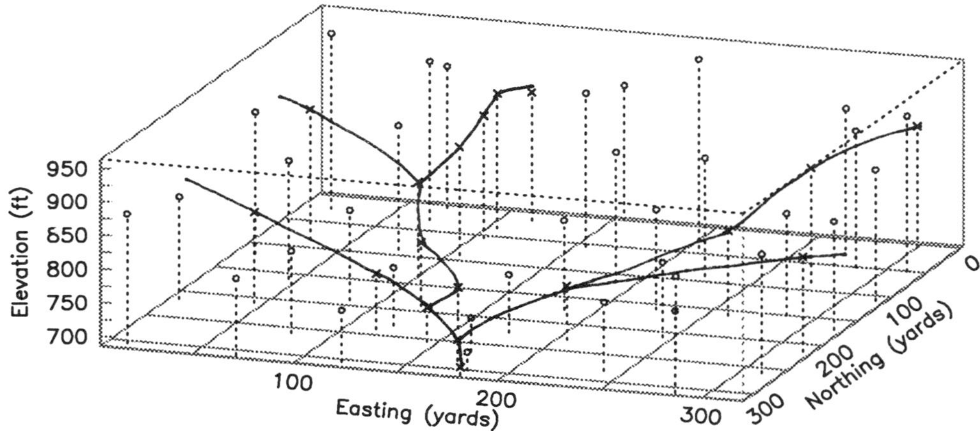

FIGURE 1: Elevations of the north face of a hill from Davis (1973), Table 5.11. The aspect ratio in the figure is spatially correct. The solid lines indicate the small streams running down the hill. Elevations on the streams are indicated by $\times$ and those not on the streams are marked by $\circ$.

We conclude that in many practical situations this uncertainty has a large impact on the estimated uncertainty of the prediction and a lesser effect on the predicted value itself. If there is little information about the model in the data the approach guards against gross error. In situations where substantial previous knowledge of the phenomena exists, the approach allows the information to be incorporated easily.

Bayesian analyses of kriging procedures are relatively new. Except for the work of Omre (1987), Omre and Halvarsen (1989) and Woodbury (1989) there appears to be no work from within the geostatistical community using the Bayesian perspective. Omre and Halvarsen (1989) describe a Bayesian approach to predicting the depth of geologic horizons based on seismic reflection times. They note the Bayesian interpretation of ordinary kriging and utilize prior information about the mean function only, not accounting for uncertainty in the covariance structure. Of course, the situation is a direct extension of standard Bayesian work in linear models where, for example, Box and Tiao (1973) §2.7 and Zellner (1971) §7 are textbook references. Here we will focus on a parametric representation of the covariance structure as its direct interpretation is of interest. Non-parametric approaches have been developed by Le and

Zidek (1992) and Pilz (1991). An alternative non-bayesian approach to covariance parameter uncertainty is given by Switzer (1984). What is novel about this paper is the spatial setting with irregularly observed locations and the general treatment of parameters in the covariance structure other than location and scale.

The framework of prediction is developed in Section 2. In Section 3 the assumption that the covariance structure is known is relaxed and the Bayesian formulation is developed. The focus is the evaluation of the performance of the traditional 'plug-in' kriging procedure. Section 4 illustrates this evaluation using topographical data from Davis (1973).

2. METHODOLOGY

2·1 Prediction using kriging

In this section we present the traditional kriging procedure as the basis for the later developments. Suppose $Z(x)$ is a real-valued stationary Gaussian random field on R with mean

$$E\{Z(x)\} = f(x)'\beta,$$

where $f(x) = \{f_1(x), \ldots, f_q(x)\}'$ is a known vector-valued function and β is a vector of unknown regression coefficients. Furthermore, the covariance function is represented by

$$\mathrm{cov}\{Z(x), Z(y)\} = \alpha K_\theta(x, y) \qquad \text{for } x, y \in R$$

where $\alpha > 0$ is a scale parameter, $\theta \in \Theta$ is a $p \times 1$ vector of structural parameters and Θ is an open set in $\mathbb{R}^p$. The division is purely formal as θ may also determine aspects of scale. In the general case, we observe $\{Z(x_1), \ldots, Z(x_n)\}' = Z$ and will focus on the prediction of $Z(x_0)$. The kriging predictor is the best linear unbiased predictor of the form $\widehat{Z}_\theta(x_0) = \lambda(\theta)'Z$; that is, the unbiased linear combination of the observations that minimizes the variance of the prediction error. It is straightforward to show that the corresponding weight vector $\lambda(\theta)$ defining $\widehat{Z}_\theta(x_0)$ is given by

$$\lambda(\theta)' = b'_\theta(F'K_\theta^{-1}F)^{-1}F'K_\theta^{-1} + k'_\theta K_\theta^{-1}, \tag{2.1}$$

where

$$\begin{aligned} F &= \{f_j(x_i)\}_{n\times q}, \\ k_\theta &= \{K_\theta(x_0, x_i)\}_{n\times 1}, \\ K_\theta &= \{K_\theta(x_i, x_j)\}_{n\times n}, \\ b_\theta &= f(x_0) - F'K_\theta^{-1}k_\theta. \end{aligned}$$

In the example $x = (x^1,\ x^2)$ and we can take $f_1(x) = x^1$ and $f_2(x) = x^2$, the northing and easting of the survey locations, respectively. A third component of the mean will be added in Section 4. The covariance function represents the covariance between the elevation at the survey locations $x = (x^1,\ x^2)$ and $y = (y^1,\ y^2)$.

2·2 Assessing Uncertainty in kriging

The quality of the prediction is determined by the distribution of the prediction error, $e_\theta(x_0) = Z(x_0) - \widehat{Z}_\theta(x_0)$. Note that the prediction weights $\lambda(\theta)'$ do not depend on α or β. Under our Gaussian model, for fixed α, β, and θ, the conditional distributions of $Z(x_0)$ and $e_\theta(x_0)$ given Z are

$$Z(x_0) \mid Z \sim N\Big(k_\theta' K_\theta^{-1} Z + b_\theta'\beta,\ \alpha\{K_\theta(x_0,x_0) - k_\theta' K_\theta^{-1} k_\theta\ \}\Big)$$

$$e_\theta(x_0) \mid Z \sim N\Big(b_\theta'(\beta - \hat\beta(\theta)),\ \alpha\{K_\theta(x_0,x_0) - k_\theta' K_\theta^{-1} k_\theta\ \}\Big)$$

where $\hat\beta(\theta) = (F'K_\theta^{-1}F)^{-1}F'K_\theta^{-1}Z$ and $N\big(\cdot,\cdot\big)$ denotes the Gaussian distribution. The sampling (or unconditional) distribution for $e_\theta(x_0)$ is

$$e_\theta(x_0) \ \sim N\Big(\ 0,\ \alpha V_\theta\ \Big) \tag{2.2}$$

where $V_\theta = K_\theta(x_0,x_0) - k_\theta' K_\theta^{-1} k_\theta + b_\theta'(F'K_\theta^{-1}F)^{-1} b_\theta$ and αV_θ is the usual prediction error variance as given in Ripley (1981).
Note that the underlying kriging procedure is motivated by sampling considerations, producing point predictions and associated measures of uncertainty for those predictions both based on sampling distributions unconditional on the observed Z. However kriging, when the mean is of known regression form, can be given a Bayesian interpretation. Traditionally, it is assumed that the covariance function is known exactly and the investigator has little knowledge about β prior to analyzing the data. The underlying kriging approach usually presumes ignorance about β and the unrelatedness of β to the behavior of the covariance function. This latter philosophy will be followed throughout the paper. Under these assumptions, an appropriate prior distribution has $\mathrm{pr}(\beta \mid \alpha, \theta)$ locally uniform. The posterior distribution of β is then

$$\beta \mid \alpha, \theta, Z \sim N_q\Big(\ \hat\beta(\theta),\ \alpha(F'K_\theta^{-1}F)^{-1}\Big)$$

The posterior distribution of the prediction error is then

$$\mathrm{pr}(e_\theta(x_0) \mid \alpha, \theta, Z) \ \ \propto \ \ \int_\beta \mathrm{pr}(e_\theta(x_0) \mid \alpha, \beta, \theta, Z)\mathrm{pr}(\beta \mid \alpha, \theta, Z) d\beta,$$

which is, by direct calculation,

$$e_\theta(x_0) \mid \alpha, \theta, Z \sim N\Big(\ 0,\ \alpha V_\theta\ \Big) \tag{2.3}$$

the same as the sampling distribution (2.2). Similarly we have

$$Z(x_0) \mid \alpha, \theta, Z \sim N\Big(\ \widehat{Z}_\theta(x_0),\ \alpha V_\theta\ \Big) \tag{2.4}$$

where $\widehat{Z}_\theta(x_0) = k_\theta' K_\theta^{-1} Z + b_\theta'\hat\beta(\theta)$ is the usual kriging point predictor. These distributions form the basis for all inferential statements about the prediction and prediction error. Hence, except for the usual differences in interpretation, we end up with the same analysis as the traditional approach. This comparison may be loosely stated as: ordinary kriging is 'Bayesian' with the non-informative prior for the mean parameter.

3. Kriging with unknown covariance

In this section, the assumption that the covariance function is known exactly is relaxed to allow the covariance function to be unknown, but still a member of the parametric class Θ.

In traditional kriging, one estimates α and θ by either likelihood methods or various *ad hoc* approaches. The likelihood approach to the estimation of the covariance structure was first applied in the hydrological and geological fields following Kitanidis and Lane (1985) and Hoeksema and Kitanidis (1985). Mardia and Marshall (1984) is a standard reference in the statistical literature. Usually the predictor and the behavior of the prediction error are themselves estimated by 'plugging-in' the estimates into (2.1) and (2.2). If θ is known so that only the location parameter β and the scale parameter α are uncertain then we are in a standard generalized least-squares setting. The distinction between the generalized least squares setting and the random field setting is the uncertainty in the structural parameter θ. While the restriction to a parametric class is a significant assumption, it still allows great latitude.

As β is a location parameter we expect that our prior opinions about β bear no relationship to those about α and *a priori* might expect α and β to be independent, leading to the use of Jeffreys's prior. Partly for convenience, the form of the prior used here will be

$$\mathrm{pr}(\alpha,\ \beta, \theta) \propto \mathrm{pr}(\theta)/\alpha$$

It easily follows from Zellner (1971) that the predictive distribution of $Z(x_0)$ conditional on θ and Z is

$$Z(x_0) \mid \theta, Z \sim t_{n-q}\left(\ \widehat{Z}_\theta(x_0),\ \ \frac{n}{n-q}\widehat{\alpha}(\theta)V_\theta\right), \tag{3.1}$$

a shifted t distribution on $n-q$ degrees of freedom.

The marginal posterior distribution of θ can be shown to be

$$\mathrm{pr}(\theta \mid Z) \propto \mathrm{pr}(\theta) \cdot |K_\theta|^{-1/2}|F'K_\theta^{-1}F|^{-1/2}\widehat{\alpha}(\theta)^{-(n-q)/2} \tag{3.2}$$

The Bayesian predictive distribution for $Z(x_0)$ is

$$\mathrm{pr}(Z(x_0) \mid Z) \propto \int_\Theta \mathrm{pr}(Z(x_0) \mid \theta, Z) \cdot \mathrm{pr}(\theta \mid Z) d\theta$$

where the integrand is given by (3.1) and (3.2). As the dependence of K_θ on θ is not specified this expression can not be simplified and further exploration will in general require numerical computation. If prior information is available it may be directly incorporated into (3.2), although additional numerical integration may be necessary if prior dependencies among (α, β, θ) are envisaged.

Suppose we use an estimation procedure to select the parameters $(\tilde{\alpha}, \tilde{\theta})$ of a covariance structure. These may be arrived at by any procedure, although the usual methods are maximum likelihood, weighted least squares or derived from empirical correlation functions. The distribution that an investigator would use as a basis for inference about $Z(x_0)$ would be

$$Z(x_0) \mid \tilde{\alpha}, \tilde{\theta}, Z \sim N\left(\ \widehat{Z}_{\tilde{\theta}}(x_0),\ \tilde{\alpha}V_{\tilde{\theta}}\ \right),$$

plugging in $(\tilde{\alpha}, \tilde{\theta})$ for (α, θ) in (2.4).

Depending on the influence of θ on the spread and location of $\mathrm{pr}(Z(x_0) \mid \theta, Z)$, the Bayesian predictive distribution might be wider or narrower than the plug-in predictive distribution. The location of the plug-in predictive distribution may also be quite different from the Bayesian predictive distribution. Typically the Bayesian predictive distribution will have no simple analytic form and must be determined numerically. The difference between the plug-in and Bayesian predictive distributions represents the difference in inference between the traditional kriging approach and the full Bayesian approach.

Note the plug-in prediction error, $\tilde{e}_\theta(x_0) = Z(x_0) - \widehat{Z}_{\tilde{\theta}}(x_0)$, is just a shifted version of $Z(x_0)$, so that comparisons of performance of the plug-in estimates will be the same whether we consider $Z(x_0)$ or $\tilde{e}_\theta(x_0)$. We could interpret this as a comparison between the plug-in distribution for $\tilde{e}_\theta(x_0)$ and the actual distribution for $\tilde{e}_\theta(x_0)$ under the full Bayesian model, although the latter distribution would not be used for inference.

3·1 The Matérn Class of Covariance Functions

In this section we describe a general class of covariance functions that we feel provides a sound foundation for the parametric modeling of Gaussian random fields. The class is motivated by the smooth nature of the spectral density, the wide range of behaviors covered and the interpretability of the parameters. It will be used throughout the later sections. The properties of the covariance function directly determine the properties of the random field model. The Matérn class is characterized by the parameter $\theta = (\theta_1, \theta_2)$. $\theta_1 > 0$ is a scale parameter controlling the range of correlation. The smoothness parameter $\theta_2 > 0$ directly controls the smoothness of the random field. The Exponential class corresponds to the sub-class with smoothness parameter $\theta_2 = 1/2$, that is

$$K_E(x) = \exp(-x/\theta_1').$$

The sub-class defined by $\theta_2 = 1$ was introduced by Whittle (1954) as a model for two dimensional fields. It is commonly used in hydrology (see e.g. Jones (1989)). As $\theta_2 \to \infty$, $K_\theta(x) \to \exp(-x^2/\theta_1^2)$, often called the 'Gaussian' covariance function. We shall refer to it as the Squared Exponential model. This model forms the upper limit of smoothness in the class and will rarely represent natural phenomena as realizations from it are infinitely differentiable.

The isotropic correlation functions have the general form

$$K_\theta(x) = \frac{1}{2^{\theta_2 - 1}\Gamma(\theta_2)} \cdot \left(\frac{x}{\theta_1'}\right)^{\theta_2} \mathcal{K}_{\theta_2}\left(\frac{x}{\theta_1'}\right)$$

where $\theta_1' = \theta_1/(2\sqrt{\theta_2})$ and $\mathcal{K}_{\theta_2}$ is the modified Bessel function of order θ_2 discussed in Abramowitz and Stegun (1964), §9.

A field with this covariance function is $\lceil \theta_2 - 1$ times (mean-square) differentiable where $\lceil$ is the integer ceiling function. The realizations will have continuous $\lceil \theta_2 - 1$ derivatives if $\theta_2 > \lceil \theta_2 - \frac{1}{2}$. If the field is Gaussian the realizations will have continuous $\lceil \theta_2 - 1$ derivatives (almost certainly) (See Cramèr and Leadbetter (1967), §4.2, §7.3, and §9.2-9.5).

All calculations of $\mathcal{K}_{\theta_2}$ in this work use the RKBESL algorithm from the SPECFUN library available free from NETLIB. A general treatment is given in the seminal work by Matérn (1986).

4. An analysis of Davis's topographical data

In this section we analyze the data introduced in Section 1 and originally from Davis (1973). It has been studied by Ripley (1981, pp. 58-72), and subsequently by Warnes (1986), Ripley (1988, pp. 15-21) and Mardia and Watkins (1989). The original data were scaled so that 50 yards in location corresponds to one map unit. We will use the more natural units of yards, although the later references continued the original scaling. The survey locations are recorded to two significant figures and the elevations to three significant figures.

The major assumptions implicit in the model are stationarity of the Gaussian random field, isotropy of the correlations and the correct specification of the mean. These are interdependent so that checking them individually is usually not the best approach. There are available methods to test if the marginal distribution of the observations is Gaussian. However it is difficult to determine if the joint distribution of the observations is Gaussian in the presence of an unknown correlation structure. In particular the marginal distribution of the observations is little guide to the joint distribution. The realizations of the random field can be assumed to be smooth, at least continuous and maybe even differentiable. Given the nature of the data and the measurement procedure it will be assumed that the measurement error is small so that the (observed) field is continuous.

The Exponential model, while providing a reasonable initial covariance class, does not allow the field to have differentiable realizations. Given that *a priori* the form of the covariance is unknown it is unreasonable to exclude the possibility of smoother random fields.

As indicated in the Section 2, the mean function should clearly include the Northing and Easting of the survey locations. In addition, there is information in the locations of the streams that should be taken into account. One crude way is to include, as $f_3(x)$, the horizontal distance of the survey point to the closest stream.

4·1 Posterior Knowledge Based on a Flat Mean

Initially we will entertain the model with a constant mean. The marginal posterior for the smoothness parameter based on the uniform prior for the smoothness and range parameters is given in Figure 2.

The mass of the distribution is between $\theta_2 = 0.5$ and $\theta_2 = 1.5$. The mode is slightly below $\theta_2 = 1$, which corresponds to Whittle's covariance function (Whittle (1954, 1962)). Interestingly, Whittle regarded this model as the natural extension of the Exponential model, $\theta_2 = 1/2$, from one to two dimensions. It corresponds to a random field with continuous realizations that are on the margin of mean-square differentiability. For $\theta_2 > 1$ the field is mean-square differentiable. It is interesting to note that the ratio of the density at the mode to the density at the Exponential model is about $5 : 1$, so that the Exponential appears too rough for this field. Such posterior densities are a useful tool for describing and understanding the behavior of the phenomena underlying the data.

Of course, alternative prior distributions can easily be used. One could express prior knowledge about (β, α) by taking the marginal prior of (β, α) to be the usual Gaussian-Gamma conjugate prior from generalized least squares. An informative prior for θ_2 could deemphasize smoothnesses less than a half or much greater than two, the rationale being that we do not expect the realizations to be discontinuous or much smoother than twice differentiable. Such a prior distribution would have little effect on the predictive distribution as the likelihood places little weight on smoothness values in that range.

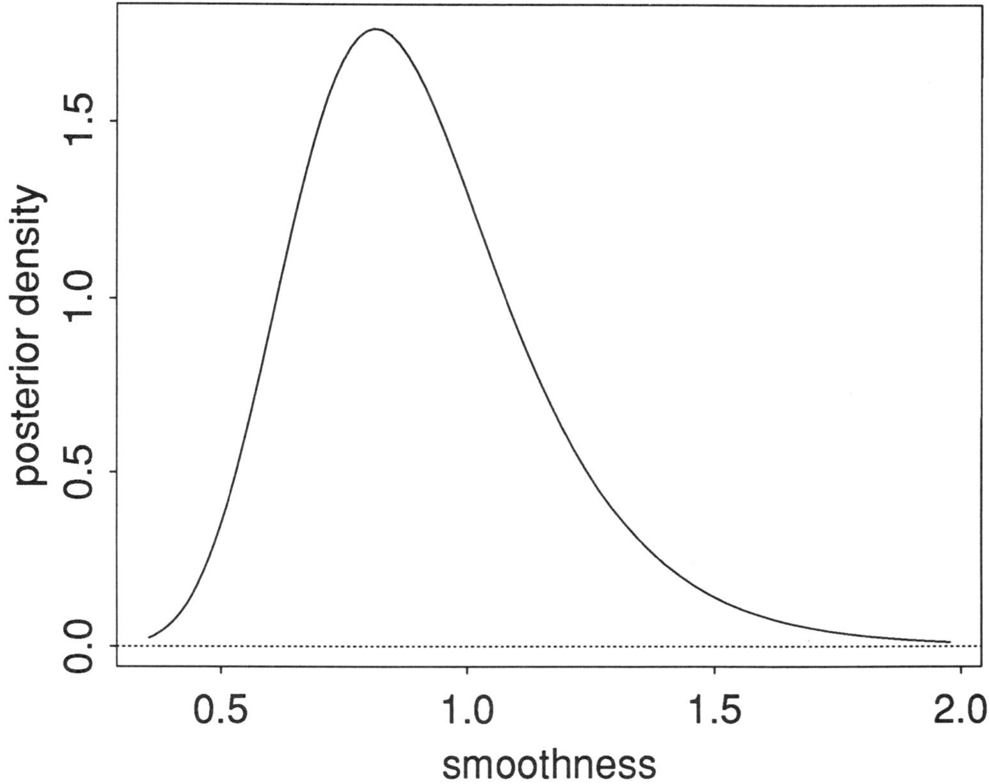

FIGURE 2: Posterior distribution for the smoothness parameter based on the Matérn model with constant mean.

4·2 Incorporating Additional Information in the Mean

The model with a constant mean may be inadequate as compared to the models including the survey locations and distance to streams as regressors because of non-stationarity in the mean (Mardia and Watkins (1989)). The location chosen to be predicted at is the surveyed location closest to the most northern junction of the stream (at (180, 300) on Figure 1). It was chosen to be reasonably close to the other survey locations. The models will be developed without this location and the elevation there will be used as a check on the predictions.

Figure 3 is the profile log-likelihood surface under this model. Figure 4 presents the predictive densities based on the Matérn model with this more sophisticated mean function and a uniform prior on the smoothness and range parameters. The plug-in predictive distribution based on the maximum likelihood estimate, $(\widehat{\alpha}, \widehat{\theta}) = (955, 68, 7.8)$, is Gaussian centered at 703 feet with a standard deviation of about 2.1 feet. The effect of the additional regressors is to substantially reduce the variability of the predictive distributions.

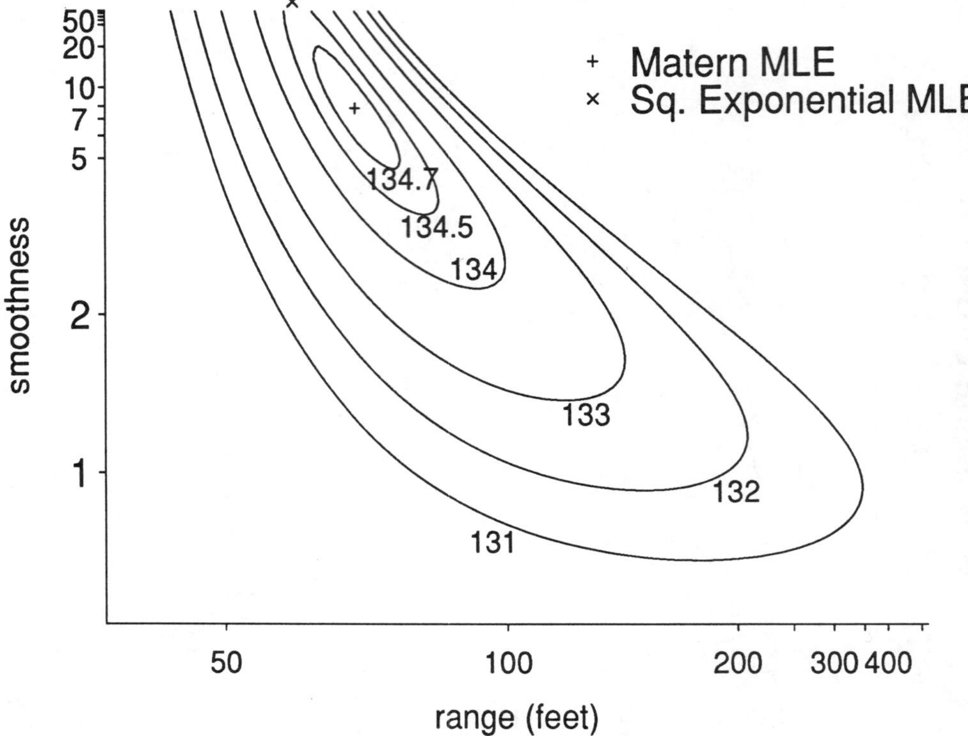

FIGURE 3: Profile log-likelihood for the Matérn model with mean based on the northing and the distance to closest stream.

Also represented are three alternate plug-in predictive distributions based on the maximum likelihood estimators under the Exponential and Squared Exponential classes and the maximum *a posteriori* Matérn value. Probability intervals based on this plug-in predictive distributions will markedly differ from those based on the Bayesian predictive distribution. The latter is a better reflection of the uncertainty in the covariance structure and should be regarded as a superior reference for inference. Smoother estimated models correspond to less perceived uncertainty in the prediction. For example, the Bayesian 95% prediction interval has nominally 99.99% probability under the plug-in predictive distribution for the maximum likelihood Matérn model. Alternatively the nominally 95% interval for this plug-in predictive distribution actually has 73% probability.

4·3 Sensitivity to Prior Specification

How sensitive is our inference to the choice of prior distributions? In these examples a prior distribution uniform on the positive values the smoothness parameter is used. Alternatively one could use the prior

$$\text{pr}(\theta_2) \quad = \quad \frac{1}{(1+\theta_2)^2}$$

reflecting the belief that larger smoothness values are *a priori* less likely than smaller values. Physically the belief is that the field is more likely to be one or two times differentiable, rather than, say, 101 times. This prior is uniform for $\theta_2/(1+\theta_2)$ on [0, 1].

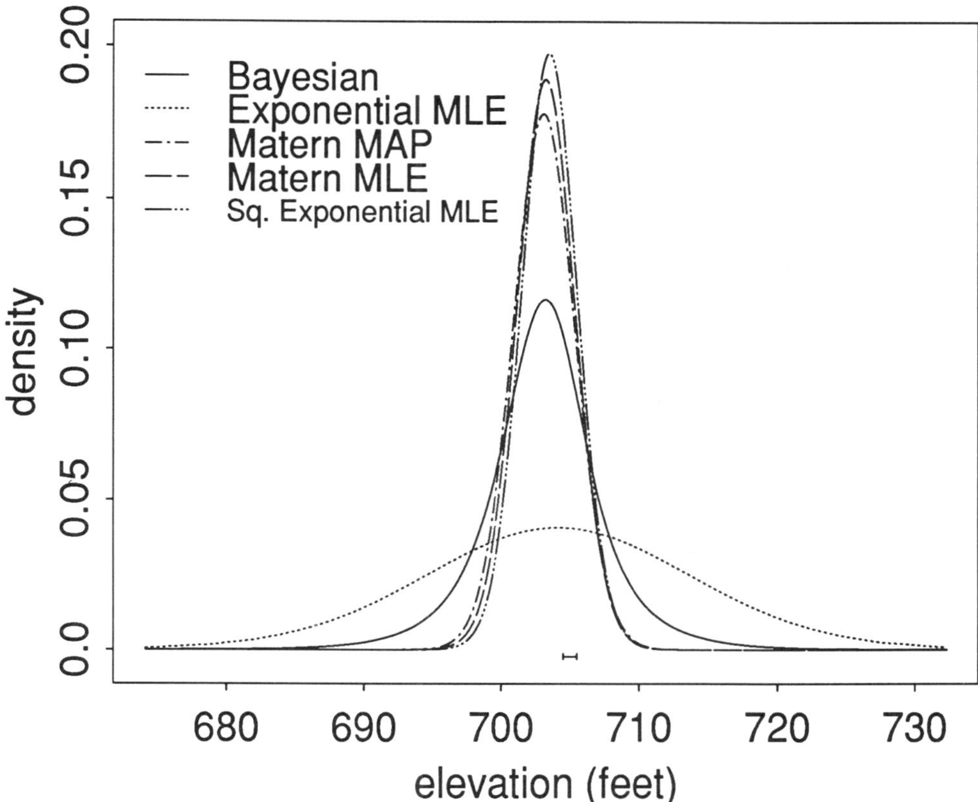

FIGURE 4: Predictive distributions based on the Matérn model with mean using the northing and the distance to closest stream. The observed elevation at this location is represented by the small horizontal bar reflecting the recording accuracy.

The effect of using this prior to deemphasize larger smoothness values relative to the uniform prior is to increase the uncertainty in the prediction, in line with the less smooth models. The inference appears to be insensitive to moderate changes in the prior for θ_2. Based on figures not presented here, we find that using a uniform prior for α instead of the usual $1/\alpha$ results in a predictive distribution with slightly thinner tails and that the predictive distribution is insensitive to changes in the prior for β. While the maximum likelihood estimate is a good representative value, the overall flatness of the likelihoods would suggest against choosing any particular member as the 'truth'. Clearly we need additional information before we can choose between members of the same class. The same comments apply to the choice of mean model. It is tempting to base the decisions on the changes in log-likelihood. It is still an open question as to the validity of this procedure in the face of the interdependence of the mean and covariance structures.

5. Conclusion

The kriging procedure is often described as optimal (Matheron (1965)) because it produces optimal predictions when the covariance structure of the random field is known. If the covariance structure needs to be estimated, then this primary motivation for kriging is in question. It is then necessary to assess the effect of the fact that the model is estimated, rather than known, on the prediction and the associated prediction uncertainty. In this paper we have seen that the Bayesian paradigm provides a framework in which to analyze the performance of the estimated kriging predictor. In conclusion, a better approach would be to base inference on the Bayesian predictive distribution. This approach takes into account the uncertainty about the covariance function expressed in the likelihood surface and ignored by point estimates of the covariance function. It also allows the performance of the usual plug-in predictive distribution based on an estimated covariance structure to be critiqued within a wider framework. The results also suggest that fitting the empirical correlation function 'by eye' may lead to plug-in predictive distributions that differ markedly from the Bayesian predictive distribution. The maximum likelihood estimate may be the best single representative available, but this reduction itself can be detrimental to the inference (See Figure 4).

References

Abramowitz, M. and Stegun, I. A. (1965), Handbook of Mathematical Functions, New York: Dover.

Box, G. E. P. and Tiao, G. C. (1973), Bayesian Inference in Statistical Analysis, Reading, Mass.: Addison-Wesley.

Cramér, H. and Leadbetter, M. R. (1967), Stationary and Related Stochastic Processes, New York: Wiley.

Davis, J. C. (1973), Statistics and Data Analysis in Geology, New York: Wiley.

Handcock, M. S. (1989), Inference for Spatial Gaussian Random Fields when the Objective is Prediction. Ph.D. thesis, Department of Statistics, University of Chicago, Chicago, Illinois.

Hoeksema, R. J. and Kitanidis, P. K. (1985), Analysis of the Spatial Structure of Properties of Selected Aquifers, Water Resources Research **21**, 563-572.

Jones, R. H. (1989), Fitting a Stochastic Partial Differential Equation to Aquifer Head Data, Stochastic Hydrology and Hydraulics **3**, 100-105

Kitanidis, P. K. and Lane, R. W. (1985), Maximum Likelihood Parameter Estimation of Hydrologic Spatial Processes by the Gauss-Newton Method, Journal of Hydrology **17**, 31-56.

Le, N. D. and Zidek, J. V. (1992), Interpolation with Uncertain Spatial Covariances: A Bayesian Alternative to Kriging, J. Multivariate Analysis **43**, 2, 351-374.

Mardia, K. V. and Marshall, R. J. (1984), Maximum likelihood estimation of models for residual covariance in spatial regression, Biometrika **71**, 135-146.

Mardia, K. V. and Watkins, A. J. (1989), On Multimodality of the Likelihood in the Spatial Linear Model, Biometrika **76**, 289-95.

Matérn, B. (1986), Spatial Variation Second Ed, Lecture Notes in Statistics, **36,** Berlin: Springer-Verlag.

MATHERON, G. (1965), Les Variables Régionalisées et leur Estimation Paris: Masson.
OMRE, H. (1987), Bayesian Kriging - Merging Observations Qualified Guesses in Kriging, Mathematical Geology **19**, 25-39.
OMRE, G. M. and HALVARSEN, D. F. (1989), A Bayesian Approach to Kriging, In M. Armstrong (ed.), Proceedings of the Third International Geostatistics Congress I, pp. 49-68, Dordrecht: Academic Publishers.
PILZ, J. (1991), Bayesian Estimation and Experimental Design in Linear Regression Models, New York: Wiley.
RIPLEY, B. D. (1981), Spatial Statistics, New York: Wiley.
RIPLEY, B. D. (1988), Statistical Inference for Spatial Processes, Cambridge: Cambridge University Press.
SWITZER, P. (1984), Inference for Spatial Autocorrelation Functions, in G. Verly et al (eds.), Geostatistics for Natural Resources Characterization, NATO ASI Series C Vol. 122, pp. 127-140. Reidel Publishing Co., Boston, MA.
WARNES, J. J. (1986), A Sensitivity Analysis for Universal Kriging, Mathematical Geology **18**, 653-676.
WHITTLE, P. (1954), On Stationary Processes in the Plane, Biometrika **41**, 434-449.
WHITTLE, P. (1962), Topographic Correlation, Power-Law Covariance Functions, and Diffusion, Biometrika **49**, 305-314.
WOODBURY, A. D. (1989), Bayesian Updating Revisited, Mathematical Geology **21**, 285-308.
ZELLNER, A. (1971), An Introduction to Bayesian Inference in Econometrics, Reprinted 1987, New York: Wiley.

"FUZZY" GEOSTATISTICS - AN INTEGRATION OF QUALITATIVE DESCRIPTION INTO SPATIAL ANALYSIS

MAREK KACEWICZ
ARCO Exploration and Production Technology
2300 W. Plano Parkway
Plano, TX 75075
USA

Cognitive processes in the earth sciences, in their early stages, are characterized by vague data that can be very well described using fuzzy set formalism. Fuzzy sets have been designed by Zadeh (1965) to handle qualitative and intuitive descriptions and incorporate them into reasoning processes. Recent attempts to put together fuzzy descriptions and geostatistics bring us closer to spatial analysis of vague data (not necessary under sampled). This paper summarizes "fuzzy" approaches in geostatistics and outlines directions of future research. New defuzzification and fuzzy indicator methods are presented which allow one to use standard geostatistical methods in spatial analysis of fuzzy data.

INTRODUCTION

A widespread understanding of limitations of spatial statistics in handling ambiguous and imprecise earth science data has stimulated new developments in geostatistics. The result is in a number of new and interesting approaches e.g., indicator kriging (Journel, 1983), soft kriging (Journel, 1986), Bayesian approaches (Kitanidis, 1986; Omre, 1987), fuzzy kriging (Bardossy, et.al., 1988; Diamond, 1989). This paper will focus on fuzzy techniques in spatial analysis of data.

Fuzzy set theory presented for the first time by Zadeh (1965) has arisen from a widespread discussion originated in philosophy in the late nineteenth and early twentieth centuries concerning so called antinomies. The antinomy is defined as a fundamental and apparently unresolvable conflict or contradiction. British philosopher Bertrand Russell discussed the problem of antinomy in mathematics and natural languages and gave some nice examples, among them the so-called barber's dilemma. This concerns a barber who was allowed to shave only those people who do not shave themselves. Should the barber shave himself? The barber would be in conflict with his orders each time after starting to shave. It's impossible to give a unique answer to the problem if two-valued logic is applied. The only way to fulfill the barber's personal needs is to use a metalanguage based on multivalued logic in which other than yes-no answers are allowed. The multivalued logic formalism has been developed by Jan Lukasiewicz in 1920's. In fuzzy logic, which is one multivalued approach, yes-no (called later, "crisp") is replaced by more-or-less (called later, "fuzzy"). It means that

R. Dimitrakopoulos (ed.), Geostatistics for the Next Century, 448–463.

instead of using only black and white, 0-1, or yes-no, we can use for example black-gray-white, 0-1/2-1 or yes-more/or/less-no.

In the earth sciences (especially in exploration) we often face the barber's dilemma. We are expected to give one answer to hydrocarbon migration problems, to put 'sharp' limits between two stratigraphic units, to assign one permeability number to a grid cell, etc., having no or 'fuzzy' data. How can we avoid conflicts with our standards which suggest neither answer is appropriate, but at the same time satisfy exploration needs? One solution is to use mathematical formalisms which easily handle ambiguous and imprecise data, qualitative descriptions, etc., and give some information about the risk. Advantages and disadvantages of fuzzy kriging, which is one of such attempts, will be discussed in this paper.

ELEMENTS OF THE FUZZY SET THEORY

Let X denote a space of objects. In the fuzzy set theory it is often called a universe. Intuitively, a fuzzy set is a class where partial membership is allowed. By definition a fuzzy set $A \subset X$ is a set of ordered pairs $A=\{(x,\mu_A(x)), x \in X\}$, where $\mu_A : X \longrightarrow [0,1]$. Let us now show some useful operations on fuzzy sets

containment: $A \subset B \iff \mu_A(x) \leq \mu_B(x) \quad \forall x$ (1)

equality: $A=B \iff \mu_A(x) = \mu_B(x) \quad \forall x$ (2)

complement: $A' \iff \mu_{A'}(x)=1-\mu_A(x) \quad \forall x$ (3)

union: $A \cup B \iff \mu_{A \cup B}=\mathrm{Max}[\mu_A(x),\mu_B(x)] \quad \forall x$ (4)

intersection: $A \cap B \iff \mu_{A \cap B}=\mathrm{Min}[\mu_A(x),\mu_B(x)] \quad \forall x$ (5)

product: $AB \iff \mu_{AB}=\mu_A(x)\mu_B(x) \quad \forall x$ (6)

sum: $A \oplus B \iff \mu_{A \oplus B}=\mu_A(x)+\mu_B(x)-\mu_A(x)\mu_B(x) \quad \forall x$ (7)

Definition 2.1 A fuzzy set A is said to be Borel measurable if $A_\theta=\{x:\mu_A(x)\leq\theta, 0\leq\theta\leq 1\}$ are Borel measurable.

Let us notice that if $\mu_A(x)$ and $\mu_B(x)$ are Borel measurable, so are $\mathrm{Max}[\mu_A(x),\mu_B(x)]$, $\mathrm{Min}[\mu_A(x),\mu_B(x)]$, $\mu_A(x)\mu_B(x)$, and $\mu_A(x)+\mu_B(x)$. From that we can assert that Borel fuzzy sets form a σ-field with respect to operations (3), (4), and (5).

Definition 2.2 A fuzzy subset of A for which $\mu_A(x)\geq\alpha$, $\forall x$ is called a α-level set.

Definition 2.3 A fuzzy set is called normal if there exists at least one such x for which $\mu_A(x)=1$.

Definition 2.4 A fuzzy set A is convex if

$$\mu_A[\alpha x+(1-\alpha)y]\geq \min[\mu_A(x),\mu_A(y)] \quad \forall x,y\in A \text{ and } \forall \alpha\in <0,1>$$

Definition 2.5 A fuzzy number is a convex normal fuzzy set of the real line **R** such that

1. There exists exactly one $x_0\in \mathbf{R}$ for which $\mu_A(x_0)=1$
2. $\mu_A(x)$ is piecewise continuous.

Sometimes in the definition of a fuzzy number the assumption number 1 is replaced by 1' which says that there exists at least one $x_0\in \mathbf{R}$ such that $\mu_A(x_0)=1$. The fuzzy set theory provides a tool that allows one to calculate functions on fuzzy sets. The so called extension principle will be used later to derive fuzzy variograms, as an example.

Definition 2.6 The extension principle.
Suppose that f is a mapping from X to Y and A is a fuzzy subset of X expressed as $A=\{(x,\mu_A(x)),x\in X\}$. Then the extension principle asserts that $f(A)=\{(f(x),\mu_A(x)),x\in X\}$. Let an n-ary function be a mapping from the Cartesian product $X_1 x X_2 x \ldots x X_n$ to a universe Y such that $y=f(x_1,x_2,\ldots,x_n)$, and $A_1,A_2,\ldots,A_n$ are fuzzy sets in $X_1,X_2,\ldots,X_n$.respectively, characterized by membership functions $\mu_{A_1}(x),\mu_{A_2}(x),\ldots,\mu_{A_n}(x)$, then

$$\mu_F(y)= \sup_{\substack{x_1,x_2,\ldots,x_n \\ y=f(x_1,x_2,\ldots,x_n)}} \operatorname{Min}[\mu_{A_1}(x_1),\mu_{A_2}(x_2),\ldots,\mu_{A_n}(x_n)] \tag{8}$$

and

$$\mu_F(y)=0 \text{ if } f^{-1}(y)=\emptyset,$$

where F is a fuzzy subset of Y.

The extension principle is used to define arithmetic operations on fuzzy numbers

$$\mu_{A\blacklozenge B}(y)= \sup_{y=x_1\blacklozenge x_2} \operatorname{Min}[\mu_A(x_1),\mu_B(x_2)] \tag{9}$$

where ♦ denotes +, -, *, or / .

FUZZY SETS: FROM WHERE?

One of the main problems encountered in fuzzy set applications is how to obtain them. There is no consensus among scientists in the matter of which algorithm, recipe, or methodology should be used. A vast variety of approaches makes fuzzy sets extremely flexible, but dangerous in some situations. In spatial applications, the way fuzzy sets are prepared depends on the qualitative and the quantitative character of data. We will distinguish between approaches based on descriptive languages (e.g. natural languages), a formalism using existing mathematical functions, and manipulation of data by an expert. Another approach based on fuzzy tolerance approximation can be found in Kacewicz (1991).

Descriptive linguistic approach
Having only a rough idea about new basins (frontier areas) one tends to use qualitative terms such as LOW POROSITY, MORE-OR-LESS HIGH PERMEABILITY, and later arbitrarily translate them to fuzzy sets containing numbers. Kacewicz (1993) shows examples of fuzzy sets describing permeability in one such case. Low permeability sandstone (LPS) and high permeability sandstone (HPS) have been defined by the following sets:

$$LPS = \{\frac{1.0}{0.1},\frac{0.8}{10},\frac{0.6}{30},\frac{0.4}{50},\frac{0.1}{100}\}$$ <— fuzzy membership <— permeability in millidarcies

$$HPS = \{\frac{0.1}{50},\frac{0.3}{100},\frac{0.6}{300},\frac{0.9}{500},\frac{1.0}{1000}\}$$

Very low, very high, more-or-less low, more-or-less high, and medium permeability can be defined as following

$$\text{VERY LPS} = \text{LPS}^m$$
$$\text{VERY HPS} = \text{HPS}^m$$
$$\text{MORE-OR-LESS LPS} = \text{LPS}^{1/2}$$
$$\text{MORE-OR-LESS HPS} = \text{HPS}^{1/2}$$
$$\text{MEDIUM PERMEABILITY SANDSTONE} = \sim\text{LPS} \cap \sim\text{HPS}$$

where

$$A^p = \{\ x \in A:\ \mu_{A^p}(x) = \mu_A^p(x)\ ,\ p \in R\ \} \tag{10}$$

Fuzzy sets defined by mathematical functions
Fuzzy sets can be described more rigorously using mathematical functions such as S-type, L-R-type, and T-type. These and other functions can be found e.g. in Zimmerman (1984), Kandel (1986), Kruse and Meyer (1987), Dubois and Prade (1980). The following function, which is very useful to define terms such as LOW, HIGH,etc. is called the S-type function

$$S\ (x;\alpha,\beta,\gamma) = \begin{cases} 1 & \text{for } x \leq \alpha \\ 1-2\left(\dfrac{x-\alpha}{\gamma-\alpha}\right)^2 & \text{for } \alpha \leq x \leq \beta \\ 2\left(\dfrac{x-\gamma}{\gamma-\alpha}\right)^2 & \text{for } \beta \leq x \leq \gamma \\ 0 & \text{for } x \geq \gamma \end{cases} \tag{11}$$

where α,β,γ are parameters to be selected by the user. This function was applied by Kacewicz (1993) to define permeability of the Travis Peak Formation (East Texas) sandstones. Dubois and Prade (1978) suggested a special representation of fuzzy numbers which is called L-R-type. They first define a reference function f of a fuzzy number. A function f is a reference function iff

1. $f(-x)=f(x)$
2. $f(0)=1$

3. f is decreasing on $[0,\infty]$.
By definition a fuzzy number A is of L-R-type if there exist reference functions L (for left) and R (for right) and scalars $a>0$, $b>0$ such that

$$\mu_A(x) = \begin{cases} L\left(\frac{m-x}{\alpha}\right) & \text{for } x \leq m \\ R\left(\frac{x-m}{\beta}\right) & \text{for } x \geq m \end{cases} \tag{12}$$

where for example $L(x)=\frac{1}{1+x^2}$ and $R(x)=\frac{1}{1+2|x|}$. Symbolically the L-R-type fuzzy number (10) is denoted by a triple

$$(m,\alpha,\beta)_{LR} \tag{13}$$

Dubois and Prade (1978) show that arithmetic operations on fuzzy numbers based on the extension principle (9) can be rewritten in a simpler form. Let $A=(m,\alpha,\beta)_{LR}$ and $B=(n,\gamma,\delta)_{LR}$ be two numbers of L-R type, then

[1] $(m,\alpha,\beta)_{LR} + (n,\gamma,\delta)_{LR} = (m+n,\alpha+\gamma,\beta+\delta)_{LR}$

[2] $-(m,\alpha,\beta)_{LR} = (-m,\alpha,\beta)_{LR}$

[3] $(m,\alpha,\beta)_{LR} - (n,\gamma,\delta)_{LR} = (m-n,\alpha+\gamma,\beta+\delta)_{LR}$

[4] $(m,\alpha,\beta)_{LR} \bullet (n,\gamma,\delta)_{LR} = (mn,n\alpha+m\gamma,n\beta+m\delta)_{LR}$, for A and B positive

$(m,\alpha,\beta)_{LR} \bullet (n,\gamma,\delta)_{LR} = (mn,n\alpha-m\delta,n\beta-m\gamma)_{LR}$, for A negative and B positive

$(m,\alpha,\beta)_{LR} \bullet (n,\gamma,\delta)_{LR} = (mn,-n\beta-m\delta,-n\alpha-m\gamma)_{LR}$, for A and B negative.

Operation [1] to [4] are used by Diamond (1989) in his version of fuzzy kriging.
Let us notice that if L and R are of the form

$$T(x)=\begin{cases} 1-|x| & 0\leq x\leq 1 \\ 0 & \text{otherwise} \end{cases} \tag{14}$$

we obtain a triangular function (T-type) which is frequently used in practice.

Manipulation of data by an expert

Applications of spatial methods to petroleum exploration on a basin scale encounter different data-related situations. Sometimes a large numbers of good quality data are clustered in some areas, but there is nothing between those clusters. Clusters may correspond to production fields which are characterized mainly by high values of parameters such as matrix or fracture porosity and permeability. High values may be related to limited diagenetic processes due to long-time preservation of hydrocarbons. Predictions based only on field data would lead to overestimation between the fields. In order to avoid those situations, gaps should be filled using additional information such as seismic data. In frontier areas, where seismic data is sparse or does not exit, a general knowledge about geology, analogs, etc. should be used. The final fuzzy set will result from a series of questions and answers. One may, for example, decide that the rock is characterized more by type A features than by B and C, but the second and the third one should not be excluded. In the resulting fuzzy set, parameters corresponding to type A will have higher membership values than parameters corresponding to types B and C.

FUZZY KRIGING

We will show here two independent approaches to define kriging with fuzzy data. Although both of them use the definition of fuzzy numbers, they are slightly different. The first is restricted to triangular membership functions; the second assumes fuzzy numbers in a general form.

Fuzzy numbers in a triangular form

In his approach, Diamond (1989) assumes that all fuzzy numbers are triangular (14) and are represented in the LR form (13). We omit here LR and use the notation A=(x,a,b), where a, b, and m stand for the supremum, infimum, and modal values of A respectively. For two fuzzy numbers A=(x,a,b) and B=(y,c,d) the distance is defined as following

$$d^2(A,B)=(x-y-(a-c))^2+(x-y+(b-d))^2+(x-y)^2 \quad (15)$$

A triangular fuzzy number-valued regionalised variable on the universe V is a quantitity w(x) at $x \in V$ whose value is a fuzzy number $(w_m, w_l(x), w_r(x))$. The regionalized variable w(x) can be considered as a realization of a fuzzy-valued random function W(x) defined on V. In Diamond (1989) it is shown that

[1] the expectation E{W(x)} exists iff $E\{d^2(W(x),0)\}$ exists and is a triangular fuzzy number.

[2] E{A+B}=E{A}+E{B} and $E\{\lambda W\}=\lambda E\{W\}$.

The variance of W is defined to be

$$\text{Var } W=E\{d^2(A,B)\} \quad (16)$$

and is a crisp non-negative number.

Definition

A fuzzy-valued fuzzy number function W(x) with mode $W_m(x)$, and support $[W_l(x),W_r(x)]$ is said to be second order stationary if

(i) $E\{W(x)\}=r=(r_m,\rho_l,\rho_r)$ exist and is independent of x.

(ii) there exist modal, upper and lower covariance functions $C_m(h)$, $C_l(h)$, and $C_r(h)$, independent of x, such that

$$\begin{aligned} &E\{W_m(x+h)W_m(x)\}-r_m^2=C_m(h)\\ &E\{W_l(x+h)W_l(x)\}-r_l^2=C_l(h) \quad (17)\\ &E\{W_r(x+h)W_r(x)\}-r_r^2=C_r(h) \end{aligned}$$

The estimate y_0 is a weighted average of data

$$y_0^*=\sum_{i=1}^{N}\lambda_i\, w(x_i) \quad (18)$$

Let a random function corresponding to y_0 be denoted by Y_0. The corresponding estimator is

$$Y_0^*=\sum_{i=1}^{N}\lambda_i\, W(x_i) \quad (19)$$

The weights λ_i are calculated assuming that

[1] Y_0^* is unbiased: $E\{Y_0^*\}=E\{Y_0\}=E\{W(x)\}$

[2] $Ed^2\{Y_0^*,Y_0\}=E\{(Y_{0m}^*-Y_{0m})^2+(Y_{0l}^*-Y_{0l})^2+(Y_{0r}^*-Y_0)^2\}$ is to be minimized

[3] $\lambda_i \geq 0$, $i=1,2,\ldots,N$

The proof of the following theorem can be found in Diamond (1989).

Theorem

Let Y_0^* be given by (19) and conditions [1],[2], and [3] are satisfied. The matrix defined as following $\Gamma_{ij}=C_m(x_i-x_j)+C_l(x_i-x_j)+C_r(x_i-x_j)$ for $i,j=1,2,\ldots,N$ is assumed to be strictly positive definite. There exists a unique, linear, and unbiased fuzzy-valued estimator $Y_0^* = \sum_{i=1}^{N} \lambda_i^+ W(x_i)$, which is best in the sense that it minimizes $Ed^2\{Y_0^*,Y_0\}$.

The weights λ_i^+ satisfy the system

$$\begin{aligned}
&\sum_{i=1}^{N} \Gamma_{ij} \lambda_i - L_j - \mu = C_m(x_j,V)+C_l(x_j,V)+C_r(x_j,V) \quad, j=1,2,\ldots,N \\
&\sum_{i=1}^{N} \lambda_i = 1 \\
&\sum_{i=1}^{N} L_i \lambda_i = 0 \\
&L_i \lambda_i \geq 0, \text{ for } i=1,2,\ldots,N
\end{aligned} \tag{20}$$

The residual is $\sigma_K^2 = C_m(V,V)+C_l(V,V)+C_r(V,V) +\mu$

$$- \sum_{i=1}^{N} \lambda_i^+ (C_m(x_i,V)+C_l(x_i,V)+C_r(x_i,V))$$

Fuzzy numbers in a general form

Another approach to krige fuzzy data was published by Bardossy, et.al.(1988). The authors first show how the extension principle works for fuzzy variograms. Next, they focus on intervals defined by α-level fuzzy sets (Definition 2.2). The following two fuzzy arithmetic operations were used

[1] the sum of two α-level intervals: $[a_\alpha,b_\alpha]+[c_\alpha,d_\alpha]=[a_\alpha+c_\alpha,b_\alpha+d_\alpha]$

[2] fuzzy number multiplied by a real number: $B_\alpha=\lambda A_\alpha$ and

if $\lambda>0$, then $c_\alpha=\lambda a_\alpha$ and $d_\alpha=\lambda b_\alpha$

if $\lambda<0$, then $c_\alpha=\lambda b_\alpha$ and $d_\alpha=\lambda a_\alpha$

where $A_\alpha=[a,_\alpha b_\alpha]$ and $B_\alpha=[c_\alpha,d_\alpha]$, and $\lambda\in\mathbf{R}$. For each $\alpha\in(0,1)$, we get two fuzzy experimental variograms- higher $\gamma^+_\alpha(h)$ and lower $\gamma^-_\alpha(h)$, where

$$\gamma_\alpha(h)=[a_\alpha,b_\alpha]$$
$$\gamma^+_\alpha(h)=b_\alpha \qquad (21)$$
$$\gamma^+_\alpha(h)=a_\alpha$$

The next step is to fit a crisp theoretical variogram to (21) and use it to solve a standard kriging system

$$\sum_{i=1}^{n}\lambda_i\,\gamma(x_i-x_j)=\gamma(x-x_j)\quad \text{for } j=1,2,\ldots,n$$
$$\sum_{i=1}^{n}\lambda_i=1$$

The crisp weights λ_i , i=1,2,...,n are used to calculate the following linear combination of fuzzy numbers

$$z^*(x)=[a_\alpha,b_\alpha]=\sum_{i=1}^{n}\lambda_i\,[a_{\alpha i},b_{\alpha i}]$$

where

$$a_\alpha=\sum_{\lambda_i>0}\lambda_i\,a_{\alpha i}+\sum_{\lambda_i<0}\lambda_i\,b_{\alpha i}$$
$$b_\alpha=\sum_{\lambda_i>0}\lambda_i\,b_{\alpha i}+\sum_{\lambda_i<0}\lambda_i\,a_{\alpha i}$$

PROBABILITY OF FUZZY EVENTS AND DEFUZZIFICATION

Suppose that randomness and fuzziness appear simultaneously. Let Z(x) be a second-order random function with expectation m, covariance C(h) and semivariogram $\gamma(h)$. Let us assume that the set $A=\{z(x_1), z(x_2), \ldots, z(x_n)\}$ contains measurements - known values. Fuzzy sets $\mathbf{B}_j$ which are defined on B_j j=1,2,...,k_j as subsets of A

$$\mathbf{B}_j=\left\{(\mu_j(z(x)),z(x)):\ z(x)\in A\right\} \qquad (22)$$

will be called fuzzy random sets, where $\mu(\cdot)$ is considered to be deterministic. $z(x_C)$ will refer to random variables which are elements of the fuzzy set C. Zadeh (1968) was the first one to introduce a rigorous definition of probability measures of fuzzy events (see also Kandel, 1986). He discussed inexact or fuzzy concepts within the confines of an extension of probability theory. In ordinary probability theory we define

$$P(X=x)=f(x) \qquad (23)$$

where f(x) is the probability density of a regional random variable X. The cumulative distribution function $F(x)=P\{X\le x\}$ is defined as

$$F(x) = \int_{-\infty}^{x} f(y)\, dy \tag{24}$$

Substituting dP(x) for f(x) dx the above equation can be written as

$$F(x) = \int_{-\infty}^{\infty} dP(y) \tag{25}$$

It is held that

$$\lim_{x \to \infty} F(x) = \int_{-\infty}^{+\infty} dP(y) = 1 \tag{26}$$

For a fuzzy random variable $z(x_A)$, by analogy with the above equation we write

$$P\left\{ z(x_A) = x \right\} = \mu_A(x)\, f(x) \tag{27}$$

Definition

Let (R^n,σ,P) be a probability space in which S is the σ-field of Borel sets in R^n and P is a probability measure over R^n. Then fuzzy event in R^n is a fuzzy set A in R^n whose membership function, $\mu_A: R^n \to [0,1]$ is Borel measurable.
The probability of the fuzzy event A is defined as follows

$$\lim_{x \to \infty} P(A,x) = \lim_{x \to \infty} \int_{-\infty}^{x} \mu_A(y)\, dP(y) \tag{28}$$

Very similarly we can define the probability of the fuzzy set A in n dimensions. Let us notice that

$$P(A) = E\left\{ \mu_A \right\} \tag{29}$$

In discrete cases the probability is defined as

$$P(A) = \sum_{i=1}^{n_A} \mu_A(z(x_i))\, p(x_i) \tag{30}$$

where n_A is a number of elements in A.
Mean value and variance can be defined as

$$m_p(A) = \frac{1}{P(A)} \sum_{i=1}^{n_A} z(x_i)\mu_A(z(x_i))\, p(z(x_i)) \tag{31}$$

$$v_p^2(A) = \frac{1}{P(A)} \sum_{i=1}^{n_A} (z(x_i)-m_p(A))^2 \mu_A)(z(x_i))\, p(z(x_i)) \tag{32}$$

where P(A) is defined by (30).
By analogy to (30), (31), (32) we can write the following averages

$$P'(A) = \frac{1}{n_A} \sum_{i=1}^{n_A} \mu_A(z(x_i)) \tag{33}$$

$$m'(A) = \frac{1}{\theta_A} \sum_{i=1}^{n_A} z(x_i)\mu_A(z(x_i)) \tag{34}$$

$$v_p^{'2}(A) = \frac{1}{\theta_A} \sum_{i=1}^{n_A} (z(x_i)-m'(A))^2 \mu_A(z(x_i)) \tag{35}$$

where

$$\theta_A = \sum_{i=1}^{n_A} \mu_A(z(x_i)) \tag{36}$$

Formulas (34) and (35) can be viewed as weighted averages of $z(x_i)$ and $(z(x_i)-m'(A))^2$ respectively. Weighting values are provided by the fuzzy membership function $\mu_A(\cdot)$. Under the stationarity condition

$$E\{m'_p(A)\} = \frac{1}{\theta_A} \sum_{i=1}^{n_A} E\{z(x_i)\}\ \mu_A(z(x_i)) = m \tag{37}$$

This means that the expected value of the fuzzy mean defined by (34) is equal to the expected value of the random function.
Expected values of (35)

$$E\{v_p^{'2}(A)\} = \frac{1}{\theta_A} \sum_{i=1}^{n_A} [E\{(z(x_i)^2\}-2E\{z(x_i)\ m'(A)\}+E\{m'(A)^2\}]\ \mu_A(z(x_i)) \tag{38}$$

where

$$E\{(z(x_i)^2\} = C(0)+m^2 \tag{39}$$

$$E\{z(x_i)\ m'(A)\} = \frac{1}{\theta_A} \sum_{j=1}^{n_A} C(x_i,x_j)\ \mu_A(z(x_j)) + m^2$$

$$E\{m'(A)\ m'(A)\} = \frac{1}{\theta_A^2} \sum_{j=1}^{n_A} \sum_{i=1}^{n_A} C(x_i,x_j)\ \mu_A(z(x_i))\ \mu_A(z(x_j)) + m^2$$

Finally

$$E\{v_p^{'2}(A)\} = C(0) - \frac{1}{\theta_A^2} \sum_{j=1}^{n_A} \sum_{i=1}^{n_A} C(x_i,x_j)\ \mu_A(z(x_i))\ \mu_A(z(x_j)) \tag{40}$$

Let us notice that if the definition (35) is replaced by

$$v_p^{'2}(A) = \frac{1}{\theta_A} \sum_{i=1}^{n_A} (z(x_i)-m)^2 \mu_A(z(x_i)) \tag{41}$$

The expected value

$$E\{v_p^{'2}(A)\} = \frac{1}{\theta_A} \sum_{i=1}^{n_A} E\{z^2(x_i)-2z(x_i)m+m^2\}\ \mu_A(z(x_i)) = C(0)$$

The estimator $z^*(x_0)$ considered is a linear combination of fuzzy mean values defined by (34)

$$z^*(x_0) = \sum_{i=1}^{n} \lambda_i \, m'(A_i) \tag{42}$$

From (37) and (42) the unbiasedness condition

$$E\{z^*(x_0) - \sum_{i=1}^{n} \lambda_i \, m'(A_i))\} = m - \sum_{i=1}^{n} \lambda_i \, m$$

and

$$\sum_{i=1}^{n} \lambda_i = 1 \tag{43}$$

The estimation variance can be represented as follows

$$\sigma_E^2 = E\{(z(x)-z^*(x_0))^2\} = E\{z(x)^2\} - 2\,E\{z(x)\, z^*(x_0)\} + E\{z^*(x_0)^2\}$$

where

$$E\{z(x)^2\} = C(0) + m^2$$

$$E\{z(x)\, z^*(x_0)\} = \sum_{i=1}^{n} \lambda_i \frac{1}{\theta_A} \sum_{j=1}^{n_{A_i}} C(x,x_j)] \, \mu_{A_i}(z(x_j)) + m^2$$

$$E\{z^*(x_0)^2\} = \sum_{i=1}^{n} \sum_{j=1}^{n} \lambda_i \lambda_j \frac{1}{\theta_{A_i}} \frac{1}{\theta_{A_j}} \sum_{k=1}^{n_{A_i}} \sum_{l=1}^{n_{A_j}} C(x_k,x_l) + m^2$$

To find the minimum of σ_E^2 we calculate n derivatives

$$\frac{\delta(E\{(z(x)-z^*(x_0))^2\} - 2\,\phi \sum_{i=1}^{n} \lambda_i)}{\delta \lambda_j}$$

and set them to zero

$$-2 \frac{1}{\theta_{A_i}} \sum_{k=1}^{n_{A_i}} C(x,x_k)\, \mu_{A_i}(z(x_k)) + 2 \sum_{j=1}^{n} \lambda_j \frac{1}{\theta_{A_i}} \frac{1}{\theta_{A_j}} \sum_{k=1}^{n_{A_i}} \sum_{l=1}^{n_{A_j}} C(x_k,x_l)$$

$$\mu_{A_i}(z(x_k))\, \mu_{A_j}(z(x_l)) - 2\,\phi = 0$$

Finally the kriging system

$$\begin{cases} \sum_{j=1}^{n} \lambda_j \, C'(A_i,A_j) - \phi = C'(x,A_i) & \forall\ i=1,2,\ldots,n \\ \sum_{j=1}^{n} \lambda_j = 1 \end{cases} \tag{44}$$

where

$$C'(A_i,A_j) = \frac{1}{\theta_{A_i}} \frac{1}{\theta_{A_j}} \sum_{k=1}^{n_{A_i}} \sum_{l=1}^{n_{A_j}} C(x_k,x_l)\, \mu_{A_i}(z(x_k))\, \mu_{A_j}(z(x_l)) \quad (45)$$

$$C'(x,A_i) = \frac{1}{\theta_{A_i}} \sum_{k=1}^{n_{A_i}} C(x,x_k)\, \mu_{A_i}(z(x_k)) \quad (46)$$

The minimum estimation variance

$$\sigma^2_{min} = C(0) - \sum_{i=1}^{n} \lambda_i\, C'(x,A_i) + \phi \quad (47)$$

Comparing (40), (45), and (46) it is seen that local fuzzy variances are included in formula (47).

FUZZY INDICATOR APPROACH

There are different ways to define a fuzzy indicator function. We will consider a fuzzy indicator to be a union of a Borel measurable (user defined) fuzzy number AROUND(z_0) and the indicator function $i(x,z_0)$

$$\mu_I(z_0)=\max[\mu_{AROUND}(z_0),i(x,z_0)] \quad (48)$$

$\mu_I(z_0)$ is also Borel measurable. Figure 1 shows an example of the fuzzy indicator function. Let us notice that μ_I are normal fuzzy sets. If AROUND(z_0) is a crisp number

$$\mu_I(z_0)=\max[\delta(z_0),i(x,z_0)]=i(x,z_0) \quad (49)$$

where $\delta(z_0)$ is the Dirac function. Another definition, providing less obvious links between fuzzy and crisp indicators can be based on S and L-R-shape functions (11) and (12). Similarly to cutoffs in classic indicator approaches we define fuzzy cutoff sets. Let $\mu(x)$ be Borel measurable fuzzy membership function defined at $x \in D$. Fuzzy cutoff $\mu(x,z)$ is defined as an intersection of a classic indicator function $i(x,z)$ and $\mu_I(x)$

$$\mu(x,z)=\min[\mu_I(x),i(x,z)] \quad (50)$$

which is equivalent to multiplication

$$\mu(x,z)=\mu_I(x)\ i(x,z). \quad (51)$$

Since, $\mu_I(x)$ and $i(x,z)$ are Borel measurable, $\mu(x,z)$ is also Borel measurable. Figure 2 shows an example cutoff fuzzy set.

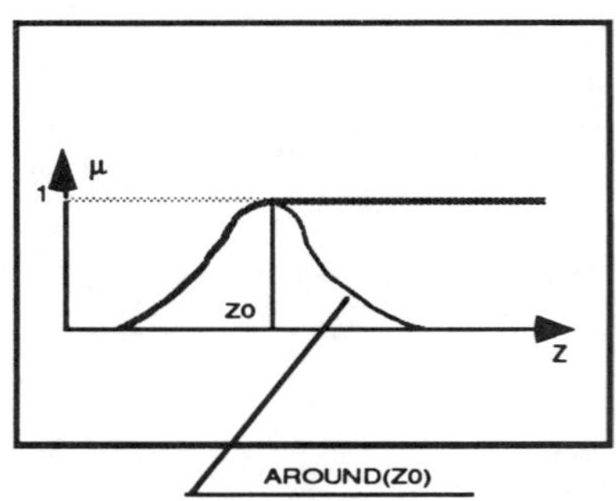

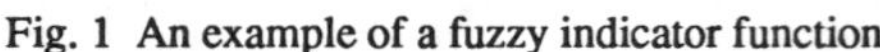

Fig. 1 An example of a fuzzy indicator function

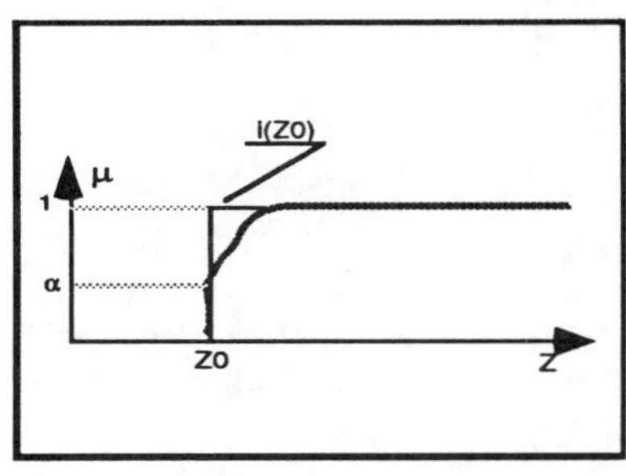

Fig. 2 An example of a fuzzy cutoff

An approach based on Zadeh's definition

Fuzzy cumulative distribution function is defined as an expected value of $\mu(x,z)$

$$F_{\mu}[z|\text{Fuzzy information}]=E\{\mu(x,z)\}=\int_D \mu(x,z)dP \tag{52}$$

Similarily to Journel (1983) we can define some useful functions.
Non-centered covariance between two fuzzy random variables separated by a vector h

$$K_{\mu}(h,z)=E\{\mu(x+h,z)\mu(x,z)\} \tag{53}$$

Centered covariance

$$C_{\mu}(h,z)=K_{\mu}(h,z)-F_{\mu}^{2}(z) \tag{54}$$

and semivariogram

$$\gamma_{\mu}(h,z)=\frac{1}{2}E\{[\mu(x+h,z)-\mu(x,z)]^2\}=\frac{1}{2}E\{\mu^2(x+h,z)\}-$$

$$E\{\mu(x+h,z)\mu(x,z)\}+\frac{1}{2}E\{\mu^2(x,z)\}=K_{\mu}(0,z)-K_{\mu}(h,z) \tag{55}$$

Fuzzy indicator kriging estimates the posterior distribution $F_{\mu}[z|\text{Fuzzy information}]$ as a linear combination of fuzzy cutoff functions

$$\mu^*(x_i,z) = \sum_{i=1}^{n} \lambda_i(x_i,z)\, \mu(x_i,z) +[1-\sum_{i=1}^{n} \lambda_i(x_i,z)]\, F_{\mu}(z) \tag{56}$$

Formula (51) suggests that (56) can be a linear combination of crisp and fuzzy indicators. Similarily as for crisp version of simple indicator kriging

$$\sum_{i=1}^{n} \lambda_i(x,z)\, C_{\mu}(x_i-x_j,z)=C_{\mu}(x-x_j,z) \tag{57}$$

for j=1,2,...,n where n is a number of fuzzy indicator variables.

Level sets defined from cutoffs

Let us assume that for each location x and the corresponding fuzzy membership function $\mu(x,z)$, α_x-level sets are obtained from z_0-cutoff, as shown in Fig. 2 . It means that α_x is considered to be a fuzzy membership value for $i(x,z_0)$. We will first calculate the simple kriging estimate of the indicator transform $i(x,z_0)$

$$i^*(x,z_0) = \sum_{i=1}^{n} \lambda_i(x,z_0)\, i(x_i,z_0) + [1-\sum_{i=1}^{n} \lambda_i(x,z_0)]\, F(z_0) \qquad (58)$$

where $\lambda_i(x_i,z_0)$ for i=1,2,...,n are the simple kriging weights corresponding to cutoff z_0, which are calculated from a simple kriging system

$$\sum_{i=1}^{n} \lambda_i(x,z)\, C_I(x_i-x_j,z_0)=C_I(x-x_j,z_0) \qquad (59)$$

where j=1,2,...,n and $C_I(h,z_0)$ is a regular indicator covariance at cutoff z_0. Next, a membership value is assigned to $i^*(x_i,z_0)$ as follows

$$\mu(x,z_0)=\mathrm{Min}(\alpha_{x_1},\alpha_{x_2},...,\alpha_{x_n}) \qquad (60)$$

The same procedure is performed for K cutoff values z_k. Finally, we obtain the following fuzzy set

$$I^*_\mu=\{(i^*(x,z_j),\mu(x,z_j)) \text{ for } j=1,2,...,k\} \qquad (61)$$

Methods based on fuzzy integration

Since formulas (52)-(57) contain operation of integration, we must discuss this problem. Intuitively, the fuzzy integral of a fuzzy function should result also in a fuzzy set. We will follow the concept of a fuzzy integral presented by Dubois and Prade (1979). Let us notice that α-level set of a fuzzy indicator can be written as $\alpha\cdot i(z_\alpha)$. An indicator fuzzy set can be alternatively defined by the formula

$$\mu_I(z_0) = \bigcup_{\alpha\in[0,1]} \mu_{\alpha I}(z_0) = \bigcup_{\alpha\in[0,1]} \alpha\cdot i(z_\alpha) \qquad (62)$$

where $\mu_{\alpha I}(z_0)$ is an α-level subset of $\mu_I(z_0)$, and z_α is such that $\mu_I(z) \geq \alpha$ for $z\geq z_0$. We will assume that $E\{\mu(x,z)\}$ is a fuzzy set of $\int_D i(z_\alpha)\, dP$ with fuzzy memberships α

$$E\{\mu(x,z)\} = \{(\int_D i(z_\alpha)\, dP,\alpha)\} \qquad (63)$$

for $\alpha\in[0,1]$.

Similarily we can define $E\{\mu(x,z)\mu(x+h,z)\}$

$$E\{\mu(x,z)\mu(x+h,z)\}=\{(\int_D i(z_\alpha)\, i(z_\beta)dP,\mathrm{Min}(\alpha,\beta))\} \qquad (64)$$

for $\alpha,\beta\in[0,1]$. The fuzzy indicator kriging problem defined using (63) and (64), requires a fuzzy least squares method, or a system of fuzzy linear equations to be solved first.These will be discussed in future.

DISCUSSION

Two approaches presented in section 4 define kriging in terms of fuzzy sets. The first method, published by Diamond (1989), is restricted to fuzzy triangular functions, which appears to be a fairly severe assumption for most geological applications. This drawback is outweighed, to some degree, by computational advantages and the fact that the approach is mathematically very consistent, i.e. all kriging equations are derived using fuzzy triangular numbers. In order to better describe geological information, LR fuzzy numbers defined by nonlinear L and R functions seem to be a good choice. Diamond (1989) shows no method or explanation of how field data of the fuzzy triangular type can be obtained. Instead, he fuzzifies published data. In our view, the fact that the data have to be fuzzified suggests that other kriging methods could be applied with similar success. Fuzzy kriging presented by Diamond (1989) is very similar to his interval kriging (Diamond, 1988). The second method (Bardossy, et.al., 1988) assumes that fuzzy numbers are of a general form, what is closer to our geologic intuition. First, they calculate upper and lower fuzzy experimental variograms using arithmetic operations on fuzzy α-intervals. Next, classic theoretical variograms are being fitted. The method allows one the iterative use of classic kriging solvers for each α-level. It looks as if there is a discrepancy between starting information - fuzzy numbers, and the transition from fuzzy experimental variogram to crisp variogram. It would appear to be better to apply Diamond's interval kriging to α-level intervals, and assign fuzzy membership α to the estimated interval. After having performed a similar procedure on all level intervals, the resulting, estimated intervals combined with the fuzzy membership values will result in a fuzzy distribution function.

In the section 5, it has been shown how geostatistics can handle fuzzy sets built on the "what-if" basis (explained in section 3.3). The approach is based on the Zadeh's definition of the probability of fuzzy events. It is applicable to basin-scale problems, where no hard data exist or data is sparsely distributed and biased by existing production. In such cases, expert knowledge is almost all what we have. Expert systems seem to be an appropriate area for applying this approach.

Zadeh's definition of the probability of fuzzy events (28) suggests that the concept of indicator can be extended to the fuzzy set environment. The following three approaches have been analyzed. The first one defines fuzzy indicator covariance and semivariogram. Calculations are performed assuming classic definitions of integrals. Derivations of fuzzy membership values from fuzzy cutoffs have been shown in the second approach. These values are assigned to classic indicator kriging results. The third method uses the definition of fuzzy integrals. A fuzzy least squares method or an algorithm solving a system of fuzzy linear equations are required to solve the kriging problem.

As has been shown, the fuzzy set theory-based geostatistics has a potential for future research. It can be applied to basin scale problems, where expert knowledge plays a significant role. Some concepts, such as indicators, can be generalized in terms of multivalued logic, which will result in a new class of geostatistical methods.

REFERENCES

Bardossy, A., Bogardi, I., and Kelly, W.E., (1988) "Imprecise (fuzzy) information in geostatistics", Math. Geol., 20, pp. 189-203.

Bardossy, A., Bogardi, I., and Kelly,W.E. (1990) "Kriging with imprecise (fuzzy) variograms. I: theory", Math. Geol .22, pp. 63-79.

Diamond, P. (1989) "Fuzzy kriging", FSS, 33, pp. 315-332.

Diamond, P. (1988) "Interval-valued random functions and the kriging of intervals", Math. Geol, 20, pp. 145-165.

Dubois, D., Prade, H., (1978) "Operation on fuzzy numbers", Int. J. Syst. Scie., 9, pp. 613-626.

Dubois, D., and Prade, H., (1979) "Fuzzy real algebra: Some results" FSS, 2, pp. 327-348.

Dubois, D., Prade, H., (1980) "Fuzzy Sets and Systems: Theory and Applications", Academic Press, New York.

Dubrule, O., and Kostov, C., (1986) "An interpolation method taking into account inequity constrains", Math. Geol., 18, pp. 269-286.

Journel, A.G., (1983) "Non-parametric estimation of spatial distribution", Math. Geol., 15, pp. 445-468.

Journel, A.G., (1986) "Constrained interpolation and qualitative information - the soft kriging approach", Math. Geol., 18, pp. 269-286.

Kacewicz, M., (1991) "Shape prediction with a fuzzy uncertainty measure", Math. Geol., 23, pp. 289-295.

Kacewicz, M., (1993) "Mathematics between source and trap: uncertainty in hydrocarbon migration modeling", in: New Developments in Mathematical Geology (in press), Oxford University Press.

Kandel, A., (1986) "Fuzzy Math. Techniques with Applications", Addison-Wesley, 274 pp.

Kitanidis, P.K., (1986) "Parameter uncertainty in estimation of spatial functions: Bayesian analysis", Water Resoures. Res., 22, pp. 499-507.

Kruse, R., and Meyer, K.D., (1987) "Statistics with Vague Data", D. Reidel, pp. 279.

Omre, H., (1987) "Bayesian kriging - merging observations and qualified guesses in kriging", Math. Geol., 19, pp. 25-39.

Yager, R.R., (1979) "A note on probabilities of fuzzy events", Inf. Sci., 18, pp. 113-129.

Zadeh, L.A., (1965) "Fuzzy sets", Inform. Contr., 8, pp. 338-353.

Zadeh, L.A., (1968) "Probability measures of fuzzy events", J. Math. Anal., 23, pp. 421-427.

Zimmerman, H.J, (1984) "Fuzzy Set Theory - and Its Appplications", Kluwer-Nijhoff, 363 pp.

ROBUST BAYES LINEAR PREDICTION OF REGIONALIZED VARIABLES

Jürgen Pilz
TU Bergakademie Freiberg
Department of Geosciences
D-09599 Freiberg, Germany

The Bayes linear kriging approach assumes that the first and second order moments of the prior distribution of the trend model parameters are known exactly. Usually, however, only partial prior knowledge about these moments is available. To cope with this uncertainty we consider both local and global Bayesian robustness measures, in the sense of Berger (1984). We derive in explicit form a globally Bayes robust linear predictor, which turns out to be a hierarchical Bayes linear predictor with a least favourable covariance matrix for the uncertain prior mean of the trend parameter vector chosen at the last stage of the hierarchy. The results are illustrated for the special case of Bayes ordinary kriging when only interval bounds for the prior expectation and variance of the mean of the regionalized variable under consideration are known.

1 INTRODUCTION

This paper is concerned with the prediction of regionalized variables when prior knowledge about the mean function (not necessarily constant) is available. Omre (1987), Omre and Halvorsen (1989) and Omre, Halvorsen and Berteig (1989) develop a Bayes linear approach to kriging which admits partial prior knowledge about the coefficients of the trend function. In the Bayesian approach, the unknown trend parameters are random and follow some a priori distribution. The restriction to linear predictors allows the analysis to be carried out in terms of first and second order moments of the prior distribution and the assumed covariance function of the regionalized variable.
A major drawback of the Bayesian kriging approach developed in the above cited papers is the assumption of *exact* knowledge of the first and second order prior moments. Omre and Halvorsen (1989) initiate a study of the effects of varying precision of the prior specification on the resulting posterior parameter distribution. In this paper we develop a systematic robust Bayesian approach, requiring only approximate knowledge of the prior moments, to make the "orthodox" Bayesian approach more flexible.

R. Dimitrakopoulos (ed.), Geostatistics for the Next Century, 464–475.

In Section 2 we reconsider the Bayesian model and the Bayes linear kriging predictor. Then we introduce Bayesian robustness measures, in particular local and global Bayesian robustness as advocated by Berger (1984). We give, in an explicit form, globally Bayes robust predictors for the case where we only have approximate knowledge of the first and second order prior moments. In the limiting cases of noninformative and exact prior knowledge about the trend parameter vector these predictors coincide with the well-known universal and simple kriging predictor, respectively.

2 THE BAYES LINEAR KRIGING PREDICTOR

Consider a regionalized variable, $\{Z(x): x \in D\}$, defined over some region $D \subset R^d$ (usually, d=2 or d=3, respectively) and having expectation

(1) $$E\, Z(x) = m(x).$$

The trend function will be modelled as

(2) $$m(x) = \theta_1 f_1(x) + \ldots + \theta_r f_r(x)$$

where $f(x)=(f_1(x),\ldots,f_r(x))^T$ is a vector of specified functions, assumed to be linearly independent over D, and $\theta=(\theta_1,\ldots, \theta_r)^T$ is an unknown vector of trend parameters. We assume second order covariance stationarity; the covariance function will be denoted by

(3) $$C(h) = \mathrm{Cov}\,(R(x+h),R(x)) = E\,\{R(x+h)R(x)\}\ ;\ x,x+h \in D$$

where $R(x)= Z(x) - m(x)$. The user-specified covariance function (3) is assumed to be independent of θ. Like Omre and Halvorsen (1989), we suppose a priori that

(4) $$E\,\theta_i = \mu_i\ ,\quad \mathrm{Cov}\,(\theta_i,\theta_j) = \phi_{ij}\ ;\ i,j= 1,\ldots,r\ .$$

Here the μ_i's can be interpreted as prior guesses for the unknown trend parameters, the ϕ_{ii}'s reflect the uncertainties about the guesses and the ϕ_{ij}'s, $i \neq j$, account for the interrelations between the (random) trend parameters. Note that

$$E\, Z(x) = E\,\{\, E\,(Z(x) \mid \theta)\,\} = \mu_1 f_1(x) + \ldots + \mu_r f_r(x) = \mu^T f(x)$$

and that the total (marginal) covariance function is given by

(5) $$C_0(h) = E\,\{C(h)\} + \mathrm{Cov}\,\{\, E(Z(x+h)|\theta),\, E(Z(x)|\theta)\,\} = C(h) + f(x+h)^T \Phi f(x).$$

Our goal is to predict the value of $Z(x_0)$ at a predetermined point $x_0 \in D$ on the basis of observations $Z=(Z(x_1),...,Z(x_n))^T$ at given locations $x_1,...,x_n \in D$, using an (inhomogeneous) linear predictor of the form

$$\hat{Z}(x_0) = w_1 Z(x_1) + \ldots + w_n Z(x_n) + w_0 = w^T Z + w_0.$$

The weights w_0, $w=(w_1,...,w_n)^T$ are chosen to minimize the total mean square error of prediction (TMSEP),

$$E\{ Z(x_0) - \hat{Z}(x_0) \}^2 = \mathrm{Var}\{Z(x_0) - \hat{Z}(x_0)\} + \{ E(Z(x_0)-\hat{Z}(x_0)) \}^2 . \tag{6}$$

This is just the Bayes risk of $\hat{Z}(x_0)$ with respect to squared error loss. With

$$c_0 = (C(x_0-x_1),...,C(x_0-x_n))^T, \quad K = (C(x_i-x_j))_{i,j=1,...,n} \quad \text{and} \quad F = (f(x_1),...,f(x_n))^T$$

one easily verifies that

$$\mathrm{Var}\{Z(x_0) - \hat{Z}(x_0)\} = C(0) + f(x_0)^T \Phi f(x_0) - 2w^T(c_0 + F\Phi f(x_0)) + w^T(K + F\Phi F^T)w$$

and

$$E(Z(x_0) - \hat{Z}(x_0)) = \mu^T f(x_0) - \mu^T F^T w - w_0 .$$

Whatever be w, we may choose $w_0 = \mu^T(f(x_0) - F^T w)$ to minimize (6). This implies

$$E\{Z(x_0) - \hat{Z}(x_0)\}^2 = C(0) + f(x_0)^T \Phi f(x_0) - 2w^T(c_0 + F\Phi f(x_0)) + w^T(K + F\Phi F^T)w \tag{7}$$

From (7) we readily obtain the equation for the optimum choice of w:

$$(K + F\Phi F^T) * w = c_0 + F\Phi f(x_0) . \tag{8}$$

Together with the above choice for w_0, this leads us to the Bayes linear predictor

$$\hat{Z}_B(x_0) = (c_0 + F\Phi f(x_0))^T (K + F\Phi F^T)^{-1} (Z - F\mu) + \mu^T f(x_0) . \tag{9}$$

The Bayes linear predictor (9) is the total mean, $E\,Z(x_0) = \mu^T f(x_0)$, adjusted by a linear combination of the prior expected residual vector, $Z - F\mu$ (see also Omre and Halvorsen (1989)).

We now draw a link to Bayes linear regression theory. For the Bayes linear regression model

$$E(Z|\theta) = F\theta \quad \text{with} \quad E\,\theta = \mu, \ \mathrm{Cov}\,\theta = \Phi,$$

the (optimum) Bayes estimator of θ within the class of linear estimators of the form $\hat{\theta}= AZ + a$, with some $(r\times n)$- matrix A and vector $a\in R^n$, is given by

$$\hat{\theta}_B = \Phi F^T(F\Phi F^T + K)^{-1}(Z - F\mu) + \mu \tag{10}$$

see, e.g. Pilz (1991). In case that K and Φ have full rank, it can be rewritten as $\hat{\theta}_B= (F^TK^{-1}F + \Phi^{-1})^{-1}(F^TK^{-1}Z + \Phi^{-1}\mu)$, which is just a matrix weighted average of the best linear unbiased estimator (BLUE) $\hat{\theta}= (F^TK^{-1}F)^{-1}F^TK^{-1}Z$ of θ and its prior mean μ. Using the matrix identity

$$(K + F\Phi F^T)^{-1} = K^{-1} - K^{-1}F(F^TK^{-1}F + \Phi^{-1})^{-1}F^TK^{-1} \tag{11}$$

the Bayes linear predictor may be written in the compact form

$$\hat{Z}_B(x_0) = f(x_0)^T\theta_B + c_0^TK^{-1}(Z - F\theta_B). \tag{12}$$

Thus, it has the same structure as the well-known universal kriging predictor

$$\hat{Z}_{UK}(x_0) = f(x_0)^T\hat{\theta} + c_0^TK^{-1}(Z - F\hat{\theta}), \tag{13}$$

with the BLUE $\hat{\theta}$ replaced by the Bayes linear estimator $\hat{\theta}_B$.
Some of the remarkable features of the Bayes linear predictor have already been listed in Omre and Halvorsen (1989), notably it works even where $n<r$ and it is numerically more stable than the universal kriging predictor. Moreover, $\hat{Z}_B$ is simple to implement on a computer, since we can use existing Kriging and/or regression packages and since $\hat{\theta}_B$ can be represented in the form of an augmented BLUE

$$\hat{\theta}_B = (\underline{F}^T\underline{K}^{-1}\underline{F})^{-1}\,\underline{F}^T\underline{K}^{-1}\underline{Z},$$
$$\text{where } \underline{F}^T = [F^T : I_r],\ \underline{K} = \mathrm{diag}\,[K,T],\ \underline{Z}^T = [Z^T, \mu^T]$$

and I_r stands for the identity matrix of order r.
From (7) and (8) we deduce that

$$\begin{aligned} E\,\{Z(x_0) - \hat{Z}_B(x_0)\}^2 = {} & C(0) + f(x_0)^T\Phi f(x_0) \\ & - (c_0+F\Phi f(x_0))^T(K+F\Phi F^T)^{-1}(c_0+F\Phi f(x_0)) . \end{aligned} \tag{14}$$

Clearly, this total risk of the Bayes linear predictor is smaller than the corresponding TMSEP of $\hat{Z}_{UK}(x_0)$. However, even the MSEP (frequentist risk) of $\hat{Z}_B(x_0)$ is smaller than that of $\hat{Z}_{UK}(x_0)$ for all parameters, θ, contained in a neighbourhood of μ of appreciable size, see Pilz (1991). The sum of the first two terms on the right-hand side of (14) equals the total (marginal) variance of $Z(x_0)$, $\mathrm{Var}\,\{Z(x_0)\} = C(0) + f(x_0)^T\Phi f(x_0)$. Observing that for the augmented observation vector

$$Z_0 = (Z(x_0), Z(x_1), \ldots, Z(x_n))^T = (Z(x_0), Z^T)^T$$

the marginal covariance matrix reads

$$\mathrm{Cov}\,Z_0 = \begin{pmatrix} f(x_0)^T\Phi f(x_0) + C(0) & \vdots & (c_0 + F\Phi f(x_0))^T \\ \cdots\cdots & \vdots & \cdots\cdots \\ c_0 + F\Phi f(x_0) & \vdots & F\Phi F^T + K \end{pmatrix}$$

we see that the TMSEP of $\hat{Z}_B(x_0)$ equals the Schur complement of $\mathrm{Var}\,\{Z(x_0)\}$ with respect to the marginal covariance matrix of the observations, $\mathrm{Cov}\,Z = F\Phi F^T + K$.

3 BAYESIAN ROBUSTNESS MEASURES

The Bayes linear predictor, $\hat{Z}_B$, depends on the basic prior inputs μ and Φ and, like any other linear predictor, on the specified covariance structure expressed by K and c_0 (via C(h)). We are interested in the robustness of our Bayes predictor against misspecifications of these basic model ingredients. Diamond and Armstrong (1984), Bardossy (1988), Posa (1989) and Stein (1989) explore the robustness of the universal kriging predictor against variogram (or related covariance function) misspecification. In the Bayes case we have to account for additional uncertainty with respect to the first and second order prior moments μ and Φ, respectively. More generally, Bayesian robustness means robustness of prediction with respect to a varying joint distribution P_{θ,Z_0} of the trend parameter θ and augmented observation vector Z_0, respectively. The uncertainty with respect to this distribution is simply expressed by allowing P_{θ,Z_0} to vary over some family, Π_0, encompassing all plausible distributions.
Berger (1984) introduces local and global Bayesian robustness to aim at inference robustness and procedure robustness, respectively. In our context, the local Bayesian robustness measure rests on the posterior expected squared error of prediction (posterior expected loss)

$$(15)\qquad l(P,\hat{Z}) = E\,\{\,[Z(x_0) - \hat{Z}(x_0)]^2 \mid Z{=}z\,\}$$

for a given realization z of the observation vector Z, i.e. the expectation is taken w.r.t. the posterior distribution of $(\theta, Z(x_0))$ given Z=z. Then an arbitrary predictor $\hat{Z}^*(x_0)$ for $Z(x_0)$ is called <u>d-locally Bayes robust</u> for some real number d > 0 iff

$$l_0(\hat{Z}^*) = \sup_{P \in \Pi_0} | l(P,\hat{Z}^*) - \inf_{\hat{Z}} l(P,\hat{Z}) | \leq d .$$

If θ and Z_0 are jointly normal then this amounts to measuring the numerical stability of $\hat{Z}_B$ with respect to varying the first and second order moments of (θ, Z_0), i.e. the quantities μ, Φ, c_0 and K, within the ranges associated with the family Π_0. To be specific, let K_0 denote the covariance matrix of the conditional distribution of Z_0 partitioned as

$$K_0 = \mathrm{Cov}\,(Z_0|\theta) = \begin{pmatrix} C(0) & \vdots & c_0^T \\ \cdots & \cdots & \cdots \\ c_0 & \vdots & K \end{pmatrix}$$

and consider the following general class of distributions

$$(16) \qquad \Pi_0 = \{ P_{\theta,Z_0} : E\,\theta \in \mathbb{M}, \mathrm{Cov}\,\theta \in \mathbb{T}, \mathrm{Cov}\,(Z_0|\theta) \in \mathbb{K}\}$$

where $\mathbb{M}$ is some subset of R^r and $\mathbb{T}$ and $\mathbb{K}$ are subsets of the sets of positive definite matrices of order r and n+1, respectively. Then the following result obtains.

<u>Proposition 1</u>: If $P_{\theta,Z_0} \in \Pi_0$ is normal we have for any predictor $\hat{Z}^*(x_0)$ of $Z(x_0)$ that

$$l_0(\hat{Z}^*) = \sup \{ | \hat{Z}^*(x_0) - \hat{Z}_B(x_0) |^2 : \hat{Z}_B \in \mathbb{Z} \}$$

where

$$\mathbb{Z} = \{ \hat{Z}_B(x_0) = f(x_0)^T \hat{\theta}_B + c_0^T K^{-1}(Z - F\hat{\theta}_B) : \mu \in \mathbb{M}, \Phi \in \mathbb{T}, K_0 \in \mathbb{K}\}.$$

This result follows from the fact that, under the normality assumption, $l(P, \cdot)$ is minimized by the Bayes linear predictor $\hat{Z}_B$, which is based on the posterior mean $\hat{\theta}_B$ of θ.

Proposition 1 tells us that the local Bayes robustness measure of a predictor is equivalent to its maximum distance from the class of Bayes linear predictors $\mathbb{Z}$. When restricting attention to linear predictors then, the normality assumption is no longer needed and the evaluation of local Bayes robustness comes down to measuring the effect of misspecification of the moments μ, Φ and K_0. In this respect, expressions for the maximum distance can be derived in explicit form from results of Das Gupta and Studden (1988). On the other hand, the results of Diamond and Armstrong (1984) and Bardossy (1988) carry over directly to the Bayes case and can also be used to evaluate the numerical stability of a Bayes linear predictor at hand.

The global Bayesian robustness criterion rests on the usual Bayes risk, i.e. on the TMSEP defined by (6) and obtained from $l(P,\hat{Z})$ after integration w.r.t. the marginal distribution of Z. Correspondingly, a predictor $\hat{Z}^*$ is called d-globally Bayes robust iff

$$\sup_{P\in\Pi_O} \left| TMSEP(P,\hat{Z}^*) - \inf_{\hat{Z}} TMSEP(P,\hat{Z}) \right| \le d$$

for some d>0. The search for an optimal predictor with respect to this robustness criterion leads us to minimax regret, which is mathematically intractable. Alternatively, we may look for a predictor $\hat{Z}^*$ satisfying

$$(17) \qquad \sup_{P\in\Pi_O} TMSEP(P,\hat{Z}^*) = \inf_{\hat{Z}} \sup_{P\in\Pi_O} TMSEP(P,\hat{Z})$$

i.e., a predictor which minimizes the maximum possible Bayes risk (TMSEP) among all linear predictors $\hat{Z}$ where the maximum is taken with respect to all plausible probability distributions $P=P_{\theta,Z_0}$. This criterion is also known as the criterion of restricted minimaxity (or Π_0-minimaxity or Gamma-minimaxity, respectively).

4 RESTRICTED MINIMAX LINEAR PREDICTION

For simplicity, assume the covariance structure of the random field $\{Z(x): x\in D\}$ is known, i.e., c_0 and K are fixed. This is the usual assumption underlying all the "classical" kriging procedures: once a theoretical covariance (or variogram) function model has been fitted to the empirical covariance (or variogram) function, prediction proceeds as if this model were the true underlying model.
If c_0 and K are specified we can restrict our attention to the subfamily Π_θ of plausible prior distributions of θ, and thus we can immediately use the results of Pilz (1991a; Ch. 6.3, 15.5) concerning (globally) robust Bayes linear estimation of the regression parameter θ. To be specific, consider the following class

$$(18) \qquad \Pi_\theta = \{ P_\theta : \ E\,\theta = \mu \in \mathcal{M}, \ \ Cov\,\theta = \Phi \le \Phi_0 \}$$

of plausible priors, where $\mathcal{M}$ is some subset of R^r and Φ_0 is a given positive definite matrix of order r serving as an upper bound for the covariance matrix of θ. By an "upper bound" we mean that $\Phi_0 - Cov\,\theta$ is positive semidefinite. For example, if the trend parameters are considered uncorrelated a-priori, which is the case in Omre and Halvorsen (1989), then we may choose $\Phi_0 = diag(\phi_1,\ldots,\phi_r)$ where ϕ_i is an upper bound for the variance $Var\,\theta_i$; $i=1,\ldots,r$.

Now, consider the globally Bayes robust predictor with respect to the above family, i.e. the Π_θ - minimax predictor. We distinguish between two cases. First, if $E\theta = \mu$ is specified, i.e. $\mathcal{M}$ is a single-element subset, then in the Bayes linear estimator we can replace Φ by its upper bound Φ_0. This is proved showing that the TMSEP of $\hat{Z}_B$ is

$$(19) \qquad \text{TMSEP}(P_\theta, \hat{Z}_B) = C(0) - c_0^T K^{-1} c_0 + (f(x_0) - F^T K^{-1} c_0)^T (F^T K^{-1} F + \Phi^{-1})^{-1} (f(x_0) - F^T K^{-1} c_0)$$

and observing that this expression attains its maximum for $\Phi = \Phi_0$. Secondly, in case that μ is unspecified but can be assumed to lie in some compact set $\mathcal{M}$ symmetric around some center point $\mu_0 \in R^r$, for example an ellipsoid or a hyperrectangle, then $\hat{Z}_B$ is to be replaced by the globally Bayes robust predictor

$$(20) \qquad \hat{Z}_B^*(x_0) = f(x_0)^T \hat{\theta}_B^* + c_0^T K^{-1}(Z - F\hat{\theta}_B^*) \quad \text{where}$$
$$\hat{\theta}_B^* = (\Phi_0 + M_q^*) F^T [\, F(\Phi_0 + M_q^*) F^T + K \,]^{-1} (Z - F\mu_0) + \mu_0 \, .$$

Here, M_q^* is a positive semidefinite matrix which maximizes the functional

$$(21) \qquad G(M_q) = f(x_0)^T \{ F^T K^{-1} F + (\Phi_0 + M_q)^{-1} \}^{-1} f(x_0)$$

defined on the set of centered second order moment matrices $M_q = \int_{\mathcal{M}} (\mu - \mu_0) \times (\mu - \mu_0)^T q(d\mu)$ generated by the class of all probability distributions q over $\mathcal{M}$.

<u>Proposition 2</u>: The Π_θ-minimax linear predictor for $Z(x_0)$ with respect to the family Π_θ from (18) is given by

$$\hat{Z}_B^*(x_0) = f(x_0)^T \hat{\theta}_B^* + c_0^T K^{-1}(Z - F\hat{\theta}_B^*)$$

with $\hat{\theta}_B^*$ defined by (20) and M^* maximizing the functional $G(\cdot)$.
For a proof see Pilz (1991a), Ch. 6.3.

The maximization of $G(\cdot)$ amounts to solving a convex optimization problem. For a compact and symmetric region $\mathcal{M}$ as considered above we can give a solution in explicit form:

$$(22a) \qquad M^* = (\mu^* - \mu_0)(\mu^* - \mu_0)^T \qquad \text{where } \mu^* \text{ maximizes}$$

$$(22b) \qquad g(\mu) = \{f(x_0)^T(\mu - \mu_0)\}^2 / \{1 + (\mu - \mu_0)^T F^T K^{-1} F (\mu - \mu_0)\}$$

and, additionally, satisfies

$$(22c)\qquad \sup_{\mu\in\mathcal{M}} \{f(x_0)^T V(V+\Phi_0+M^*)^{-1}(\mu-\mu_0)\}^2 = \{f(x_0)^T V(V+\Phi_0+M^*)^{-1}(\mu^*-\mu_0)\}^2$$

For an arbitrary compact $\mathcal{M}$ one has, in general, to use iterative procedures to obtain approximately optimal solutions (see Pilz (1991a), Ch. 12, 15).
From the structure of $\hat{Z}_B$ it is seen that this predictor is the Bayes linear predictor for $Z(x_0)$ when θ follows a prior distribution with the moments $E\,\theta = \mu_0$ and $Cov\,\theta = \Phi_0 + M^*$. Thus, uncertainty w.r.t. the prior location parameter $\mu = E\,\theta$ causes an "inflation" of the prior covariance matrix $Cov\,\theta$. This is equivalent to adding a further stage to the Bayes model assuming that the hyperparameter μ follows a prior distribution with the moments $E\,\mu = \mu_0$ and $Cov\,\mu = M^*$. Thus, we obtain

<u>Proposition 3:</u> Π_θ- minimax linear prediction with respect to the family (18) with a symmetric and compact region $\mathcal{M}$ is equivalent to Bayes linear prediction in the three-stage-hierarchical Bayes model

$$E\,Z(x) = f(x)^T\theta\,,\quad Cov\,(Z(x_i),Z(x_j)) = C(x_i-x_j);\ i,j=0,1,\ldots,n$$
$$E\,\theta = \mu\,,\quad Cov\,\theta = \Phi_0$$
$$E\,\mu = \mu_0,\ Cov\,\mu = M^*$$

where μ_0 is the center point of $\mathcal{M}$ and M^* maximizes the Bayes risk among all covariance matrices generated by the (hyper) priors defined over $\mathcal{M}$.

Consider the two extreme cases of noninformative and exact knowledge of θ, respectively. In case of complete prior ignorance about θ we have $\mathcal{M} = R^r$ and $(M^*+\Phi_0)^{-1} = 0$ and thus $\hat{Z}_B^*$ coincides with the UK predictor. Conversely, if we have exact knowledge of θ then $\mathcal{M}=\{\mu_0\}$, $\Phi_0=M^*=0$ and $\hat{Z}_B^*$ coincides with the simple kriging predictor. Usually, we will be faced with a situation of partial prior knowledge between these two extremes. Then the gain in efficiency obtainable when using $\hat{Z}_B^*$ instead of the UK predictor is given by

$$\begin{aligned} \text{eff} &:= TMSEP(\hat{Z}_{UK}) - TMSEP(\hat{Z}_B^*) \\ &= (f(x_0)-F^TK^{-1}c_0)^T V\{V+M^*+\Phi_0\}^{-1} V(f(x_0)-F^TK^{-1}c_0) \end{aligned}$$

The preceding results can be generalized, to some extent, to the case when there is also uncertainty about the covariance structure of the observations. To this end, consider the following situation. Assume that

$$Cov\,(Z|\theta) \le K^* \text{ and } c_0 = (C(x_0-x_1),\ldots,C(x_0-x_n))^T = 0,$$

i.e. we have no correlations between $Z(x_0)$ and the actual observations. This will be the case, for example, when we are faced with an extrapolation problem where the distance of the point x_0 from the observation sites is large enough to neglect the spatial dependence. Then the preceding results still hold with K replaced by K^*, since in the case $c_0 = 0$ the TMSEP (19) is maximized by setting $K=K^*$.

Finally, a few words about the upper bound matrices, Φ_0 and K^*. For Toeplitz-type covariance structures, the rsults of Grenander and Szegö (1958) may be used to derive such upper bounds. When the error structure admits a first order spatial autoregressive or moving average process, upper bounds in explicit form may be derived from results due to Oman (1982).

5 ROBUST BAYES ORDINARY KRIGING

Let us now consider the special case where

$$E\,Z(x) = \theta\text{ , i.e. } f(x)=1;\ E\,\theta \in \mathcal{M} = [\mu_l, \mu_u];\ \mathrm{Var}\,\theta \le \phi_0$$

Then $F=(1,\dots,1)^T$, a vector of n ones, and the Bayes kriging predictor reads

$$\hat{Z}_B = (1 - c_0{}^T K^{-1} F)\hat{\theta}_B + c_0{}^T K^{-1} Z = (1 - \sum c_{0i} k^{ij})\hat{\theta}_B + \sum c_{0i} k^{ij} Z(x_j),$$

where

$$\hat{\theta}_B = (\sum k^{ij} Z(x_j) + \mu/\phi)/ (\sum k^{ij} + 1/\phi);\ \ K^{-1} = (k^{ij})$$

and the summation is over $i,j=1,\dots,n$. Having specified $\mu=\mu_1$ and $\phi=\phi_1$, the resulting predictor has local Bayes robustness measure

$$l(\hat{Z}_B{}^*) = (1-\sum c_{0i} k^{ij})^2 \times \sup \{g(\mu,\phi):\ \mu\in\mathcal{M},\ \phi\in(0,\phi_o]\ \}$$

where

$$g(\mu,\phi) = \{(a+\mu_1/\phi_1)/(b+1/\phi_1) - (a+\mu/\phi)/(b+1/\phi)\}^2 \text{ and } a= \sum k^{ij} z(x_j),\ b= \sum k^{ij}.$$

It can be seen that the maximum of $g(\mu,\phi)$ is attained at one of the two corner points (μ_l,ϕ_o), (μ_u,ϕ_o); so $l(\hat{Z}_B{}^*)$ can be easily computed.

Now, finding a globally Bayes robust predictor is not difficult either. Rewriting $\mathcal{M}$ as $\mathcal{M} = \{\mu\in R^1\!:\ |\mu-\mu_0|\le (\mu_u - \mu_l)/2\}$ where $\mu_0 = (\mu_u + \mu_l)/2$ is the symmetry point of $\mathcal{M}$, the solution to (22a) - (22c) is given by

$$M^* = (\mu^* - \mu_0)^2 = (\mu_u - \mu_l)^2/ 4.$$

This is the largest possible variance. Hence, the globally Bayes robust ordinary kriging predictor takes the form

$$\hat{Z}_B{}^0 = \sum c_{0i}k^{ij}Z(x_j) + (1 - \sum c_{0i}k^{ij})\,\hat{\theta}_B{}^0$$

where

$$\hat{\theta}_B{}^0 = (\sum k^{ij}Z(x_j) + \mu_0/(\phi_0+M^*))/(\sum k^{ij} + 1/(\phi_0+M^*)).$$

The gain in efficiency over the ordinary kriging predictor is given by

$$\text{eff} = (1 - \sum k^{ij}c_{0j})^2\,(\sum k^{ij})^2 / (\sum k^{ij} + M^* + \phi_0).$$

In the limiting case $(\mu_u-\mu_l)\to\infty$, $\phi_0\to\infty$, which means that our prior knowledge about θ is noninformative (θ can be anywhere), we have $(M^*+\Phi_0)^{-1}= 0$ and $\hat{Z}_B{}^0$ coincides with the OK predictor. On the other hand, if $(\mu_u-\mu_l)\to 0$, $\phi_0\to 0$ which implies that $\theta_B=\mu_l=\mu_u=\mu$ with probability one and $M^*=0$, we obtain the simple linear kriging predictor.

CONCLUSIONS

Our robustness investigations are limited to studying the effects of uncertainty with respect to the trend model. This type of modelling is well suited to situations where the trend coefficients have some physical meaning, which is the case, for example, in the majority of models including an external drift. Clearly, there is a need for extensions to modelling uncertainty with respect to the covariance structure of the regionalized variable. In a parametric Bayes setting, this would mean to consider a parameterized covariance function, $C(h) = C_\beta(h)$, say, and then to express uncertainty with respect to C through a prior distribution P_β for the vector of covariance parameters β. But the basic difference to the above approach is that common types of covariance functions are *nonlinear* with respect to the components of β, at least with respect to the range component; so we cannot hope for results in an explicit form. What can be done, however, is to start with the numerical computation of Bayes estimators using prior distributions for the covariance components and then to study the effects of a possible mis-specification, as proposed by Handcock and Stein (1993). This requires the specification of a sufficiently rich and tractable class of plausible priors for β as well as the elaboration on noninformative priors P_β which could be used as standard or reference priors in case of vague prior knowledge. An alternative route would be the derivation of robust Bayes analogues to the REML estimates based on the nonlinear regression model and covariance structures considered in Christensen (1993).

Summarizing, future efforts in this area of research should be concentrated on the extension of the concept of global Bayes robustness to work with flexible and tractable families $\Pi_{(\theta,\beta,Z_0)}$ of probability distributions, that is to work with families of plausible likelihood functions for Z_0 and prior densities for the trend and covariance parameters involved.

REFERENCES

Bardossy, A. (1988) "Notes on the robustness of the kriging system", Math. Geology 20, 189-203.

Berger, James O. (1984) "The robust Bayesian viewpoint", in J.B. Kadane (ed.), Robustness of Bayesian Analysis, Elsevier Science Publ., New York, 63-124.

Christensen, R. (1993) "Quadratic covariance estimation and equivalence of predictions", Math. Geology 25, 541-558.

Das Gupta, A. and Studden, W.J. (1988) "Robust Bayesian analysis and optimal experimental designs in normal linear models with many parameters- I", Tech. Report #88-14, Dept. of Statistics, Purdue Univ.

Diamond, P. and Armstrong, M. (1984) "Robustness of variograms and conditioning of kriging matrices", Math. Geology 16, 809-822.

Grenander, U. and Szegö, G. (1958) Toeplitz Forms and Their Applications, Univ. of California Press, Berkeley and Los Angeles.

Handcock, Mark S. and Stein, Michael L. (1993) "A Bayesian analysis of kriging", this volume.

Oman, Samuel D. (1982) "Contracting towards subspaces when estimating the mean of a multivariate normal distribution", J. Multivar. Anal. 12, 270-290.

Omre, H. (1987) "Bayesian kriging - merging observations and qualified guesses in kriging", Math. Geology 19, 25-39.

Omre, H. and Halvorsen, Kjetil B. (1989) "The Bayesian bridge between simple and universal kriging", Math. Geology 21, 767-786.

Omre, H., Halvorsen, Kjetil B. and Berteig, V. (1989) "A Bayesian approach to kriging", in M. Armstrong (ed.), Geostatistics, Vol. 1, Kluwer Academic Publishers, Dordrecht, 109-126.

Pilz, J. (1990) "Bayes estimation of variograms and Bayesian collocation", in Proceedings 22nd Symposium APCOM, TUB Dokumentation Heft 51, Vol. II, Berlin, 565-576.

Pilz, J. (1991) Bayesian Estimation and Experimental Design in Linear Regression Models, J. Wiley & Sons, Chichester, New York.

Pilz, J. (1992) "Zur Verwendung von a-priori-Kenntnissen in geostatistischen Modellen", in G.J. Peschel (ed.), Beiträge zur Mathematischen Geologie und Geoinformatik, Band 3, Verlag Sven von Loga, Köln, pp. 2-11.

Posa, D. (1989) "Conditioning of the stationary kriging matrices for some well-known covariance models", Math. Geology 21, 755-765.

Schaffrin, B. (1992) "Biased kriging on the sphere", in A. Soares (ed.), Geostatistics Troia, Vol. 1, Kluwer Academic Publishers, Dordrecht, 121- 131.

Stein, Michael L. (1989) "The loss of efficiency in kriging prediction caused by misspecifications of the covariance structure", in M. Armstrong (ed.), Geostatistics, Vol. 1, Kluwer Academic Publishers, Dordrecht, 273-282.

Yakowitz, S. and Szidarovszky, F. (1985) "A comparison of kriging with non-parametric regression methods", J. Multivar. Anal. 16, 21-53.

A FRACTAL-MULTIFRACTAL APPROACH TO GEOSTATISTICS

C. E. PUENTE
Hydrologic Science
Department of Land, Air and Water Resources
University of California, Davis CA 95616
USA

A new approach for the description of geophysical data in space (either along a line or over the plane) is introduced. The procedure consists of using fractal interpolating functions (either, from the line to the line or from the line to the plane) to transform arbitrary multinomial multifractal measures defined over a closed interval. Specifically, and for the most general case, let X be a multinomial multifractal measure on a closed interval I with parameters $p_1, p_2, \ldots, p_N$, and let $f_\Theta : I \rightarrow R^2$ be a deterministic continuous fractal interpolating function, as introduced by Barnsley (1988), which interpolates a set of data points $\{(x_n, y_n, z_n); x_0 < \ldots < x_N, n = 0, 1, \ldots, N\}$, and which has parameters Θ. Then, we show via simulations that measures $Y = f_\Theta(X)$, provide good models for a variety of geophysical phenomena. A deterministic connection between turbulence-related measures and the Gaussian distribution is shown as a limiting case when the graph of the fractal interpolating function $f_\Theta(X)$ is space-filling. It is suggested that the evolution of complex geophysical patterns may be approached via this geometric description.

INTRODUCTION

With the advent of the modern theory of turbulence (chaos) and fractal geometry, a new vision has been gained for understanding the intricacies of physical phenomena. In this relation, three important paradigms have appeared:

(a) Details that were thought to be unimportant may play crucial roles in our ability to predict; e.g. Lorenz (1963), Moon (1987).

(b) What appears unpredictable and "random" at a local scale could be explained as part of a global deterministic process; e.g. Lorenz (1963), Mandelbrot (1983), Meneveau and Sreenivasan (1987).

(c) Very complicated processes, which were thought to require partial differential equations for their description (e.g. convection, fully-developed turbulence), may be accurately described by means of very simple deterministic models; e.g. Libchaber (1982), Meneveau and Sreenivasan (1987), Feigenbaum (1980).

The framework provided by these paradigms constitutes the basis herein.

R. Dimitrakopoulos (ed.), Geostatistics for the Next Century, 476–487.

Proper description (interpolation, estimation, prediction, simulation, etc.) of geophysical data (spatial patterns) are crucial for the understanding and quantification of geophysical phenomena and its consequences. A common trait of these data sets is that they are *complex*. They exhibit "heterogeneities", "anisotropies", and "intermittencies" which typically preclude a simple mathematical description. In order to describe these intricate (and seemingly random) patterns, several procedures have been introduced in the geostatistics literature. For good overviews on spatial statistics see for instance Cressie (1991) and NRC (1991).

Among the procedures that have been developed the most are the ones based on stochastic methods (e.g. Mantoglou and Wilson, 1982; Borgman et al.,1984; Davis, 1987; Christakos, 1987). Instead of concentrating on data sets per-se, these procedures center their attention on "relevant statistics" of the patterns at hand (i.e. spatial correlations, extremes, etc). Assumptions like stationarity and ergodicity are often used to aid in the development of statistical theories, even though it is common to have geophysical phenomena which provide but a single spatial pattern. The typical assumption made is that what one sees are just realizations of underlying ergodic random fields (processes). Despite substantial progress throughout the years, the problem of "spatial variability" is still an important limitation in several geophysical disciplines, and the development of new methods is needed, see NRC (1991). Although state of the art stochastic methods provide a viable representation of complex data, more often than not (given their assumptions) they result in "smoothed," "distorted" and/or "unreal" representations of observed patterns. Notice that these approximate representations may be unacceptable if details matter.

Descriptions of natural phenomena via fractal geometry have been developed and popularized by Mandelbrot (1983, 1989). The underlying assumption here is that nature possesses a geometry that repeats when looked at increasingly smaller scales. This self-similarity (self-affinity) represents a plausible way for bridging the gap across alternative scales, and has been found of use in several disciplines in Physics, e.g. Mandelbrot (1983), Feder (1988). Lately, these ideas have played an important role in the ever intriguing field of turbulence. In fact, the use of multifractal measures, e.g. Meneveau and Sreenivasan (1987) and Sreenivasan (1991), appears to provide an adequate framework to study the very intermittent and highly heterogeneous behavior seen in turbulence. It is worth mentioning that even though the power of fractal ideas is unquestionable, a couple of common objections have been that: (i) the use of the most simple mathematical fractals gives rise to self-similarity (self-affinity) ad-infinitum, a property hardly observed with real (geophysical) data; and (ii) the use of fractal techniques sometimes is not properly linked with the "Physics" of the underlying phenomena.

This work introduces a new procedure for the quantification of spatial variability of geophysical phenomena. The idea is to reproduce such "complex", "jagged", and "intricate" patterns as deterministic fractal transformations of distributions found of relevance to study turbulence (multifractal probability measures). The transformations used belong to the family of deterministic continuous fractal interpolating functions, as introduced by Barnsley (1986, 1989). It will be shown, via simulations, that this representation provides a compact quantification of observed data, which may lead to better estimation of the underlying phenomena, and which may result in improved dynamic representations of geophysical processes.

The organization of this paper is as follows. First, we review fractal interpolators in two dimensions, and their subsequent derived distributions, and show examples in which such distributions resemble geophysical realizations along a line. Then, we discuss three dimensional fractal interpolating functions and show examples of bivariate measures which exhibit behavior similar to what is found with geophysical records over the plane. The article concludes with a summary and final remarks.

FRACTAL INTERPOLATION IN 2 DIMENSIONS

This section follows closely Barnsley (1988). Start with a set of $N+1$ data points on the plane, $\{(x_n, y_n), x_0 < x_1 < \cdots < x_N, n = 0, 1, \ldots, N\}$, and a set of N affine maps of the form

$$w_n \begin{pmatrix} x \\ y \end{pmatrix} = \begin{pmatrix} a_n & 0 \\ c_n & d_n \end{pmatrix} \begin{pmatrix} x \\ y \end{pmatrix} + \begin{pmatrix} e_n \\ f_n \end{pmatrix}; \qquad n = 1, \ldots, N \tag{1}$$

subject to

$$w_n \begin{pmatrix} x_0 \\ y_0 \end{pmatrix} = \begin{pmatrix} x_{n-1} \\ y_{n-1} \end{pmatrix}, \quad w_n \begin{pmatrix} x_N \\ y_N \end{pmatrix} = \begin{pmatrix} x_n \\ y_n \end{pmatrix}; \; n = 1, \ldots, N. \tag{2}$$

This leads to 4 equations in 5 unknowns for each $n \in \{1, \ldots, N\}$ enabling the solution of 4 affine mapping parameters in terms of the remaining one and the data point coordinates. As in Barnsley (1988), we choose d_n, the vertical scalings, as the free variables. When these quantities satisfy $0 \leq |d_n| < 1$, all the mappings become contractile on the plane, and in virtue of fixed-point theorems they lead to a unique (deterministic) set $G = \cup_{n=1}^{N} w_n(G)$ which is the *attractor* of the set of mappings. Barnsley (1986,1988) showed that this attractor is the *graph* of a continuous function, f, which passes through each data point, i.e. $G = \{(x,\ f(x)) | x \in [x_0,\ x_N]\}$. Points on G may be obtained by progressively iterating the contractile affine mappings, starting at any given data point. For this reason, the set of mappings together with the space in which they are defined, $\{\mathcal{R}^2, w_1, \ldots w_N\}$, are called an iterated function system (IFS). Once all parameters are fixed, the fractal dimension of G gives $D = \min(1, D_0)$ where D_0 is the unique solution of

$$\sum_{n=1}^{N} |d_n| a_n^{D_0 - 1} = 1. \tag{3}$$

The graph G can be constructed following simple deterministic recursive (geometric) rules (Barnsley, 1988; Puente et al., 1993). As previously mentioned, it may be found successively iterating the affine mappings starting at a given data point. When all possible calculations are carried, this leads to an infinite N-ary tree: the first N nodes of such a tree are the images of the chosen data point, obtained using the affine mappings $w_1, \ldots, w_N$; the tree continues by applying the mappings to the newly acquired points on G, and so on. Another simple way of creating the graph, G, is via a Monte-Carlo approach called the "chaos game", in which successive points are found recursively through images of (an arbitrary) data point employing the maps w_n according to a (fixed) set of weights p_n ($\sum_{n=1}^{N} p_n = 1, p_n > 0, n = 1, \ldots, N$). The chaos game does not compute the whole tree but only a branch of it. Due to an ergodic theorem, this method gives (with probability 1) the same outcome G irrespective of the weights p_n, see Barnsley (1988). Due to its convenience in computer simulations, the chaos game is used herein.

By counting the number of times image points fall into small intervals in either the x or y axis, the methods above induce unique stationary measures in x and y, X and Y respectively (Barnsley, 1986). These measures, which depend on the method of construction, can be scaled and thought of as probability distributions so that $Y = f(X)$ is the derived distribution of X via f. Given the structure of the first component of the affine mappings (see equation (1)), the unique measures in x induced by the chaos game are the deterministic multifractals with parameters p_n and length scales given by the data spacings. These measures could be described deterministically as follows. Start with a uniform bar over $[x_0,\ x_N]$ and cut it into N pieces of sizes $[x_0,\ x_1], \ldots, [x_{N-1},\ x_N]$ such that on the nth piece there is $p_n\%$ of the mass. The final multinomial multifractal measure is obtained repeating the same procedure ad infinitum each time on each of the (N) pieces employing sub-pieces with same scales as at the beginning and partitions given by the p_n's. Binomial multifractal measures ($N = 3$) appear in the description of fully developed turbulence, e.g. Meneveau and Sreenivasan (1987) ($p_1 = 0.7$), and in the classical problem of the gambler's ruin in relation to the strategy of bold play, e.g. Feller (1968), and Billingsley (1983).

Due to the structure of the affine mappings, the measures in x appear irrespective of the shape (and fractal dimension) of G. Puente (1992) reported on the structure of the derived measures in y. They could be either multifractal (as the parent measure in x) or absolutely continuous (as the fractal dimension of G increases). In the limit, when G fills up the plane (as $D \to 2$), the limiting derived distribution is Gaussian (Puente et al., 1993).

Figure 1 illustrates graphically the construction of derived distributions for the case when of $N = 3$. As is seen, the distribution in y (Y, denoted dy in the figure) may be interpreted as a weighted projection of the fractal function f via the multifractal measure X (denoted by dx in the graph). Figure 2, shows a couple of examples that illustrate that "physically reasonable" outcomes are attainable. The one-dimensional graph on the left (y vs. dy) may describe a geophysical phenomenon with marked intermittency, and with an exponential distribution (when a histogram of the measure is found). It was obtained using a fractal interpolating function which passes by the points $\{(0,0),\ (0.5,1),\ (1,0)\}$, such that the two needed affine mappings have vertical scalings $d_1 = -d_2 = 0.5$ and are iterated according to the weights $p_1 = 0.3$, $p_2 = 0.7$. The graph was found playing the chaos game making 1 million iterations, and computing relevant histograms over 512 equally-spaced bins. The graph on figure 1(b) shows a log-normal one-dimensional pattern similar to what is commonly observed in several geophysical applications. This graph was found using the same interpolating points as before with $d_1 = -d_2 = 0.7$ and $p_1 = 0.35$, $p_2 = 0.65$. Contrary to what is assumed in many applications, the derived distributions obtained by this procedure can not have themselves Gaussian distributions (histograms). This is because the measures in y have always positive values. Observe that in this regard the limiting Gaussian measure produces a uniform histogram.

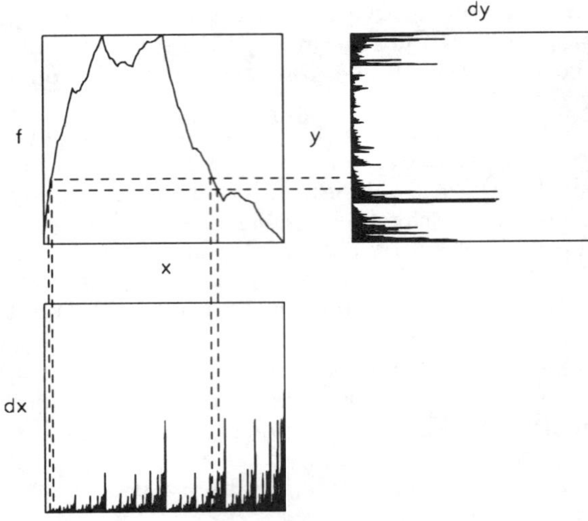

Figure 1: Schematic construction of derived measures. dx and dy are measures in x and y, respectively.

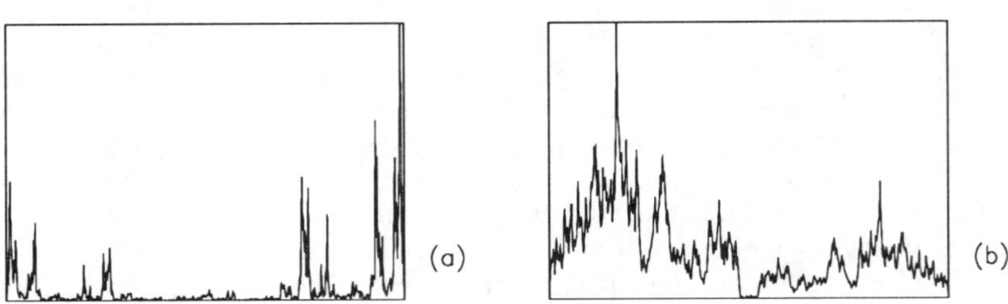

Figure 2: Examples of derived measures along a line. (a) exponential, (b) log-normal.

FRACTAL INTERPOLATION IN 3 DIMENSIONS

Analogous to the two-dimensional case, we consider $N+1$ data points, $\{(x_n, y_n, z_n) : x_0 < \cdots < x_N; n = 0, 1, \ldots, N\}$, and a set of N contractile maps of the form

$$w_n \begin{pmatrix} x \\ y \\ z \end{pmatrix} = \begin{pmatrix} a_n & 0 & 0 \\ c_n & d_n & h_n \\ k_n & l_n & m_n \end{pmatrix} \begin{pmatrix} x \\ y \\ z \end{pmatrix} + \begin{pmatrix} e_n \\ f_n \\ g_n \end{pmatrix}, n = 1, \ldots, N \tag{4}$$

such that

$$A_n = \begin{pmatrix} d_n & h_n \\ l_n & m_n \end{pmatrix} \tag{5}$$

has L_2-norm less than 1 ($\|A_n\|_2 = \sqrt{\lambda_{\max}(A_n^T A_n)} < 1$) and

$$w_n \begin{pmatrix} x_0 \\ y_0 \\ z_0 \end{pmatrix} = \begin{pmatrix} x_{n-1} \\ y_{n-1} \\ z_{n-1} \end{pmatrix}, \quad w_n \begin{pmatrix} x_N \\ y_N \\ z_N \end{pmatrix} = \begin{pmatrix} x_n \\ y_n \\ z_n \end{pmatrix}; \quad n = 1, \ldots, N. \tag{6}$$

Now, we obtain a set, G, which is the unique attractor of the IFS, $\{\mathcal{R}^3; w_1, \ldots, w_N\}$, and the graph of a continuous deterministic function from $[x_0,\ x_N]$ to the yz-plane. Our construction leaves now four free parameters per map, with a_n, c_n, e_n, f_n, g_n, and k_n all determined in terms of d_n, h_n, l_n, m_n and the $N+1$ data points. Transforming to polar coordinates, we can redefine the submatrix A_n

$$A_n = \begin{pmatrix} d_n & h_n \\ l_n & m_n \end{pmatrix} = \begin{pmatrix} r_n^{(1)} \cos\theta_n^{(1)} & -r_n^{(2)} \sin\theta_n^{(2)} \\ r_n^{(1)} \sin\theta_n^{(1)} & r_n^{(2)} \cos\theta_n^{(2)} \end{pmatrix} \tag{7}$$

The entries on these matrices and the interpolating data points constitute the parameter set Θ mentioned in the abstract of the paper.

If the magnitude of the scaling parameters $r_n^{(i)}$ are all less than one, then we are guaranteed the existence of a unique attracting set $G = w_1(G) \cup \cdots \cup w_N(G)$, which is the graph of a continuous function from the interval $[x_0, x_N]$ to the yz-plane and containing the original data points. Moreover, the attractor G possesses fractal properties as the magnitude of the scaling parameters increases towards one (Barnsley, 1988). Analogous to the two-dimensional case, a formula may be derived for the fractal dimension of G, Puente and Klebanoff (1993).

For illustration purposes, figure 3 shows several contour plots of joint derived measures that were obtained transforming binomial multifractal measures via a 3-dimensional fractal interpolating function. The parameters (Θ) that were used are given in table 1. In all cases, three interpolating points were used, with the first and last being $(0,0,0)$ and $(1,0,0)$, and with $p_1 = 0.3, p_2 = 0.7$. All graphs were found via the chaos game employing 5 million iterations, storing bivariate histograms in 128 x 128 equally-spaced bins. Notice the wide variety of geometries that could be captured by means of the *"fractal-multifractal"* procedure. Observe that depending on the parameter choice these figures reflect situations that have both highly "intermittent" (multifractal) and anisotropic behavior, or "smooth" and seemingly isotropic behavior. Notice that some of these figures reflect data sets that are recognizable

Table 1: Set of data and parameters for figure 3. (angles in degrees)

	Middle Point Coordinates	First transformation				Second transformation			
		$r_1^{(1)}$	$\theta_1^{(1)}$	$r_1^{(2)}$	$\theta_1^{(2)}$	$r_2^{(1)}$	$\theta_2^{(1)}$	$r_2^{(2)}$	$\theta_2^{(2)}$
(a)	(0.5,0.2,1)	0.5	30	0.5	30	0.5	180	0.995	180
(b)	(0.5,0.2,1)	0.995	30	0.5	30	0.995	180	0.7	180
(c)	(0.5,0.5,1)	0.5	0	0.5	0	0.7	0	0.995	15
(d)	(0.5,0.2,1)	0.5	0	0.5	0	0.9	90	0.995	90
(e)	(0.5,-0.7,1)	0.995	30	0.5	30	0.995	270	0.7	270
(f)	(0.5,0.2,1)	0.995	0	0.5	0	0.995	45	0.9	45
(g)	(0.5,0.5,1)	0.995	65	0.995	45	0.995	85	0.995	45
(h)	(0.5,0.5,1)	0.995	90	0.995	270	0.995	270	0.995	90

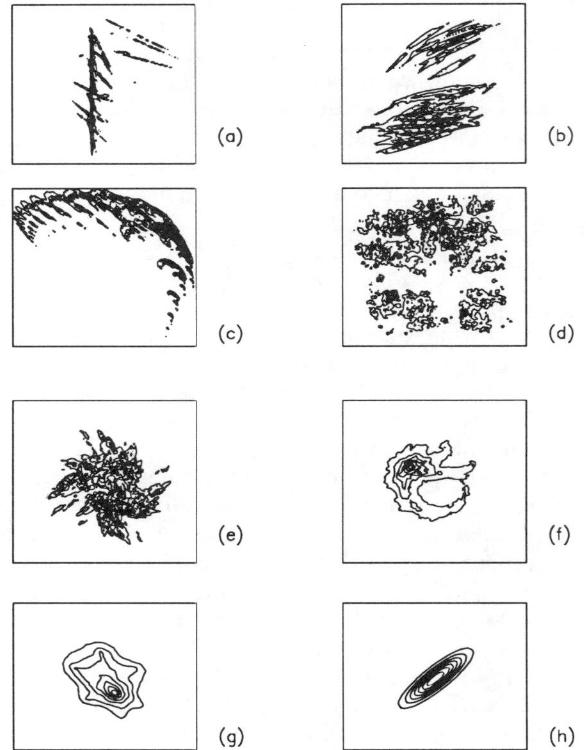

Figure 3: Examples of bivariate derived measures for parameters in table 1.

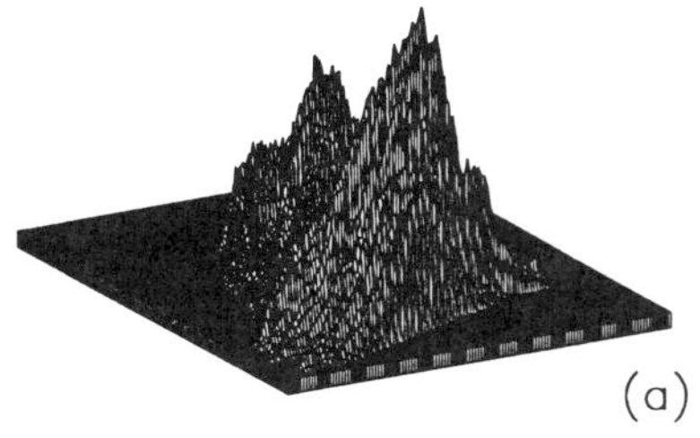

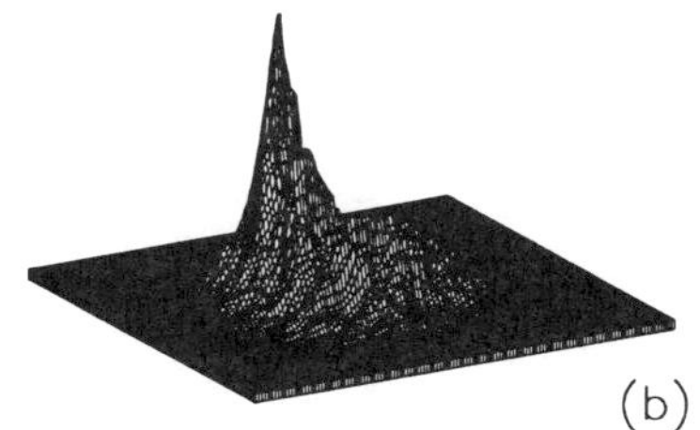

Figure 4: Perspective views of derived measures. (a) fig. 3(b), (b) fig. 3(f).

as appearing in geophysical applications. Observe that even though some of these figures may be considered "fractal", they do not exhibit mathematical self-similarity (self-affinity). Because the affine functions being used are continuous in their parameters, small changes on the shape of the fractal interpolating function lead to small changes in the corresponding derived measures. A similar behavior holds in regards to the multifractal parameters p_n. Figure 4 shows a couple of the measures in figure 3, as seen from a three-dimensional perspective.

As in the 2 dimensional case, in the limit when the r_n's tend to 1 we obtain joint bivariate Gaussian distributions, Puente and Klebanoff (1993), see figure 3(h). Figure 5 shows a more complete example which explains how Gaussians with different correlations may be constructed. All graphs are such that fractal interpolators pass by the set of points $\{(0,0,0),(1/2,H,1),(1,0,0)\}$, and have parameters $r = r_1^{(1)} = -r_1^{(2)} = -r_2^{(1)} = r_2^{(2)} = 0.995$ and $\theta_1^1 = \theta_1^2 = \theta_2^1 = \theta_2^2 = \pi/2$. H (the interpolating height in y) varies in the graph from 1, 0.5, 0, -0.5, -1 from left to right, and from top to bottom. When $H = 1$, due to symmetries in the maps w_n, the fractal projections xy and xz are identical and the joint density in yz is Gaussian with perfect positive correlation. As H is decreased to 0.5, the xz fractal has "shifted" with respect to the xy fractal, which still maintains a clear symmetry with respect to the first and second halves of its domain. An elliptical pattern on the joint density now appears, indicating a bivariate Gaussian whose correlation coefficient is approximately 0.8. When H equals 0, the xz fractal has "shifted" completely with respect to the xy fractal. As H is further reduced, the xz fractal appears shifted again, but now upside-down with respect to the xy fractal. When $H = -0.5$, a bivariate Gaussian with correlation about -0.8 is obtained. When H takes the value -1, the coefficient of correlation equals -1. These graphs show that the coefficient of correlation between Gaussians may be explained in terms of geometric similitudes between the $x - y$ and $x - z$ projections of a suitable fractal interpolating function, an unexpected and intriguing result.

It is important to stress at this time the main differences between the procedure just described and classical statistical procedures currently in use. First of all, the new procedure is entirely deterministic: both the parent multifractal measure and

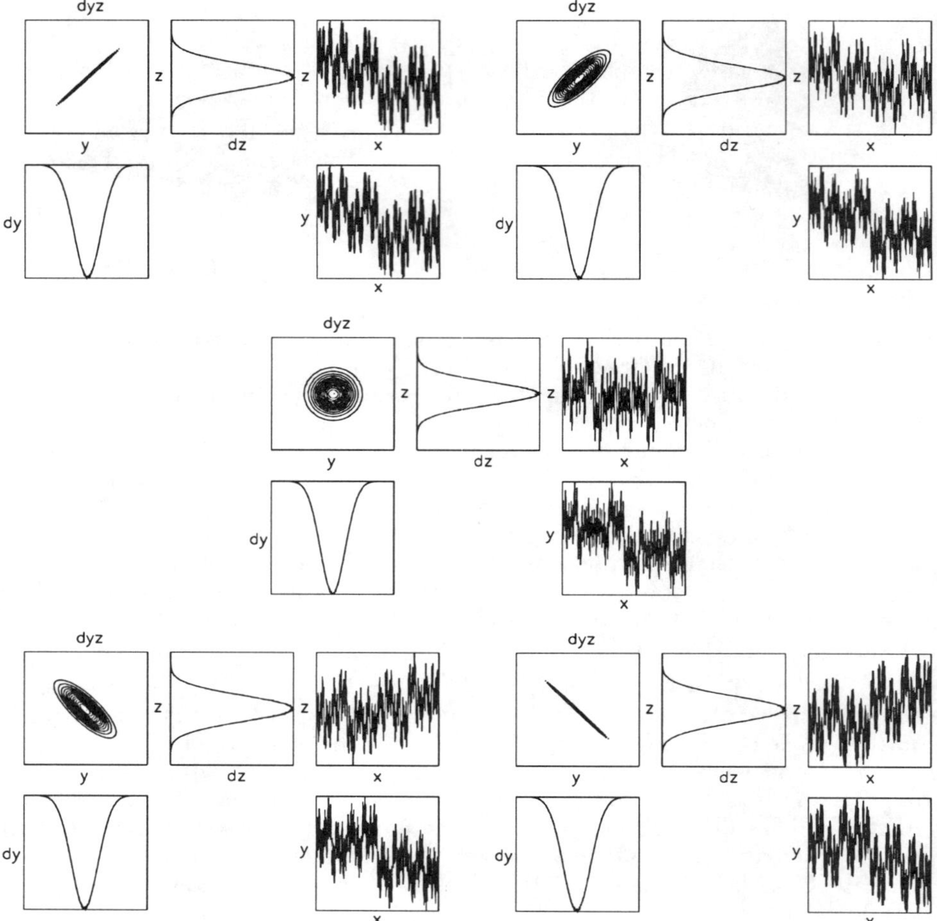

Figure 5: *Data:* $\{(0,0,0),(1/2,H,1),(1,0,0)\}$. *Parameters:* $r = r_1^{(1)} = -r_1^{(2)} = -r_2^{(1)} = r_2^{(2)} = 0.995$. $\theta_1^1 = \theta_1^2 = \theta_2^1 = \theta_2^2 = \pi/2$. H varies from 1, 0.5, 0, -0.5, -1 from left to right, top to bottom. Note the apparent "shifts" in the graphs of the fractal interpolating functions as the correlation ρ changes. The distribution in x (not shown) is uniform, i.e. $p_1 = p_2 = 0.5$.

the transforming mapping may be uniquely obtained via simple recursive procedures. The data at hand is interpreted as a "normalized" distribution, a probability measure, which is encoded via a parent (deterministically generated) multifractal measure and a unique (and hence deterministic) fractal interpolating function. Second, instead of focusing on the statistics of the actual realization or realizations at hand, interest is in the actual description of the patterns as we see them. It appears that the fractal-multifractal procedure just described allows a very parsimonious description of natural patterns. Third, the fractal-multifractal procedure gives a variety of distributions which are not restricted to Gaussian-related fields.

As mentioned in the introduction, multifractal measures are being increasingly identified as relevant models of physical phenomena, especially in circumstances related to turbulence, Sreenivasan (1991). This connection and the fact that fractals are of increasing importance to model nature (e.g. Mandelbrot, 1982; Kaye, 1989) bring a "physical" explanation to the procedure. The idea is that what we see may be just a "reflection" of turbulence via a fractal transformation of an appropriate dimension.

FINAL REMARKS

We have introduced a new framework for the description of complex geophysical data. The method relies on the use of fractal-geometric techniques and in particular on the combination of fractal interpolation functions and multinomial multifractal measures. Even though the procedure appears to be just geometrical, this new framework may have physical significance. This is because the two components that make up the method are being found of importance to understand nature. The physical realism of the procedure is stressed by the unexpected connection that plane-filling and space-filling fractal interpolating functions provide between "disorder" (turbulence) and "harmony"(Gaussian behavior).

For a long time it was believed that small distortions had little or no effect on our ability to predict. Recently, however, with the advent of chaos theory (e.g. Moon, 1987; Rasband, 1990), it has been recognized that details may be quite important. Given that most geophysical phenomena are in principle described by sets of nonlinear partial differential equations, it appears that details could be quite important in the overall description of these phenomena. In this regard, the fractal-multifractal procedure provides a framework suitable for description of complexity as we see it.

In the spirit of the paradigms mentioned in the introduction, it is conceivable that a dynamic description may be developed based on geometric encodings of actual data. This means no use of (stochastic) partial differential equations for relevant physical processes. Instead, it may be that "relevant" physics can be captured by focusing on accurate geometric descriptions of observed patterns. Rather than arriving at complex mathematical formulas for the patterns, we may concentrate on finding suitable deterministic mechanisms that generate them. Then, we may follow such mechanisms and use "surrogate" geometric parameters to account for dynamics. If the set of such surrogates is small, one may literally see how patterns evolve without the need of partial differential equations. Given the intricate appearance of most geophysical patterns, it appears that the fractal-multifractal procedure may result in compact dynamic descriptions of geophysical phenomena.

Research is continuing on the development of efficient algorithms to properly describe real geophysical data sets. Of course, the practical merit of this work will rest on our ability to choose the proper model (fractal-multifractal combination) for a given situation. Clearly, efficient solutions of inverse problems are required, specially when dealing with dynamical applications. A complete statistical description of the patterns that may be obtained via the fractal-multifractal procedure is being studied and will be reported elsewhere.

Acknowledgements

The editorial work by Mr. Aldo Romero and Dr. Aaron Klebanoff is appreciated.

References

Barnsley, M. F. (1986) "Fractal Functions and Interpolation", *Constr. Approx.* **2**, 303-329.

Barnsley, M. F. (1988) *Fractals Everywhere,* Academic Press, New York.

Barnsley, M. F. (1989) "Hidden variable fractal interpolation functions", *SIAM J. of Math. Anal.* **20(5)**, 1218-1242.

Billingsley, P. (1983) "The singular function of bold play", *Am. Sci.* **71**, 392-397.

Borgman, L. E., M. Taheri and R. Hagan (1984) "Three dimensional, frequency-domain simulations of geologic variables", in G. Verly, M. David, A. G. Journel (eds.), *Geostatistics for Natural Resources Characterization, Part 1*, Reidel, Dordrecht, pp. 517-541.

Christakos, G. (1987) "Stochastic simulation of spatially correlated geoprocesses", *Math. Geol.* **19**, 807-831.

Cressie, N. (1991) *Statistics for Spatial Data,* John Wiley and Sons, New York.

Davis, M. W. (1987) "Production of conditional simulations via the LU triangular decomposition of the covariance matrix", *Math. Geol.***19**, 91-98.

Feder, J. (1988) *Fractals,* Plenum Press, New York.

Feigenbaum, M. J. (1980) "Universal behavior in nonlinear systems", *Los Alamos Science,* Summer, 4-27.

Feller, W. (1968) *An Introduction to Probability Theory and Its Applications,* vol 1, 3rd edition, John Wiley, London.

Kaye, B. H. (1989) *A Random Walk Through Fractal Dimensions,* Verlagsgesellschaft, Weinheim.

Libchaber, A. (1982) "Convection and turbulence in liquid helium I", *Physica B* **109** & **110**, 1583-1589.

Lorenz, E. N. (1963) "Deterministic nonperiodic flow", *Journal of the Atmospheric Sciences* **20**, 130-141.

Mandelbrot, B. B. (1983) *The Fractal Geometry of Nature,* Freeman, San Francisco.

Mandelbrot, B. B. (1989) "Multifractal measures especially for the geophysicist", in C. H. Scholz and B. B. Mandelbrot (eds.), *Fractals in Geophysics,* Birkhauser Verlag, Basel, pp. 5-42.

Mantoglou, A. and J. L. Wilson (1982) "The turning bands method for simulation of random fields using line generation by a spectral method", *Water Res. Res.* **18**, 1379-1384.

Meneveau, C. and K. R. Sreenivasan (1987) "Simple multifractal cascade model for fully developed turbulence", *Phys. Rev. Lett.* **59**, 1424-1427.

Moon, F. C. (1987) *Chaotic Vibrations,* Wiley, New York.

NRC, National Research Council (1991) *Spatial Statistics and Digital Image Analysis,* National Academy Press, Washington, D. C..

Puente, C. E. (1992) "Multinomial multifractals, fractal interpolators, and the Gaussian distribution", *Phys. Lett. A* **161**, 441-447.

Puente, C. E., M. M. López, J. E. Pinzón and J. M. Angulo (1993) "The Gaussian distribution revisited", Submitted to *Journal of Applied Probability.*

Puente, C. E. and A. Klebanoff (1993) "Gaussians everywhere", In press *Fractals.*

Sreenivasan, K. R. (1991) "Fractals and multifractals in fluid turbulence", *Annu. Rev. Fluid Mech.* **23**, 539-600.

PROBABILITY KRIGING OF SUB-SEISMIC FAULT THROWS WITH MULTIFRACTAL DISTRIBUTIONS

Renjun Wen and Richard Sinding-Larsen

Department of Geology and Mineral Resource Engineering
Norwegian Institute of Technology (NTH),
University of Trondheim, N-7034, Trondheim, Norway

ABSTRACT

Faults with throws less than the resolution limit of seismic data (*ca.* 15 meters) are numerous in faulted hydrocarbon reservoirs, but are not observable on a reservoir scale. Estimation of sub-seismic fault throws from seismically observable fault throws can be made using various methods under different assumptions. The purpose of this paper is to evaluate the efficiency of the probability kriging approach in estimating multifractally distributed fault throws.

A fractal distribution may be fitted to fault-throw data across many scales of magnitude (10^{-3} to 10^{3} m) for faults observed on cores, seismic sections, and outcrops. Such a distribution however only characterizes the bulk property of fault throws in an area. As the fractal dimension of fault-throws are spatially dependant, a multifractal model therefore appears to be a more realistic description of fault throws.

A synthetic data base was simulated, based on a multifractal model whose parameters were estimated from real data. Using samples from the synthetic data, we estimated the conditional probability of having a magnitude of sub-seismic fault-throws less than selected threshold values using probability kriging by taking into account both the fractal distribution and biased sampling of throws. The results indicated that a multifractal distribution of fault throws can be used as a theoretical basis for probability kriging in the estimation of the local distribution of fault-throws.

KEY WORDS: *fault throws, fractal, multifractal, probability kriging, sub-seismic faults.*

INTRODUCTION

Seismic data are the major data source for mapping subsurface faults in hydrocarbon reservoirs. However, small-scale faults with throw values less then *ca.* 15 meters are not observable due to the resolution limit of the data, except at very few well locations where faulted cores may be detected. An interpreted fault map or a section represents a smoothed picture of the subsurface reality (Figure 1). Geologists and reservoir engineers all would like to know the spatial distribution of small faults that have not been mapped by seismic data (i.e.. so called sub-seismic faults). For example, if a well is to be drilled at location **A**

R. Dimitrakopoulos (ed.), Geostatistics for the Next Century, 488–497.

(Figure 1), it is interesting to know what is the likelihood of encountering sub-seismic faults in the target horizon. The presence of unmapped faults may lead to a dry well resulting in missing the target. From an engineering point of view, small faults are known to have important influence on the behavior of fluid flow in the reservoirs. Small-scale faulting may increase the effective permeability of the reservoirs by generating small cracks which connected pores. On the other hand, faults may hinder the horizontal connectivity by juxtaposing reservoirs with sealing layers. Horizontal permeability may therefore be reduced as a result of the formation of shale smears on fault planes which create vertical barriers for horizontal flows (Abu-Elbashar, *et al.*, 1991). Information about the distribution of sub-seismic faults in reservoirs is therefore needed before undertaking the drilling of production wells or the simulation of reservoir performances.

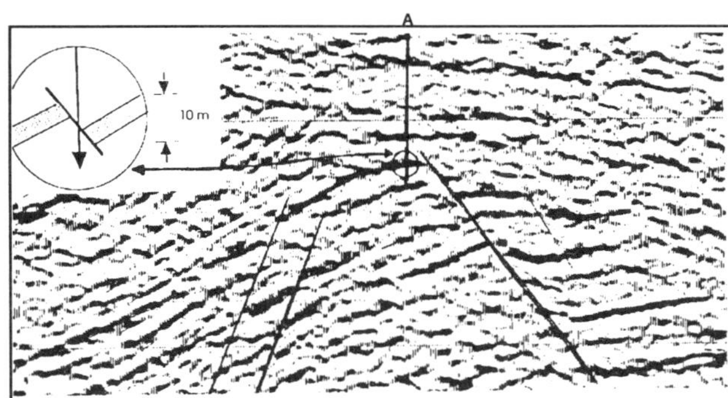

Figure 1. An interpreted geological section is a smoothed picture of reality due to the resolution limit of seismic data. Faults with throws less than *ca.* 15 meters are not observable on the seismic section. An exploration well may be ruined by an unidentified sub-seismic fault. 'A' denotes the well position and the circular caption shows a dry well scenario due to faulting of a reservoir.

In this paper, we introduce a methodology based upon a multifractal model and probability kriging (PK) for predicting sub-seimic fault throws from seismically observable fault throws. In the next section the formulation of the problem is presented, with a discussion of possible solutions. A multifractal modelling procedure of fault-throw is then presented afterwards. The estimation of likely throw values of sub-seismic faults using PK subject to a change in support is described. Finally an example using simulated fault-throw data is presented with concluding remarks.

FORMULATION OF THE PROBLEM AND POSSIBLE SOLUTIONS

Regionalized Variables With Varying Support and Resolution Limit

A fault is a discontinuous planar surface in a 3D rock volume. It can be parameterized by four parameters: length, throw, dip, and orientation (or dip azimuth). Faults are never idealized planer surfaces due to inhomogeneties of the rock media and faulting processs. Fault parameters are functions of spatial coordinates, both for a single fault and a fault population in a region.

A fault plane is uniquely defined if fault-throw values are specified along the fault. Hence we choose to study exclusively fault-throw. This is also due to the fact that throw is the most important fault attribute that characterizes faulting conditions in a reservoir. For example, the ratio between throw and reservoir thickness determine the likelihood that the reservoir will juxtaposed with sealing layers; the largest throw value in an area may be used as an indicator for the regional intensity of tectonic activities, reflecting the density of small cracks resulting from tectonic stresses in the reservoir.

In this paper fault-throw within two-dimensional cells are regarded as a regionalized variable (RV). New RVs are defined by changing cell cells or supports (Olea, 1991). We deal only with square shaped supports (cells). The side-length of a cell is the only parameter of the square support. Let C denote a set of n_C resolution limit values,

$$C = \{c_k\}, k = 1, 2, \ldots n_C. \tag{1}$$

where n_C is the number of resolution limit values. In this paper we use resolution limit as the fault-throw value that can be mapped. For the spatial resolution, we use cell size. Capital letters are used for denoting random processes (RF) and sets, small letters for one realization of RF or elements of sets.

Let $\Re^2$ denote a square spatial domain with a side length L, and S_ξ denote a square support with side length ξ, $0 < \xi \le L$. Given that we can divide the area $\Re^2$ into $n(\xi)$ square cells, s_ξ, we have

$$\begin{aligned} S_\xi &= \{s_\xi\} \\ \cap s_\xi &= \varnothing \\ n(\xi)\xi^2 &= L^2 \end{aligned} \tag{2}$$

Hence, fault-throw is a set defined on sets C, S_ξ, with the spatial coordinates set $X = \{x_i\}$, $i = 1, \ldots n(\xi)$.

$$t(c_k, s_\xi, x_i) \in T(C, S_\xi, X) \tag{3}$$

It must be emphasized that whenever we refer to fault throw data, implicitly we are concerned with data observed above a resolution limit on a finite support. This is why we define our variable relative to the sets C and S. Fault-throw data are strongly dependant on the resolution limit of the observation tool, for example, 2D seismic data may detect faults with throws about 25 meters, and 3D seismic data can detect faults with throw values of ca. 15 meters. On cores and outcrops we can identify micro-faults with centimeters throws. One fault cannot be characterized by a single throw value. Defining variable T on the set S_ξ instead of a fault set represents a convenient way of addressing the problem of presence or absence of small faults.

Statement of the Problem

The objective is to predict sub-seismic fault throws from seismically observable fault throws in a reservoir. Given a set of fault-throw values within a range $[a, b]$ that have been observed in an area $\Re^2$ that is partitioned into $n(\xi)$ disjoint cells, where ξ is the side-length of a cell, The problem is to estimate the likely distribution of fault-throw values within each cell when the resolution limit is reduced to c, $c < a$, or its local cdf, $F[t(c, s_\xi, x_i)]$.

Comments

(1) The distribution of fault-throw in one cell is a function of at least three parameters: (i) spatial location of the cell; (ii) the area reference (support), fault throws within different supports will be different; and (iii) resolution of calibration data.

(2) Small faults with throw values less than *ca.* 15 meters are numerous in reservoirs as can be seen from observations on outcrops and cores. They are genetically related to seismically resolvable faults. Therefore it appears possible to infer the throw values of small faults from those of large faults.

Possible Solutions

A Simple Fractal Approach

One possible solution may be based on the fractal distribution property of fault throws. This is appealing because it means scale-invariance, i.e., each level is an up-sized or down-sized version of the levels below or above it (Mandelbrot, 1982). We may predicate sub-seismic fault-throw based on a fractal distribution. Based on the observed fractal nature of fault throws Walsh *et al.* (1992) suggested that a power-law of fault throws may be used to describe variations from one millimeter to one kilometer as indicated in Figure 2. If a power-law relationship can be extended to throw values below 15 m., the proportion of small fault throws in the sampled area can be calculated. In fact, this is only a special case of equation (3) when $n(\xi) = 1$ or $\xi = L$.

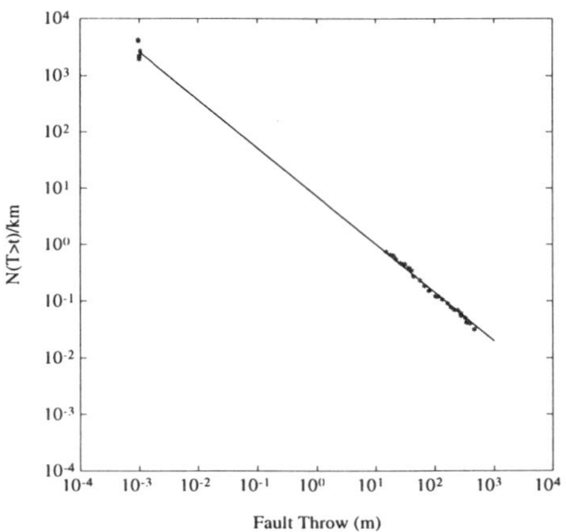

Figure 2. Power-law (equivalent to fractal) distribution of fault-throw values within several orders of magnitude. Data used to plot this figure were digitized from figure 1 of Walsh *et al.* (1992).

Stochastic Simulation

Stochastic simulations may be used to generate equi-probable realizations of fault-throw values. The simulation can either be model-based, or data-driven. In the model-based approach, the Gaussian process model or its transforms are commonly used (Journel and Huijbregts, 1978). For example, Miller and Borgman (1985) simulated the fracture parameters (dip, dip azimuth, and fracture spacings) using a Fourier transform algorithm. Chilés (1988) simulated fractures in a three-dimensional rock volume by using a nonscaling fractal model with variable similarity dimension and a parent-daughter model with a regionalized density. Because fault-throw data show strong non-Gaussian properties, a fault simulation based on the Gaussian process or its transforms may not fully capture the real nature of faults. A nonparametric approach using multiple-indicator conditional simulation (Journel, 1989) has therefore been applied by Sinding-Larsen *et al.* (1992) for simulating fault throws in a North Sea oilfield. The work of Omre *et al.* (1990) represents another non-Gaussian approach for fault simulations. They applied the marked-point-process model for simulating small faults within a fault zone. The model parameters are difficult to estimate from available fault data, and were usually specified heuristically based on expert experience.

Kriging

Conventional kriging methods, like simple kriging (SK), ordinary kriging (OK), and universal kriging (UK), provide smoothed estimations. However, fault-throw values show very irregular spatial behaviors and a long tail in frequency distribution. Extreme values in sampled data make it difficult to estimate reliable variogram models from original data. Therefore, SK, OK and UK cannot directly be used in estimating sub-seismic fault throws. Transformed fault-throw values may be better for variogram estimation, but call for multi-Gaussian distribution in kriging (Journel, 1984). Probability kriging (PK) may be a better choice for the local estimation of fault-throw values. In this method, the kriged parameter is not the original variable, but its local probabilities are conditioned by the available data. The original parameter is transferred into indicator variables at several threshold values. PK has superior properties. Firstly variogram estimation is robust for indicator variables; secondly, qualitative information, such as geologist's judgments can be used through indicator kriging; thirdly, probabilities greater or less than certain given threshold values are estimated instead of a single unique value. Sinding-Larsen *et al.* (1992) estimated local probability images of fault-throws in a North Sea oilfield using the PK approach, but without considering the change in support.

Comments

(1) The demonstrated fractal distribution describing fault-throw across several orders of magnitude should be considered in predicating sub-seismic faults. Simulations generate equi-probable realizations, whereas probability kriging estimates conditional probabilities from available data.

(2) The change of support or cell-size must be solved properly in order to use information from seismically resolvable fault-throw data when using stochastic simulations or the probability kriging.

With regard to comments (2) and (3), both the simulation and probability kriging methods could be formulated specifically for predicating sub-seismic faults. In this paper, we only consider the probability kriging approach.

MULTIFRACTAL MODELLING FOR THE PROBABILITY DISTRIBUTION OF FAULT-THROW

The fault map shown in Figure 3 is used to introduce the concept of a multifractal model for the spatial distribution of fault-throw. Suppose we have fully sampled all fault-throw data on the fault map for a small support s. All measured fault-throw values can then be plotted as a cumulative frequency plot, a power-law or fractal distribution can be fitted to the data, i.e.,

$$N(T > t) \propto t^{-D} \quad (4)$$

where $N(t)$ is the number of fault-throw data larger than a throw value t, D is called fractal dimension (Turcotte, 1989). Note that there is usually a range of t: $[t_a, t_b]$, in which equation (4) is valid. Equation (4) is also called the Pareto distribution (Johnson and Kotz, 1970) and may be written in the probability form as,

$$p(t) = Prob(T \geq t) = t_0 \cdot t^{-D} \quad (5)$$

Where t_0 is a constant of proportionality. The cumulative probability function (cdf) is,

$$F(t) = 1 - t_0 \cdot t^{-D} \quad (6)$$

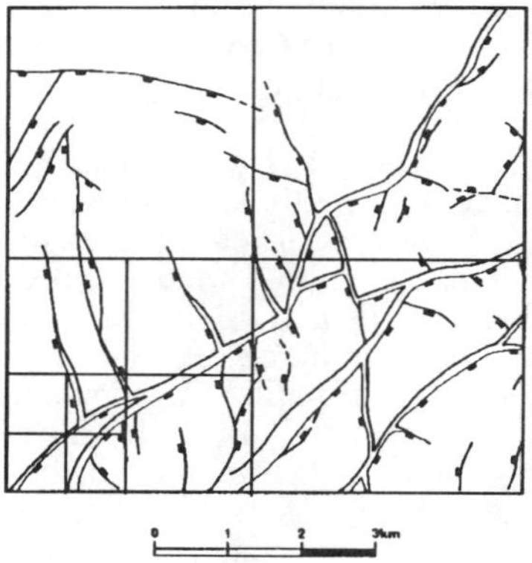

If we subdivide the map into four sub-regions (cells) (Figure3), and plot the fault-throw data of each cell on separate plots, fractal relationships (equation 4 or 7) hold for describing the throw data of each cell, but with different D values. The fractal relationship still hold as we continue to subdivide the map into smaller cells. Again, the D values depend on both spatial locations and cell size. i.e.,

Figure 3. A fault map is subdivided into n^2 cells step by step, $n = 1, ..., n(\xi)$. Fault-throw values within cells of different size have power-law or fractal distribution, but with different fractal dimensions, depending on spatial locations and cell-size. This type of spatial distribution can be conveniently modeled as a multifractal. The progressively finer divisions are only illustrated in one corner of the figure.

$$p(x, s_\xi, t) = Prob(T \geq t | x, s_\xi) \propto t^{-\alpha(x)} \quad (7)$$

It is clear that a single fractal dimension is not sufficient for describing fault-throw distributions in an area. Fractal dimensions of fault-throw are both spatially dependant and support-size dependant. Such fractal distributions are called multifractal distributions. Strictly speaking, they are multifractal measures (Feder, 1988). A multifractal probability model of fault-throws in an area is related to a two-dimensional set S, which is the union of subsets S_ξ.

$$S = \bigcup_{\xi} S_\xi \tag{8}$$

The fault-throw values in each set in the union is fractal with its own fractal dimension, depending on ξ. This is one reason for the term *multifractal* (Feder, 1988).
In the multifractal literature, the Lipschitz-Hölder exponent, α, is regarded as more useful than ξ (e.g., Mandelbrot, 1982). α is defined by the equation,

$$\mu_\xi \propto \delta^{-\alpha} \tag{9}$$

where μ_ξ is the increment of probability when the throw value has an increment of δ. There is a one-to-one correspondence between parameter ξ and α (see chapter 6 of Feder (1989) for more details). In this paper we are content with using ξ because it has a clear physical meaning in terms of the side-length of cells.
For each sub-set S_ξ with a fractal dimension $f(\xi)$, the probability of fault-throw changing from t to t+δ in a cell with side-lengh ξ is expressed as,

$$p(s_\xi, \delta) = \rho(\xi) d\xi \delta^{-f(\xi)} \tag{10}$$

where $\rho(\xi)d\xi$ is the number of sets from S_ξ to $S_\xi + d\xi$.
In a multifractal, fractal dimension varies as a function of spatial location and support-size. This captures the most important feature of a fault-throw distribution in an area. In fact many complex phenomena have been demonstrated to be of multifractal nature, such as turbulent flows (Sreenivasan *et al.*, 1989), spatial distribution of copper ores (Mandelbrot, 1989), fracture surfaces (Nolte, *et al.*, 1989), and the spatial distribution of seismic epicenters (Geilikman *et al.*, 1990).
As stated before, our objective is to estimate sub-seismic fault-throw in cells with smaller size than the seismic resolution size. How the estimation procedure is affected if fault-throws are multifractally distributed is discussed in the following section.

PK OF SUB-SEISMIC FAULT-THROW WITH MULTIFRACTAL DISTRIBUTION

Due to the limited resolution of seismic data , average value of fault-throw can only be measured from data down to lower resolution limit t_c and cell-size limit. l_c. The probability kriging (PK) may however be used to estimate throw below these limits. PK was formulated for estimating conditional probability in the framework of nonparametric geostatistics (Journel, 1984). In this section we suggest that the inclusion of prior information regarding the fractal distribution may further improve the estimator.

Indicator Variables with Varying Supports

Indicator variables of fault-throw can be defined for different cell-size and cutoff values of throws. For a cell size s_ξ, the indicator function is,

$$i(s_\xi, t, x_i) = \begin{cases} 0 & if\ t \le t(x_i) \\ 1 & if\ t > t(x_i) \end{cases} \tag{11}$$

Using this indicator function, multiple indicator variables are derived from the original fault-throw data for different cell-size and threshold values of throw.

PK for Varying Supports

PK was originally formulated as a co-kriging estimator for conditional probabilities of a regionalized variable from indicator variables and transformed variables of raw data (Journel, 1984). In this study we use PK as an estimator for conditional probability of fault-throw from multiple indicator variables.

$$F[s_\psi, t, x_j | (n, n_\xi)] = i(s_\psi, t; x_j)^* = \sum_{\xi}^{n_\xi} \sum_{j=1}^{n} a_{\xi,j}(s_\xi, t, x_j) \cdot i(s_\xi, t, x_j) + \sum_{j=1}^{n} b_j(t; x_j) \cdot u(x_j) \quad (12)$$

where: $F[s_\psi, t; x|(n, n_\xi)]$: kriged conditional probability of a cell with size s_ψ at location x

$[i(s_\psi, t; x)]^*$: same as $F(s_\xi, t; x|(n))$.

$i(s_\xi, t; x_i)$: the indicator value for a cell with size s_ξ, at location x_i for threshold value t.

t : a threshold value of interest.

x : spatial location of a kriged cell.

n : number of neighboring cells whose indicator variables are used for kriging.

n_ξ : number of cell-sizes (supports) whose indicator variables are used for kriging.

$u(x_i)$: transformed variable of $t(x_i)$.

$a_{\xi,i}(s_\xi, t; x_i)$: weighting coefficients.

$b_i(t; x_i)$: weighting coefficient.

The weighting coefficients $a_j(z; x_j)$ and $b_j(z; x_j)$ are calculated from a cokriging system. Despite the fact that we have included more variables in equation (12), the algorithm of finding weighting coefficients is similar to the original PK .
Three types of variogram models need to be provided as input to a PK calculation.

Indicator semivariogram,

$$\gamma_i(s_\xi, h; t) = \frac{1}{2} E\{[i(s_\xi, x+h; t) - i(s_\xi, x; t)]^2\} \quad (13)$$

$\gamma_i(h; z)$ is inferred following the similar procedure as that for conventional semivariograms.

Semivariogram of $u(x_j)$.

$$\gamma_u(h; t) = \frac{1}{2} E\{[u(x+h; t) - u(x; t)]^2\} \quad (14)$$

Cross-semivariogram between $i(s_\xi, x, t)$ *and* $u(x_j)$,

$$\gamma_{i,u}(s_\xi, h, t) = \frac{1}{2} E\{[i(s_\xi, x+h, z) - i(s_\xi, x, z)][u(x+h, t) - u(x, t)]\} \quad (15)$$

$$\gamma_{s_k, s_j}(h,\ t) = \frac{1}{2} E\{[i(s_k, x+h, z) - i(s_k, x, z)][i(s_j, x+h, z) - i(s_j, x, z)]\} \quad (16)$$

The cross-seimivariogram of $\gamma_{i,\ u}(.)$ is equivalent to the conditional variogram of order 1 (Journel, 1984; Sullivan, 1984). $\gamma_{s_k, s_j}(h,t)$ is the cross-semiarograms between indicator variables with different supports.

Transform of $t(x_j)$ to $u(x_j)$

Because of large scaling differences, original values of $t(x_j)$ cannot be used directly in a PK calculation. A transformed variable $u(x_j)$ must be introduced to normalize $t(x_j)$ into the range between 0 and 1. Several schemes have been proposed. For example, $u(x_j)$ can be defined to be a uniformly distributed variate between 0 and 1 through a rank order transform (Journel, 1984; 1989; Olea, 1991),

$$u(x_i) = \frac{1}{n} \mathrm{r}[t(x_i)] \qquad (17)$$

where r[·] is the rank number of $t(x_j)$ in the n samples. Sullivan (1984) transformed $t(x_j)$ through its experimental cumulative distribution function $F[t(x_j)]$. i. e.,

$$u(x_j) = F[(t(x_j)] \qquad (18)$$

Using equation (18) seems to be more logical than (17). Furthermore, PK is interpreted as cokriging a local probability distribution from indicator variables and the global distribution, $F[t(x_j)]$. As for fault throws, the experimental cumulative histogram cannot be used directly because of bias sampling (Heffer and Bevan, 1990).

Figure 4. One realization of a multifractally distributed fault-throw in a simulated reservoir. This realization was used as test data for probability kriging.

A Simulated Example

A numerical algorithm was designed to simulate random fields with multifractal properties expressed in equations 7 - 10. A data-base of fault-throw was simulated based on multifractal parameters that were estimated from North sea seismic data. The data-base has 64 × 64 cells, each cell 10 × 10 m^2. Figure 4 shows one realization. We use this simulated data set to test the efficiency of PK in estimating the magnitude of sub-seismic fault-throws.

The simulated fault-throw data has a global fractal distribution (Equ. 4) with t_0 = 15507, and D_0 = 1.1. Indicator variables of fault-throw were derived from 100 simulated throw data for four threshold values: 15, 20, 25 and 30 meters, and three cell-sizes with side-lengths being 10, 20, and 40 meters. Models were fitted to semivariograms of $i(s_\xi, t; x_j)$ and $u(x_j)$ and cross-semivariograms between $i(s_k, t; x_j)$, $i(s_p, t; x_j)$,and $u(x_j)$.

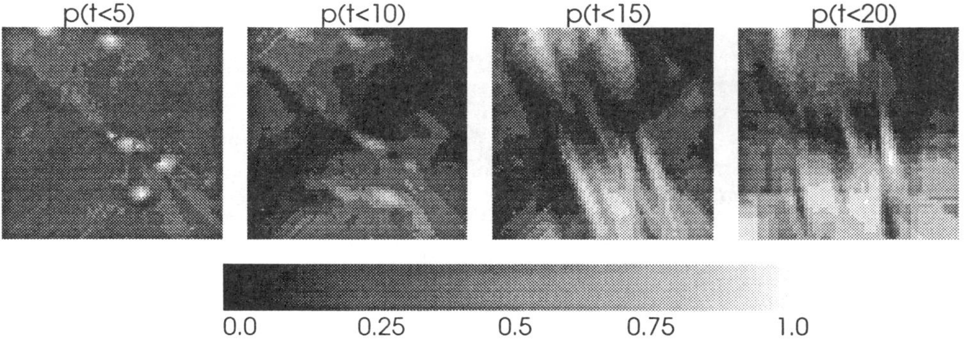

Figure 5. Probability images of fault-throw larger than six threshold values in the simulated reservoir.

After solving the co-kriging equation for the four threshold values of fault-throw, conditional probability of fault-throw for cell with size being 10×10 m^2 were calculated. Probabilities of sub-seismic fault-throw are obtained by extending the distribution curve of each cell to the lower value of faults in the manner shown on Figure2. Four probability images are shown in Figure 5 representing throw threshold values less than 5, 10, 15, and 20 meters for a cell-size 10×10 m^2. From Figure 5, we can observe that the large whitish area in the south with high probability for fault throws less than 20 m shrinks as we consider smaller and smaller throws.

CONCLUDING REMARKS

In this paper we have formulated a methodology for predicting sub-seismic fault-throw from seismically observable faults using probability kriging. Spatial probability of fault-throw is modelled as a multifractal distribution. A simulated data set has been used to test the efficiency of PK in estimating sub-seismic fault-throw. The following comments are made regarding the results presented in this paper.

(1) The fractal dimension of fault-throw is spatially dependant and heterogeneous, hence a single power-law relationship will fail to capture the spatial variability of the data. The multifractal model allows for a more realistic description.

(2) Probability kriging can be adopted for estimating sub-seismic fault throws from larger cell-size and larger threshold values. A fixed cell-size may be used if sufficient data are available (Sinding-Larsen, *et al.*, 1992).

(3) Parameters of a global fractal distribution may be used in the PK by transforming the original variables into normalized variables (probability normalization). PK is then interpreted as the result of co-kriging between indicator variables having one or multiple supports and global probabilities.

(4) Efficient estimation of cross-variograms between indicator variables with different supports remains a problem and is under study.

ACKNOWLEDGMENTS

This work was supported by the Resource Geology Group of the Department of Geology, Norwegian Institute of Technology (NTH). The authors thank Dr. F. Agterbert and an anonymous reviewer for constructive criticisms. The paper has benefited from the suggestions and comments of the reviewers.

REFERENCES

Abu-Elbashar, O. B., S. T. Daltaban, C. G. Wall, and J. S. Archer (1991) Effect of stochastic faults on reservoir permeabilities in presence of stochastic shales. SPE paper 21395.

Chilés, J. P. (1988) Fractal and geostatistical methods for modeling of a fracture network. *Mathematical Geology*. Vol. 20. No. 8. Pages 631-654.

Childs, C., J. J. Walsh,. and J. Watterson (1990) A method for estimation of the density of fault displacements below the limit of seismic resolution in reservoir formations. In *North Sea Oil and Gas Reservoirs II* (edited by A. T. Buller, E. Berg, O. Hjelmeland, J. Kleppe, O. Torsæter and J. O. Aasen), Graham & Trotman, London, pages 309-318.

Heffer, K. and Bevan, T. (1990) Scaling relationship in natural fractures - data, theory and applications. SPE paper 20981.

Feder J. (1988). *Fractals*. Plenum Press, New York and London. pages 283.

Geilikman M. B., T. V. Golubeva and V. F. Pisarenko (1990) Multifractal patterns of seismicity. *Earth and Planetary Science Letters* Vol. 99, p. 127-132.

Johnson N. L. and S. Kotz (1970) *Continuous univariate distributions -1.* Houghton Mifflin Company, Boston, pages 300.

Journel, A. G. (1984) The place of non-parametric geostatistics, in *Geostatistics for Natural Resources Characterization.* (Editors: Verly, G., M. David, A. G. Journel, and A. Marechal). NATO ASI Series, C. Vol. 122. D. Reidel Publishing Company, Dordrecht, The Netherlands, Part 1, p. 307-335.

Journel, A. G. (1989) *Fundamentals of Geostatistics in Five Lessons*, American Geophysical Union, pages 40.

Journel, A. G. and Huijbregts, Ch. J. (1978) *Mining geostatistics.* Academic Press, London. pages 600 .

Mandelbrot B. B. (1982) *The Fractal Geometry of Nature.* W. H. Freeman, New York. pages 648.

Mandelbrot B. B. (1989) The multifractal measures, especially for the geophysicist. *Pure and Applied Geophysics.* Vol. 131, No. 1/2, p. 5-42.

Miller, S. M. and L. E. Borgman. (1985) Spectral-type simulation of spatially correlated fracture set properties. *Mathematical Geology*, Vol. 17, No. 1, p. 41 - 51.

Nolte, L. J., L. R. Myer, and D. D. Nolte (1992) Fractures: finite-size scaling and multifractals. *Pure and Applied Geophysics* Vol. 138, No. 4, p. 679 - 706.

Olea, R. A. (1991) (Editor). *Geostatistical Glossary and Multilingual Dictionary.* International Association for Mathematical Geology Studies in Mathematical Geology No. 3. Oxford University Press.

Omre, H., K. Sølna and B. Tørudbakken; 1990. Stochastic modelling and simulation of fault zones. *Proceedings of the CODATA Conference on Geomathematics and Geostatistics.* Leeds, September 11-14, 1990.

Sinding-Larsen, R., R. Wen and G. Caillet (1992) Nonparametric geostatistical studies of fault parameters in the Snorre Area, North Sea. In *North Sea Oil and Gas Reservoirs - III* (Editors: J. O. Aasen, E. Berg, A. T. Buller, O. Hjelmeland, J. Kleppe and Torsæter O.) (in press).

Stanley H. E. and P. Meakin (1988) Multifractal phenomena in physics and chemistry. *Nature* Vol. 335, p. 405-409.

Sreenivasan, K. R., R. R. Prasad, C. Meneveau and R. Ramshankar (1989) The fractal geometry of interfaces and the multifractal distribution of dissipation in fully turbulent flows, *Pure and Applied Geophysics.* Vol. 131, No. 1/2, p. 43 - 60.

Sullivan, J. (1984) Conditional recovery estimation through probability kriging - theory and practice. in *Geostatistics for Natural Resources Characterization.* (Editors: Verly, G., M. David, A. G. Journel, and A. Marechal). NATO ASI Series, C. Vol. 122. D. Reidel Publishing Company, Dordrecht, The Netherlands, Part 1, p. 365-384.

Turcotte, D. L. (1989) Fractals in geology and geophysics, *Pure and Applied Geophysics.* Vol. 131, No. 1/2, p. 171- 196.

Walsh, J. J., Watterson, J. and Yielding, G. (1990) Determination and interpretation of fault populations: procedures and problems. In *North Sea Oil and Gas Reservoirs - III* (Editors: J. O. Aasen, E. Berg, A. T. Buller, O. Hjelmeland, J. Kleppe and Torsæter O.) (in press).

Yielding, G., J. Walsh and J. Watterson (1992) The prediction of small-faulting in reservoirs. *First Break.* Vol. 10. No. 12. p. 449-460.

Quantitative Geology and Geostatistics

---

1. F. M. Gradstein, F. P. Agterberg, J. C. Brower and W. S. Schwarzacher: *Quantitative Stratigraphy*. 1985 ISBN 90-277-2116-5
2. G. Matheron and M. Armstrong (Eds.): *Geostatistical Case Studies*. 1987 ISBN 1-55608-019-0
3. *Cancelled*
4. M. Armstrong (Ed.): *Geostatistics*. Proceedings of the 3rd International Geostatistics Congress, held in Avignon, France (1988), 2 volumes. 1989 Set ISBN 0-7923-0204-4
5. A. Soares (Ed.): *Geostatistics Tróia '92*, 2 volumes. 1993 Set ISBN 0-7923-2157-X
6. R. Dimitrakopoulos (Ed.): *Geostatistics for the Next Century*. 1994 ISBN 0-7923-2650-4